Handbook of Organic Waste Management

Handbook of
Organic Waste Management

Editor: Ollie Hall

Callisto Reference
www.callistoreference.com

Callisto Reference,
118-35 Queens Blvd., Suite 400,
Forest Hills, NY 11375, USA

Visit us on the World Wide Web at:
www.callistoreference.com

ISBN: 978-1-64116-262-3 (Hardback)

The publisher's policy is to use permanent paper from mills that operate a sustainable forestry policy. Furthermore, the publisher ensures that the text paper and cover boards used have met acceptable environmental accreditation standards.

Printed in China.

Cataloging-in-Publication Data

Handbook of organic waste management / edited by Ollie Hall.
p. cm.
Includes bibliographical references and index.
ISBN 978-1-64116-262-3
1. Organic wastes--Management. 2. Compost--Management. 3. Refuse and refuse disposal.
4. Recycling (Waste, etc.). I. Hall, Ollie.
TD804 .H36 2020
628.445--dc23

Table of Contents

Preface

The main aim of this book is to educate learners and enhance their research focus by presenting diverse topics covering this vast field. This is an advanced book which compiles significant studies by distinguished experts in the area of analysis. This book addresses successive solutions to the challenges arising in the area of application, along with it; the book provides scope for future developments.

Organic waste refers to the waste that can be broken down into carbon dioxide, methane, water or simple organic molecules by the action of micro-organisms or other living things. This is possible through the processes of anaerobic digestion, aerobic digestion or composting. Some common examples of organic waste are manure, sewage, slaughterhouse waste, human waste, green waste, food waste, sewage sludge, biodegradable plastics and paper waste. Organic waste materials along with other wastes are disposed in landfills and incinerators. Since they are biodegradable they can also be used for land application and composting. Certain organic substances such as cardboard and paper are used for recycling. Biogas can also be produced from organic waste and used as a source of heat and electricity. This book studies, analyzes and upholds the pillars of organic waste management and its utmost significance in modern times. It is compiled in such a manner, that it will provide in-depth knowledge about the theory and practice of this area of study. It aims to serve as a resource guide for students and experts alike and contribute to the growth of the discipline.

It was a great honour to edit this book, though there were challenges, as it involved a lot of communication and networking between me and the editorial team. However, the end result was this all-inclusive book covering diverse themes in the field.

Finally, it is important to acknowledge the efforts of the contributors for their excellent chapters, through which a wide variety of issues have been addressed. I would also like to thank my colleagues for their valuable feedback during the making of this book.

Editor

1

Valorization of various sources of rabbit manure in agro-piscicultural system in Benin (West Africa): dynamics and effect of mineralization upon quality of fresh water

Richard Adande[1] · Hervé Kouessivi Janvier Bokossa[1] · Mouhamadou Nourou Dine Liady[1] · Emile Didier Fiogbé[1]

Abstract

Purpose Rising demand of fish generates various pressures on aquatic ecosystems, causing its deterioration and the loss of its biodiversity. With the aim of a better management and fish production in a salubrious environment, this study proposes the utilization of organic fertilizers from rabbit manure.

Methods The treatments consisted of manure of rabbits fed with: manufactured diet (T_1), improved diet with *Ipomea aquatic* (T_2), improved diet with *Elaeis guineensis* (T_3), and improved diet with *Panicum maximum* (T_4). Indeed, experimental design was a completely randomized block design with three replications. 6 g in 10 L of water from the different manures from each diet was followed during 50 days. Various nutrients were measured.

Results After the mineralization process, the best manure which released more nutrients was T_4 with measured proportions of N-NO_2^-, N-NO_3^-, N-NH_3, P-PO_4^{3-}, Ca^{2+}, Mg^{2+}, K^+, and suspended matter being 0.06 ± 0.0, 0.38 ± 0.1, 0.85 ± 0.1, 8.27 ± 0.9, 12.11 ± 0.8, 4.47 ± 0.3, 3.02 ± 0.2, and 140.78 ± 18.6 mg/L, respectively. The highest content of phosphorus and nitrogen was recorded in T_4 and T_2, respectively. The principal component analysis showed a positive correlation between P-PO_4^{3-}, Ca^{2+}, Mg^{2+}, MES, conductivity, and salinity during mineralization.

Conclusion Manure from rabbits fed by an improved diet with *P.* maximum (T_4) was recommended for agro-piscicultural production system.

Keywords Rabbit manure · Mineralization · Water · Nutrient · Agro-piscicultural

Introduction

Today, aquaculture is increasingly developing in the world (e.g., over 60% of the total fish production in China) despite the serious environmental problems that may occur (Li 2002; Lazard 2014). Although it remains under developed in most parts of Africa (Lazard 2014), its production has increased in recent years, compared to other livestock such as pigs, cows, and poultry (FAO 2010). Similarly, the strong fish demand has increased with population growth (FAO 2012). However, such demand has led to the deterioration of natural resources, especially quality, with the loss of its biodiversity (Moura E Silva et al. 2016). Fish farming could play a key role in facing the challenge of rising demands of animal proteins (Kaushik 2014). Strengthening the sector depends upon the efficient use of organic fertilizers of animal origin for an optimal primary production in fish conventional systems (Goulding et al. 2008). Indeed, the burst of optimal and sustainable fish farming must necessarily use a dynamic resource available such as organic fertilizers (FAO 2012). Thus, organic fertilizer use is a suitable alternative for agricultural activities to improve fish productivity and yields (Pucher et al. 2012; Bokossa et al. 2014a). Steinbronn (2009) showed that aquaculture production contributes significantly to food security, generates income, and plays a significant role in farmers' livelihood strategies. According to Wilkins (2008), the practice of integrated systems (animal and plant

✉ Richard Adande
richard_adande@yahoo.fr

[1] Laboratory of Research in the Wetlands (LRZH), Faculty of Science and Techniques (FAST), University of Abomey-Calavi (UAC), 01 BP 526, Cotonou, Benin

production) would give an economic and ecological efficiency in operations. Likewise, Wilkins (2008) and Lemaire et al. (2013) revealed that this practice would reduce the loss of nitrogen, phosphorus, and carbon to the environment and might limit the use of chemical fertilizers. Although animal manures have been investigated in zooplankton production in West African countries (e.g., Akodogbo et al. 2015), to our best knowledge, no study has dealt with the nutrients released in water from manure of rabbits.

The aims of our study were to (i) determine the effect of the diet on manure quality and (ii) assess the water quality and nutrient dynamics during the mineralization process.

Materials and methods

Experimental site

The experiment was conducted from June to August 2016 at a site (6°24′53.3″ N; 2°20′18.5″E; altitude 9 m above sea level) located at the University of Abomey-Calavi Benin. The area is characterized by a sub-equatorial climate with two rainy seasons (March to end of July and mid-September to November) and two dry seasons (August to mid-September and December to March).

Rabbits (*Oryctolagus cuniculus*) feeding and manure collection

The rabbit manure was obtained from 36 8-month-old local breed young rabbits. Four triplicates including a control were formulated. They were placed in 12 galvanized iron cages of 1 m length, 0.5 m wide, and 0.5 m height each. The cages were randomly placed in a hangar. Each cage was equipped by a manger and a trough. Nets of mesh (0.1 mm) were suspended to the bottom of the cages to collect rabbit manures. Rabbit manures were stored in bags of 25 kg.

Four diets were involved during the study period: classic manufactured (T_1), improved diet with *Ipomea aquatic* (T_2), improved diet with *Elaeis guineensis* (T_3), and improved diet with *Panicum maximum* (T_4). T_1 was manufactured locally by a veterinarian society service. Its chemical composition comprised: proteins (raw material) 17%, calcium ≥1.14%, total phosphorus ≥0.8%, lysine ≥0.75%, methionine + cystine ≥0.6%, crude fat ≥2%, crude fiber ≥14%, and ≥0.1% flavomycine.

The compositions of the three other diets were:

- T_2 Tuber cassava 2%, born bran 20%, palm kernel meal 10%, soy bran 10%, cottonseed meal 15%, shell 2%, distillers grains 20%, *Azolla filiculoïdes* 10%, brewer's yeast 5%, *Ipomoea aquatica* 15%, and salt 1%.
- T_3 Cossette cassava 2%, corn bran 25%, palm kernel meal 15%, soy bran 5% cottonseed meal 10%, shell 2%, distillers grains 15%, *Azolla filiculoïdes* 5%, brewer's yeast 10%, *Elaeis guineensis* 10%, and salt 1%.
- T_4 Cossette cassava 2%, corn bran 30%, palm kernel meal 10%, soy bran 10%, cottonseed meal 5%, shell 2%, distillers grains 10%, *Azolla filiculoïdes* 5%, brewer's yeast 15%, *Panicum maximum* 10%, and salt 1%.

Experimental design and water enrichment process

The experimental design was a completely randomized block design with three replications. The treatments consisted of the different types of manure collected from the experimental cages. The rabbit manures were grounded into nylon mosquito toils with a mesh size of 1 mm. For each treatment, 12 nylon mesh bags containing 6 g of dried manures were tied and suspended in 10 L of water (Mascha et al. 2010; Agadjihouèdé et al. 2011; Qiu et al. 2012; Bokossa et al. 2014a). A total of 150 plastic buckets (10 lots × 15 buckets) were used in the experiment. Each manure nylon mesh bags were placed randomly within 12 buckets. Three control (no fertilizer) buckets per lot were used. They were used for calibration purpose (i.e., correct the values form fertilized buckets). Half of the ten lots were sampled every 3 days, while the remaining was sampled every 7 days to take into account the labile and resistant fractions, respectively. At the sampling date, 12 nylon bags containing manures were collected, as well as 15 water samples of 500 ml each (including three controls). Water samples were then sent to the laboratory for analysis.

In all the buckets, temperature, pH, conductivity, dissolved oxygen, total dissolved solids (TDS), and salinity were measured ''in situ'' at the sampling days with a multiparameter probe (softer version/2015 2138 SN-ODEON CALYPSO; ±0.1 °C sensitivity).

Mineralization process and water analysis

Water samples were stored in a refrigerator at 4 °C to evaluate the released nutrients during the mineralization process. Analyses included the determination of suspended solids, $N\text{-}NO_3^-$, $N\text{-}NO_2^-$, $N\text{-}NH_3$, $P\text{-}PO_4^{3-}$, Mg^{2+}, K^+, and Ca^{2+} contents. Their measurement was performed using a molecular absorption spectrophotometer (DR 2800). After homogenization of the mixture, the suspended solids were measured at a wavelength of 810 nm. The ammonia combines with chlorine to form monochloramine which reacts with salicylate to form 5-aminosalicylate. $N\text{-}NH_3$ content was obtained at 655 nm. Regarding

Table 1 Chemical characteristics of rabbit manure regarding the composition of the diet used to feed the animal

Treatment	Nutrient content (mg/L)			
	N	P	K	N:P
T_1	12.33 ± 0.01c	1.03 ± 0.00c	1.66 ± 0.05b	11.97 ± 0.07bc
T_2	16.34 ± 0.03a	1.04 ± 0.00bc	2.19 ± 0.23a	15.66 ± 0.15a
T_3	14.02 ± 0.58b	1.06 ± 0.01b	0.79 ± 0.04c	13.23 ± 0.64b
T_4	15.02 ± 0.58ab	1.26 ± 0.00a	0.84 ± 0.02c	11.89 ± 0.44c
CV (%)	14.43	1.09	1.37	13.19

CV coefficient of variation

Means followed by the same letter are not significantly different ($p > 0.05$) according to the LSD. Treatments' description: T_1 manufactured diet, T_2 improved diet with *Ipomea aquatica*; T_3 improved diet with *Elaeis guineensis*, and T_4 improved diet with *Panicum maximum*

$N\text{-}NO_3^-$, the cadmium reduces nitrate to nitrite sample. The nitrite ion reacts with sulfanilic acid to form an intermediate diazonium salt. The nitrite present in the sample reacts with sulfanilic acid to form an intermediate diazonium salt. This combines with chlorotropique acid to produce a pink-colored complex, whose intensity is directly proportional to the nitrite concentration in the solution. The reading was obtained at 507 nm. $P\text{-}PO_4^{3-}$ was measured by the colorimetric method using blue ammonium molybdate reagent; measurements were realized at a wavelength of 880 nm. K^+, Ca^{2+}, and Mg^{2+} concentrations were determined by atomic absorption spectrophotometry (Rodier et al. 2012).

Data Analysis

The SAS software (SAS Version 9.2, SAS Institute Inc., Cary, NC, USA) was used for data analysis. The ANOVA was performed on the nutrient released in water during the decomposition process using time and treatments as factors. Main effects and interactions between treatments were assessed for significance levels, and separation was conducted using Fisher's protected LSD test. All levels of significance were set at p value <0.05. A principal component analysis (PCA) was also performed to assess the correlation between different parameters. This was done using the STATISTICA software (Statsoft inc., Tulsa, OK, USA).

Results

Influence of the diet on manure quality

The nutrient content from the manure from rabbits fed with the different diets are presented in Table 1. The potassium, phosphorus, and nitrogen contents were significantly different ($p < 0.05$) between treatments. N:P:K ratios from T_4, T_3, T_2, and T_1 were, respectively, 18:2:1, 18:1:1, 8:1:1, and 7:1:1. For nitrogen content, treatments can be classified as $T_2 > T_4 > T_3 > T_1$, while for the phosphorus content, the classification was $T_4 > T_3 > T_2 > T_1$. Regarding N:P ratios, the order was $T_2 > T_3 > T_1 > T_4$ (Table 1). Overall, the manures from rabbits fed with improved diet with *Panicum maximum* had significantly high phosphorus content ($p < 0.05$). Whereas manures from rabbits fed with improved diet with *Ipomea aquatica* recorded the highest nitrogen content.

Water quality and nutriment dynamics

The ANOVA (Table 2) showed that the diet had significant influence on nutrient contents including $N\text{-}NO_2^-$, $N\text{-}NO_3^-$, $N\text{-}NH_3$, $P\text{-}PO_4^{3-}$, Mg^{2+}, K^+, and suspended solid ($p < 0.0001$), and Ca^{2+} ($p < 0.05$). However, no significant effect was noticed for the in situ parameters ($p > 0.05$) (Table 2). The time significantly influenced all the measured parameters ($p < 0.05$), thus showing that nutrient and "in situ" parameters changed with time. The different nutrient dynamics depended on the diet: the treatment–time interaction was significant for all nutrients ($p < 0.0001$), except $P\text{-}PO_4^{3-}$, whereas it was not for the "in situ" parameters.

The physicochemical quality of water was affected by the release of the various fertilizing elements from rabbit manures (Table 3). The increase of conductivity, temperature, pH, dissolved oxygen, dissolved total solids, and salinity was found for all the treatments during the process of the enrichment of water following the decomposition of rabbit manures during 50 days of the experimentation in a progressive way. However, no significant difference between the treatments was found at the end of the experiment (Fig. 1; Table 3). Indeed, the temperature ranged from 29.66 ± 0.3 to 29.81 ± 0.3 in all the treatments. The conductivity was 240.85 ± 14.5, 239.72 ± 14.0, 238.98 ± 13.5, and 237.58 ± 14.1 μS/cm in treatments T_4, T_2, T_3, and T_1, respectively. A high increase of conductivity was recorded in the treatments resulting from

Table 2 Value of F and level significance of variance analysis (ANOVA), various modes on days, and physicochemical parameters

Sources of variation	ddl	Fisher's values													
		NO_2^-	NO_3^-	NH_3	PO_4^{3-}	Ca^{2+}	Mg^{2+}	K^+	SS	T °C	02	pH	Cond	TDS	Sal
Repetitions	2	0.14 ns	1.67 ns	0.16 ns	11.04***	1.23 ns	3.40*	0.73 ns	5.24***	3.38*	0.40 ns	10.06***	2.08 ns	2.21 ns	3.77*
Diet	3	8.46***	2.52 ns	158.80***	18.47***	3.01*	9.55***	3926***	48.19***	1.07 ns	0.28 ns	0.17 ns	0.19 ns	0.05 ns	0.19 ns
Days	9	249.10***	34.69***	3.65*	108.72***	65.66***	42.15***	825***	13526***	261.02***	88.51***	367.1***	243.8***	265.4***	130.4***
Diet*days	27	4.27***	2.84***	4.54***	1.19 ns	2.26*	3.29***	5.04***	6.97***	0.78 ns	0.66 ns	0.31 ns	0.55 ns	0.74 ns	0.83 ns

ns $p > 0.05$

* $p < 0.05$; ** $p < 0.01$; *** $p < 0.0001$

tested diets (T_4, T_3, and T_2) compared to those from the standard diet (T_1). That could be explained by a higher release of ortophosphate, salinity, suspended solids, calcium, and magnesium in the improved diets (Table 3). There was no significant difference between treatments for pH, dissolved oxygen, TDS, and salinity. This result can be explained by similar experimental condition and microorganism activity in the buckets.

The different types of manure have induced significant increase in N fractions (N-NH_3, N-NO_3^-, and N-NO_2^-) at the end of experiment. The K^+, Ca^{2+}, Mg^{2+}, P-PO_4^{3-}, and suspended solid contents in the water at the end of the decomposition process of the rabbit manures fed with different diets had also significantly increased ($p < 0.05$) as indicated by the large amount of released nutrients.

According to the observed nutrient dynamics (Table 3), nitrites increased during the first week before decreasing in the last weeks of the experimentation in all the treatments generally. Such a dynamic was noticeable particularly in T_1, T_2, and T_3. Mean NO_2^- values in the different treatments increased from 0.0 ± 0.0 mg/L (T_1) to 0.1 ± 0.0 mg/L (T_4) during 50 days. These variations can be explained by the microbial activity which influences the permanent availability of nitrites. The reductions of nitrates in the treatments were revealed from the 9th days. These reductions were related to the ambient temperature. Indeed, when temperature became high, biochemical reactions were very important affecting nitrogen transformation or bacteria denitrification. The highest N-NO_3^- concentrations were obtained in T_2 (0.38 ± 0.1 mg/L) followed by T_4 (0.33 ± 0.0 mg/L). During the experiment, the average N-NH_3 content ranged from 0.43 ± 0.0 mg/L in T_1 to 1.62 ± 0.0 mg/L in T_4. However, significant decreases by 3.67, 1.9, and 1.26 times were noticed in ammonium concentrations in the manure from treatments T_1, T_2, and T_3, respectively, compared to T_4. This is related to the equilibrium constant which can be favourable to a massive loss of nitrogen as ammonia gas or favourable to storage in water as NH_4^+. In the rabbit manures, the volatilization process of ammonia consisted in the transfer of ammonia gas in the immediate atmosphere from ammonia in manure decomposing in water during experimentation. The concentration of orthophosphate, which was 10.83 ± 1.0 mg/L in the T_4 treatment, was 1.33, 1.30, and 1.2 times higher, respectively, than in treatments T_1, T_2, and T_3 ($p < 0.05$). Tendency similar trend was practically noticed with the other elements such as Ca^{2+}, Mg^{2+}, and K^+. This result revealed the richness of T_4 compared to the others treatments. We also noticed a rapid increase of K^+ concentrations in T_2 contrary to T_1; while T_3 and T_4 indicated during the first 12 days of experiment a rapidly mineralizable labile fraction of organic matter even from this time to the end of the experiment, a readily degradable fraction was found in rabbit manure.

Table 3 Effect of nutrients released from the different manure on water quality

Treatment	Days	Nutrient content							
		$N\text{-}NO_2^-$ (mg/L)	$N\text{-}NO_3^-$ (mg/L)	$N\text{-}NH_3$ (mg/L)	$P\text{-}PO_4^{3-}$ (mg/L)	Ca^{2+} (mg/L)	Mg^{2+} (mg/L)	K^+ (mg/L)	SS (mg/L)
T_1	3	0.01 ± 0.00c	1.01 ± 1.01a	0.83 ± 0.83a	3.33 ± 3.33b	5.87 ± 5.87d	2.72 ± 2.72ef	1.74 ± 1.74ba	5.23 ± 5.22d
	6	0.14 ± 0.01a	1.12 ± 1.12a	0.89 ± 0.89a	4.14 ± 4.14b	6.89 ± 6.89dc	2.84 ± 2.84ef	1.84 ± 1.84ba	19.59 ± 19.58d
	9	0.15 ± 0.00a	0.33 ± 0.02b	0.71 ± 0.04b	5.57 ± 7.71b	10.10 ± 0.14cb	3.55 ± 0.16edf	1.03 ± 0.10b	82.71 ± 0.94c
	12	0.08 ± 0.07ab	0.13 ± 0.01dc	0.07 ± 0.05e	2.84 ± 0.46b	7.68 ± 0.65cbd	2.27 ± 0.26f	1.06 ± 5.56b	117.00 ± 0.06c
	15	0.01 ± 0.00c	0.20 ± 0.04c	0.22 ± 0.01d	3.25 ± 0.51b	8.53 ± 0.32cbd	4.43 ± 0.15bdc	2.71 ± 0.39a	92.67 ± 19.67c
	22	0.01 ± 0.00c	0.14 ± 0.00dc	0.30 ± 0.03dc	11.50 ± 0.45a	14.79 ± 2.27a	4.44 ± 0.26bdc	2.39 ± 9.70a	184.33 ± 0.52b
	29	0.03 ± 0.00c	0.06 ± 0.04de	0.33 ± 0.02c	12.64 ± 1.85a	15.86 ± 1.29a	4.02 ± 0.47edc	1.69 ± 0.59ba	218.67 ± 10.91ba
	36	0.00 ± 0.00c	0.01 ± 0.00e	0.34 ± 0.02c	12.64 ± 1.85a	10.92 ± 1.71b	6.23 ± 1.06a	1.76 ± 13.66ba	237.33 ± 0.61a
	43	0.00 ± 0.00c	0.05 ± 0.04de	0.33 ± 0.02c	12.44 ± 1.85a	16.06 ± 1.29a	5.35 ± 0.92bac	1.69 ± 0.59ba	246.67 ± 24.06a
	50	0.00 ± 0.00c	0.06 ± 0.04de	0.33 ± 0.02c	12.64 ± 1.85a	15.86 ± 1.29a	5.68 ± 0.74ba	1.69 ± 0.59ba	248.67 ± 24.06a
	Means T_1	0.04 ± 0.01C	0.31 ± 0.07AB	0.43 ± 0.04D	8.10 ± 0.86B	11.25 ± 0.77B	4.15 ± 0.27A	1.76 ± 0.14D	145.28 ± 16.95A
T_2	3	0.01 ± 0.00d	1.01 ± 0.03c	1.05 ± 0.00bac	2.71 ± 0.29c	6.26 ± 0.41c	2.27 ± 0.26e	2.23 ± 0.05cd	10.60 ± 1.78e
	6	0.18 ± 0.02b	1.23 ± 0.16a	1.27 ± 0.11a	3.83 ± 0.23cb	9.56 ± 0.90b	3.09 ± 2.34ed	2.22 ± 0.05cd	10.68 ± 0.52ed
	9	0.24 ± 0.01a	0.34 ± 0.05a	1.24 ± 0.15ba	5.49 ± 0.60b	8.74 ± 0.18b	2.40 ± 0.11e	1.44 ± 0.25d	66.19 ± 3.59e
	12	0.12 ± 0.01c	0.38 ± 0.38ab	0.76 ± 0.76bdc	3.05 ± 3.05c	7.96 ± 7.96cb	4.35 ± 4.35cd	6.10 ± 6.10a	63.33 ± 63.33cd
	15	0.01 ± 0.00d	0.12 ± 0.02c	0.43 ± 0.02d	2.37 ± 0.66c	7.96 ± 0.46cb	2.98 ± 0.34ed	3.78 ± 0.86b	133.67 ± 34.43ed
	22	0.01 ± 0.00d	0.18 ± 0.08c	0.42 ± 0.06d	13.76 ± 0.68a	16.10 ± 1.01a	5.45 ± 0.38cb	3.37 ± 0.27cb	152.33 ± 21.69cb
	29	0.02 ± 0.00d	0.11 ± 0.07c	1.10 ± 0.14bac	12.67 ± 1.22a	16.07 ± 1.01a	5.83 ± 0.88b	2.67 ± 0.14cbd	235.33 ± 29.47b
	36	0.00 ± 0.00d	0.24 ± 0.06c	0.61 ± 0.34dc	13.67 ± 1.22a	16.09 ± 1.02a	7.39 ± 0.49a	3.05 ± 0.17cb	175.00 ± 13.07a
	43	0.00 ± 0.00d	0.11 ± 0.07c	1.17 ± 0.30ba	12.47 ± 1.22a	16.27 ± 1.01a	5.83 ± 0.88b	2.67 ± 0.14cbd	281.00 ± 14.64b
	50	0.01 ± 0.00d	0.11 ± 0.07c	0.50 ± 0.05d	12.67 ± 1.22a	16.07 ± 1.01a	5.16 ± 0.58cb	2.67 ± 0.14cbd	279.67 ± 9.76cb
	Means T_2	0.06 ± 0.01A	0.38 ± 0.07A	0.85 ± 0.07C	8.27 ± 0.93B	12.11 ± 0.78AB	4.47 ± 0.33A	3.02 ± 0.24B	140.78 ± 18.64A
T_3	3	0.03 ± 0.00d	0.80 ± 0.35a	0.77 ± 0.51b	5.07 ± 0.88c	8.20 ± 0.86b	1.67 ± 0.14e	2.16 ± 0.16b	10.95 ± 1.15d
	6	0.18 ± 0.02b	0.42 ± 0.26ab	1.14 ± 0.18ba	5.61 ± 0.41c	9.89 ± 0.22b	2.72 ± 0.11d	2.28 ± 0.14b	30.93 ± 3.08dc
	9	0.26 ± 0.03a	0.19 ± 0.04b	1.32 ± 0.05a	5.59 ± 0.64c	9.09 ± 0.47b	2.98 ± 0.04dc	1.44 ± 0.20c	44.49 ± 9.57cbd
	12	0.10 ± 0.00c	0.13 ± 0.00b	1.47 ± 0.06a	4.43 ± 0.84c	8.57 ± 0.36b	3.17 ± 0.16dc	2.28 ± 0.14b	48.00 ± 5.13cbd
	15	0.01 ± 0.00d	0.15 ± 0.01b	1.42 ± 0.06a	4.67 ± 0.48c	9.22 ± 0.85b	1.66 ± 0.28e	1.24 ± 0.20c	62.67 ± 14.74cb
	22	0.01 ± 0.00d	0.29 ± 0.05b	1.44 ± 0.02a	12.12 ± 0.30ba	14.40 ± 1.44a	5.05 ± 0.26a	3.31 ± 0.22a	77.00 ± 6.65b
	29	0.01 ± 0.00d	0.13 ± 0.11b	1.47 ± 0.03a	10.39 ± 2.01b	15.76 ± 0.69a	3.70 ± 0.50bc	3.41 ± 0.12a	153.33 ± 18.80a
	36	0.01 ± 0.00d	0.26 ± 0.04b	1.40 ± 0.05a	14.39 ± 1.69a	15.76 ± 0.47a	4.14 ± 0.28ba	3.36 ± 0.29a	170.67 ± 20.16a
	43	0.00 ± 0.00d	0.13 ± 0.11b	1.47 ± 0.03a	13.19 ± 1.69ba	15.96 ± 0.69a	4.92 ± 0.47a	3.41 ± 0.12a	151.33 ± 21.53a
	50	0.01 ± 0.00d	0.13 ± 0.11b	0.97 ± 0.02ba	13.39 ± 1.69ba	14.87 ± 0.69a	5.03 ± 0.41a	3.18 ± 0.02a	146.67 ± 20.33a
	Means T_3	0.06 ± 0.01A	0.26 ± 0.05A	1.28 ± 0.06B	8.88 ± 0.80B	12.17 ± 0.63AB	3.50 ± 0.24B	2.61 ± 0.15C	89.60 ± 11.11C

Table 3 continued

Treatment	Days	Nutrient content							
		N-NO_2^- (mg/L)	N-NO_3^- (mg/L)	N-NH_3 (mg/L)	P-PO_4^{3-} (mg/L)	Ca^{2+} (mg/L)	Mg^{2+} (mg/L)	K^+ (mg/L)	SS (mg/L)
T_4	3	0.01 ± 0.00d	0.81 ± 0.16a	1.41 ± 0.12e	3.51 ± 0.42e	6.25 ± 0.74f	1.72 ± 2.67e	3.04 ± 0.32bcd	10.67 ± 0.59 h
	6	0.35 ± 0.03a	0.44 ± 0.06bc	1.54 ± 0.05edc	4.88 ± 0.41de	8.74 ± 0.61fe	2.88 ± 0.05dc	3.47 ± 0.36bc	60.92 ± 5.38 g
	9	0.25 ± 0.01b	0.57 ± 0.02ba	1.65 ± 0.06bdc	7.30 ± 0.71c	10.32 ± 0.46de	3.55 ± 0.23c	2.07 ± 0.27d	78.85 ± 5.30ef
	12	0.12 ± 0.00c	0.19 ± 0.00cd	1.85 ± 0.04a	6.26 ± 0.31dc	9.21 ± 0.16e	4.68 ± 0.22b	2.78 ± 0.31dc	97.66 ± 4.09d
	15	0.01 ± 0.00d	0.15 ± 0.09d	1.74 ± 0.06ba	6.32 ± 0.23dc	12.06 ± 0.76dc	2.26 ± 0.07de	2.8 ± 0.45dc	76.00 ± 1.15f
	22	0.01 ± 0.00d	0.28 ± 0.08cd	1.87 ± 0.05a	16.40 ± 0.83ba	16.14 ± 1.12ba	4.58 ± 0.18b	4.22 ± 0.05ba	88.33 ± 4.63ed
	29	0.02 ± 0.00d	0.19 ± 0.09cd	1.72 ± 0.05bac	14.96 ± 0.80b	14.28 ± 1.08bc	3.64 ± 0.31c	3.84 ± 0.40bac	130.33 ± 4.40c
	36	0.01 ± 0.00d	0.32 ± 0.03cd	1.52 ± 0.04ed	17.76 ± 0.60a	18.24 ± 0.91a	6.42 ± 0.31a	4.10 ± 0.70ba	146.66 ± 4.63b
	43	0.00 ± 0.00d	0.19 ± 0.09cd	1.35 ± 0.02e	15.55 ± 0.81b	14.48 ± 1.08bc	6.34 ± 0.33a	3.84 ± 0.40bac	185.00 ± 1.15a
	50	0.02 ± 0.00d	0.19 ± 0.09cd	1.53 ± 0.04ed	15.42 ± 0.82b	14.28 ± 1.08bc	6.01 ± 0.03a	4.68 ± 0.11a	137.00 ± 2.88cb
	Means T_4	0.08 ± 0.02A	0.33 ± 0.04AB	1.62 ± 0.03A	10.83 ± 1.00A	12.40 ± 0.69A	4.21 ± 0.30A	3.48 ± 0.17A	101.14 ± 8.83B

For each treatment, mean values with the same lower case letter in a column indicate not statistically significant differences ($p > 0.05$) according to LSD test. For each column, mean values followed by the same capital letter indicate not statistically significant differences between treatments ($p > 0.05$) according to LDS test

SS suspended solid

Overall, T_4 released the higher concentrations of nutrients being able to be used for fish production systems by its contribution for primary production.

Classification of treatments according to their nutrient profile

Figure 2 depicts the results of the PCA. All the variables can be classified according to the main components which explained 66.13% of the information (51.61% for F1 and 14.52% for F2). The most correlated variables to F1 were: conductivity, orthophosphate, suspended solid, total dissolved solids (TDS), salinity, magnesium, and nitrate (Table 4). Among these variables, those who contribute most to F1 were two different groups: for the first group conductivity, orthophosphate, suspended solid, salinity, and magnesium which are highly correlated between each other and negatively to F1, since they define eigenvectors same direction, and for the second group composed only by the nitrate which was positively correlated with F1 (Fig. 2a). On F2, the most correlated variables are temperature, pH, and oxygen. However, some variables (potassium and ammonium) are near to centre of the factorial design, and their correlation is not certainly very strong. These variables are probably better explained by others minor components. Taking into account all the information provided by F1 and F2 (Table 4), the physicochemical parameters belonging to the axis F1, only orthophosphate was high and was at the origin of the mineralization, explaining the increase in conductivity, suspended solids, TDS, and salinity. The biochemical activity of degradation of the organic matter by microorganisms depends on the physical and chemical conditions of the environment. A temperature rise could increase P-PO_4^{3-}, Ca^{2+}, Mg^{2+}, SS concentrations, and salinity. Likewise, a rise in the temperature and pH is followed by an increase of nitrate concentration. Indeed, the fluctuation of temperature, dissolved oxygen, and pH modulate mineralization rate because of their key role on the microbial activity on organic matter.

Discussion

Diet and manure's nutrient content

The efficient uses of organic fertilizers necessitate a good knowledge of their nutrient contents. Nutrients released from organic manure constitute the key of the primary production. However, their excessive concentrations can negatively impact the aquatic environment (e.g., eutrophication). Therefore, it is essential to quantify their concentration in the aquatic environments to avoid such negative

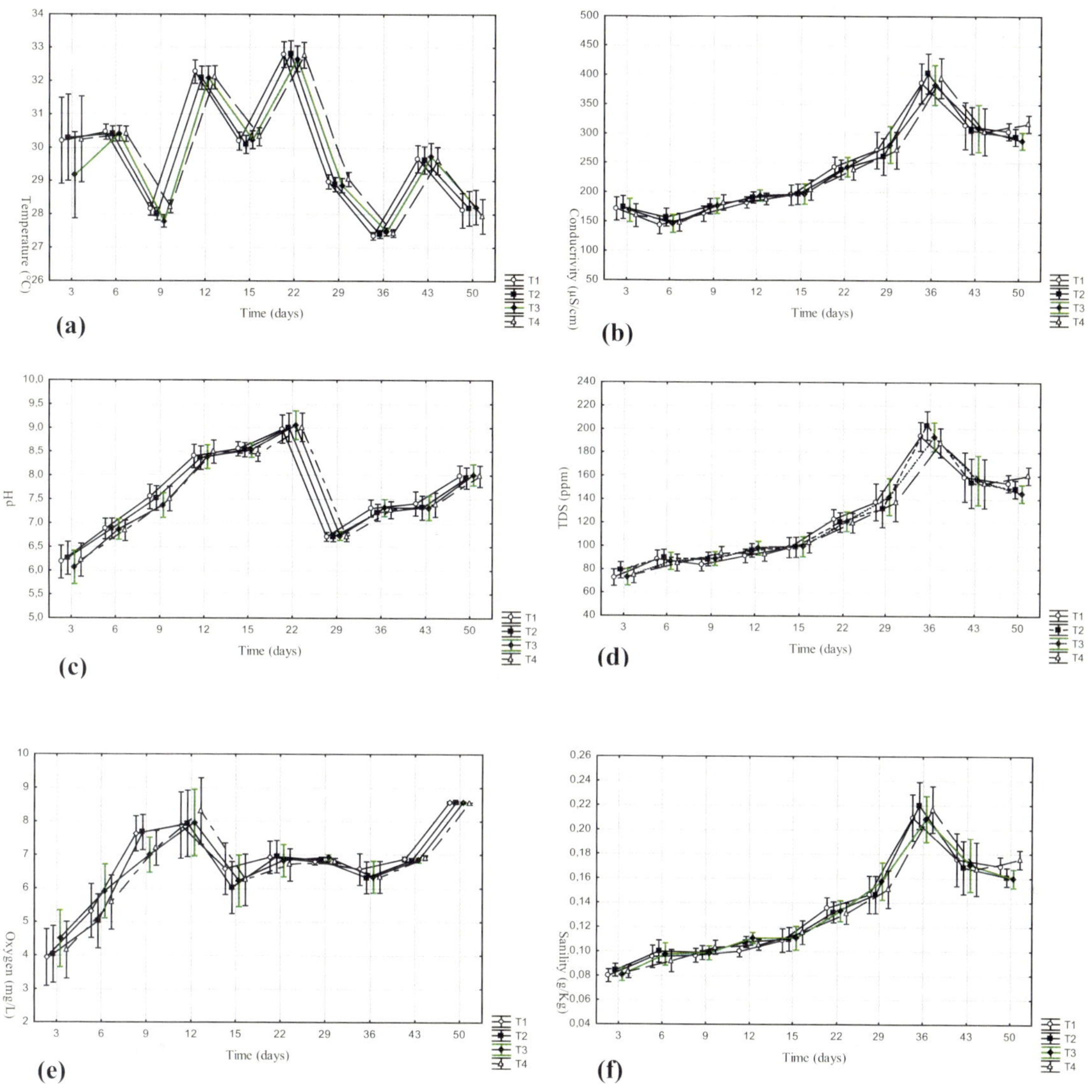

Fig. 1 Physico-chemical parameter of the various treatments according to time **a** temperature; **b** conductivity; **c** pH; **d** total dissolved solid (TDS); **e** oxygen; and **f** salinity. Treatments' description: T_1 manufactured diet, T_2 improved diet with *Ipomea aquatica*, T_3 improved diet with *Elaeis guineensis*, and T_4 improved diet with *Panicum maximum*

impacts or mitigate them. Our results showed that the nutrient content from the different types of rabbit manure varied qualitatively and quantitatively according to the tested diet. This result could be explained by the diet composition (nitrogen levels, phosphorus, potassium, calcium, magnesium, etc.). This is in line with Martinez-Ballestra et al. (2008) who noted that the concentrations of these nutrients in animal waste strongly depend on the composition of food rations. Rations are influenced by factors such as the species, type of digestion, chemical composition, and presence of secondary compounds (Mackie 2002; Michelland et al. 2012). The concentrations of 1.26 mg/L of phosphorus for T_4 and 16.34 mg/L of nitrogen for T_2 were higher than those found by Vodounnou et al. (2016), which were 0.011 mg/L for phosphorus and 0.025 mg/L for nitrogen in rabbit manures. This difference could be explained by the type of tested diets and rabbit species used in one hand, and the conservation process of these rabbit manures in the other hand. However, these results are lower than those obtained by Bokossa et al. (2014a) on manure from pigs fed with *Azolla* and rice bran. This difference is due to the composition of diets and digestibility of rations applied to monogastric animals. Indeed, rabbits have a complex digestive physiology compared to other herbivores and omnivores.

The value of the N:P ratio indicates which one between N and P is the limiting factor. According to rough nutrient balance estimation, the best ratio for sustainable aquaculture production was that obtained with treatment T_2 (N:P = 15:1 and N:P:K = 18:1:1), but N:P ratio was

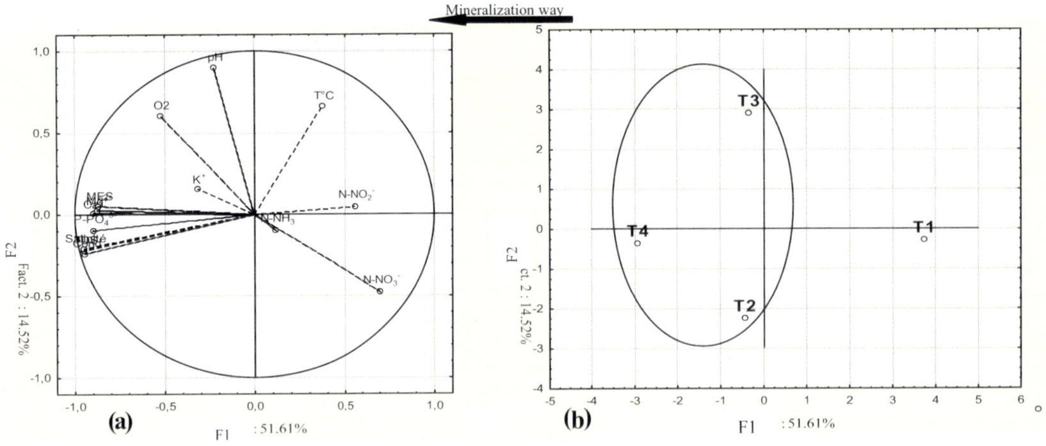

Fig. 2 Principal component analyses of treatments and their effect on water quality. **a** Correlation between chemical parameters; **b** classification of treatments

Table 4 Correlation coefficients between variables and axis principal

	Factorial axe 1	Factorial axe 2
$N\text{-}NO_2^-$	0.55	0.04
$N\text{-}NO_3^-$	0.69	−0.47
$N\text{-}NH_3$	0.11	−0.09
$P\text{-}PO_4^{3-}$	−0.90	−0.09
Ca^{2+}	−0.89	0.00
Mg^{2+}	−0.87	0.02
K^+	−0.31	0.15
SS	−0.87	0.04
T °C	0.37	0.66
O_2	−0.52	0.60
pH	−0.22	0.89
Cond	−0.94	−0.23
TDS	−0.94	−0.21
Salinity	−0.95	−0.20

between 10:1 and 20:1 for all treatments (Borchardt 1996). This ratio from our study confirmed the enrichment of T_4 treatment in phosphorus. These results are similar to those obtained by Güsewell and Verhoeven (2006) and Guckland et al. (2009), where N:P ratio was 16:1 during the decomposition of leaves and 11:1 in the soil, respectively.

Panicum maximum at 3 week age contains 32.3% calcium, 46.6% phosphorus, and 71% of magnesium (Oyenuga 1960). These high contents of magnesium and phosphorus in the form of orthophosphate contribute to their high growth, density, biomass, and net primary production (Rejmánková 2001; Cody et al. 2010). The high phosphorus content recorded in treatment T_4 would play a crucial role in phytoplankton production in fish net.

Effect of rabbit manure on nutrient dynamics and water quality

The experiment took place during the dry season with temperature fluctuating between 28 and 30 °C. It had played an important role in organic matter mineralization, but also induced a good solubility and diffusion of gases such as oxygen in water (Belghiti et al. 2013; Bokossa et al. 2014b). The speed of oxygen diffusion depends for a large part on the contact area between air and water. The quantitative importance of this process depends on the temperature, light, and richness in nutrients. After the water fertilization was achieved, 3 day latency time (enzymatic adaptation) was observed during which values of chemical parameters did not varied a lot. The pH remained around 6 in all the treatments. Microbial activity started thus after these 3 days and showed a fast variation of these parameters (Table 3); pH and $P\text{-}PO_4^{3-}$ increased, whereas nitrogen forms showed great fluctuation (nitrification process) providing a favourable environment for phytoplankton production (Fig. 1e), and consequently that of zooplankton when phytoplankton biomass is not so high to cause growth inhibition (Liady et al. 2015 and references there in). Indeed, the extent of pH preferences for fish is usually 5–7 to 7–9 (Peterson et al. 1989). Our findings were like those obtained by Bokossa et al. (2014a). The high content of suspended solids can be correlated with all inorganic and organic substances contained in a liquid, the total dissolved substances in water and salinity in the mineral environment released from rabbit manures during mineralization in T_4 and T_2. This corroborates the results of Guibaud et al. (2013) on the effect of mineralization in fresh water with waste and with various anthropogenic activities in Morocco (Nechad et al. 2014). The organic fertilizer decomposition is assured by heterotrophic microorganisms which release various minerals while consuming oxygen (Mérigout 2006). The different values of nitrite recommended in fish farming ponds range from 0.5 to 2 mg/L (Dwamd 1994). This tendency which was found in T_1, T_2, and T_3 around the 36th day of the experiment confirms the results obtained by Grunert et al. (2016) during the conversion of nitrogen in the growth of horticultural plants. This decline could be explained by the low phosphorus content in rabbit manures in our experiments. Phosphorus content was important in T_4 and explains that the degradation of organic matter continues under its influence. In fact, phosphorus allows microorganism's longer life time than nitrogen (Durand et al. 1983). The orthophosphate content from T_4 was higher than those obtained by Bokossa et al. (2014a), whose maximum value was 2.80 mg/L. This difference could be explained by the type of organic fertilizer, animal species, and tested diet. The phosphorus richness in T_4 can be a reliable source of mineralization for the development of aquaculture. The nutrient concentrations in the rabbit manure can be used to a good production of microalgae in aquaculture systems, namely, because of their role on microalgae's growth in freshwater (Martin et al. 1985). The high concentrations of K^+, Ca^{2+}, and Mg^{2+} in T_4, T_3, and T_2 proved the presence of released nutrients in water and justify the high conductivity obtained. Suspended solids induced degradation of aquatic ecosystems by reducing light penetration in water, damage the gills of fish, total or partial closure of habitats, and spawning leading depletion oxygen and create a hostile environment for the life of aquatic organisms (Massa 2000; Foster and Charlesworth 1996), but from our results, the low value of suspended solid in T_3 and T_4 treatments is a worthy asset for aquaculture production.

Correlation between physicochemical variables and nutrient dynamics

The use of PCA clarifies the relationships between variables and phenomenon behind these correlations (Fig. 2a). It is used in many water dynamic studies and facilitates aquatic ecosystem studies (Ladhar et al. 2014). PCA showed that the mineralization was related mainly to orthophosphate content and to some extent to ammonium, calcium from which increasing conductivity. These results are different from those obtained by Amadou et al. (2014) in physicochemical and bacteriological analyses of wastewater. The most important effect of the mineralization in our study was characterized by high levels of orthophosphate in T_4 which is a major source of plant nutrients (Fig. 2b). Similar results were observed by Diallo et al. (2014), and highlight the inseparable link between orthophosphates and the growth of plants and even point out its essential role in eutrophication process. The increase in conductivity is correlated with increased nutrient levels (Belghiti et al. 2013; Rodier et al. 2012; Bokossa et al. 2014a). Indeed, T_2 and T_4 released nutrients such as nitrogen, phosphorus, potassium, and calcium, which are the cause of the increase in conductivity. We noticed a nitrogen loss during the mineralization process. It would be necessary to offer to agriculture and fish farmers a contribution on a regular way of use of rabbit manures to fill the gap.

Conclusion

Following the results obtained from the mineralization process of different rabbit manures in fresh water, we can conclude that it can serve as enrichment for fresh water in agro-piscicultural systems. Nutrients released from organic wastes (as rabbit manure) can provide a good yield in such systems at a lower cost compared to chemical fertilizers. This study of enrichment of water by rabbit manure resulting from different diets including classic diet (T_1), improved diet with *I. aquatica* (T_2), improved diet with *E. Guineensis* (T_3), and improved diet with *P.* Maximum (T_4) showed that the mineralization moved towards orthophosphates, calcium, and magnesium. Treatments T_4, T_2, and T_3 had good physicochemical characteristics offering favourable conditions for fish production by the means of algae production (phytoplankton), contrary to the treatment T_1. Moreover, T_4 treatment because of its high phosphorus content offers more optimal conditions for the production of microalgae. Water quality in terms of minerals content varies according to the diets.

Acknowledgements We express our gratitude to the Ministry for the Higher Education and Scientific Research of Benin, which set up the project "support at PhD student" in which this study of research was made.

References

Agadjihouèdé H, Montchowui E, Chikou A, Lalèyè PA (2011) Libération comparée de sels dans l'eau par la minéralisation de l'azolla, la bouse de vache, la fiente de volaille et les sons de riz et de maïs utilisés en pisciculture. Int J Biol Chem Sci 5:1883–1897. doi:10.4314/ijbcs.v5i5.10

Akodogbo HH, Bonou CA, Adande R, Sossou DS, Fiogbé ED (2015) Optimization of zooplankton production from pig dung optimal dose: renewed medium. Agric Adv 4:15–21. doi:10.14196/aa.v4i2.1804

Amadou H, Laouali MS, Manzola A (2014) Analyses physico-chimiques et bactériologiques des eaux de trois aquifères de la région de Tillabéry: application des méthodes d'analyses statistiques multi variées. Lar J 20:25–41

Belghiti ML, Chahlaoui A, Bengoumi D, Moustaine REL (2013) Etude de la qualité physico-chimique et bactériologique des eaux souterraines de la nappe plio-quaternaire dans la région de meknès (Maroc). Lar J 14:21–36

Bokossa HKJ, Saïdou A, Fiogbé ED, Kossou D (2014a) Decomposition rate of pig's manures and nutrient release pattern in wetland condition. Agric For Fish 3:271–278. doi:10.11648/j.aff.20140304.19

Bokossa HKJ, Saïdou A, Sossoukpe E, Fiogbé DE, Kossou D (2014b) Decomposition and mineralization effect of various sources of Pig manure on water quality and nutrients availability for agro-Fish System in Benin. Agric Sci 5:1194–1206. doi:10.4236/as.2014.512129

Borchardt MA (1996) Nutrients. In: Stevenson RJ, Bothwell ML, Lowe RL (eds) Algal ecology: freshwater benthic ecosystems. Ac Press, San Diego, pp 183–227

Cody RJ, Luecke C, Whalen SC, Evans MA (2010) Direct and indirect effects of fish on pelagic nitrogen and phosphorus availability in oligotrophic Arctic Alaskan lakes. NRC Sci 67(10):1635–1648. doi:10.1139/F10-085

Diallo AD, Ibnonamr K, N'diaye AD, Garmes H, Kankou M, Wane O (2014) L'intérêt des méthodes d'analyses statistiques dans la gestion du suivi de la qualité physicochimique de l'eau de la rive droite du fleuve Sénégal. Lar J 17:101–114

Durand M, Beaumatin P, Dumay C (1983) Estimation in vitro à l'aide du phosphore radioactif des besoins en phosphore des microorganismes du rumen. Reprod Nutr Dév 23:727–739

Dwamd (1994) Division of water ambient monitoring database, 18th edn p 325

FAO (2010) (Food and Agriculture Organization). Statistical databases. rome: fao [online]. http://www.fao.org. Accessed 09 Oct 2016

FAO (2012) (Food and Agriculture Organization). The State of world fisheries and aquaculture, 2000, p 230

Foster IDL, Charlesworth SS (1996) Hearvy metals in the hydrological cycle: trends and explanation. Hydr Processes 10(2):227–261

Goulding K, Jarvis S, Whitmore A (2008) Optimizing nutrient management for farm systems. Philos Trans R Soc B 363:667–680. doi:10.1098/rstb.2007.2177

Grunert O, Reheul D, Labeke MCV, Perneel M, Sanabria EH, Siegfried EV, Boon N (2016) Growing media constituents determine the microbial nitrogen conversions in organic growing media for horticulture. Micro Biotechol 9:389–399. doi:10.1111/1751-7915.12354

Guckland A, Brauns M, Flessa H, Thomas FM, Leuschner C (2009) Acidity, nutrient stocks and organic matter content in soils of a temperate deciduous forest with different abundance of European beech (*Fagus sylvatica L.*). J Plant Nutr Soil Sci 172: 500–511

Guibaud G, Charriau A, Bordas F, Rabiet M (2013) Evaluation, through the measurements of electric conductivity and mineral compound contents in waters and soils, of the impact of an old municipal landfill on a wetland (peat bog). Eur J Water Qual 44:13–38

Gusewell S, Verhoeven JTA (2006) Litter N: p ratios indicate whether N or P limits the decomposability of graminoid leaf litter. JTA Plant Soil 287:131. doi:10.1007/s11104-006-9050-2

Kaushik S (2014) L'apport de la pisciculture à l'alimentation de l'homme. Cah Agric 23:18–23. doi:10.1684/agr.2014.0679

Ladhar C, Tastard E, Casse N, Denis F, Ayadi H (2014) Strong and stable environmental structuring of the zooplankton communities in interconnected salt ponds, p15. Hydro. doi:10.1007/s10750-014-1998-y

Lazard J (2014) La diversité des piscicultures mondiales illustrée par les cas de la Chine et du Nigeria. Cah Agric 23:24–33. doi:10.1684/agr.2014.0680

Lemaire G, Franzluebbers A, Carvalho PCF, Dedieu B (2013) Integrated crop-livestock systems: strategies to achieve synergy between agricultural production and environmental quality. Agric Ecosyst Environ 190:4–8. doi:10.1016/j.agee.2013.08.009

Li SF (2002) Aquaculture research and its relation to development in China. In: Zhang LX, Liu J, Li SF, Yang NS, Gardiner PR (eds) Agricultural development and the opportunities for aquatic resources research in China. World Fish Center, Penang, p p58

Liady MND, Tangou TT, Fiogbe ED, Cauchie HM, Vasel JL (2015) About the interest of a zooplankton compartment in pond systems: methodology to study the growth of *D. pulex*. Wat Sci Technol 71:1436–1443

Mackie RI (2002) Mutualistic fermentative digestion in the gastrointestinal tract: diversity and evolution. Int Comp Biol 42:319–326

Martin C, Picard G, de la Noüe J (1985) World Epuration biologique du lisier de porc par production de biomasses d'algues unicellulaires. J Microbiol Biot 1:173. doi:10.1007/BF01742582

Martinez-Ballestra MC, Lopez-Perez L, Hernandez M, Lopez-Berenguer C, Fernandez-Garcia N, Carvajal M (2008) Agricultural practices for enhanced human health. Phyto Chem Rev 7:251–260. doi:10.1007/s11101-007-9071-3

Mascha J, Karin V, Andrea P, Frank MT (2010) Leaf litter decomposition in temperate deciduous forest stands with a decreasing fraction of beech (*Fagus sylvatica*). Ecol 164:1083–1094. doi:10.1007/s00442-010-1699-9

Massa F (2000) Sédiment, physico-chimie du comportement interstiel et développement larvaire de la truite commune (Salmo-trutta) : Etude en milieu naturel antropisée et en condition contôlée, Thèse de l'INA-P-G, p. 199

Mérigout P (2006) Etude du métabolisme de la plante en réponse à l'apport de différents fertilisants et adjuvants culturaux. Influence des phytohormones sur le métabolisme azoté. Sciences of the Universe [physics]. INAPG (AgroParisTech), 2006. HAL Id: pastel-00002273 https://pastel.archivesouvertes.fr/pastel-00002273. Submitted on 22 Mar 2007

Michelland RJ, Combes S, Monteils V, Bayourthe C, Cauquil L, Enjalbert FJC, Kimse M, Troegeler-Meynadier AZA, Gidenne T, Fortun-Lamothe L (2012) Analyse comparée des écosystèmes digestifs du rumen de la vache et du caecum du lapin. INRA Prod Anim 25:395–406

Moura E Silva MS, Graciano TS, Losekann ME, Luiz AJ (2016) Assessment of benthic macroinvertebrates at Nile tilapia production using artificial substrate samplers. Braz J Biol 76(3):735–742. doi:10.1590/1519-6984.02815

Nechad I, Fadil K, Fadil F (2014) Qualité physicochimique des eaux des sources ain regrag et ain Sidi Bouali dans la région de Sefrou (moyen atlas, Maroc). Lar J 20:127–146

Oyenuga AV (1960) Effet of stage of growth and frequency of cutting on the yield and chemical composition of some Nigerian fodder grasses-*Panicum maximum* Jacq. With 10 text-figures printed *in* Great Britain. J Agric Sci 55:339. doi:10.1017/S00218596 00023212

Peterson RH, Coombs K, Power J, Paim U (1989) Responses of several fish species to pH gradients. Rev Can de Zool 67:1566–1572. doi:10.1139/z89-222

Pucher J, Steinbronn S, Mayrhofer R, Schad I, El-Matbouli M, Focken U (2012) Improved sustainable aquaculture systems for small-scale farmers in Northern Vietnam. Environ Sci and Eng, pp 4–8. doi 10.1007/978-3-642-33377

Qiu S, Mc Comb AJ, Bell RW (2012) Leaf litter decomposition and nutrient dynamics in woodland and wetland conditions along a forest to wetland hill slope. Int Sch Res Netw ISRN Soil Sci 12:8

Rejmánková E (2001) Effect of experimental phosphorus enrichment on oligotrophic tropical marshes in Belize, Central America. Plant Soil 236:33. doi:10.1023/A:1011953715153

Rodier JBL, Merlet N, Coll (2012) Analyse de l'eau. 9e édition, DUNOD, p 20225

Steinbronn S (2009) A case study: fish production in the integrated farming system of the Black Thai in Yen Chau district (Son La province) in mountainous North-Western Vietnam—current state and potential. Ph. Dissertation, University of Hohenheim, Stuttgart

Vodounnou DSJV, Kpogue DNS, Tossavi CE, Mensah GA (2016) Effect of animal waste and vegetable compost on production and growth of earthworm (*Eisenia fetida*) during vermiculture. IRWA 5:87–92. doi:10.1007/s40093-016-0119-5

Wilkins RJ (2008) Eco-efficient approaches to land management: a case for increased integration of crop and animal production systems. Philos Trans R Soc B Biol Sci 363:517–525. doi:10.1098/rstb.2007.2167

2

Chrysosporium queenslandicum: a potent keratinophilic fungus for keratinous waste degradation

Tarun Kumar Kumawat[1] · Anima Sharma[2] · Seema Bhadauria[3]

Abstract

Purpose Keratinous wastes are the solid environmental pollutant generated from poultry farms, slaughterhouses and barber's shops. The aim of the present study is the degradation of keratinous wastes in an eco-friendly way by biological methods, which should further be helpful to reduce the wastes and recycled into valuable feed and fertilizers.

Methods Degradation of keratinous substrates was assessed by highly potent keratinophilic fungi, namely *Chrysosporium queenslandicum* TKKASb Apinis and R.G. Rees. This chicken feather degrading fungal strain previously isolated and identified by morphological and 18 s rDNA sequencing in laboratory, was used in the present study. *Ch. queenslandicum* was inoculated into the basal salt medium (BSM) with keratinous substrates for 12 and 24 days at 28 ± 2 °C to observe degradation. The rate of degradation was expressed as weight loss of keratin substrate over incubation days and pH variation.

Results In this study, *Ch. queenslandicum* showed maximum degradation on chicken feathers followed by human nail clippings, animal hair and human hair. The degradation rate on chicken feather was 38.40 ± 0.80 and 46.40 ± 2.50% after 12 and 24 days, respectively. The basal salt medium's pH was increased over incubation time. Scanning electron microscopy (SEM) examination also demonstrated the degradation of chicken feathers.

Conclusions The results suggest that *Ch. queenslandicum* possess the potential biotechnological applications which can be used in the hydrolysis of keratinous waste and recycling of poultry waste for environmental protection. The hydrolyzed keratinous material can also be utilized as the source of fertilizers for plants and feed for animals.

Keywords *Chrysosporium queenslandicum* · Keratinous wastes · Environmental pollutant · Degradation · Fertilizers

Introduction

Environmental wastes are originated in enormous amount from many countries and contain a considerable amount of protein and carbon compounds (Park and Son 2009). The keratinous wastes are considered as the environmental pollutants (Werlang and Brandelli 2005; Kumawat et al. 2016a). Food industry, poultry farm, slaughterhouse, leather industries and wool industries are constantly producing million tons of keratinous wastes. The major producers include USA, Brazil and China, account for more than 40 million tons per year (Sharma and Gupta 2016). According to a current report, India's part alone is 350 million tons (Agrahari and Wadhwa 2010; Sousa et al. 2015). The keratin-containing wastes are dumped, buried, used for land filling, or incinerated (Fellahi et al. 2016). These activities cause the soil, water and air pollution. Discarded feathers cause various human ailments, including chlorosis and fowl cholera (Williams et al. 1991).

Keratin protein is a tough, fibrous and the third most abundant polymer in the environment after cellulose and

✉ Anima Sharma
sharmaanima6@gmail.com

[1] Department of Biotechnology, JECRC University, Jaipur, Rajasthan, India

[2] ICAR-Central Sheep and Wool Research Institute, Arid-Region Campus, Bikaner, Rajasthan, India

[3] Department of Microbiology, JECRC University, Jaipur, Rajasthan, India

chitin (Lange et al. 2016). This protein provides an exoskeleton covering of animals in the form of horns, hooves, hair, and feathers (Gupta and Nayak 2015). The mammalian keratin proteins are categorized into two groups based on their structure and function: (a) hard keratin and (b) soft keratin (Kakkar et al. 2014). Due to the strength and stability of keratin protein, very few microorganisms are able to degrade it down. Keratinophilic fungi have the capability to decompose the keratinous wastes with the production of the keratinase enzyme (Sharma and Rajak 2003; Kumawat et al. 2013; Sharma et al. 2015a). Soils are the best place for the growth and incidence of keratinophilic fungi, because of the presence of organic materials and keratinous substrates (Ulfig 2006; Sharma et al. 2015b).

Keratin-rich by-products, i.e., chicken feathers, horns and hooves are the resource of nutrient for animals (amino acids) and plants (N, S) (Kornillowicz-Kowalska and Bohacz 2011). The preparation of compost fertilizers from the keratinous wastes was also abandoned in the European countries, and from biochemical point of view, the keratinous wastes which contain about 20–25% protein should represent superior source of nitrogen for plants (Gousterova et al. 2003).

Chrysosporium queenslandicum belongs to the family Onygenaceae and isolated first time from the feathers of domestic fowl Cunnumala, Southern Queensland, Australia (Apinis and Rees 1976). Antifungal compound "Queenslandon", and antibacterial compounds "chrysoqueen (1) and chrysolandol (2)" related to members of the dihydronaphthoquinone group was produced by *Chrysosporium queenslandicum* IFM 51121 (Hoshino et al. 2002; Ivanova et al. 2002).

The focus of this paper is to evaluate the keratinous waste decomposition by *Chrysosporium queenslandicum*. Reduction of keratinous wastes by *Ch. queenslandicum* showed the biodegrading potential as well as management of keratinous wastes. The degraded keratinous wastes can be used as the source of feed and fertilizers.

Materials and methods

Fungal culture

The fungal culture of *Chrysosporium queenslandicum* TKKASb ((GenBank accession number KU560575) was isolated from poultry farm house soil of Rajasthan by the hair baiting technique, was earlier confirmed by morphological and 18s rDNA sequencing (molecular characterization) in the laboratory (Kumawat et al. 2016b). The strain was maintained on Sabouraud's dextrose agar (SDA).

Degradation of different keratinous waste substrates

The degradation of keratinous substrates (chicken feathers, human hair, animal hair and human nail clippings) was assessed at different incubation period, i.e., 12 and 24 days by weight loss method (Singh 2003). The keratinous substrates degradation from *Chrysosporium queenslandicum* was tested in basal salt medium (NaCl, 0.5; $MgCl_2{\cdot}6H_2O$, 0.1; $CaCl_2$, 0.06; KH_2PO_4, 0.5; and K_2HPO_4 0.5; keratinous substrates; distilled water, 1 L; pH 7.0) (Sharma et al. 2012). *Chrysosporium queenslandicum* was inoculated into the flask containing 100 ml of BSM with 250 mg of keratinous substrates, and incubated at 28 °C on an orbital shaking incubator (Remi CIS-24 Plus) at 70 rpm for 12 and 24 days. A control, having the keratin substrates without keratinophilic fungal suspension, was also run along with the test flasks. The degradation by *Chrysosporium queenslandicum* was expressed as percentage weight loss of keratinous waste substrates. The experiment was performed in triplicates.

Determination of percentage of degradation

After the incubations days, the keratinous substrates fragments were collected on preweighed Whatman filter paper No. 1 and dried at 40 °C for 48 h and weighed for final weight. The keratin degradation by each keratinophilic species was expressed as percentage weight loss. The percent of keratinous substrates degradation was determined by weight loss method and calculate percentage weight loss using the following equation (Kim et al. 2001):

$$\text{Percentage weight loss } (\%) = (\text{IW} - \text{FW}) \times 100/\text{IW}$$

where IW is the initial weight, FW is the final weight.

pH variation

A clear-cut confirmation to assess the progress of keratinous substrates degradation was obtained by observing the changes in hydrogen ion concentration of the basal salt media. Changes in the pH of the BSM after incubation time were also determined using digital pH meter (Model 181, Electronics India).

Scanning electron microscopy (SEM)

To characterize the degradation pattern of the chicken feather degradation by *Chrysosporium queenslandicum* at microscopic level, scanning electron microscopy (SEM) was used. For examination of a degraded chicken feathers under a scanning electron microscope, the degraded chicken feather sample from basal salt medium (after incubation days) was first washed using distilled water, then treated with acetone for dehydrating followed by air-drying. The dried samples were then mounted over the stubs with double-sided carbon tape. A thin layer of gold was coated over the samples using an automated sputter coater QUARUM-Q15OAS (Quorum Technologies, UK). The specimens and preparations were observed using a scanning electron microscope EVO-18 SEM (Carl Zeiss, Germany) at an accelerating voltage of 20 kV (Mehta et al. 2014).

Statistical analysis of data

Results are given as mean ± Standard error (SE) of n observations taken in three replicates ($n = 3$). Data obtained were statistically analyzed with one-way analysis of variance (ANOVA) and the observations were considered significant when P value was less than 0.05 ($P < 0.05$).

Results

The degradation of various keratinous waste substrates from *Chrysosporium queenslandicum* was accessed on the percent weight loss of substrates during treatment. The degradation percentage (%) after 12 days of incubation revealed significant differences ($P < 0.05$). The percentage of degradation of chicken feathers, human hair, animal hair and human nail clippings treated was found to be 38.40 ± 0.80, 08.80 ± 2.12, 13.60 ± 1.06 and 22.40 ± 1.44%, respectively, after 12 days of incubation. The results of feather degradation of *Chrysosporium queenslandicum* were shown in figure 1. After 24 days incubation, the results of keratinous substrates degradation revealed significant differences ($P < 0.05$). The percentage weight loss of chicken feathers, human hair, animal hair and human nail clippings after 24 days incubation was 46.40 ± 2.50, 22.40 ± 1.74, 19.20 ± 0.40 and 30.80 ± 1.20%, respectively. It was very clear that keratinous waste substrates degradation was highest in sample subjected to long time incubation (Fig. 1).

In the present study, changes in pH of the culture medium due to the keratinous substrates degradation process were observed after 12 and 24 days of incubation. The pH of the control cultures did not change during the incubation periods. Keratinous substrates degradation was accompanied by alkalization of the culture medium with elevation of pH from 7.67 and 7.93 for human hair, 7.90 and 7.73 for animal hair, 7.95 and 8.00 for human nail clippings and 8.11 and 8.16 for chicken feathers, respectively, after 12th and 24th days of incubation (Fig. 2). Statistical analysis showed a significance difference ($P < 0.05$) among tested treatments. Such an alkalinisation of the medium may be due to excretion of excess nitrogen via deamination and ammonium excretion.

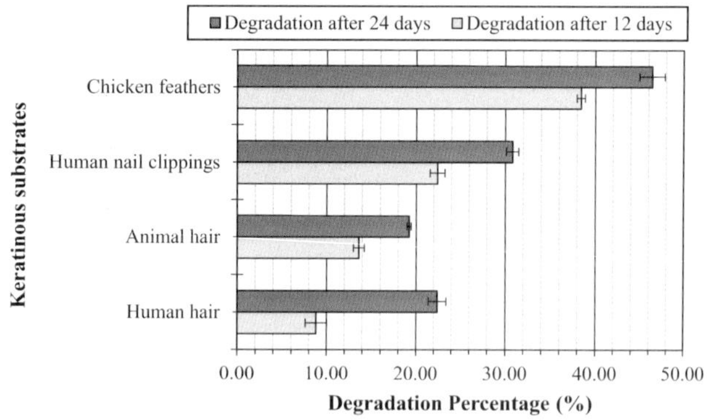

Fig. 1 Degradation percentage (%) of keratinous substrates by *Chrysosporium queenslandicum*; data means and error bars at each point indicate ± SE. Statistical analysis showed a significant difference ($P < 0.05$) among tested treatments

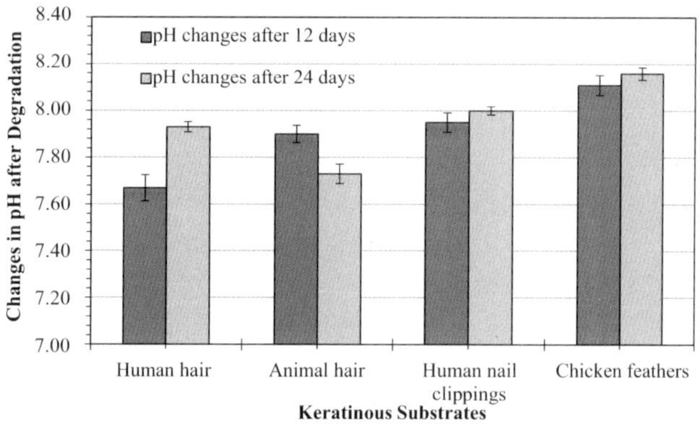

Fig. 2 Changes in pH of basal salt medium after keratin degradation by *Chrysosporium queenslandicum*; Data means and error bars at each point indicate ± SE. Statistical analysis showed a significant difference ($P < 0.05$) among tested treatments

Keratinous substrates from control flasks showed no signs of degradation (Fig. 3a). In contrast, keratin substrates from experimental flasks showed varying degrees of digestion after incubation days (Fig. 3b). Scanning electron microscopy (SEM) was applied to visualize and confirm the degradation of chicken feather by *Chrysosporium queenslandicum*. Image of micrographs clearly shows the degradation of chicken feather shaft and barbs and shows the breakdown and weakening of the chicken feather structure by *Ch. queenslandicum*, thus indicating keratinolysis (Fig. 4a). The shaft of feathers with barbs (Fig. 4b) could be clearly observed in SEM micrographs of uninoculated chicken feathers.

Discussion

Keratinophilic fungi are able to colonize different keratinous substrates and decompose them in low molecular weight (Maruthi et al. 2011). Several keratinophilic fungi

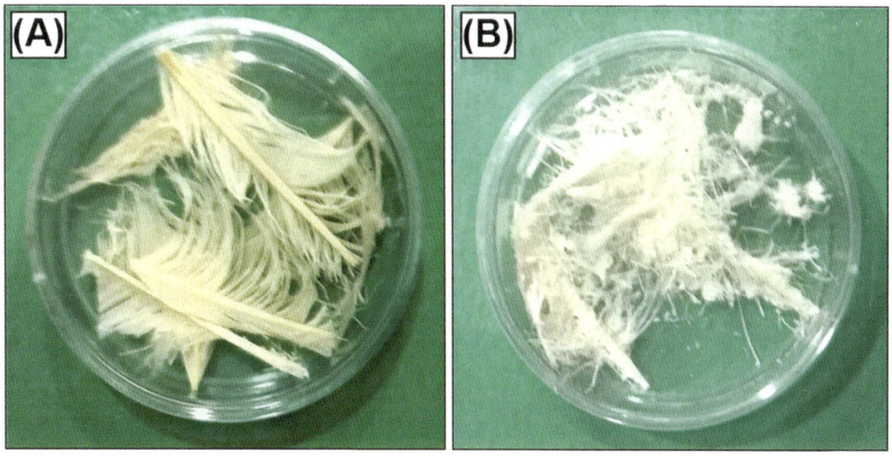

Fig. 3 Degradation of chicken feathers; **a** control chicken feathers, **b** degraded chicken feathers by *Chrysosporium queenslandicum*

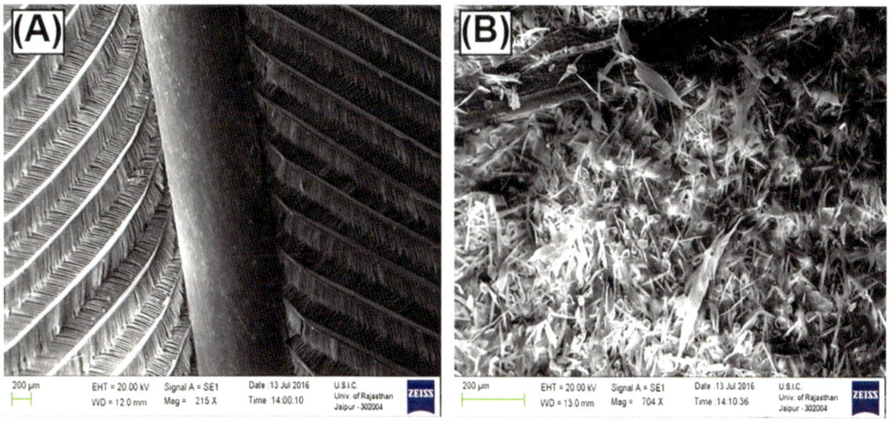

Fig. 4 Scanning electron microscopic (SEM) images of **a** undigested and **b** degraded chicken feathers by *Chrysosporium queenslandicum* at ×704

use keratinous substrates as source of carbon and nitrogen (Gopinath et al. 2015). Keratinous waste degradation in BSM with microbial cultures is one of the most capable techniques to utilize keratin protein and to obtain important by-products (Laba and Rodziewicz 2014). According to Syed et al. (2009), microbiological degradation of keratin waste is also an ecologically and economically safe method of utilization of such waste.

We have isolated and identified a potent keratinophilic fungus, *Chrysosporium queenslandicum*. This isolate showed the degradation of various keratinous substrates. The present study describes the degradation of various keratinous substrates in basal salt medium, changes of pH and scanning electron microscopy. In this study, *Ch. queenslandicum* showed maximum degradation on chicken feathers followed by human nail clippings, animal hair and human hair after 12 days of incubation. After 24 days of incubation, *Ch. queenslandicum* showed maximum degradation on chicken feathers followed by human nail clippings, human hair and animal hair. Muhsin and Hadi (2002) reported that chicken feather was highly degraded by *Aspergillus flavus* (32%). Kumawat et al. (2016a) observed that the keratinous waste degradation was the highest on the chicken feathers (39.2%) and the pH increased from the initial 7.0–8.15 by colonization of *Arthroderma multifidum.*

In the present study, percentage of degradation of human hair was found to be 08.80 ± 2.12 and 22.40 ± 1.74%, after 12 and 24 days, respectively. Muhsin and Hadi (2002) observed the rate of degradation in which human hair had the highest degradation rate *Chrysosporium pannicola* 62% and *Microsporum gypseum* 48%. Deshmukh and Verekar (2014) studied that all isolated 13 species viz. *Aphanoascus durus*, *Arthroderma corniculatum*, *Auxarthron umbrinum*, *Chrysosporium evolceanui*, *Chrysosporium indicum*, *Chrysosporium tropicum*, *Chrysosporium zonatum*, *Chrysosporium* states of *Arthroderma tuberculatum*, *Chrysosporium* state of *Ctenomyces serratus*, *Gymnascella dankaliensis*, *Microsporum gypseum*, *Myriodontium keratinophilum* and *Trichophyton mentagrophytes* were decomposed human hair (18.4–40.2%) after four weeks of incubation.

The degradation percentage for the human nail clippings was 22.40 ± 1.44 and 30.80 ± 1.20% after 12 and 24 days incubation, respectively. Oyeka and Gugnani (1998) reported that *Scytalidium hyalinum* degraded 40%, whereas

Scytalidium japonicum degraded 38% of human nail clippings. In the similar research, in another case *Fusarium solani* degraded the human nail clippings (43%). Sharma et al. (2011) studied that *Microsporum gypseum* (49.34%) and *Trichophyton verrucosum* (49.34%) showed the greatest degradation of animal hair. These keratin substrates exhibited low degradation by *Fusarium oxysporum* (16.66%). Bohacz (2017) reported the ability of strains of *Aphanoascus fulvescens* and *Chrysosporium articulatum* isolated from soil (phaesol) to degrade native feather keratin. According to them, keratinophilic fungi exhibit the keratinolytic activity against native feather keratin.

In this study, the changes in pH were assessed after the degradation of keratinous wastes by *Ch. queenslandicum*. The pH of the medium-containing keratinous substrates was increased after 12 and 24 days incubation in the experimental flask. Kaul and Sumbali (1999) stated that alkalinisation of the medium may be due to excretion of excess nitrogen via deamination and ammonium excretion. Godheja and Shekhar (2014) reported that there was a gradual increase of pH in to the alkaline phase for feathers till 25 day incubation during the process of biodegradation.

Scanning electron microscopy (SEM) was employed to examine structural changes during the chicken feather degradation. Koutb et al. (2012) reported that the chicken feathers were completely degraded after 7 days of incubation by *Aspergillus niger*. Mehta et al. (2014) studied the scanning electron microscopy of chicken feather degradation after seven days by *Bacillus sonorensis* at 8000× magnification. Kshetri and Ningthoujam (2016) reported the degradation of chicken and other bird feathers by *Bacillus* sp. MBRL 575. Degradation of feather barbules was apparent from 12 h of incubation onwards and the complete degradation (98% weight loss) was observed after 48 h at 30 °C. Scanning electron micrograph of degraded chicken feather by *Bacillus* sp. MBRL 575 after 12 and 24 h of incubation were observed by scanning electron microscope.

Conclusion

The keratinophilic fungus isolate, *Chrysosporium queenslandicum,* degraded the various keratinous wastes (chicken feathers, hairs) effectively and showed the keratinolytic activity. The keratinous waste degradation by *Ch. queenslandicum* is not only economical but also a possible process for better management of keratinous wastes. The isolate could be used for biotechnological application in recycling of poultry waste for environmental protection (production of nitrogenous fertilizer and animal feed) and its fermentation broth could be useful in leather industry and textile industry, etc.

Acknowledgement We thank to Director, School of Sciences for encouragement and Head, Department of Biotechnology, JECRC University, Jaipur and Department of Medical Microbiology, PGIMER, Chandigarh (India) for providing the opportunity and facilities to complete the work.

Conflict of interest The authors have no conflict of interest to declare.

References

Agrahari S, Wadhwa N (2010) Degradation of chicken feather a poultry waste product by keratinolytic bacteria isolated from dumping site at Ghazipur poultry processing plant. Int J Poult Sci 9:482–489. doi:10.3923/ijps.2010.482.489

Apinis AE, Rees RG (1976) An undescribed keratinophilic fungus from southern Queensland. Trans Br Mycol Soc 67:522–524. doi:10.1016/S0007-1536(76)80188-X

Bohacz J (2017) Biodegradation of feather waste keratin by a keratinolytic soil fungus of the genus Chrysosporium and statistical optimization of feather mass loss. World J Microbiol Biotechnol 33:13. doi:10.1007/s11274-016-2177-2

Deshmukh SK, Verekar SA (2014) Isolation of keratinophilic fungi from selected soils of Sanjay Gandhi National Park, Mumbai (India). J Mycol Med 24:319–327. doi:10.1016/j.mycmed.2014.08.004

Fellahi S, Chibani A, Feuk-Lagerstedt E, Taherzadeh MJ (2016) Identification of two new keratinolytic proteases from a *Bacillus pumilus* strain using protein analysis and gene sequencing. AMB Express 6:42. doi:10.1186/s13568-016-0213-0

Godheja J, Shekhar SK (2014) Biodegradation of keratin from chicken feathers by fungal species as a means of sustainable development. J Bioremed Biodeg 5:232. doi:10.4172/2155-6199.1000232

Gopinath SCB, Anbu P, Lakshmipriya T et al (2015) Biotechnological aspects and perspective of microbial keratinase production. Biomed Res Int 2015:140726. doi:10.1155/2015/140726

Gousterova M, Nustorova I, Goshev P, Christov D, Braikova K, Tishinov T Haertle, Nedkov P (2003) Alkaline hydrolysate of waste sheep wool aimed as fertilizer. Biotechnol Biotechnol Equip 17:140–145. doi:10.1080/13102818.2003.10817072

Gupta P, Nayak KK (2015) Characteristics of protein-based biopolymer and its application. Polym Eng Sci 55:485–498. doi:10.1002/pen.23928

Hoshino Y, Ivanova VB, Yazawa K, Ando A, Mikami Y, Zaki SM, Karam AZ, Youssef YA, Grafe U (2002) Queenslandon, a new antifungal compound produced by *Chrysosporium queenslandicum*: production, isolation and structure elucidation. J Antibiot (Tokyo) 55:516–519. doi:10.7164/antibiotics.55.516

Ivanova VB, Hoshino Y, Yazawa K, Ando A, Mikami Y, Zaki SM, Grafe U (2002) Isolation and structure elucidation of two new antibacterial compounds produced by *Chrysosporium queenslandicum*. J Antibiot (Tokyo) 55:914–918. doi:10.7164/antibiotics.55.914

Kakkar P, Madhan B, Shanmugam G (2014) Extraction and characterization of keratin from bovine hoof: a potential material for biomedical applications. Springerplus 3:596. doi:10.1186/2193-1801-3-596

Kaul S, Sumbali G (1999) Production of extracellular keratinase by keratinophilic fungal species inhabiting feathers of living poultry birds (*Gallus domesticus*): a comparison. Mycopathologia 146:19–24. doi:10.1023/A:1007086720237

Kim JM, Lim WJ, Suh HJ (2001) Feather-degrading Bacillus species from poultry waste. Process Biochem 37:287–291. doi:10.1016/S0032-9592(01)00206-0

Kornillowicz-Kowalska T, Bohacz J (2011) Biodegradation of keratin waste: theory and practical aspects. Waste Manag 31:1689–1701. doi:10.1016/j.wasman.2011.03.024

Koutb M, Morsy FM, Bagy MMK, Hassan EA (2012) Optimization of extracellular keratinase production by *Aspergillus Terreus* isolated from chicken's litter. J Adv Lab Res Biol 3:210–216

Kshetri P, Ningthoujam DS (2016) Keratinolytic activities of alkaliphilic *Bacillus* sp. MBRL 575 from a novel habitat, limestone deposit site in Manipur, India. Springerplus 5:595. doi:10.1186/s40064-016-2239-9

Kumawat TK, Sharma V, Seth R, Sharma A (2013) Diversity of keratin degrading fungal flora in industrial area of Jaipur and keratinolytic potential of *Trichophyton mentagrophytes* and *Microsporum canis*. IJBBR 4:359–364

Kumawat TK, Sharma A, Bhadauria S (2016a) Biodegradation of keratinous waste substrates by *Arthroderma multifidum*. Asian J Appl Sci 9:106–112. doi:10.3923/ajaps.2016.106.112

Kumawat TK, Sharma A, Bhadauria S (2016b) Effect of culture media and environmental conditions on mycelium growth and sporulation of *Chrysosporium queenslandicum*. Int J ChemTech Res 9:271–277

Laba W, Rodziewicz A (2014) Biodegradation of hard keratins by two bacillus strains. Jundishapur J Microbiol 7:e8896. doi:10.5812/jjm.8896

Lange L, Huang Y, Busk PK (2016) Microbial decomposition of keratin in nature-a new hypothesis of industrial relevance. Appl Microbiol Biotechnol 100:2083–2096. doi:10.1007/s00253-015-7262-1

Maruthi YA, Lakshmi KA Rao SR, Chaitanya DA (2011) *Chrysosporium tropicum*—a potential feather/hair waste degrading keratinophilic fungi. E3 J Environ Res Manag 2:014–018

Mehta RS, Jholapara RJ, Sawant CS (2014) Isolation of a novel feather-degrading bacterium and optimization of its cultural conditions for enzyme production. Int J Pharm Pharm Sci 6:194–201

Muhsin TM, Hadi RB (2002) Degradation of keratin substrates by fungi isolated from sewage sludge. Mycopathologia 154:185–189. doi:10.1023/A:1016335623534

Oyeka CA, Gugnani HC (1998) Keratin degradation by *Scytalidium* species and *Fusarium solani*. Mycoses 41:73–76. doi:10.1111/j.1439-0507.1998.tb00381.x

Park GT, Son HJ (2009) Keratinolytic activity of *Bacillus megaterium* F7-1, a feather-degrading mesophilic bacterium. Microbiol Res 164:478–485. doi:10.1016/j.micres.2007.02.004

Sharma S, Gupta A (2016) Sustainable management of keratin waste biomass: applications and future perspectives. Braz Arch Biol Technol 59:e16150684. doi:10.1590/1678-4324-2016150684

Sharma R, Rajak RC (2003) Keratinophilic fungi: nature's keratin degrading machines! Their isolation, identification and ecological role. Reson 8:28–40. doi:10.1007/BF02837919

Sharma M, Sharma M, Rao VM (2011) *In vitro* biodegradation of keratin by dermatophytes and some soil keratinophiles. Afr J Biochem 5:1–6

Sharma A, Chandra S, Sharma M (2012) Difference in keratinase activity of dermatophytes at different environmental conditions is an attribute of adaptation to parasitism. Mycoses 55:410–415. doi:10.1111/j.1439-0507.2011.02133.x

Sharma V, Kumawat TK, Sharma A, Seth R, Chandra S (2015a) Distribution and prevalence of dermatophytes in semi-arid region of India. Adv Microbiol 5:93–106. doi:10.4236/aim.2015.52010

Sharma V, Kumawat TK, Sharma A, Seth R, Chandra S (2015b) Dermatophytes: diagnosis of dermatophytosis and its treatment. Afr J Microbiol 9:1286–1293. doi:10.5897/AJMR2015.7374

Singh CJ (2003) Optimization of an extracellular protease of *Chrysosporium keratinophilum* and its potential in bioremediation of keratinic wastes. Mycopathologia 156:151–156. doi:10.1023/A:1023395409746

Sousa M, Souza O, Maciel M, Cruz R, Rego MG, Magalhaes O, Pessoa-Junior A, Porto A, Souza-Motta C (2015) Keratinolytic potential of fungi isolated from soil preserved at the Micoteca URM. Eur J Biotechnol Biosci 3:10–15

Syed DG, Lee JC, Li WJ, Kim CJ, Agasar D (2009) Production, characterization and application of keratinase from *Streptomyces gulbargensis*. Bioresour Technol 100:1868–1871. doi:10.1016/j.biortech.2008.09.047

Ulfig K (2006) Sludge liming decreases the growth of keratinolytic and keratinophilic fungi. Polish J Environ Stud 15:341–346

Werlang PO, Brandelli A (2005) Characterization of a novel feather degrading *Bacillus* sp. strain. Appl Biochem Biotechnol 120:71–79. doi:10.1385/ABAB:120:1:71

Williams CM, Lee CG, Garlich JD, Shih JCH (1991) Evolution of bacterial feather fermentation product, feather lyaste, as a feed protein. Poult Sci 70:85–94. doi:10.3382/ps.0700085

3

Agro-environmental assessment of composting plants in Southwestern of Morocco (Souss-Massa Region)

K. Azim[1,2] · S. Komenane[2] · B. Soudi[2]

Abstract

Purpose The region of Souss-Massa generates huge quantities of organic horticultural wastes estimated up to 1,307,465 tons·year^{-1}. Fruit and vegetable producers are asked to process the organic wastes because of the GlobalGAP certification requirements. As a result, there is a need to assess the existing composting plants at regional scale, to improve their efficiency, produce a high-quality soil amendment, free from any pathogens, and weed seeds.
Methods The survey consisted of a diagnosis of the quantitative and qualitative status of the organic horticultural wastes (axillary buds; pruned leaves), and an agro-environmental assessment of nine composting units.
Results The analysis of macronutrients (nitrogen, phosphorus, and potassium) in the organic wastes reveals that 13 million euro worth of N, P, and K can be generated by composting all horticultural wastes (12,000 tons of nitrogen, 6000 tons of phosphorus and 14,000 tons of potassium · year^{-1}), constituting for soils an important source of fertilization backup. Results of field investigations showed that 55% of the composting plants have an area equal to one hectare. The proximity between mature piles and raw materials in 55.5% of cases could highly contaminate them by root knot nematode inoculum. Only 11% of composting units were operating on impermeable surface. The test of circular chromatography showed that 80% of sampled composts are immature, even with high operational costs.
Conclusion Much effort is required to control and optimize the maturation process. The composting activity should be organized through a professional skilled organization, supported by scientific research and government subsidies.

Keywords Tomato by-products · Organic wastes · Composting · Environment · Pollution · Horticulture

Introduction

There is no doubt that sustainable agricultural practices can supply synergies that limit serious effects of the climate change. The report of the FAO stated that sustainable agriculture helps to counteract climate change by restoring soil organic matter content as well as reducing soil erosion and improving soil physical structure (Scialabba and Hattam 2002). Furthermore, global warming increases the rate of soil organic matter mineralization. The evidence for increased carbon sequestration in organic soils seems clear: the addition of manures, compost, mulches, and cover crops restored organic matter (Mae-Wan and Li Ching 2008). Enormous quantities of vegetable wastes are produced annually from the horticultural industries. Composting can be a feasible treatment to stabilize horticultural wastes and, thus, to improve their properties for use as organic fertilizers (Gavilanes et al. 2016). The organic agricultural waste from animal breeding and horticultural production consist mainly of (1) cattle manure and (2) crop residues. The crop residues are considered as plant by-products (Ballerini 2006). In the Souss-Massa region (South Western Morocco), the production of such plant by-

✉ K. Azim
azim.khalid@yahoo.fr

1 Research Unit of Integrated Crop Production, Regional Center of Agronomy Research, INRA-CRRA, B.P.124, CP: 80350 Inzegane, Agadir, Morocco

2 Department of Environment and Natural Resources, IAV Hassan II-Rabat, B.P. 6202, Madinat Al Irfane, Morocco

product has reached 406,045 tons·year^{-1} (ORMVA/SM 2009). The development of the agricultural production can cause several types of pollution (Elasri and Afilal 2016). Those agricultural wastes, which are not well managed, can lead to serious sanitary and environmental complications, such as soil, air, and groundwater pollution. The good management of those wastes would be an effective means of organic matter restoration through carbon restitution of the depleted soils through organic amendments.

Composting of organic waste products to agricultural land is considered as being one of the most economical, practical, and environmentally beneficial management options (Topal et al. 2016). As a rule, the process consists of the natural biological decomposition of organic waste components and involves diverse species of microorganisms. The final product of such decomposition is a humus-like substance, which can be used successfully either as an agent for soil enrichment or as an organic fertilizer (Neklyudov et al. 2006). In this way, composting could be considered as a feasible treatment for recycling solid organic wastes, since it transforms them into organic amendments and fertilizers that can be used in agricultural soils. However, the application of unstable or immature compost into soils may produce several negative effects, such as phytotoxicity, competition for oxygen between microbial biomass and plant roots/seeds, or plant nitrogen deficiency (Gómez-Brandón et al. 2008). Consequently, compost maturity is the most essential criterion both in recycling organic waste, and in its marketing and utilization in agriculture as organic amendment (Lasaridi et al. 2006). Maturity and stability indicators that have been used in other composting studies include carbon–nitrogen (C/N) ratio, microbial activity, germination index, cation exchange capacity (CEC), humic substances content, compost concentration of water soluble carbon (WSC), dissolved organic matter, NH^{4+}–N and NO^{3-}–N; ratios of NH^{4+}–N = NO^{3-}–N, WSC/TN, and WSC/organic-N (Harada and Inoko 1980; Zucconi et al. 1981; Iglesias-Jiménez and Pérez-García 1992).

To deal with the significant quantities of organic wastes generated locally by greenhouse cropping, farmers in the Souss-Massa region try to use a simple composting technology, such as the dynamic pile (aboveground) method to process animal manure and crop residues. This method is implemented with varying efficiency and has not been fully evaluated. In spite of the increasing interest in this type of organic waste processing, few, if any, papers have been published relating to the composting process evaluation in the Sous Massa region. Besides, there has not been, until now, any regulation that could organize composting activity to prevent a negative impact on the environment. In the light of the situation, the objectives of this study were to (1) evaluate the quantitative and qualitative potentials of crop residues and to (2) compare the performance of composting method used by farmers in terms of both agronomic and economic issues.

Materials and methods

Climate and soil conditions of the studied area

The High-Atlas Mountains in the north and the Anti-Atlas Mountains in the southeast delineate the Souss-Massa plain. Its proximity to the Atlantic Ocean generates a favorable climate for winter production, with mild winters and relatively hot and dry summers. The study was carried out in an irrigated area 30 km to the south of Agadir (latitude = 30.6; longitude = 9.36; altitude = 75 m). The region is characterized by an arid climate with climatic mean values are as follows (Table 1).

Survey method

Data were collected, whilst visiting nine compost plants selected randomly from a total of 13 in the studied region. The nine appointed compost plants were named, respectively, A, B, C, D, E, F, G, H, and I; had been visited (Table 2) and using a questionnaire that was completed by the responsible person in charge.

The questionnaire (Table 3) included questions relating to the origin and the nature of the composted raw materials, mixing ratios, sizing of the plants, and the techniques used for the preservation of natural resources such as soil, water, and air quality with respect to the international standards. At the end of the survey for each composting plant, a qualitative rating was used for indicating environmental impact, process handling, mechanization, and quality of compost. The investigator used four levels of importance: low, medium, high, and very high.

Composting plants description

The configuration of the surveyed plants is more or less variable. They are equipped with storage area for raw wastes and with matured compost area, while the maturing area and the fermentation area are generally confused. These different configurations revealed at the level of the compost plants can cause problems with the quality of the products. Indeed, the very close arrangement between the compost piles and the fermentation piles or even the inorganic refuse pool can lead to mixtures of materials and thus interfere with the normal evolution of the composting process. For instance, composting plant A is configured with a distance of 5 m between the compost pile (which is assumed to be mature) and the inorganic refuse pool. In the composting plant D, it was

Table 1 Souss-Massa plain: climatic mean values (INRA-CRRA Compact Meteo station 2013)

Parameter	Temperature (°C min)	Temperature (°C max)	Humidity (max %)	Humidity (min %)	Insolation (hrs $year^{-1}$)	Rainfall (mm $year^{-1}$)
Values	11.5	24.8	85	55	3600	173

Table 2 Souss-Massa: GPS location of the composting plants

Compost plant	Latitude (°, min, s)	Longitude (°, min, s)
A	30°10′36.9019″N	9°28′07.8298″W
B	30°07′35.8684″N	9°26′58.7508″W
C	30°07′29.9302″N	9°27′12.5878″W
D	30°03′45.0941″N	9°30′44.9514″W
E	30°05′47.1638″N	9°21′57.1954″W
F	30°07′40.8626″N	9°31′02.0095″W
G	30°02′00.0755″N	9°30′17.1227″W
H	30°11′04.3552″N	9°29′57.7241″W
I	30°11′18.4859″N	9°25′43.0473″W

Table 3 Detail of the questionnaire

Nature of the question	Number of questions
Description of the sourcing farms (units that generate organic wastes)	11
Description of the organic wastes generated	14
Composting plant description and sizing	72
Waterproofness	3
Reception, sorting and control area	12
Raw material storage area	10
Fermentation area (thermophile area)	20
Maturation area (curing area)	11
Sifting area	10
Drainage system	6
Operational details	10
Certification (GlobalGAP…)	8
Total of questions	115

noticed that the fermentation piles are almost in the middle of crop wastes and decaying fruits and vegetables. This proximity could lead to the presence of phytopathogens and nematodes inoculum in the final product. Vegetables and especially tomato fruits wastes are disposed for free decaying and not managed at all.

Sampling method

During the survey, samples were taken from the raw wastes intended for the composting, and also at the level of the produced compost. Concerning raw material, different samples were taken from tomato wastes (leaves, axillary buds, and the entire end cycle plant), other vegetables (leaves and stems), citrus (spring pruning buds), banana (post-harvest trunk), and sheep manure. A composite sample of 2 kg of each origin was taken, and three sub-samples of produced compost were taken from at least 70 cm depth at the shaded pile side to constitute a composite sample. All the samples were put in plastic bags, weighed for moisture content after drying at 105 °C 24 h in the oven. Dried samples from raw material and matured compost served for laboratory analysis. Sampling from compost plants C and F were not carried out because of the absence of raw material and matured compost.

Laboratory analysis

Oven dried samples of raw material and compost were sieved ($\varnothing = 2$ mm) and tested for pH in 1:10 soil/water (w/v) suspensions using a glass electrode pH meter (Bio-Block pH meter Microprocessor 99621) at 25 °C (Rhoades 1982). Electrical conductivity (EC) (dS/m) was measured using an electrical conductivity meter (WTW Inolab-Cond Level-2P, Germany, at T°25 °C) in 1:5 soil/water (w/w) solutions. Total Organic Matter content (TOM) was determined by calcination of a dry sample with known moisture content, at 480 °C for 6 h. Total nitrogen content analysis was performed using the standard Kjeldahl procedure (Gerhardt: Kjeldatherm-Vapodest 20, Germany) as described by (Jackson 1967). C/N ratio was calculated by dividing total carbon (issued from TOM) on total nitrogen. Total phosphorus (TP) and total potassium were determined as described by Mehlich (1953). Calcium and magnesium were analyzed by shaking 1 g of compost sample with sodium acetate solution for 1 h prior to atomic absorption spectroscopy detection (Varian Spectra AA 220FS, USA). Qualitative tests of compost samples were performed using the circular chromatography test with silver nitrate as reagent (Brinton 1983).

Results and discussion

Assessment of the organic wastes generated at regional scale

Investigations at farm level compiled with the previous studies (ORMVA/SM 2009; Heck and El Harrak 2010),

Table 4 Estimation of potential macronutrients in the organic wastes produced in Souss-Massa region in 2011

Crops	Quantities of organic wastes produced (T)	Dry Matter content (T)	Percentage of dry organic wastes	Macronutrients content (% w/w of DM)			Estimation of potential macronutrient in the organic waste (T)		
				N	P_2O_5	K_2O	N	P_2O_5	K_2O
Citrus	495,418	435,968	42.41	0.69	1.05	0.95	3008	4578	4142
Other vegetables crops (except tomato wastes)	458,045	290,126	28.22	2.38	0.12	0.5	6905	348	1451
Tomatoes	337,431	297,648	28.95	0.7	0.31	3.01	2084	923	8959
Banana	14,618	2997	0.29	1.17	0.31	1.8	35	9	54
Sheep manure	1953	1325	0.13	1.19	0.31	1.11	16	4	15
Total	1,307,465	1,028,063	100	–	–	–	12,048	5862	14,620

Table 5 Estimation of the economic value of organic agricultural wastes

Macronutrient	Nutrient content of the organic agricultural wastes (tons·year^{-1})	Average price of the corresponding fertilizer (k€/ton)	Economic value of the organic waste (k€)	Percentage (%)
Nitrogen (N)	12,048	0.38	4578.24	33.32
Phosphorus (P)	5862	0.44	2579.28	18.78
Potassium (K)	14,620	0.45	6579.00	47.90
Total Economic Value (k€)			13,736.52	–

Table 4 indicate the quantities of potential macronutrients in the organic wastes. About 25% of the biomass is generated from tomato greenhouses. Tomato wastes represent the major source of interest not only because of the quantities produced, but also due to the high level of water content (71%) which reduces direct processing via on-site composting. Citrus (wood and leaves from autumn and spring pruning) and other vegetable crop residues constitute more than 70%, and are generally not recycled in the soil but presented as fodder for cattle, or in case of abundance, not composted but thrown outside the farms. Workers for cooking and heating purposes generally take woody residues from citrus plantations. Cow manure production represents only 0.13% of the organic wastes; this can be explained by its direct soil amendment by farmers in their citrus orchards, since most of them have dairy production. Hence, actual quantities of produced cow manure are not recorded. On the other hand, farmers use sheep manure in organic fertilization rather than cow manure because of its low price and its availability from other regions.

In terms of macronutrients, potential nitrogen from other vegetables crop wastes seems to be important. About 6900 tons of nitrogen that represents 57.2% of the total potential nitrogen can be valorized from vegetables crop residues (except tomato residues) by composting. Similarly, potential potassium from tomato wastes is estimated up to 9000 tons and represents about 61% of the total potential potassium source in the region.

According to the analysis of the macro-element (nitrogen, phosphorus and potassium) of the organic waste samples (Table 5), 13 million euro worth of N, P, and K can be generated through composting of organic residues from different horticultural crops. Farmers commonly take into consideration the fertilizer uptake of the crops during the cropping cycle when setting fertilization program. Unfortunately, they do not consider the representation of macronutrients concealed in the organic waste generated during the cultivation cycle. Potential potassium worth is highly significant and represents about 48% due to its high content particularly in tomato residues.

Composting plant description and sizing

Description

Collected data of this first part are given in Table 6, and represent the description and the positioning of the compost plants within their environments.

The assessment showed that 67% of the composting plants have an area equal or higher that 1 ha (average of 1.24 ha), and are completely separated from the production farm, reflecting the awareness of the producers of contamination risks due to inappropriate composting technics.

Size

The size of the composting units did not respect the logical workflow of the composting process, mainly due to

Table 6 Description of the composting plants environment

Compost plant	Total area of the farm (ha)	Compost plant surface	Distance from the well (m)	Distance from the habitation	Soil type
A	1	1 ha	5	≈ 10 m	Sandy loam
B	1	1 ha	>500	≈ 50 m	Sandy loam
C	6	2000 m^2	>500	>500 m	Sandy loam
D	4	150 m^2	>500	>500 m	Sandy loam
E	220	1 ha	>500	>500 m	Sandy
F	1	1 ha	≈ 500	≈ 25 m	Sandy loam
G	1	1 ha	>500	>500 m	Sandy loam
H	22	700 m^2	≈ 5	≈ 25 m	Sandy loam
I	45	600 m^2	0	≈ 1.5 km	Sandy loam

confusion between fermentation and maturing phase. The proximity between the raw material and maturation piles in 55.5% of cases could result in a high degree of contamination by phytopathogens and nematode inoculum. In addition, only 11% of the surveyed units had impermeable surface and drainage facilities to collect compost leachates. Two management approaches have been identified:

- *High-quality compost production* 56% of composting plants are mixing sheep manure and olive mill wastes to produce compost;
- *Organic waste management* 44% of composting plants convert tomato waste into an acceptable organic amendment. The produced compost is used as an amendment to soil of citrus orchard to avoid potential inoculation with root knot nematodes that could not be suppressed by composting due to the low level of sanitization.

Operational details as illustrated in Table 7, data set reveals that only 37.5% of compost plants are treating tomato wastes, whilst the rest are composting manure and olive mill wastes. The lack of scientific and technical data of tomato wastes composting could be the origin of this result. Based on survey responses, mixing ratios in all the compost plants are based on empirical experience instead of scientifically based recommendations. Composting plants that manage tomato wastes do not use any additional raw material to decrease the C/N ratio to 30 as recommended by literature. Tomato wastes, in general, consist of the whole plant (main stem, leaves, and some small immature fruits) and have a C/N ratio about 47 making the composting process too long (7 months). Roots are not composted, but incinerated to avoid root knot nematode contamination to amended soils.

Agronomic performance of the material and the final compost

Table 8 reveals that C/N ratios of different final composts are above the optimal value of 12–15, and it is mostly due to the mixing ratio of different raw materials that result in the initial C/N ratio above the optimal value of 30. As a result, 40% of final composts are not mature and still need to be processed for an extra maturation phase. Total nitrogen of final compost OMW + SM and SM are better when compared to the compost of TW. This is due mainly to the fact that composting TW only is not optimal with regard to its high C/N ratio and low moisture content. Concerning the agronomic value, nutrient content of all final compost complies with the recommendations of the Bulletin of Transfer of Technology in Agriculture (Soudi 2005).

With regard to the nutrient content from the final compost samples, results showed that for 100 g of compost, average of nutrients content can be estimated as follows: nitrogen = 0.82 g; phosphorus = 0.5 g; potassium = 0.958 g. The Total Organic Carbon (TOC) is high in all composts and exceeds the threshold given by the World Health Organization (WHO). The high levels of the TOC are due to the initial TOC of the raw material, as well as there was no additional raw material added with high TN to decrease the C/N ratio of the mixture. High levels of TOC will need more time (3 months) to be decomposed, thus increasing operational costs of the produced compost. Regarding the other standards, the produced compost complies with the acceptance thresholds. The composts have a moisture content ranging from 30 to 50% and a rather alkaline pH ranging from 8.4 to 9.8. Electrical Conductivity (EC) measurements show values ranging from 2.6 to 4.8 dS/m. These values are in conformity with threshold given by Francou (2003).

Qualitative test of final compost

This test of maturity (reagent silver nitrate) was used to assess the maturity of various composts. Among five tested composts, only 20% presented an advanced degree of maturity whereas 80% were immature. It is likely that immature compost could spread phytopathogens and

Table 7 Operational details of the surveyed composting plants

Compost unit	Parameters										
	Process duration (months)	Quantity of treated wastes (T $year^{-1}$)	Nature of waste	Origin of wastes	Cost (wastes + transportation) €/T	Mixing ratio	Fermentation area (S)	Fermentation phase (months)	Maturing phase (months)	Height of pile (m)	Width of pile (m)
A	4	7000	Sheep manure	Local market	41	50%	2500 m^2	2	2	1.5	2
			Olive mill wastes	Marrakech region	18	50%					
B	6	2160	Tomato waste	On farm (54 ha)	0	100%	–	2	4	2	2
C	3	–	Sheep manure	Beni Mellal region	45	33%	100 m^2	2	1	2	5
			Cow manure			33%					
			Olive mill wastes	Marrakech region	18	33%					
D	4–7	20,880	Tomato waste	On farm (240 ha)	0	100%	300 m^2	4	Variable	3	3
E	3	2400	Sheep manure	Local market	13	100%	0.5 ha	2	1	2.5	2
F	5–7	4698	Tomato waste	On farm (70 ha)	0	28% leaves	0.5 ha	4	Variable	4	4
						70% stems					
						2% fruits					
H	5	744	Cow manure	Local market	40	85%	210 m^2	1	4	1.6	3
			Olive mill wastes	Marrakech region	23	15%					
I	5	276	Sheep manure	Local market	40	66%	400 m^2	1	4	1.5 m	2.5 m
			Olive mill wastes	Marrakech region	33	33%					

Table 8 Physical and chemical analyses of raw material and final compost

Parameters	Input material						Compost						WHO standards (1993)
	OMW + SM		SM		TW		OMW		TW		SM		Min–Max
	Average	SD	Average	SD	Average	SD	Average	SD	Average	SD	Average	SD	–
Moisture	36%	17%	26%	16%	40%	18%	37%	6%	47%	3%	37%	4%	–
pH	9.40	0.28	7.69	1.05	8.14	0.48	9.40	0.582	8.93	0.06	8.63	0.20	6–9
EC (dS/m)	5.86	0.08	2.98	2.51	2.85	1.17	3.77	0.287	2.92	0.32	4.32	0.56	2–5
TOC	67.01	1.41	71.63	23.30	77.55	9.11	63.25	5.596	60.25	1.32	56.47	2.74	30–40
TMM	32.99	1.41	28.37	23.30	22.45	9.11	36.75	5.596	39.75	1.32	43.53	2.74	–
C/N Ratio	39.79	9.60	50.24	11.65	29.92	7.18	23.92	0.767	22.94	0.44	32.41	3.33	20–30
TN	0.87	0.19	0.78	0.41	1.36	0.37	1.33	0.145	1.31	0.03	0.88	0.08	0.10–1.80
P_2O_5	2.50	1.84	2.00	0.10	1.21	1.35	1.62	0.801	0.73	0.05	1.61	1.09	0.10–2.30
K_2O	1.74	0.25	0.99	0.56	1.19	0.12	1.58	0.454	1.57	0.08	0.98	0.76	0.10–1.70
Na_2O	0.44	0.41	0.27	0.26	0.24	0.20	0.29	0.259	0.16	0.01	0.09	0.05	–
CaO	6.78	1.66	4.58	3.12	3.77	1.35	5.30	2.403	7.67	0.67	5.35	3.27	–
MgO	0.79	0.32	0.42	0.14	0.51	0.14	0.81	0.378	0.74	0.03	0.38	10	–

OMW Olive Mill Waste, *SM* Sheep Manure, *TW* Tomato Waste, *SD* Standard Deviation, *WHO* World Health Organization

Table 9 Summary of the assessment criteria of the composting plants in the Souss-Massa

Compost plants	Environmental impact	Process handling	Mechanization	Quality of compost
A	Very high	High	Low	Medium
D	Very high	High	High	Low
E	Medium	Very high	Medium	Medium
F	Very high	High	Very high	Low
G	High	High	Very high	High
H	Low	High	Very low	Medium
I	Low	High	Very low	Medium

nematodes, especially root knot nematodes that could result in up to 90% of tomato yield loss. Furthermore, the amendment of immature compost could cause asphyxia of roots because of the high level of oxygen consumption by soil microorganisms as stated by Smith and Hughes (2004). As a soil amendment, immature compost will continue to be degraded by microorganisms that consume more oxygen and consequently cause damage to the plant at rhizosphaera level.

Assessment criteria of the composting plants in Souss-Massa

The process of composting presents a positive effect on the environment, especially for the effective management of produced waste. However, there are some failures in the management of mechanization, especially in relation to the quality of the final product. Table 9 shows the assessment summary of the studied composting plants.

Environmental impacts are high because of the absence of the impermeability of the composting plant. Leachates (especially during the rainy season) can infiltrate the sandy soil and be considered as a potential pollutant of the ground water. Concerning the process handling, management of the residues is very organized to the extent of enabling a rapid workflow, consequently processing the maximum residues per year. Regarding the mechanization, there is a heterogeneity in the perception. The composting plants that turn tomato and citrus residues are more mechanized than the others that compost sheep and olive mill wastes. This can be explained by the fact that tomato and citrus residues are more important in volumes and hence need more mechanization. Quality of the compost is the most important criterion for this assessment, but could be at odds with the qualitative test mentioned above. The quality of compost in this assessment refers to the physical and chemical parameters (agronomic parameters), whilst the maturity test refers to the degree of humification of the organic matter. High levels of TOC require a long maturation phase to allow proper humification.

Conclusion

This diagnosis concludes that there is a potential of organic wastes estimated up to 1 Million of dry matter year^{-1}, and 13 million euro worth of N, P, and K can be generated by composting all horticultural wastes. Compost has not only beneficial effects on soil fertility, but also on plant nutrition. However, the lack of regulation relating to the composting activity in Morocco engenders problems of process monitoring and low quality of produced compost. This maturity not being achieved can present serious phytotoxic risks for the crops. The functioning of the composting plants raises problems of water pollution, whilst the effects on the soil need to be studied further. The management policies of wastes should favor its valorization, either in terms of recycling, or in terms of energy valuation. Composting activity should be organized through a professional association, and employs a skilled labor force to ensure impermeable composting plants. The use of computer software for the optimization of mixtures should also be encouraged. Furthermore, the optimal composting approach is the open dynamic windrow, using necessarily a compost turner to homogenize the organic material. There is no need to remind that the monitoring of temperature and oxygen uptake and carbon dioxide release are important to achieve an adequate maturity level. For this purpose, also there are many multi-parametric devices available for large composting plant. The acceptance by local farmers of the benefits of compost to the soil and crops would be a significant incentive to reduce the expensive mineral fertilizers and optimize their use. The development of the composting activities is actually limited and has to be supported by scientific research to afford a better knowledge concerning the real impact of compost on agriculture, environment, and public health.

Acknowledgements We thank without exception all persons in charge of the farms and the composting plants who helped us to complete this study. It was partially financed by the Moroccan-French cooperation in the framework of the project PRAD 13-09 N°28050 PE between INRA (National Institute of Agronomy Research), IAV (Agronomic and Veterinary Institute of Hassan II), and the APEFEL (Association of Producers and Exporters of Fruits and Vegetables).

References

Arslan Topal E, Ünlü A, Topal M (2016) Effect of aeration rate on elimination of coliforms during composting of vegetable–fruit wastes. Int J Recycl Org Waste Agric 5(3):243–249. doi:10.1007/s40093-016-0134-6

Ballerini D (2006) Les biocarburants: état des lieux. Perspectives et enjeux du développement. Editions TECHNIP

Brinton WF (1983) A qualitative method for assessing humus condition. In: Knorr D (ed) Sustainable food systems. AVI Publishing, Westport, p 416

Elasri O, El amin Afilal M (2016) Potential for biogas production from the anaerobic digestion of chicken droppings in Morocco. Int J Recycl Org Waste Agric 5(3):195–204. doi:10.1007/s40093-016-0128-4

Francou C (2003) Stabilisation de la matière organique au cours du compostage de déchets urbains: influence de la nature des déchets et du procédé de compostage—Recherche d'indicateurs pertinents. Dissertation, Institut national agronomique Paris-Grignon

Gavilanes-Terán I, Jara-Samaniego J, Idrovo-Novillo J, Bustamante M, Moral R, Paredes C (2016) Windrow composting as horticultural waste management strategy—a case study in Ecuador. Waste Manag 48:127–134

Gómez-Brandón M, Lazcano C, Domínguez J (2008) The evaluation of stability and maturity during the composting of cattle manure. Chemosphere 70(3):436–444

Harada Y, Inoko A (1980) Relationship between cation-exchange capacity and degree of maturity of city refuse composts. Soil Sci Plant Nutr 26(3):353–362

Heck P, El Harrak R (2010) Etude sur les potentiels de biomasse pour la région Souss-Massa-Drâa et la province d'Essaouira. Project final report of the case Study. GTZ Organization, p 216. https://www.giz.de/de/downloads/fr-etude-sur-les-potentiels-de-biomasse.pdf. Last Accessed 08 Jan 2017

Iglesias Jiménez E, Pérez García V (1992) Determination of maturity indices for city refuse composts. Agric Ecosyst Environ 38(4):331–343

Jackson ML (1967) Soil chemical analysis. In: Karlen DL, Varvel GE, Bullock DG, Cruse RM (1994) Crop rotations for the 21st century. Advances in Agronomy, pp 1–45, 53

Lasaridi K, Protopapa I, Kotsou M, Pilidis G, Manios T, Kyriacou A (2006) Quality assessment of composts in the Greek market: the need for standards and quality assurance. J Environ Manag 80(1):58–65

Mae-Wan H, Li-Ching L (2008) Mitigating climate change through organic agriculture and localized food systems. ISIS Report 31/01/2008. http://www.i-sis.org.uk/mitigatingClimateChange.php. Last Accessed 08 Jan 2017

Mehlich A (1953) Determination of P, Ca, Mg, K, Na and NH_4^+. North Carolina Soil Test Division. Publ., North Carolina Department of Agriculture, Raleigh, pp 1–57. http://www.ncagr.gov/AGRONOMI/pdffiles/mehlich53.pdf. Last Accessed 08 Jan 2017

Neklyudov AD, Fedotov GN, Ivankin AN (2006) Aerobic processing of organic waste into composts. Appl Biochem Microbiol 42:341. doi:10.1134/S0003683806040016

ORMVA/SM (2009) Etude de la gestion des déchets agricoles dans la zone d'action de l'ORMVA/SM comprise à l'intérieur de la province de Chtouka Ait Baha. Office Régionale de la Mise en Valeur Agricole du Souss Massa Agadir. Final Technical Report of consultation, Group EDIC-SEGU, p 83

Rhoades JD (1982) Methods of soil analysis. Part 2. Chemical and Microbiological Properties. Agron Monogr 9.2. ASA, SSSA, Madison. doi:10.2134/agronmonogr9.2.2ed

Scialabba NE-H, Hattam C (2002) Organic agriculture, environment and food security. Environment and Natural Resources Service

Sustainable Development Department FAO, Rome. http://www.fao.org/docrep/005/Y4137E/y4137e00.htm#TopOfPage. Last Accessed 08 Jan 2017
Smith D, Hughes J (2004) Changes in maturity indicators during the degradation of organic wastes subjected to simple composting procedures. Biol Fertil Soils 39(4):280–286
Soudi B (2005) The composting of cattle manure and crop residues from greenhouse production system. Bullet Transfer of Technology in agriculture. PNTTA June edition (2005) No 129. http://www.agrimaroc.net/129.pdf. Last Accessed 08 Jan 2017
Zucconi F, Pera A, Forte M, De Bertoldi M (1981) Evaluating toxicity of immature compost. BioCycle 22:54–57

4

The phytoextraction potential of selected vegetable plants from soil amended with oil palm decanter cake

A. Embrandiri[1] · P. F. Rupani[1] · M. Shahadat[2] · R. P. Singh[3] · S. A. Ismail[4] · M. H. Ibrahim[1] · M. O. Abd. Kadir[1]

Abstract

Purpose The investigation of the phytoextraction potential of three vegetable plants grown in soils amended with decanter cake.

Method Pot experiments were conducted to investigate the response of decanter cake composition on the phytoextraction of metals (Mg, Zn, Ni and Cu) by lady's finger, tomato and brinjal plants. The phytoextraction properties of these plants were determined by calculating the bioconcentration and translocation factors at different decanter cake amendments (10, 20 and 30%).

Results Results indicated that in all three plants, there was no transfer of excess metal ions from the control soil or amendments to the fruit portion as evidenced by bioconcentration factor (BCF). In addition, substantial amount of the metals was found to be accumulated in the roots and shoots, which depicts the phytoextraction ability of these vegetable plants. The translocation factors (TF) of the three plants were found to be higher than control plants. The accumulation of metal ions did not exceed the permissible standards for vegetables thus rendering the fruits safe for human consumption.

Conclusion On the basis of significant findings, lady's finger, tomato and brinjal plants were not found to be suitable for phytoextraction of metals as both BCF and TF were not greater than 1.

Keywords Decanter cake · Bioconcentration factor · Translocation factor · Phytoextraction · Palm waste · Lady's finger · Tomato · Brinjal

Introduction

The unprecedented rise in land use area for oil palm plantations in Malaysia and other palm-producing nations in the world have led to the immense quantities of wastes generated annually (Liew et al. 2015). This has lead to a big nuisance with regards to ground water pollution, infestation by rodents, spread of disease and, not to mention, the obvious odour issues. Various methods of re-utilizing the residues of the palm oil production process have been reported by a number of researchers. Almost all of the wastes have found use in one way or the other in different industries. However, with the exception of Embrandiri et al. (2012, 2013, 2016), Haron and Mohammed (2008), Razak et al. (2012), Gafar et al. (2013) and Adam et al. (2014) there is insufficient literature to show the utilization of

✉ A. Embrandiri
ashanty66@gmail.com

M. Shahadat
mdshahadat93@gmail.com

R. P. Singh
rajeevprataps@gmail.com

S. A. Ismail
sultanismail@gmail.com

M. H. Ibrahim
mhakimi@usm.my

M. O. Abd. Kadir
akomar@usm.my

[1] Environment Division, School of Industrial Technology, Universiti Sains Malaysia, 11800 Pulau Pinang, Malaysia

[2] Department of Biochemical Engineering and Biotechnology, Indian Institute of Technology, IIT Delhi, New Delhi 110016, India

[3] Institute of Environment and Sustainable Development, Banaras Hindu University, Varanasi 221005, India

[4] Eco-Science Research Foundation, Chennai 600041, India

Table 1 Analysis of soil, fresh decanter cake and mixtures (mean ± SD)

Parameter	Unamended soil (0% DC)	10% DC	20% DC	30% DC	100% DC
pH	$7.93^{a} \pm 0.03$	$6.81^{b} \pm 0.06$	$6.71^{b} \pm 0.02$	$6.64^{b} \pm 0.02$	$4.40^{c} \pm 0.01$
EC ($mScm^{-1}$)	$1.43^{b} \pm 0.02$	$1.67^{b} \pm 0.06$	$2.72^{a} \pm 0.03$	$2.87^{a} \pm 0.02$	$2.96^{a} \pm 0.02$
Cd ($mg\ g^{-1}$)	$0.13^{a} \pm 0.02$	$0.11^{a} \pm 0.01$	$0.10^{a} \pm 0.01$	$0.08^{a} \pm 0.01$	$0.05^{a} \pm 0.01$
Pb ($mg\ g^{-1}$)	$0.77^{a} \pm 0.11$	$0.67^{b} \pm 0.03$	$0.63^{b} \pm 0.04$	$0.60^{b} \pm 0.01$	$0.56^{c} \pm 0.03$
Ni ($mg\ g^{-1}$)	$0.77^{a} \pm 0.01$	$0.73^{a} \pm 0.02$	$0.66^{ab} + 0.03$	$0.65^{ab} \pm 0.01$	$0.63^{ab} \pm 0.02$
Cu ($mg\ g^{-1}$)	$0.09^{b} \pm 0.02$	$0.15^{b} \pm 0.02$	$0.18^{b} \pm 0.04$	$0.30^{b} \pm 0.02$	$0.81^{a} \pm 0.04$
Mg ($mg\ g^{-1}$)	$0.001^{b} \pm 0.01$	$0.003^{b} \pm 0.02$	$0.004^{b} \pm 0.01$	$0.005^{b} \pm 0.02$	$0.012^{a} \pm 0.03$
P ($mg\ g^{-1}$)	$0.53^{c} \pm 0.03$	$0.72^{b} \pm 0.02$	$0.86^{b} \pm 0.06$	$1.00^{a} \pm 0.09$	$1.23^{a} \pm 0.03$
N (%)	$0.95^{c} \pm 0.02$	$1.74^{bc} \pm 0.01$	$2.13^{b} \pm 0.04$	$3.34^{b} \pm 0.02$	$6.52^{a} \pm 0.03$
K ($mg\ g^{-1}$)	$0.002^{c} \pm 0.01$	$0.006^{c} \pm 0.01$	$0.008^{c} \pm 0.01$	$0.017^{b} \pm 0.01$	$0.068^{a} \pm 0.02$
Organic carbon (%)	$2.60^{e} \pm 0.03$	$10.2^{d} \pm 0.02$	$16.3^{c} \pm 0.01$	$28.8^{b} \pm 0.03$	$74.4^{a} \pm 0.02$
C:N ratio	$2.74^{d} \pm 0.03$	$5.86^{c} \pm 0.01$	$7.65^{c} \pm 0.03$	$8.62^{b} \pm 0.02$	$11.44^{a} \pm 0.02$
Bulk density ($g\ cm^{-3}$)	$0.98^{a} \pm 0.01$	$0.87^{b} \pm 0.03$	$0.83^{b} \pm 0.02$	$0.77^{c} \pm 0.01$	$0.57^{d} \pm 0.04$

Source: Embrandiri (2016)

Superscript letters indicate the significant difference at $p > 0.05$

decanter cake in agriculture, most especially with respect to plant growth. Notwithstanding, several studies have been reported to utilize organic wastes for the enhancement of plant growth. (Kelepsei et al. 2009; Ndaeyo et al. 2013; Adjei-Nsiah and Obeng 2013; Danaher and Pickens 2016; Kumar and Chopra 2016). Therefore, this work contributes to knowledge on the employment of decanter cake for the growth of vegetable plants in palm-producing countries. Although the wastes from the oil mills are expected to be organic in composition, there is unforeseen contamination of metals. This is as a result of leaks in the machinery or during washing process (Singh 2008). Due to the expected increase in oil palm acreage in the years ahead, it is eminent that serious measures have to be taken to ensure the proper management of decanter cake to address the growing environmental concerns. The aim of the present research was to investigate the uptake and accumulation of metal ions in the different plant parts from the soil and decanter cake amendments via the bioconcentration and translocation factors. The focus was on the transfer of nutrients and which parts have accumulated more metals than others and not on the potential of the plant to remediate the soil. It is generally found that when the BCF value is ≤ 1, the plant can only absorb but not accumulate metals, but BCF value of >1 denotes that plant can accumulate metals.

Methodology

Decanter cake (DC) samples sourced from a mill in Penang, Malaysia, were dried and powdered and then sieved with a 2 mm mesh sieve. Soil samples used in the experiment were collected from a depth of 10 cm also from the vicinity of the mill. It was dried, homogenized and sieved (2 mm) according to standard protocol. The experiments consisted of 3 kg of DC-amended soils in planter pots (15 cm × 15 cm) arranged in completely, randomized block design (CRBD). Amendments (0% (control), 10, 20 and 30% w/w DC) were mixed homogenously, filled into pots and left for 15 days stabilization (Singh 2008). The seeds of lady's finger, tomato and brinjal/eggplant were immersed in distilled water for 3 min prior to sowing to enhance germination. Ten seeds were sown manually at equidistant positions and seedlings were thinned to three plants per pot and kept in the nursery throughout the experiment.

For the analysis of metals in plant parts, all the samples were dried in the oven at 80 °C. A fixed amount of the ground sample (1 g) was digested in 20 mL tri-acid mixture (HNO_3:H2SO_4:$HClO_4$ in the ratio 5:1:1) till a clear solution was obtained (Allen et al. 1986). The concentrations of metal ions were determined from the digested solution using the atomic absorption spectrophotometer (PerkinElmer Analyst 200 AAS 2007).

The uptake and distribution of metals in different plant parts were calculated from the translocation and bioconcentration factor values. To evaluate the plants' ability in transferring metals from the roots to the aerial parts, the translocation factor (TF) was calculated as shown in the formula below according to Mattina et al. (2003):

$$TF = \frac{\text{Concentration of metals in aerial parts at different treatments}}{\text{Concentration of metal in root}} \times 100. \qquad (1)$$

The bioconcentration factor (BCF) was used to calculate the metal uptake capacity from soil to plant tissues, using the equation below (Zhao et al. 2003):

Table 2 Bioconcentration factors of metals in leaves, shoots and fruits of lady's finger grown in different decanter cake amendments (mean ± SD)

Plant part	DC (%)	Cu	Ni	Zn	Mg
Leaf	0	$0.914^{a} \pm 0.06$	$0.266^{c} \pm 0.02$	$0.522^{b} \pm 0.09$	$0.010^{bc} \pm 0.01$
	10	$0.604^{b} \pm 0.05$	$0.473^{b} \pm 0.03$	$0.706^{b} \pm 0.16$	$0.030^{bc} \pm 0.01$
	20	$0.200^{c} \pm 0.04$	$0.418^{b} \pm 0.02$	$0.840^{a} \pm 0.15$	$0.020^{bc} \pm 0.01$
	30	$0.269^{c} \pm 0.07$	$0.371^{c} \pm 0.01$	$0.735^{a} \pm 0.09$	$0.020^{bc} \pm 0.01$
	0	$0.839^{a} \pm 0.06$	$0.501^{b} \pm 0.01$	$0.401^{c} \pm 0.48$	$0.010^{bc} \pm 0.02$
Shoot	10	$0.751^{a} \pm 0.05$	$0.327^{c} \pm 0.01$	$0.947^{a} \pm 0.33$	$0.020^{b} \pm 0.01$
	20	$0.560^{b} \pm 0.06$	$0.339^{c} \pm 0.04$	$0.640^{b} \pm 0.16$	$0.010^{a} \pm 0.01$
	30	$0.214^{c} \pm 0.02$	$0.302^{c} \pm 0.01$	$0.715^{b} \pm 0.18$	$0.010^{ab} \pm 0.02$
	0	$1.200^{a} \pm 0.23$	$0.520^{a} \pm 0.03$	$0.400^{c} \pm 0.08$	$0.010^{bc} \pm 0.01$
Fruit	10	$0.490^{c} \pm 0.401$	$0.511^{a} \pm 0.05$	$0.240^{d} \pm 0.05$	$0.030^{bc} \pm 0.01$
	20	$0.350^{c} \pm 0.09$	$0.501^{b} \pm 0.30$	$0.271^{d} \pm 0.06$	$0.050^{a} \pm 0.04$
	30	$0.696^{b} \pm 0.19$	$0.547^{a} \pm 0.01$	$0.280^{d} \pm 0.03$	$0.040^{a} \pm 0.05$

Mean values in the same column followed by the same superscript letters are not significantly different ($p > 0.05$)

Table 3 Bioconcentration factors of metals in leaves, shoots and fruits of tomato grown at different decanter cake amendments (mean ± SD)

Plant part	DC (%)	Ni	Cu	Zn	Mg
Leaf	0	$0.42^{ef} \pm 0.01$	$1.95^{a} \pm 0.40$	$2.14^{c} \pm 0.26$	$0.00^{ab} \pm 0.01$
	10	$0.46^{f} \pm 0.02$	$0.88^{a} \pm 0.10$	$0.80^{ab} \pm 0.15$	$0.02^{c} \pm 0.02$
	20	$0.34^{bcd} \pm 0.01$	$1.53^{a} \pm 0.33$	$0.67^{ab} \pm 0.15$	$0.01^{c} \pm 0.01$
	30	$0.41^{def} \pm 0.02$	$1.23^{a} \pm 0.15$	$0.87^{ab} \pm 0.13$	$0.01^{c} \pm 0.02$
	0	$0.34^{bcd} \pm 0.03$	$0.75^{a} \pm 0.25$	$1.83^{bc} \pm 0.07$	$0.01^{ab} \pm 0.02$
Shoot	10	$0.27^{ab} \pm 0.01$	$1.16^{a} \pm 0.16$	$0.80^{ab} \pm 0.15$	$0.01^{c} \pm 0.01$
	20	$0.30^{bc} \pm 0.02$	$0.57^{a} \pm 0.15$	$0.69^{ab} \pm 0.16$	$0.01^{c} \pm 0.01$
	30	$0.22^{a} \pm 0.01$	$0.29^{a} \pm 0.03$	$0.71^{ab} \pm 0.11$	$0.01^{ab} \pm 0.03$
	0	$0.35^{cde} \pm 0.07$	$3.58^{b} \pm 0.72$	$2.07^{c} \pm 0.30$	$0.01^{ab} \pm 0.03$
Fruit	10	$0.27^{ab} \pm 0.11$	$1.19^{b} \pm 0.63$	$0.85^{ab} \pm 0.26$	$0.01^{b} \pm 0.01$
	20	$0.30^{bc} \pm 0.03$	$0.48^{b} \pm 0.08$	$0.66^{ab} \pm 0.38$	$0.01^{ab} \pm 0.02$
	30	$0.23^{a} \pm 0.03$	$0.27^{b} \pm 0.11$	$0.54^{a} \pm 0.17$	$0.03^{a} \pm 0.01$

Mean values in the same column followed by the same superscript letters are not significantly different ($p > 0.05$)

$$\mathrm{BCF} = \frac{\text{Concentration of metals in plant parts}}{\text{Concentration of metal in test soil}}. \quad (2)$$

Statistical analysis

The obtained data were subjected to one-way analysis of variance (ANOVA) using SPSS 16. Duncan's multiple range test (DMRT) was performed to test the significance of difference between the treatments at $p = 0.05$.

Results and discussion

The accumulation of metals by plants generally depends on a number of factors including the physicochemical properties of the soil, the species planted, climatic conditions and chemical speciation of metals (Alloway et al. 1991; Greger 1999). The initial physiochemical characteristics are presented in Table 1. The results of BCF revealed the highest level of Zn in the leaves of brinjal plants, which were grown in control soil (2.37 ± 0.26) as compared to other plants, whereas the BCF values of Mg BCF in all three plants were found to be in very low range (0.01–0.05) (Tables 1, 2, 3). The bioconcentration factor of Cu was highest in the fruits of lady's finger grown in control (3.58 ± 0.72), while BCF of Ni was maximum in brinjal leaf at 10% DC treatment (0.66 ± 0.02). This varying trend can be attributed to the chelation properties between the metals and amino acids (Odum et al. 2005). Kim et al. (2003) have accredited the difference in metal concentrations in various plant parts to compartmentalization and translocation through the vascular system. Plants could accrue and store metals in root and stem in nontoxic forms (Baker and Brooks 1989).

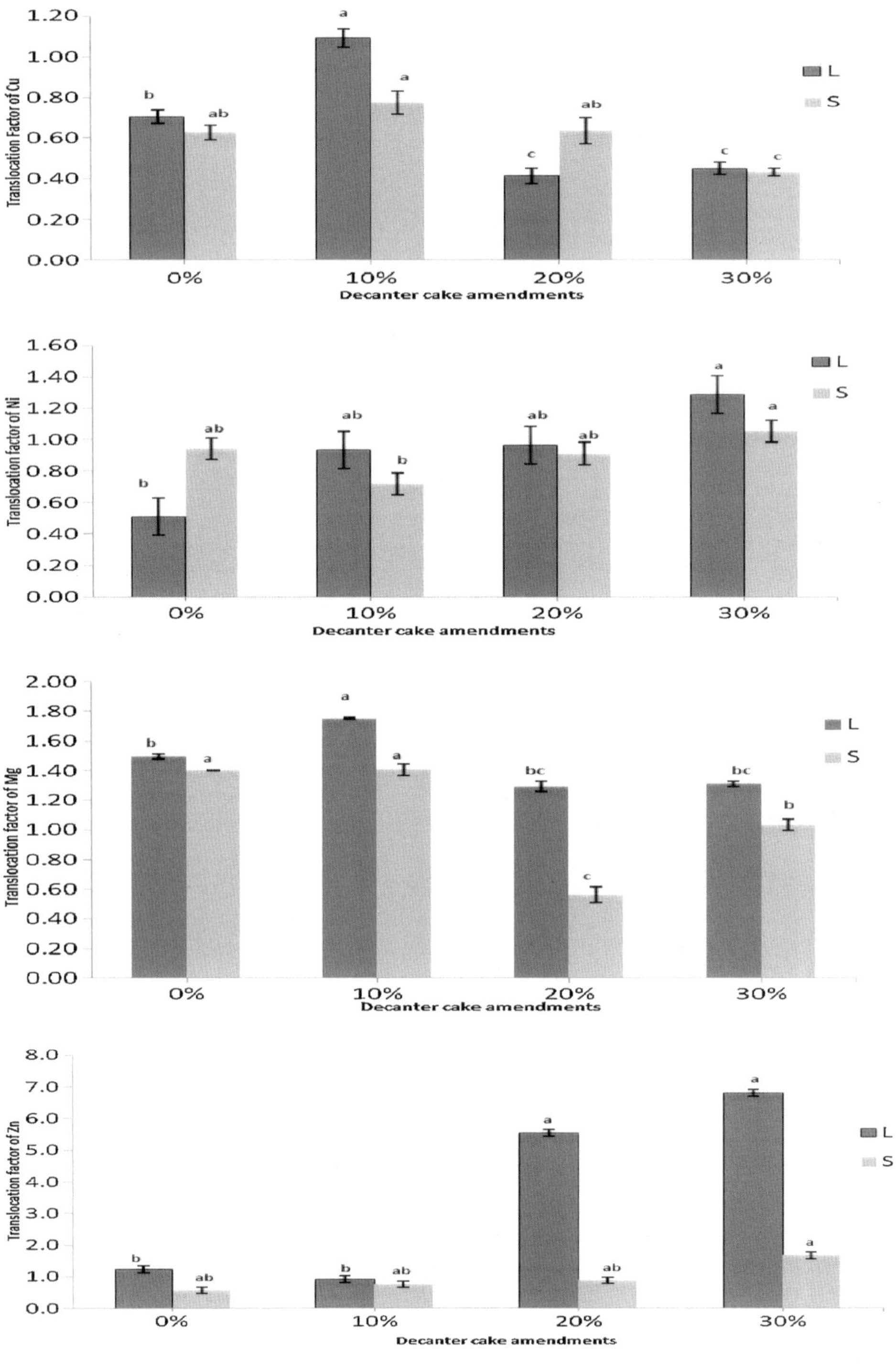

Fig. 1 Bar graphs showing the translocation factors of Cu, Ni, Mg and, Zn of the leaves (L) and shoots (S) of lady's finger plants (mean ± SD)

Lower BCF values indicate that the test plants showed difficulties to mobilize metals from the soil. The movement of metal ions from roots to shoots occurred due to speciation of elements, soil pH and soil organic matter (Kabata-Pendias and Pendias 1984). The wide variations in BCF values of heavy metals for sunflower, castor oil, alfalfa and mustard plants grown in hydroponic cultures were observed by Zhi-xin et al. (2007). The BCF values obtained for control and DC-amended plants were less than 1 (Tables 1, 2, 3). In tomato and brinjal plants, some BCF values were found to be above 1, which can also be considered safe for consumption as they do not exceed the recommended WHO limits.

The value of TF for all three plants was higher than control plants but with respect to individual metal ions and plant parts there were varying trends. TF value of Zn in the leaves of lady's finger plants was greater than 1 (Fig. 1), whereas in both shoots and leaves, translocation

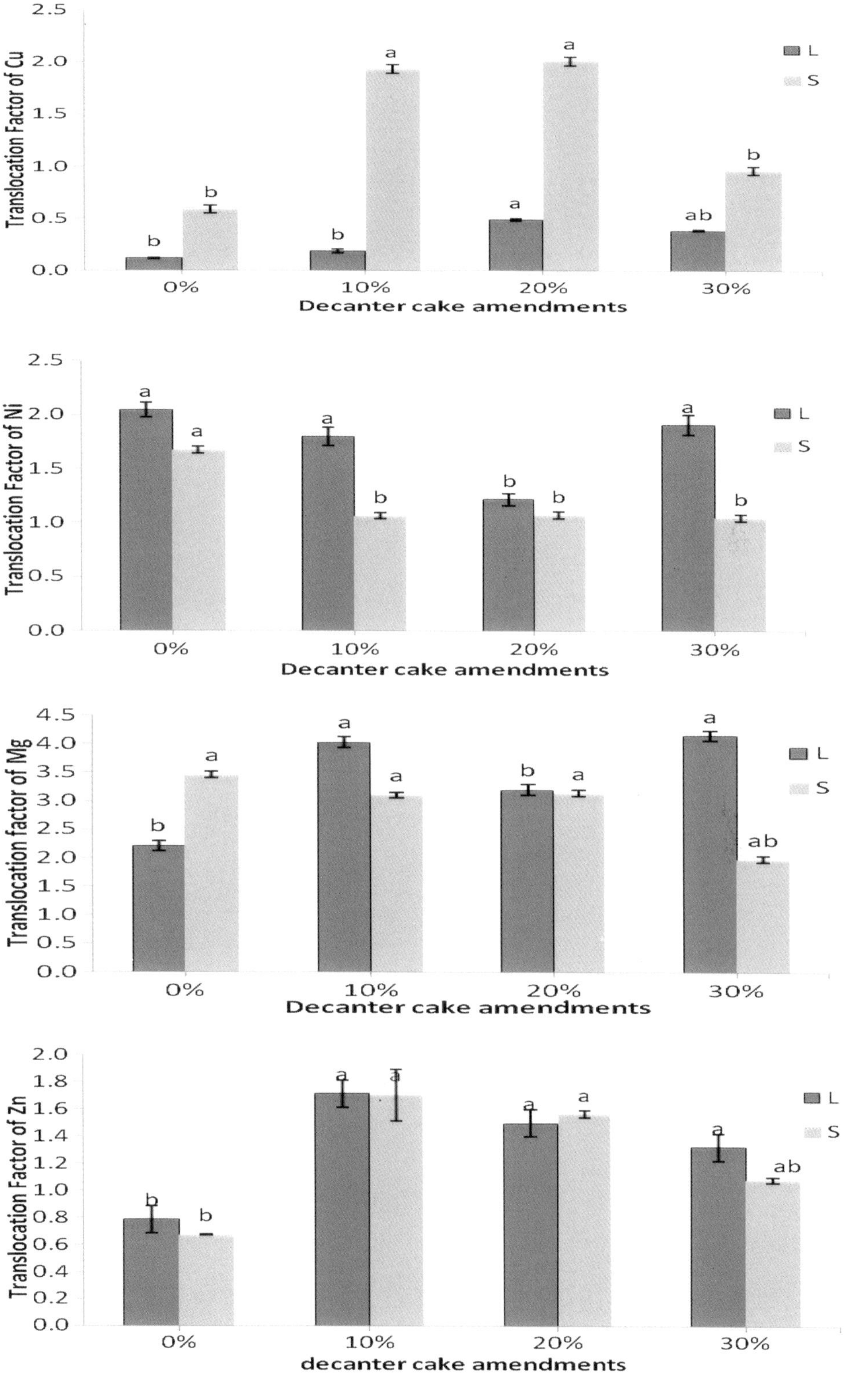

Fig. 2 Bar graphs showing the translocation factors of Cu, Ni, Mg and Zn of the leaf (L) and shoot (S) portions of tomato plants (mean ± SD)

factor of Ni was higher in only 30% DC soil. In tomato plants, all treatments showed translocation factor of Zn greater than 1, except in control, whereas for translocation factor of Mg (Fig. 2) values were greater than 1 in root and shoot portions. For brinjal plants, Ni TF was higher than 1 in leaves and shoots (Fig. 3). This varying concentration of metals is due to the difference in assimilation rate and plant physiology (Singh et al. 2007, 2009, 2012). Even though tomato and brinjal belong to the same family, brinjal has a higher rate of

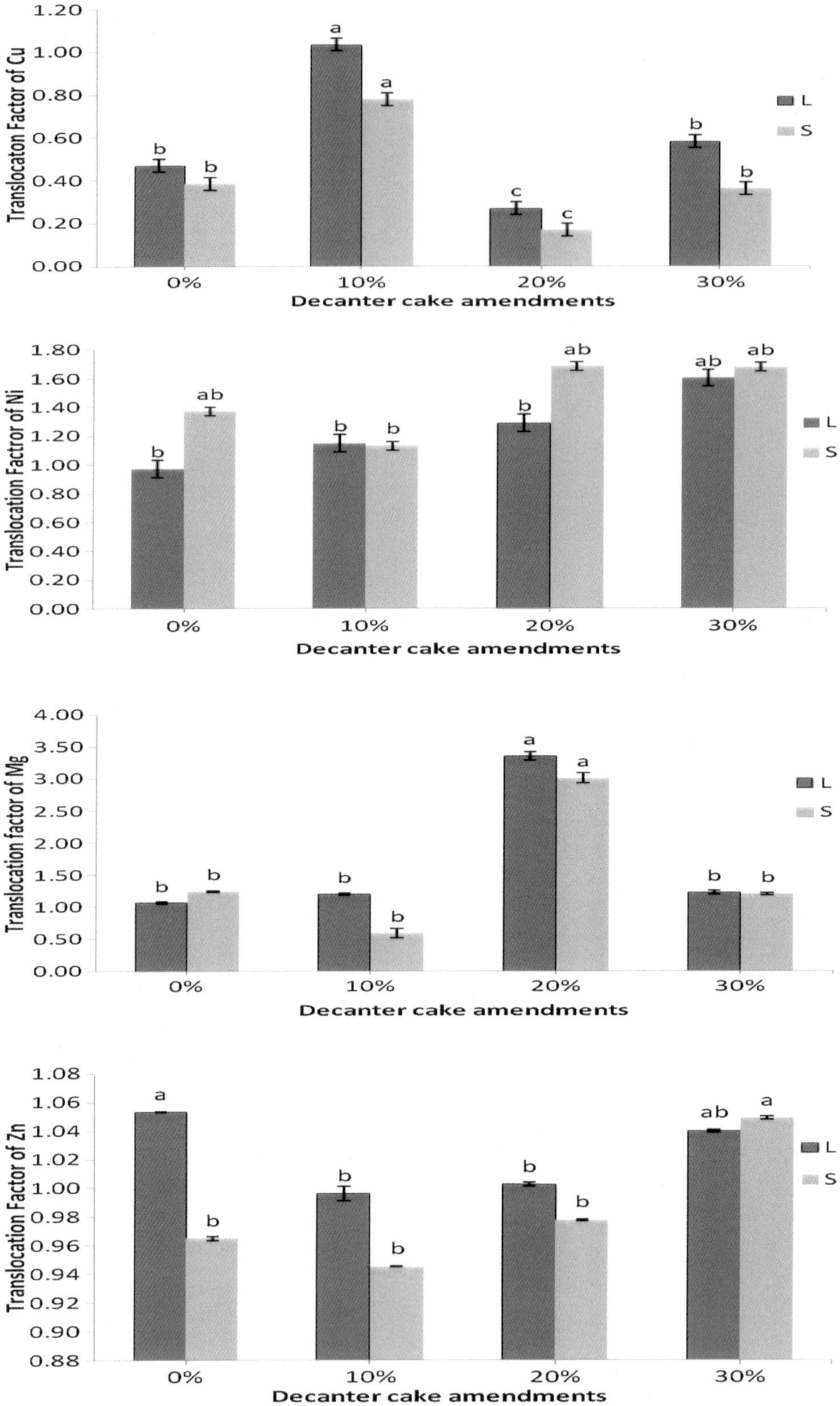

Fig. 3 Bar graphs showing the translocation factors of Cu, Ni, Mg and Zn of the leaf (L) and shoot (S) portions of brinjal plants (mean ± SD)

assimilation of nutrients while tomato and lady's finger do not. This is could be attributed to the different physiological mechanisms and graft incompatibility in solanaceous plants (Kawaguchi et al. 2008). In a similar study, the translocation factor of Fe for *Jatropha curcas* clones was higher in plants grown in the control soil than in the treatment plants (Ghavri and Singh 2010). Plants with both bioconcentration factor and translocation factor

Fig. 4 Pictures during the experiment. **a** Lady's finger fruit in 10% DC in soil. **b** Brinjal plant at 30% DC in soil. **c** Tomato plants uprooted at 10% DC. **d** Fresh decanter cake

Table 4 Bioconcentration factors of metals in leaves, shoots and fruits of brinjal plants at different decanter cake amendments (mean ± SD)

Plant part	DC (%)	Cu	Ni	Zn	Mg
Leaf	0	$1.85^{a} \pm 0.57$	$0.62^{ef} \pm 0.01$	$2.37^{c} \pm 0.26$	$0.03^{ab} \pm 0.01$
	10	$0.78^{a} \pm 0.10$	$0.66^{f} \pm 0.02$	$0.73^{ab} \pm 0.15$	$0.05^{c} \pm 0.02$
	20	$1.43^{a} \pm 0.33$	$0.54^{bcd} \pm 0.01$	$0.58^{ab} \pm 0.15$	$0.02^{c} \pm 0.02$
	30	$1.13^{a} \pm 0.15$	$0.61^{def} \pm 0.02$	$0.72^{ab} \pm 0.13$	$0.02^{c} \pm 0.01$
	0	$0.85^{a} \pm 0.54$	$0.54^{bcd} \pm 0.03$	$1.74^{bc} \pm 0.17$	$0.02^{ab} \pm 0.02$
Shoot	10	$1.26^{a} \pm 0.16$	$0.47^{ab} \pm 0.03$	$0.60^{ab} \pm 0.15$	$0.03^{c} \pm 0.03$
	20	$0.67^{a} \pm 0.15$	$0.50^{bc} \pm 0.02$	$0.59^{ab} \pm 0.16$	$0.01^{c} \pm 0.04$
	30	$0.39^{a} \pm 0.03$	$0.42^{a} \pm 0.01$	$0.61^{ab} \pm 0.11$	$0.02^{ab} \pm 0.03$
	0	$2.37^{b} \pm 0.72$	$0.55^{cde} \pm 0.07$	$2.17^{c} \pm 0.30$	$0.04^{ab} \pm 0.03$
Fruit	10	$1.04^{a} \pm 0.63$	$0.47^{ab} \pm 0.11$	$0.95^{ab} \pm 0.06$	$0.02^{b} \pm 0.01$
	20	$0.38^{a} \pm 0.08$	$0.50^{bc} \pm 0.03$	$0.76^{ab} \pm 0.38$	$0.01^{ab} \pm 0.02$
	30	$0.17^{a} \pm 0.11$	$0.53^{a} \pm 0.08$	$0.64^{a} \pm 0.07$	$0.01^{a} \pm 0.02$

Mean values in the same column followed by the same superscript letters are not significantly different ($p > 0.05$)

greater than 1 (TF and BCF > 1) have the potential to be used in phytoextraction (Yoon et al. 2006). Therefore, based on data in this study, all three tested plants did not prove to have phytoextraction potential. On the other hand, Yoon et al. (2006) also noted that plants with bioconcentration factor greater than 1 and translocation factor less than 1 (BCF > 1 and TF < 1) have the potential for phytostabilization (Fig. 3; Table 4). Figure 4 presents some of the pictures taken during the experiment.

Conclusions

This study revealed that even though there was bioconcentration and translocation of metals such as Ni, Cu, Mg and Zn from the soil to plant parts, concentrations did not transfer to the fruit portions or exceed permissible FAO limits for vegetables. There is potential of lady's finger, tomato and brinjal to be used as phtyostabilizers with thorough investigations. Oil palm decanter cake compost amendments could therefore serve as a potential fertilizer source without the risk of metal accumulation and toxicity. Characterization of decanter cake and soil is necessary prior to pot application as metal mobility increases with lowering of pH and enhancement of organic carbon in the receptor soil. Decanter cake can function as an economical source of soil amendment to farmers particularly in Malaysia and other palm-producing nations. This would, over time, reduce the dependency on synthetic chemicals that have resulted in severe damages to flora and fauna communities and by extension to man.

Acknowledgements Authors are grateful to the Universiti Sains Malaysia RUI Grant (1001/PTEKIND/81124) for the financial assistance.

References

Adam MA, Sulaiman A, Said MS, Som MA, Bahruddin AS, Mokhtar MN (2014) Preliminary study of oil palm decanter cake natural polymer composite (OPDC-NPC). Adv Mater Res 911:40–44

Adjei-Nsiah S, Obeng CB (2013) Effect of palm bunch ash application on soil and plant nutrient composition and growth and yield of garden eggs, pepper and okra. Int J Plant Soil Sci 2(1):1–15. doi:10.9734/IJPSS/2013/2039

Allen SE, Grim HM, Rowland AP (1986) Chemical analysis. In: Moore PD, Chapman SB (eds) Methods in plant ecology. Blackwell Scientific Publication, Oxford, pp 285–344

Alloway BJ, Jackson AP, Morgan H (1991) The behavior of heavy metals in sewage sludge amended soils. Sci Total Environ 100:151–176

Baker AJM, Brooks RR (1989) Terrestrial higher plants which hyper accumulate metallic elements: a review of their distribution, ecology and phytochemistry. Biorecovery 1:81–126

Danaher JJ, Pickens JM, Sibley JL (2016) Tomato seedling growth response to different water sources and a substrate partially replaced with dewatered aquaculture effluent. Int J Recycle Org Waste Agricult 5:25. doi:10.1007/s40093-016-0114-x

Embrandiri A, Ramli AA, Singh RP, Ibrahim MH (2012) Land application of biomass residue generated from palm oil processing: its potential benefits and threats. The Environmentalists 32(1):111–117

Embrandiri A, Singh RP, Ibrahim MH (2013) Biochemical, morphological and yield responses of Lady's finger plants to varying ratios of decanter cake application as a bio-fertilizer. Int J Recycle Org Waste Agricult (IJROWA) pp 2–7. doi:10.1186/2251-7715-2-7

Embrandiri A, Quaik S, Rupani PF, Srivastava V, Singh P (2015) Sustainable utilization of oil palm wastes: opportunities and challenges. In: RP Singh, Sarkar A (eds) In waste management: challenges, threats and opportunities, Nova Science Publishers, New York, pp 217–232

Embrandiri A (2016) Effects of oil palm decanter cake on growth of selected vegetables. PhD thesis, Universiti Sains Malaysia

Gafar AA, Alimon AR, Sazili AQ, Man YC, Abubakar AR (2013) Effect of varying levels of palm oil decanter cake on feed intake, growth performance and carcass characteristics of Kacang goats. IOSR J Agricult Vet Sci (IOSR-JAVS) 3(4):24-29

Ghavri SV, Singh RP (2010) Phytotranslocation of Fe by biodiesel plant *Jatropha curcas* L. grown on iron rich wasteland soil. Br Soc Plant Physiol 22(4):235–243

Greger M (1999) Metal availability and bioconcentration in plants. In: Prasad MNV, Hagemeyer J (eds) Heavy metal stress in plants-from molecules to ecosystems. Springer, Berlin, pp 1–27

Haron K, Mohammed AT (2008) Efficient use of inorganic and organic fertilisers for oil palm and potential utilisation of decanter cake and boiler ash for biofertiliser production. National Seminar on Biofertiliser. Biogas and Effluent Treatment in the Oil Palm Industry, Pp, pp 21–32

Kabata-Pendias A, Pendias H (1984) Trace elements in soils and plants, 4th edn. CRC Press, Boca Raton, FL. 85:107–129

Kawaguchi M, Taji A, Backhouse D, Oda M (2008) Anatomy and physiology of graft incompatibility in solanaceous plants. J Hortic Sci Biotechnol 83(5):581–588

Kelepsei S, Tzortzakis NG (2009) Olive mill wastes: a growing medium component for seedling and crop Production of lettuce and chicory. Int J Veg Sci 15:325–329. doi:10.1080/14620316.2008.11512427

Kim IS, Kang HK, Johnson-Green P, Lee EJ (2003) Investigation of heavy metal accumulation in *Polygonum thunbergii* for phytoextraction. Environ Pollut 126:235–243

Kumar V, Chopra AK (2016) Effects of sugarcane pressmud on agronomical characteristics of hybrid cultivar of eggplant (*Solanum melongena* L.) under field conditions. Int J Recycle Org Waste Agricult 5:149. doi:10.1007/s40093-016-0125-7

Liew WL, Kassim MA, Muda K, Loh SK, Affam AC (2015) Conventional methods and emerging wastewater polishing technologies for palm oil mill effluent treatment: a review. J Environ Manag 149:222–235. doi:10.1016/j.jenvman.2014.10.016

Mattina MJI, Lannucci-Berger W, Musante C, White JC (2003) Concurrent plant uptake of heavy metals and persistent organic pollutants from soil. Environ Pollut 124:375–378

Ndaeyo NU, IkehAO NK, Akpan EA, Udoh EI (2013) Growth and foliar yield responses of waterleaf (*Talinum triangulare* Jacq) to complementary application of organic and inorganic fertilizers in a ultisol. Am J Exp Agricult 3(2):324–335. doi:10.9734/AJEA/2013/2599

Odum E, Barret P, Gary W (2005) Fundamentals of ecology, 5th edn. Thomson Brooks/Cole, London

Razak MNA, Ibrahim MF, Yee PL, Hassan MA, Aziz SA (2012) Utilization of oil palm decanter cake for cellulase and polyoses production. Biotechnol Bioprocess Eng 17:547–555

Singh RP (2008) Land application of sewage sludge: appraisal of potential effects on plants. PhD Dissertation, Banaras Hindu University

Singh RP, Agrawal M (2007) Effects of sewage sludge amendment on heavy metal accumulation and consequent responses of *Beta vulgaris* plants. Chemosphere 67:2229–2240

Singh RP, Agrawal M (2009) Use of sewage sludge as fertilizer supplement for *Abelmoschus esculentus* plants: physiological, biochemical and growth responses. Int J Environ Waste Manag 3:91–106

Singh P, Sulaiman O, Hashim R, Peng LC, Singh RP (2012) Using biomass residues from oil palm industry as a raw material for pulp and paper industry: potential benefits and threat to the environment. Environ Dev Sustain 12(2):1–17

Yoon J, Cao X, Zhou Q, Ma LQ (2006) Accumulation of Pb, Cu, and Zn in native plants growing on a contaminated Florida site. Sci Total Environ 368:456–464

Zhao FJ, Lombi E, McGrath SP (2003) Assessing the potential for zinc and cadmium phytoremediation with the hyperaccumulator *Thlaspi caerulescens*. Plant Soil 249(1):37–43

Zhi-Xin N, Li-Na S, Tie-Heng S, Yu-Shuang L, Hing W (2007) Evaluation of phytoextracting cadmium and lead by sunflower, ricinus, alfalfa and mustard in hydroponic culture. J Environ Sci 19:961–967

5

Enhancement in the productivity of ladies finger (*Abelmoschus esculentus*) with concomitant pest control by the vermicompost of the weed salvinia (*Salvinia molesta*, Mitchell)

Naseer Hussain[1] · Tasneem Abbasi[1] · S. A. Abbasi[1]

Abstract
Purpose In a novel attempt, vermicompost derived from an intransigent and noxious weed salvinia was assessed for its fertilizer value and pest repellent properties.
Methods In outdoor experiments which simulated the way vegetables are cultivated by farmers, ladies finger (*Abelmoschus esculentus*) seeds were germinated and grown in soil supplemented with salvinia vermicompost at four levels: 0 (V_0), 2.5 (V_1), 3.75 (V_2) and 5 (V_3) t/ha. Besides assessing germination success and subsequent growth, yield, and biochemical content of the plants, the impact of pest attacks on them was also studied.
Results Salvinia vermicompost significantly enhanced germination success, growth, and yield of the plants. Maximum growth in terms of shoot length (96.2 cm), root length (48.2 cm), shoot and root dry weight (23.31, 7.96 g), stem diameter (14.04 mm), and number of leaves and branches (26.8, 4.8) was recorded in V_4 (5t/ha). Likewise, the mineral and biochemical content in vermicompost-treated plants was significantly higher than in the controls. The vermicompost also induced resistance in plants against pests and disease. Compared to the controls, vermicompost had reduced the fruit borer infection by 65, 78 and 82% in V_1, V_2 and V_3, respectively.
Conclusion The toxicity of salvinia is largely eliminated when it is vermicomposted, and the product acquires the qualities of a good organic fertilizer. The present work can potentially lead to the development of an inexpensive, sustainable and eco-friendly method of utilizing billions of tons of phytomass that is generated annually by salvinia, and which presently goes to waste.

Keywords Weed control · Organic fertilizer · Vermitechnology · Pest repellent · *Abelmoschus esculentus*

Introduction

Salvinia molesta Mitchell, a free-floating fern, is among the most widespread and environmentally harmful of the invasive plant species (Schooler et al. 2011). It is distinguished by its explosive rate of multiplication and growth, which is perhaps the highest of all plants (Crites et al. 2006). Salvinia has invaded lakes, rivers, channels and other wetlands in tropical and subtropical habitats around the world (Room 1990), covering several of them from bank-to-bank, and elbowing out most other vegetation (Luque et al. 2014; Bhat 2016). After spreading horizontally, salvinia forms dense mats of up to 1 m thickness that cuts the sunlight off from the underlying water and makes the waterways unnavigable (Coetzee et al. 2011). This also hampers the photosynthesis in the water body and the aerial replenishment of oxygen in the water lying beneath the salvinia carpet is blocked (Hussain et al. 2016b). As a result, the dissolved oxygen levels fall and the neustans and benthos are severely stressed. The stagnation of water caused by the salvinia facilitates mosquito breeding, and there is a rapid and precipitous loss of biodiversity in the invaded water body (Abbasi and Nipaney 1993). Salvinia's unabated spread is because of its ability and to regenerate vegetatively even from tiny pieces of its leaves and its

✉ S. A. Abbasi
prof.s.a.abbasi@gmail.com

[1] Centre for Pollution Control and Environmental Engineering, Pondicherry University, Pondicherry 605014, India

propensity to double its number every 3–4 days (Schooler et al. 2011).

With huge tracts of wetlands colonized by salvinia in South America, Africa, South Asia, and Australia (Abbasi and Nipaney 1993; Bhat 2016), billions of tonnes of salvinia biomass is generated every year across the world. As no method exists to utilize any sizeable fraction of this biomass (Ganeshkumar et al. 2014; Bhat 2016), it remains unharvested, causing serious harm to the wetlands (Abbasi and Nipaney 1986; Abbasi 1997). On senescence, it decays aerobically and anaerobically—mostly latter—generating global warming gases carbon dioxide and methane (Abbasi et al. 2011). As anaerobic decomposition in the anoxic epilimnion is the predominant salvinia decay pathway, it becomes a major contributor of methane emissions. With methane having molecule to molecule, 34 times greater global warming potential than carbon dioxide (Shindell et al. 2009), the harm caused by salvinia decay is immense.

The extensive and rapid spread of salvinia has been halted and repelled by the use of a few biological agents (Room et al. 1981; Sullivan et al. 2011) but such successes have been few and far between. Worse, none has led to any enduring solution to the salvinia problem. Mechanical removal of salvinia and use of weedicide have recorded even lesser success in controlling its spread. Attempts to utilize salvinia for the generation of compost, green manure or mulch (Raju and Gangwar 2004; Arthur et al. 2007; Dorahy et al. 2009), as fodder (King et al. 2004; Leterme et al. 2010), as a source of biogas (Abbasi et al. 1990; Abbasi and Nipaney 1991; Abbasi et al. 1992a, b; Abbasi and Abbasi 2010; Ganesh et al. 2005), or for making drugs (Chantiratikul et al. 2009; Choudhary et al. 2008; Li et al. 2013; Tauseef et al. 2013), have all proved uneconomical. Thus, finding a sustainable, ecologically sound, and inexpensive way with which salvinia biomass can be utilized on the scales it is generated appears to be the only recourse which can make it profitable to regularly harvest the weed, thereby keeping a check on its alarming growth.

Vermicomposting is one such option, by which the weed can be regularly harvested and converted it into beneficial biofertilizer. Studies conducted across the world on manure-based vermicompost have shown that such vermicompost is a well stabilized, aesthetically pleasing, and finely divided peat-like material with excellent physicochemical and biological characteristics (Edwards et al. 2011) that one seeks in a good organic fertilizer. In a recent study, Ganeshkumar et al. (2014) have found that salvinia can be directly and quickly converted into vermicompost, following the concept of high-rate vermicomposting earlier developed by Abbasi and co-workers (Abbasi et al. 2009, 2015). But before the process is put to large-scale use, it must be ascertained whether salvinia vermicompost is as much, or at least nearly as much, soil friendly and plant friendly as manure-based vermicompost is known to be. This question is highly significant because salvinia plants contain unusually large concentrations of polyphenols and lignin. This feature makes salvinia strongly allelopathic and hardy (Leterme et al. 2010; Hussain et al. 2016a) and it becomes essential to rule out any adverse effect that the polyphenols and lignin content of salvinia might exert on the utility value of salvinia's vermicompost. To alley this apprehension, the present field-scale study has been carried out to see how a common vegetable ladies finger (*Abelmoschus esculentus*) responds to the application of salvinia vermicompost during its germination growth, and fruit production phases.

Materials and methods

General

All analytical work was performed with analytical reagent grade chemicals, and alkali-resistant borosilicate glassware. The dilution water was deionized and double distilled before use. Dry weight determinations of all samples of soil, plants, and vermicompost were done by heating known quantities of samples at 105 °C to constant weights.

Vermicompost was generated from the whole plants of salvinia, collected from the ponds in the vicinity of Pondicherry University campus, employing earthworm species *Eisenia foetida*. Pulse-fed, high-rate, vermireactors were utilized for the purpose. The setting and operation of the vermireactors were as detailed earlier by Nayeem-shah et al. (2014) and Ganeshkumar et al. (2014). No pre-composting, manure supplementation or any other form of pre-treatment was done. The vermicompost was periodically assessed in the form of easily identifiable, separable, and quantifiable vermicast that was generated in each 20-days pulse (Abbasi et al. 2009, 2015). The total nitrogen, phosphorus and potassium (NPK) contents of the vermicompost were 17.9, 2.92 and 11.45 g/kg, respectively. The soil used in the present experiment was collected from an uncultivated area of Pondicherry University campus. This was done to prevent interference from any previous fertilizer application. The soil, which was characterized as sandy loam, had pH 6.35, and contained organic carbon, total nitrogen, phosphorus and potassium to the extent of 8.8, 2.71, 0.24, and 0.75 g/kg, respectively.

Planting of ladies finger

Germination and plant growth studies were carried out in 50-l plastic containers. In each treatment, 35 such bags were filled with soil and amended with four levels of salvinia vermicompost: V0 (control), V1-25 g, V2-37.5 g, and V3-50 g on dry weight basis. This corresponds to 0,

2.5, 3.75 and 5 t/ha VC, respectively. Five seeds of ladies finger (*Abelmoschus esculentus*) were sown in each container. The study was carried out during February–May which is the period considered ideal for growing ladies finger in the region where authors work (ICAR 2011).

Analysis

The fraction of seeds that germinated within 8 days of sowing was enumerated to determine the germination success. Thereafter, one of the healthy seedlings was retained in each bag while discarding the other four. The growth experiments were then carried out for 15 weeks, irrigating all the containers with tap water every day. 100 days after sowing, five plants from each treatment were randomly harvested. Their shoot length (cm), root length (cm), plant biomass (g), number of leaves, stem diameter (mm) and number of branches were determined. The speed and the extent of flowering were computed in terms of number of days taken by the plants to produce their first flower, and the total number of flowers produced per plant. The number of pods and the average length (cm), diameter (mm), and weight (g) of the pods per plant were recorded.

Fresh leaves of the ladies finger plants were collected for estimation of chlorophyll and carotenoids. These pigments were extracted from the leaves using *N, N*-dimethyl formamide (DMF) following a procedure of Moran and Porath (1980). The concentrations of the pigments were determined spectrophotometrically at 470, 647 and 664 nm as detailed by Wellburn (1994). The protein, carbohydrates, and mineral content of the ladies finger pods were estimated by Kjeldahl, Anthrone and dry washing methods, respectively (Nielsen 2010).

Disease incidence and pest infestation

Soon after the start of the experiments, the leaves of the ladies finger plants began getting infected by leaf miners and leaf spot diseases. It could be seen that the prevalence of infection was highly pronounced in the control plants but much lesser in the vermicompost-fertilized plants. The symptoms of the leaf miners infection, which is caused by *Liriomyza* spp, were feeding punctures and white speckles on the leaf surface (Ahmed 2000). The leaf spot disease, caused by the fungus *Alternaria alternate,* was manifested by light brown spots appearing on the leaves, which later turned into concentric dark brown spots (Tohyama et al. 2005; Hussain et al., 2017b). When the infection became particularly severe, it caused the infected leaves to turn brown and fall off (Antonijevic et al. 2007). Another form of infection, which affected only the fruits, was due to *Earias vittella* (Sharma et al. 2010; Halder et al. 2015; Hussain et al. 2017b) called 'borer infection'. It was quantified as the fraction, by weight, of the infected fruits relative to the total fruits.

Tests of significance

These were performed by one-way analysis of variance and least significant difference (Alan and David 2001; Field 2009).

Results and discussion

Seed germination and plant growth

In vermicompost treatments, the germination success was significantly ($P < 0.05$) higher than in the controls. The maximum (97%) germination was observed in V_3 treatment. It was followed by 96% and 87% success in V_4 and V_2 treatments (Fig. 1a). The controls had shown the least germination success. The augmentation of soil with salvinia vermicompost also enhanced the growth of ladies finger in terms of shoot and root length, plant biomass, number of leaves, and number of branches (Fig. 1b–h). The maximum shoot length (96.2 ± 9.4), root length (48.2 ± 4.8), shoot diameter (14.04 ± 0.7), shoot dry weight (23.31 ± 2.3), root dry weight (7.96 ± 0.6), number of leaves (26.8 ± 2.3), and number of branches (4.8 ± 0.5) were recorded in V_4 treatment. The lowest growth was recorded in the controls. With the exception of root length, the improvement in growth even with the lowest (V1) level of vermicompost augmentation in comparison to controls was dramatic and highly significant.

Past studies on the effect of vermicompost derived from animal manure on plants have shown that such vermicomposts have major, medium, and micronutrients in more bioavailable form compared to their parent substrate (Hussain et al. 2016a). The vermicompost is also endowed with much richer and diverse microflora than the parent substrate (Hussain et al. 2015, 2016b). It appears that salvinia-based vermicompost also acquires similar plant-friendly and soil friendly attributes. Additionally, manure-based vermicomposts are known to contain fulvic acids, humic acids, and phytohormones (Atiyeh et al. 2002; Ievinsh 2011; Doan et al. 2015) which rejuvenate the soil and makes it plant-friendly. Recently, Hussain et al. (2016b) have reported that salvinia loses its reproductive ability after it is ingested by earthworms. The microbial and enzymatic degradation of salvinia during its passage through the earthworm gut leads to generation of several fatty acids, alcohols, alkanes, alkenes, and nitrogenous compounds. These compounds together with the presence of growth hormones and humic acids may have contributed

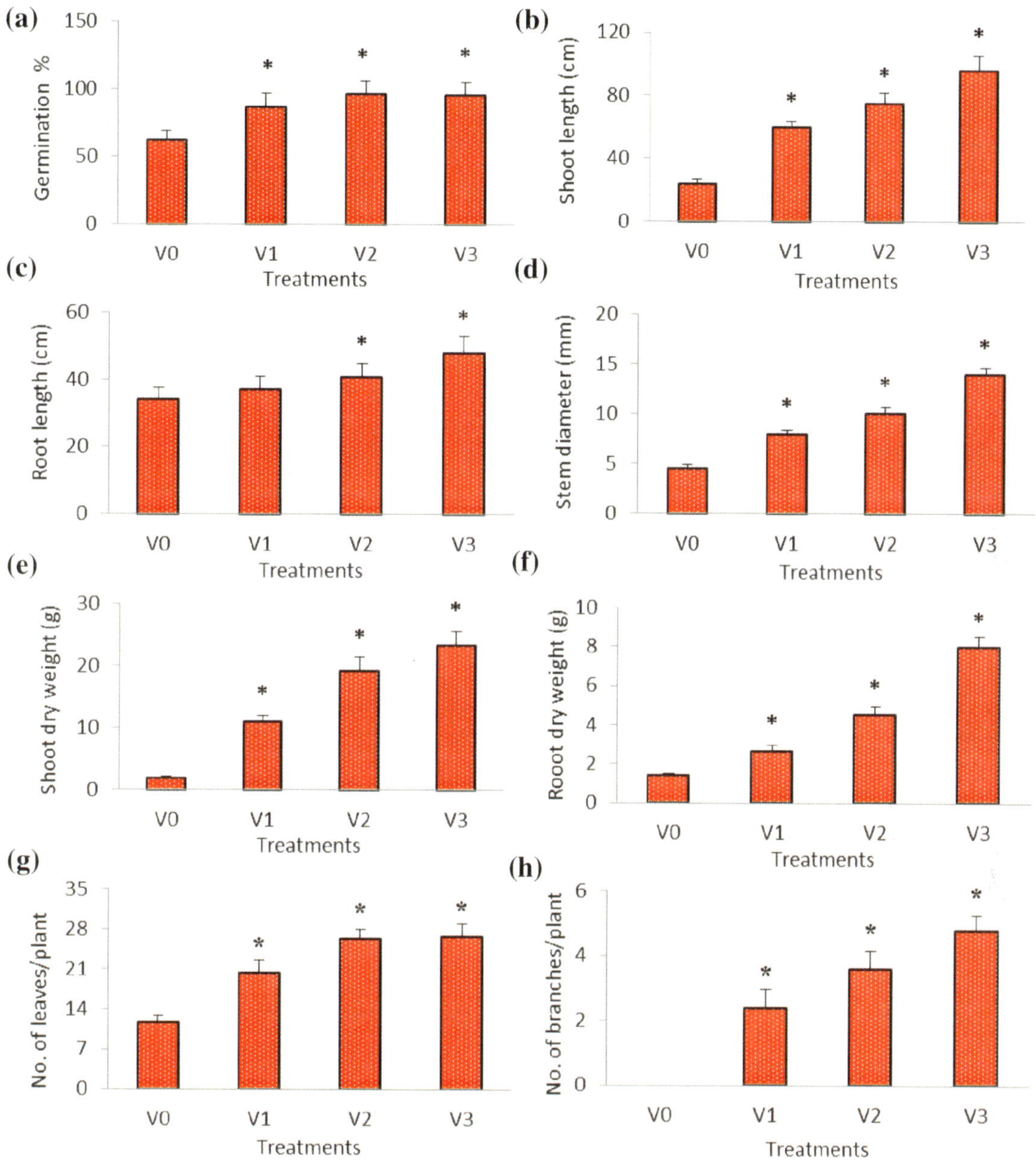

Fig. 1 Effect of salvinia vermicompost on **a** germination percentage; **b** shoot length; **c** root length; **d** stem diameter; **e** shoot dry weight; **f** root dry weight; **g** number of leaves; and **h** number of branches, of ladies finger plants. The standard deviation is indicated on the chart. The results shown in columns accompanied with an asterisk are significantly different from the controls at $P \leq 0.05$

to the better growth in VC-treated plants than in controls, as has been observed earlier in case of manure-based vermicomposts (Atiyeh et al. 2002; Doan et al. 2015; Kumar et al. 2015).

Humic acids are known to stimulate plant growth through increased cell division, enhancement of uptake of nutrients and water, and enhancement of the soil microbial population (Atiyeh et al. 2002; Chen et al. 2004; Delgado et al. 2002; García-Gil et al. 2004; Xu et al. 2014; Hussain et al. 2017a, b). It has been reported that humic acids derived from manure-based vermicompost facilitate elongation of roots and emergence of lateral roots—besides enhancing the root's H^+-ATPase activity in maize (Canellas et al. 2002). Both humic acids and plant growth hormones are probably produced by the diverse microbial population present in vermicompost (Frankenberger and Muhammad 1995). Improvement in the physical properties of the soil, especially enhancing its air and water holding capacity, have also been reported which increase the plant productivity in soil treated with manure-based vermicompost (Edwards 2004). Evidently, salvinia vermicompost possesses similar virtues to be able to generate similar effects.

Flowering and yield

Salvinia vermicompost not only induced early flowering in the ladies finger plants compared to the controls, but also resulted in the appearance of a significantly ($p < 0.05$) larger number of flowers. In all the vermicompost treatments, the yield was significantly higher than the controls. The fruit yield, reflected in the number and weight of pods per plant (10.90 ± 1.10, 105.23 ± 5.77), and the size (length and diameter) of the pods (12.35 ± 0.98; 16.17 ± 1.54), was significantly higher in VC treatments than in the controls (Fig. 2).

The findings are similar to the ones achieved with manure-based vermicompost by other authors earlier (Jouquet et al. 2011; Abduli et al. 2013; Joshi et al. 2015; Doan et al. 2015; Song et al. 2015), and indicate that the salvinia vermicompost also possesses similar virtues.

Biochemical aspects

Vermicompost application increased the concentration of photosynthetic pigments exemplified by the chlorophyll and carotenoid content in the leaves to a highly significant extent as compared to the controls (Fig. 3a–c). There was also an increase in the protein and carbohydrates content of the pods, but the increase was not statistically significant (Fig. 3f, g). On the other the total solids and ash (mineral) content of the pods grown in vermicompost-augmented soil was significantly higher than that of control soil (Fig. 3d, e). These findings indicate that salvinia-based vermicompost also has as much beneficial effect in terms of plant biochemistry as manure-based vermicomposts are known to cause (Abduli et al. 2013; Yadav et al. 2015).

Disease incidence

Vermicompost caused a reduction in the incidence of diseases in the leaves as well as the pods; the extent of this

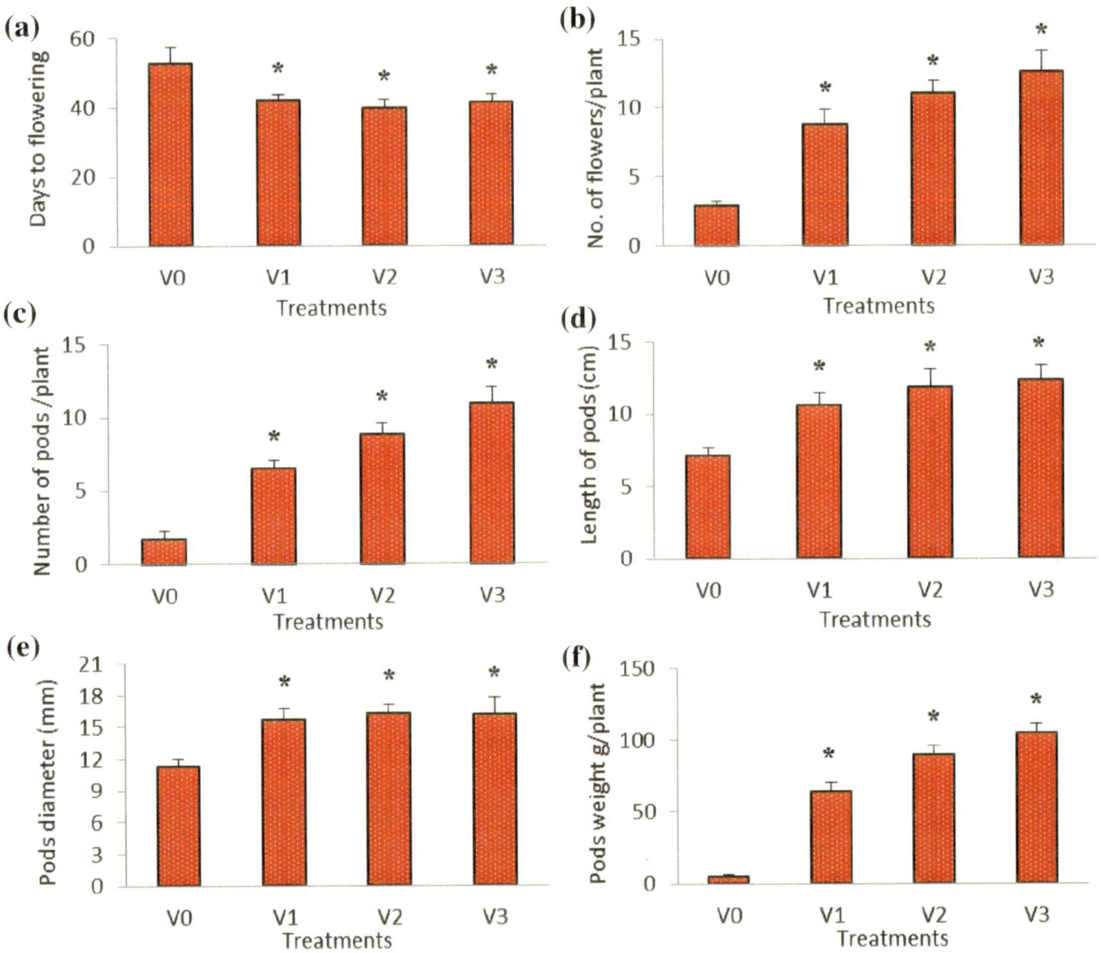

Fig. 2 Effect of salvinia vermicompost on **a** early flowering; **b** number of flowers per plant; **c** number of pods per plant; **d** length of pods; **e** diameter of pods; and **f** weight of pods per plant, of ladies finger. The standard deviation is indicated on the chart. The results shown in columns accompanied with an asterisk are significantly different from the controls at $P \leq 0.05$

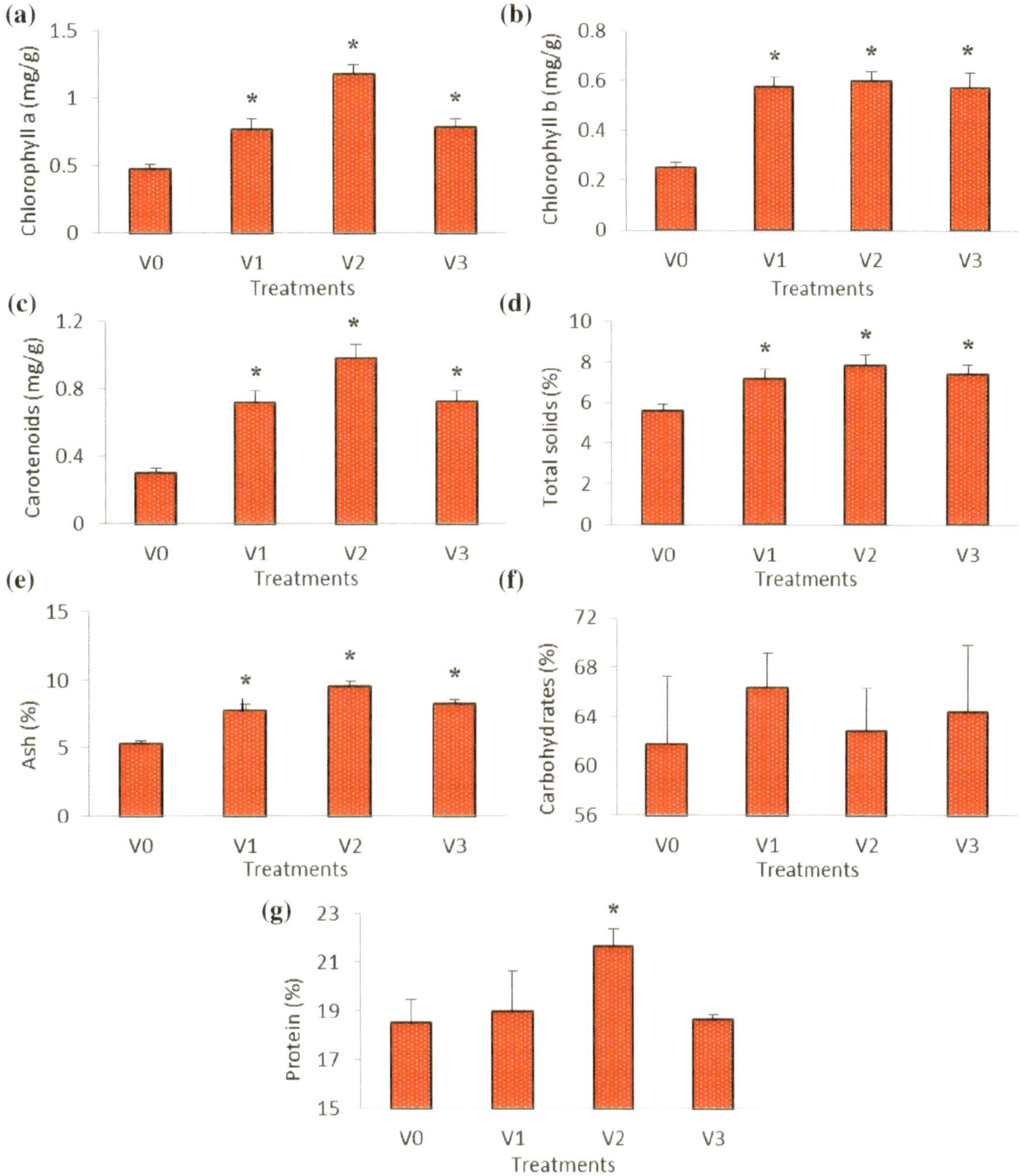

Fig. 3 Effect of salvinia vermicompost on **a** chlorophyll 'a'; **b** chlorophyll 'b'; **c** carotenoids content of ladies finger leaves and **d** total solids; **e** ash; **f** carbohydrates and **g** protein content, of ladies finger fruits. The standard deviation is indicated on the chart. The results shown in columns accompanied with an asterisk are significantly different from the controls at $P \leq 0.05$

effect increased in proportion to the VC concentration (Fig. 4). At the early stages of growth, the seedlings were seen severely infested by leaf liner, but the infection subsided as the growth increased. In this respect, also, the salvinia vermicompost seems to possess beneficial effects similar to the ones known to be possessed by manure-based vermicomposts.

There is no definite knowledge on the mechanism governing the pesticidal properties of the vermicompost (Hussain 2016). A possible cause can be that vermicompost supplies highly bioavailable nutrients which enhance the vitality of the plants, empowering them to ward off infection. Another likely reason is the presence of pathogen-destroying microorganisms in the vermicompost (Yardim et al. 2006; Aghamohammadi et al. 2016; Datta et al. 2016; Xiao et al. 2016).

Past studies have reported that vermicompost derived from animal manure possesses both these virtues (Edwards 2004; Edwards et al. 2011) and, hence, induced the resistance in plants against several plant pathogens (Szczech

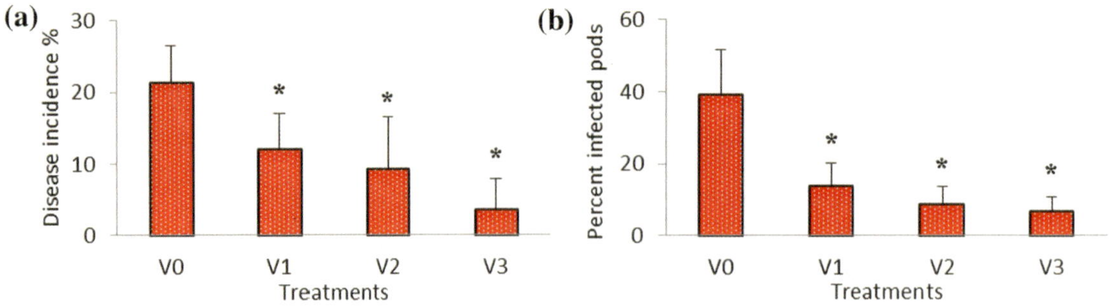

Fig. 4 Disease incidence and percentage of infected fruits in ladies finger as affected by the salvinia vermicompost. The standard deviation is indicated on the chart. The results shown in columns accompanied with an asterisk are significantly different from the controls at $P \leq 0.05$

1999; Szczech and Smolinska 2001; Choudhary et al. 2010; Nath and Singh 2011). The present study indicates that salvinia vermicompost may also be acquiring similar virtues which enable it to induce resistance against pathogens in ladies finger plants.

Conclusion

The paper reports studies which have assessed the effect of the vermicompost of the aquatic weed salvinia on the germination, growth, yield, and quality of the fruits of a common food plant ladies finger (*Abelmoschus esculentus*). The role of the vermicompost in warding off the diseases of the ladies finger leaves and pods was also witnessed. In general, a positive effect was exerted by the vermicompost in increasing germination success, plant growth, yield, and pest resistance. The degree of benefit increased in proportion to the concentration of the applied vermicompost when it was varied in the range 2.5–5 tonnes/ha. In contrast to the toxic and allelopathic nature of salvinia, its vermicompost proved totally benign, and appeared to be highly plant friendly, with virtues of an organic fertilizer similar to the ones known to be possessed by the vermicomposts derived from animal manure.

Acknowledgements SAA thanks the Council of Scientific and Industrial Research (CSIR), New Delhi, for the Emeritus Scientist grant (21(1034)/16/EMR-II). NH thanks the UGC for the Maulana Azad National Fellowship.

References

Abbasi SA (1997) Wetlands of India. Vol III: Kerala. Discovery Publishing House, New Delhi, p 276

Abbasi T, Abbasi SA (2010) Production of clean energy by anaerobic digestion of phytomass—new prospects for a global warming amelioration technology. Renewable Sustainable Ener Rev 14:1653–1659

Abbasi SA, Nipaney PC (1986) Infestation by aquatic weeds of the fern genus salvinia—its status and control. Environ Conserv 13:235–241. doi:10.1017/S0376892900036286

Abbasi SA, Nipaney PC, Schaumberg GD (1990) Bioenergy potential of eight common aquatic weeds. Biol Wastes 34(4):359–366

Abbasi SA, Nipaney PC (1991) Effect of temperature on biogas production from aquatic fern Salvinia. Indian J Technol 29:306–309

Abbasi SA, Nipaney PC (1993) World's worst weed (salvinia): its impact and utilization. International Book Distributors, Dehradun, p 226

Abbasi SA, Nipaney PC, Ramasamy EV (1992a) Studies on multiphase anaerobic digestion of Salvinia. Indian J Technol 30:483–490

Abbasi SA, Nipaney PC, Ramasamy EV (1992b) Use of aquatic weed salvinia (*Salvinia molesta,* Mitchell) as full partial feed in commercial biogas digesters. Indian J Technol 30:451–457

Abbasi T, Gajalakshmi S, Abbasi SA (2009) Towards modeling and design of vermicomposting systems: mechanisms of composting/vermicomposting and their implications. Indian J Biotechnol 8:177–182

Abbasi T, Tauseef SM, Abbasi SA (2011) Global warming and the role of wetlands. Lap Lambart Academic, Germany, p 264 **(ISBN 978-3846556009)**

Abbasi SA, Nayeem-Shah M, Abbasi T (2015) Vermicomposting of phytomass: limitations of the past approaches and the emerging directions. J Cleaner Prod 93:103–114

Abduli MA, Amiri L, Madadian E, Gitipour S, Sedighian S (2013) Efficiency of vermicompost on quantitative and qualitative growth of tomato plants. Int J Environ Res 7:467–472

Aghamohammadi Z, Etesami H, Alikhani HA (2016) Vermiwash allows reduced application rates of acaricide azocyclotin for the control of two spotted spider mite, *Tetranychus urticae* Koch, on bean plant (*Phaseolus vulgaris* L.). Ecol Eng 93:234–241. doi:10.1016/j.ecoleng.2016.05.041

Ahmed MMM (2000) Studies on the control of insect pests in vegetables (okra, tomato, and onion) in Sudan with special reference to neem-preparations. Doctoral dissertation, University of Göttingen

Shindell DT, Faluvegi G, Koch DM, Schmidt GA, Unger N, Bauer SE (2009) Improved attribution of climate forcing to emissions. Science 326:716–718. doi:10.1126/science.1174760

Song X, Liu M, Wu D, Griffiths BS, Jiao J, Li H, Hu F (2015) Interaction matters: synergy between vermicompost and PGPR agents improves soil quality, crop quality and crop yield in the field. Aust Soil Ecol 89:25–34. doi:10.1016/j.apsoil.2015.01.005

Alan CG, David SH (2001) Practical statistics and Experimental Design for Plant and Crop Sciences. Wiley, England, p 239

Antonijevic D, Fakultet P, Zemun B, Mitrovic P (2007) Leaf spot of oilseed rape Biljni lekar (Serbia). Plant Doctor 35:443–449

Arthur GD, Stirk WA, Novák O, Hekera P, van Staden J (2007) Occurrence of nutrients and plant hormones (cytokinins and IAA) in the water fern Salvinia molesta during growth and composting. Environ Exp Bot 61:137–144. doi:10.1016/j.envexpbot.2007.05.002

Atiyeh RM, Lee S, Edwards CA, Arancon NQ, Metzger JD (2002) The influence of humic acids derived from earthworm-processed organic wastes on plant growth. Bioresour Technol 84:7–14. doi:10.1016/S0960-8524(02)00017-2

Bhat MA (2016) Waste water treatment with a novel clean-green bioreactor SHEFROL. Pondicherry University, Pondicherry, p 116

Canellas LP, Olivares FL, Okorokova-Façanha AL, Façanha AR (2002) Humic acids isolated from earthworm compost enhance root elongation, lateral root emergence, and plasma membrane H^+-ATPase activity in maize roots. Plant Phys 130:1951–1957. doi:10.1104/pp.007088

Chantiratikul P, Meechai P, Nakbanpotec W (2009) Antioxidant activities and phenolic contents of extracts from *Salvinia molesta* and *Eichornia crassipes*. Res J Biol Sci 4:1113–1117

Chen Y, De Nobili M, Aviad T (2004) Stimulatory effect of humic substances on plant growth. In: Magdoff F, Weil RR (eds) Soil organic matter in sustainable agriculture, pp. 103–130. doi:10.1201/9780203496374.ch4

Choudhary MI, Naheed N, Abbaskhan A, Musharraf SG, Siddiqui H (2008) Phenolic and other constituents of fresh water fern *Salvinia molesta*. Phytochem 69:1018–1023. doi:10.1016/j.phytochem.2007.10.028

Choudhary S, Pareek S, Saxena J (2010) Management of dry root rot of greengram (*Vigna radiata*) caused by *Macrophomina phaseolina*. Indian J Agric Sci 80:988–992

Coetzee JA, Hill MP, Byrne MJ, Bownes A (2011) A review of the biological control programmes on *Eichhornia crassipes* (C. Mart.) Solms (Pontederiaceae), *salvinia molesta* D.S. Mitch. (Salviniaceae), *Pistia stratiotes* L. (Araceae), *Myriophyllum aquaticum* (Vell.) Verdc. (Haloragaceae) and *Azolla filiculoides* Lam. (Azollaceae) in South Africa. Afr Entomol 19:451–468. doi:10.4001/003.019.0202

Crites RW, Middlebrooks EJ, Bastian RK (2006) Natural wastewater treatment systems. CRC Press, Boca Raton

Datta S, Singh J, Singh S, Singh J (2016) Earthworms, pesticides and sustainable agriculture: a review. Environ Sci Pollut Res 23:8227–8243. doi:10.1007/s11356-016-6375-0

Delgado MA, Farinas JC, Ruano S (2002) Firm productivity and export markets: a non-parametric approach. J Int Econ 57:397–422. doi:10.1016/S0022-1996(01)00154-4

Doan TT, Henry-des-Tureaux T, Rumpel C, Janeau JL, Jouquet P (2015) Impact of compost, vermicompost and biochar on soil fertility, maize yield and soil erosion in Northern Vietnam: a three year mesocosm experiment. Sci Tot Environ 514:147–154. doi:10.1016/j.scitotenv.2015.02.005

Dorahy CG, Pirie AD, McMaster I, Muirhead L, Pengelly P, Chan KY, Jackson M, Barchia IM (2009) Environmental risk assessment of compost prepared from Salvinia, and Alligator weed. J Environ Qual 38:1483–1492. doi:10.2134/jeq2007.0555

Edwards CA (2004) Earthworm ecology, 2nd edn. CRC Press, Washington

Edwards CA, Arancon NQ, Sherman RL (2011) Vermiculture technology: earthworms, organic wastes, and environmental management. CRC Press, Boca Raton, pp 17–19

Field A (2009) Discovering statistics using SPSS statistics. Sage Publications, London

Frankenberger WT Jr, Muhammad A (1995) Phytohormones in soils: microbial production and function. Marcel Dekker Inc., New York, p 503

Ganesh PS, Ramasamy EV, Gajalakshmi S, Abbasi SA (2005) Extraction of volatile fatty acids (VFAs) from water hyacinth using inexpensive contraptions, and the use of the VFAs as feed supplement in conventional biogas digesters with concomitant final disposal of water hyacinth as vermicompost. Biochem Eng J 27(1):17–23

Ganeshkumar T, Premalatha M, Gajalakshmi S, Abbasi SA (2014) A new process for the rapid and direct vermicomposting of the aquatic weed salvinia (*Salvinia molesta*). Bioresour Bioprocess 1(1):26. doi:10.1186/s40643-014-0026-4

García-Gil JC, Plaza C, Senesi N, Brunetti G, Polo A (2004) Effects of sewage sludge amendment on humic acids and microbiological properties of a semiarid Mediterranean soil. Biol Fertil Soils 39:320–328. doi:10.1007/s00374-003-0709-z

Halder J, Sanwal SK, Rai AK, Rai AB, Singh B, Singh BK (2015) Role of physico-morphic and biochemical characters of different okra genotypes in relation to population of okra shoot and fruit borer, *Earias vittella* (Noctuidae: Lepidoptera). Indian J Agric Sci 85:278–282

Hussain N (2016) Gainful utilization of toxic and allelopathic weeds for the generation of biofertilizer through vermicomposting: an assessment of the fertilizer value of the resultant vermicompost. PhD thesis, Pondicherry University

Hussain N, Abbasi T, Abbasi SA (2015) Vermicomposting eliminates the toxicity of lantana (*Lantana camara*) and turns it into a plant friendly organic fertilizer. J Hazard Mat 298:46–57. doi:10.1016/j.jhazmat.2015.04.073

Hussain N, Abbasi T, Abbasi SA (2016a) Vermiremediation of an invasive and pernicious weed salvinia (*Salvinia molesta*). Ecol Eng 91:432–440. doi:10.1016/j.ecoleng.2016.03.010

Hussain N, Abbasi T, Abbasi SA (2016b) Transformation of toxic and allelopathic lantana into a benign organic fertilizer through vermicomposting. Spectrochim Acta A 163:162–169. doi:10.1016/j.saa.2016.03.019

Hussain N, Abbasi T, Abbasi SA (2017a) Toxic and allelopathic ipomoea yields plant-friendly organic fertilizer. J Clean Prod 148:826–835. doi:10.1016/j.jclepro.2017.01.176

Hussain N, Abbasi T, Abbasi SA (2017b) Detoxification of parthenium (Parthenium hysterophorus) and its metamorphosis into an organic fertilizer and biopesticide. Bioresour Bioprocess 4:26. doi:10.1186/s40643-017-0156-6

ICAR (2011) Handbook of agriculture. Indian Council of Agricultural Research, Pusa

Ievinsh G (2011) Vermicompost treatment differentially affects seed germination, seedling growth and physiological status of vegetable crop species. Plant Growth Regul 65:169–181. doi:10.1007/s10725-011-9586-x

Joshi R, Singh J, Vig AP (2015) Vermicompost as an effective organic fertilizer and biocontrol agent: effect on growth, yield and quality of plants. Rev Environ Sci Bio/Technol 14:137–159

Jouquet EP, Bloquel E, Doan TT, Ricoy M, Orange D, Rumpel C, Duc TT (2011) Do compost and vermicompost improve macronutrient retention and plant growth in degraded tropical soils? Compost Sci Util 19:15–24. doi:10.1080/1065657X.2011.10736972

King C, McIntosh D, Fitzsimmons K (2004) Giant Salvinia (*Salvinia molesta*) as a partial feed for Nile Tilapia (*Oreochromis niloticus*). In: Proc. 6th Int. Symp. Tilapia in Agric, pp 12–16

Kumar R, Singh MK, Kumar V, Verma RK, Kushwah JK, Pal M (2015) Effect of nutrient supplementation through organic sources on growth, yield and quality of coriander (*Coriandrum sativum* L.). Indian J Agric Res 49:278–281. doi:10.5958/0976-058X.2015.00045.1

Leterme P, Londoño AM, Ordoñez DC, Rosales A, Estrada F, Bindelle J, Buldgen A (2010) Nutritional value and intake of aquatic ferns (*Azolla filiculoides* Lam. and *Salvinia molesta* Mitchell.) in sows. Anim Feed Sci Technol 55:55–64. doi:10.1016/j.anifeedsci.2009.10.002

Li S, Wang P, Deng G, Yuan W, Su Z (2013) Cytotoxic compounds from invasive giant salvinia (*Salvinia molesta*) against human tumor cells. Bioorg Med Chem Lett 23:6682–6687. doi:10.1016/j.bmcl.2013.10.040

Luque GM, Bellard C, Bertelsmeier C, Bonnaud E, Genovesi P, Simberloff D, Courchamp F (2014) The 100th of the world's worst invasive alien species. Biol Invasions 16:981–985. doi:10.1007/s10530-013-0561-5

Moran R, Porath D (1980) Chlorophyll determination in intact tissues using *N*,*N*-dimethylformamide. Plant Physiol 65:478–479

Nath G, Singh K (2011) Combination of vermicomposts and biopesticides against nematode (*Pratylenchus* sp.) and their effect on growth and yield of tomato (*Lycopersicon esculentum*). IIOAB J 2:27–35

Nielsen SS (2010) Food analysis, 4th edn. Springer, London and New York. doi:10.1007/978-1-4419-1478-1

Raju RA, Gangwar B (2004) Utilization of potassium-rich green-leaf manures for rice (*Oryza sativa*) nursery and their effect on crop productivity. Indian J Agron 49:244–247

Room PM (1990) Ecology of a simple plant-herbivore system. Biological control of Salvinia. Trends Ecol Evol 5:74–79. doi:10.1016/0169-5347(90)90234-5

Room PM, Harley KL, Forno IW, Sands DP (1981) Successful biological control of the floating weed salvinia. Nature 294:78–80

Schooler SS, Salau B, Julien MH, Ives AR (2011) Alternative stable states explain unpredictable biological control of *Salvinia molesta* in Kakadu. Nature 470:86. doi:10.1038/nature09735

Shah MN, Gajalakshmi S, Abbasi SA (2014) Direct vermicomposting of vegetable waste using the concept of high-rate vermireactor operation. Int J Environ Sci Eng Res 4:59–65

Sharma RP, Swaminathan R, Bhati KK (2010) Seasonal incidence of fruit and shoot borer of okra along with climatic factors in Udaipur region of India. Asian J Agric Res 4:232–236. doi:10.3923/ajar.2010.232.236

Sullivan PR, Postle LA, Julien M (2011) Biological control of Salvinia molesta by *Cyrtobagous salviniae* in temperate Australia. Biol Control 57:222–228. doi:10.1016/j.biocontrol.2011.03.009

Szczech MM (1999) Suppressiveness of vermicompost against *Fusarium wilt* of tomato. J Phytopathol 147:155–161. doi:10.1046/j.1439-0434.1999.147003155.x

Szczech M, Smolińska U (2001) Comparison of suppressiveness of vermicomposts produced from animal manures and sewage sludge against *Phytophthora nicotianae* Breda de Haan var. nicotianae. J Phytopathol 149:77–82. doi:10.1046/j.1439-0434.2001.00586.x

Tauseef SM, Premalatha M, Abbasi T, Abbasi SA (2013) Methane capture from livestock manure. J Environ Manag 117:187–207

Tohyama A, Hayashi K, Taniguchi N, Naruse C, Ozawa Y, Shishiyama J, Tsuda M (2005) A new post-harvest disease of okra pods caused by *Alternaria alternate*. Jpn J Phytopathol 61:340–345. doi:10.3186/jjphytopath.61.340

Wellburn AR (1994) The spectral determination of chlorophylls a and b, as well as total carotenoids, using various solvents with spectrophotometers of different resolutions. J Plant Physiol 144:307–313. doi:10.1016/S0176-1617(11),81192-2

Xiao Z, Liu M, Jiang L, Chen X, Griffiths BS, Li H, Hu F (2016) Vermicompost increases defense against root-knot nematode (*Meloidogyne incognita*) in tomato plants. Appl Soil Ecol 105:177–186. doi:10.1016/j.apsoil.2016.04.003

Xu Y, Wang C, Bi Y, Zhang Y, Cheng W, Sun Z, Zhang J, Lv Z, Guo X, Jiuxianzhen YC (2014) Influence of cow manure vermicompost soil mixtures on two flowers seedling cultivation. Acta Hort 1018:583–588. doi:10.17660/ActaHortic.2014.1018.64

Yadav A, Suthar S, Garg VK (2015) Dynamics of microbiological parameters, enzymatic activities and worm biomass production during vermicomposting of effluent treatment plant sludge of bakery industry. Environ Sci Pollut Res 22:14702–14709. doi:10.1007/s11356-015-4672-7

Yardim EN, Arancon NQ, Edwards CA, Oliver TJ, Byrne RJ (2006) Suppression of tomato hornworm (*Manduca quinquemaculata*) and cucumber beetles (*Acalymma vittatum* and *Diabotrica undecimpunctata*) populations and damage by vermicomposts. Pedobiologia 50:23–29. doi:10.1016/j.pedobi.2005.09.001

6

Response of okra (*Abelmuschus esculenthus* L. Moench) to combined organic and inorganic foliar fertilizers

A. I. Afe[1] · F. Oluleye[1]

Abstract

Aim This study investigated the effects of combined organic poultry manure (PM) and inorganic foliar fertilizer (FF) on growth and yield of okra (*Abelmuschus esculenthus* L. Moench).

Materials and methods Field trials were carried out at Shao and Teaching and Research Farm of Kwara State University, Malete, respectively, in 2013 and 2014 cropping seasons. The treatment consisted of PM at 10.0 t/ha, combinations of FF with 10.0, 7.5, 5.0, or 2.5 t/ha, FF alone and control without fertilizer arranged in a randomized complete block design and replicated three times.

Results Combinations of PM and FF influenced the growth and fruit yield of okra. PM at 10.0 t/ha plus FF produced the highest plant height of 48.40 and 58.50 cm in 2013 and 2014, respectively. Also, highest number of fruits per plant and fruit yield was obtained at the same treatment over the two years. The control, without fertilizer, showed significantly longer days to flower compared with other treatments in the two years.

Conclusion The highest yield obtained at PM 10.0 t/ha +FF was not significantly different from PM 7.5 t/ha +FF. Hence, poultry manure applied at 7.5 t/ha +FF is therefore recommended.

Keywords Foliar fertilizer · Okra fruit yield · Poultry manure · Soil-applied · Yield component

✉ A. I. Afe
adeafe22@yahoo.com; ade.afe@kwasu.edu.ng

[1] Department of Crop Production, Kwara State University, Malete, Nigeria

Introduction

Sustainable soil productivity is one of the major constraints of tropical agriculture. Continuous cultivation is a common practice by poor resource farmers in the tropics. This has resulted in rapid decline of soil nutrients and unstable soil microbial population (ECA 2001). Consequently, there is decline productivity, low farmers income, and increasing poverty (FAO 2006). Hence, the use of organic manure as alternative soil amendment strategy for soil nutrient management has been advocated (Shehu et al. 1997).

Fertilizer plays a major role among cultural practices for increased crop production. However, blanket application of inorganic fertilizer to farmland soils without adequate knowledge of the nutrient status, often leads to increased soil acidity, particularly when nitrogen fertilizers are applied (Akande et al. 2010). Although, inorganic fertilizers, contain high concentrations of nutrients that are rapidly available and released for plant uptake, their use is limited due to scarcity, high cost, nutrient imbalance, and soil acidity (Akande et al. 2010).

Manure provides necessary macro- and micro-nutrients in available form, and improves the physical and biological properties of the soil (Abou El-Magd et al. 2006). Poultry manure (PM) is an excellent source of organic manure. It supplies both macro and micro-nutrients during mineralization, increases the organic matter content of the soil, and consequently enhances the texture, structure, aeration, moisture holding capacity, nutrient retention and water infiltration in the soil (Akinrinde et al. 2006; Dekissa et al. 2008). According to Garge and Bahla (2008), PM supplies phosphorous more readily to plants than other organic sources. In another study, Abd El-Kader et al. (2010) reported that manure from poultry increased okra yield and water use efficiency than composted plant residue.

Additionally, abattoir waste was found as an excellent organic fertilizer in the cultivation of vegetables and also could promote healthy environment (Roy et al. 2015).

Foliar fertilizers (FF) are applied to the foliage of the plant to boost the nutrient density in the crop. It is a convenient way of applying nutrients to the plant to supplement absorbed soil nutrients. It improves the efficiency of soil-applied nutrients and also acts as a catalyst in the uptake and use of certain macro nutrients (Philips 2004; Fageria et al. 2009). Besides being environmentally safe, it increases crop yield and quality (Fageria et al. (2009). As such, El-Aal et al. (2010), Zodape et al. (2011) have recommended the technique in integrated nutrient management. Paliwal et al. (1999) had earlier reported enhanced development of branches per plant in okra due to foliar fertilization. Similarly, Abasi et al. (2010) reported improved days to 50% flowering, plant height number of branches and fruits per plant and fruit yield in okra using combinations of foliar fertilizer and soil-applied inorganic fertilizer in okra.

Boost Xtra is one of the foliar fertilizers that have gained popularity among farmers in Nigeria. It is formulated as soluble granules containing 20% each of nitrogen, phosphorous and potassium along with certain micro-nutrients. Farmers use this fertilizer alone or in combination with soil-applied inorganic fertilizer on various crops to improve growth and yield. However, there is a paucity of information on the combinations of foliar fertilizer and organic manure usage for vegetable crop production. Therefore, this study was initiated to investigate the effects of combined PM and FF on growth and yield of okra.

Materials and methods

Field trials were carried at Shao and the Teaching and Research Farm of Kwara State University, Malete, respectively, in 2013 and 2014 cropping seasons. The two locations are in Moro Local Government Area of Kwara State, Nigeria. The PM was collected from deep liter system of poultry unit in the University Farm and the boost Xtra purchased from the agro chemical outlet in Ilorin, Nigeria. The treatments consisted of PM at 10.0 t/ha, combinations of FF with 10.0, 7.5, 5.0, or 2.5 t/ha, FF alone and control without fertilizer arranged in a randomized complete block design and replicated three times. The nutrient compositions of the PM and FF are shown in Tables 1 and 2.

Cultural practices

The land was plowed once and harrowed twice. Planting was carried out at the two locations on the flat. Each plot size measured 2.5 m × 3.5 m (8.75 m^2) with 0.5 m between plots and 1.0 m between blocks. Four seeds were planted at inter and intra row spacings of 0.6 and 0.3 m, respectively. Crops were thinned at two weeks after planting (WAP) to two plants per stand. Air-dried PM was applied uniformly at two weeks before planting. Foliar fertilizer was split applied at four and five WAP using knapsack sprayer. Pendimethalin [N-(1-ethylpropyl)-3, 4dimethyl-2, 6 dinitrobenzene amine] was applied two days after planting at the

Table 1 Nutrient composition of foliar fertilizer (Boost Xtra)

Analysis (wt/vol.)		
Nitrogen	N	20%
Phosphorous	P	20%
Potassium	K	20%
Magnesium	Mg	1.5%
Iron (EDTA)	Fe	0.35%
Manganese EDTA	Mn	0.075%
Copper EDTA	Cu	0.075%
Zinc EDTA	Zn	0.075%
Boron	B	0.0315%
Cobalt EDTA	Co	0.0012%
Molybdenum	Mo	0.0012%
pH (10% solution)	4.0–4.5	
Density	1.5SG@18C	

Table 2 Physio-chemical properties of the experimental field (0–15 cm) and nutrient composition of the poultry manure

Soil composition			Poultry manure
Physical properties	Shao	Malete	
Sand (%)	55.28	53.55	
Silt (%)	18.45	19.64	
Clay (%)	28.44	30.23	
Textural class	Sandy-loam		
Chemical properties			
pH (H_2O)	5.88	5.72	5.83
Organic carbon	1.05	1.22	7.10
Available P mg kg^{-1}	6.76	6.85	5.21
Total N	0.18	0.92	8.05
Exchangeable bases			
Ca (mg kg^{-1})	2.62	2.48	0.13
Na (mg kg^{-1})	0.55	0.39	0.22
Mg (mg kg^{-1})	1.15	2.01	0.88
Mn (mg kg^{-1})	0.92	0.88	36.58
Zn (mg kg^{-1})	2.36	2.61	28.94
Cu (mg kg^{-1})	0.33	0.29	4.10
Fe (mg kg^{-1})	3.24	2.98	3.72
K (mg kg^{-1})	0.92	0.88	

Table 3 Growth of okra as affected by the combined application poultry manure and foliar fertilizer in 2013 and 2014 cropping seasons

Treatments	2013				2014			
	Plant height (cm)	Days to 50% flowering	Days to 50% fruiting	Stem girth (cm)	Plant height (cm)	Days to 50% flowering	Days to 50% fruiting	Stem girth (cm)
PM10T +foliar	48.40a	32.77c	43.43b	4.20	58.50a	34.85b	45.62b	4.40
PM7.5t +foliar	46.67b	33.00c	43.40b	3.95	56.10a	34.72b	46.18b	4.18
PM 5.0t +foliar	42.33c	33.47c	43.57b	3.80	55.75ab	35.10b	45.74b	3.60
PM 2.5t +foliar	43,33c	32.50c	41.97b	2.95	48.03c	33.90b	43.39b	3.20
PM 10t alone	45.43b	33.50c	44.00b	3.85	53.50b	34.62b	46.10b	4.25
Foliar alone	38.80d	34.50b	43.80b	2.68	36.36d	35.20b	44.98b	2.62
No fertilizer	30.63e	40.03a	55.00a	2.59 NS	27.50e	39.10a	54.32a	2.55 NS

Values with the same letter(s) under the same column are not significantly different at 5% level of probability by Duncan's Multiple Range Test

rate of 1.5 kg active ingredients per hectare. This was followed by manual weeding at 5 WAP to keep the experimental plot weed free. Insect pests were controlled with cyalothrin [3-(2-chloro-3,3,3,-trifluoropropenyl)-2.2, dimethylclopropanecarboxylate] at the rate of 1 ml of the product per liter of water using knapsack sprayer.

Data collection

Plant height at fruiting, stem girth, number of fruits per plant, fruit length, fruit circumference, and fresh fruit weight at each harvest were measured and days to 50% flowering was observed. Fruits were harvested at four days interval for eight and nine consecutive periods in 2013 and 2014, respectively.

Statistical analysis

Data were subjected to analysis of variance using SAS (1989) package and treatment means compared using Duncan's Multiple Range Test. This was done where results were significant at 5% level of probability.

Results

Soil and poultry manure analysis

Soils at the two experimental sites were sandy-clay loam, slightly acidic with pH 5.88 and 5.72, total nitrogen, 0.18 and 0.92, available phosphorous, 6.76 and 6.85 and exchangeable potassium 0.92 and 0.88, respectively, for Shao and Malete (Tables 1, 2). The PM used was slightly richer in the plant nutrients compared to un-amended experimental soils of the study locations.

Plant growth

Growth of okra was significantly affected with the application of FF in combination with PM (Table 3). The highest plant height of 48.40 cm and 58.50, respectively, in 2013 and 2014 were observed when PM at 10.0 t/ha was combined with FF. Application of PM at 7.5 t/ha combined with FF had similar plant height compared to when PM was applied at 10 t/ha alone. Compared with other treatments, the control without fertilizer had significantly reduced plant height. Days to 50% flowering and fruiting followed similar trends in all the treatments. Where neither of the fertilizer was applied, took longer days to flower and fruit compared with other treatments. It flowered and fruit at 40.03 and 55.00 days in 2013 and 39.10 and 54.32 days in 2014. Although not statistically significant, combination of PM with FF had shorter days to flower and fruit compared with control treatment.

Yield and yield components

Okra fruit yield and yield components were significantly affected by different treatments (Table 4). Plots treated 10.0 t/ha PM mixed with FF produced the highest number of fruits per plant in 2014. Similar number of fruits were produced with application of 10.0 t/ha PM alone and 7.5 t/ha PM mixed with FF in 2014. The least number of fruits were recorded at the control treatments in the two years.

Application of 10.0 t/ha PM mixed with FF significantly produced longer fruits than other treatments, except

Table 4 Effects of combined application of poultry manure and foliar fertilizer (boost Xtra) on yield and yield components of okra in 2013 and 2014 cropping seasons

Treatments	2013				2014			
	No of fruits/plant	Fruit length (cm)	Fruit circum (cm)	Yield kg/ha	No of fruits/plant	Fruit length (cm)	Fruit circum (cm)	Yield (kg/ha^{-1})
PM10t +foliar	23.10a	11.00a	11.81a	1,892.94a	23.65a	10.76a	12.00a	1825.92a
PM7.5t +foliar	20.20a	11.10a	11.60a	1,140.00a	19.37b	9.58ab	11.33a	1549.31ab
PM5.0t +foliar	11.80b	9.30bc	10.95a	796.62b	13.22d	8.13b	10.40a	1057.70cd
PM2.5t +foliar	11.50b	8.50cd	9.61a	720.62bc	11.98d	8.15b	10.20a	958.52cd
PM10t alone	13.20b	10.90ab	11.42a	1092.12a	16.35c	8.23b	11.10a	1307.78bc
Foliar alone	10.00b	8.64cd	9.63a	632.22c	9.57e	8.06b	9.45a	958.33cd
No fertilizer	4.63c	7.10d	9.10a	283.38d	7.30f	6.38c	8.45a	637.40d

Values with the same letter(s) under the same column are not significantly different at 5% level of probability by Duncan's Multiple Range Test

combinations of 7.5 t/ha PM with FF. There was no difference in fruit length between 10.0 t/ha PM alone, as well as 7.5, 5.0, and 2.5 t/ha PM combined with FF. Shorter fruit length was observed at the control treatments.

Okra fruit yield was highest (1182.94 kg/ha in 2013 and 1825.92 kg/ha in 2014) with the integrated application of PM at 10.0 t/ha plus FF. Recorded fruit yield in this treatment was not statistically different from those obtained with application of 7.5 t/ha PM mixed with FF in the two years. Application of 7.5 t/ha PM integrated with FF had similar fresh fruit weight with applied 10.0 t/ha PM alone in the two cropping seasons. Over the two years, there was no significant different between combinations of FF with PM at 2.5 t/ha and when boost Xtra was applied alone. The yield was minimum, 283.38 kg/ha and 637.40 kg/ha in 2013 and 2014, respectively, at the control treatment without fertilizer. There was no significant difference in fruit yield between application of FF alone and the control.

Discussion

The inherent low nutrient status of the two locations suggests the need for external soil amendment for optimum crop yield. In overall, result of the study indicated that growth and yield attributes of okra were enhanced with combinations of PM and FF. The improved growth and yield observed in this study was due to balanced nutrient in the PM that was synergistically complemented with nutrient present in the FF. This was evident by the earlier commencement of flowering and fruiting in the treatments where FF was combined with high tonnage of PM. The higher values in all the parameters with the combinations of FF with PM at 7.5 t/ha compared to 10.0 t/ha PM alone also corroborated this observation. Earlier commencement of flowering and fruiting, growth, and yield parameters as observed in the study could also be due to excellent macro- and micro-nutrients in the poultry manure that was complemented with the nutrient that was supplied through the foliage. Garge and Bahla (2008) reported that poultry manure readily supply phosphorous to plants than other organic sources. Similarly, Tu et al. (2006) observed that microbial biomass and activity were higher in organically managed soils compared to conventional soil amendment with synthetic inorganic fertilizer. In a more recent study, Roy et al. (2013) reported rapid released of available plant nitrogen in organic fertilizer produced from bovine-rumen-blood-digesta mixture (BBRD).

It is possible that the combined application favored increased production of photosynthates during the growth stages and consequently partitioning and allocation of the dry matter at the developmental stages of okra. This finding is consistent with the observation of earlier researchers on the integration of organic manure with inorganic soil-applied fertilizer for okra (Akande et al. 2010; Abd El-Kader et al. (2010) and other crops such as maize (Makinde and Ayoola 2008), rice (Satyanarayana et al. 2007), sorghum (Bayu et al. 2006), and sweet maize (Uwah et al. 2011). Foliar fertilizers are fast acting because the nutrients are absorbed at the site where they are readily utilized. Apart from supplying micro-nutrients, this method of fertilizer application has also been reported to act as catalyst in the uptake and utilization of certain macronutrients (Philips 2004). In another study, Selvi and Rani (2000), found higher yield, income, and benefit cost ratio in okra when FF was mixed with inorganic NPK fertilizer than when NPK was applied alone.

Conclusion

In the present study, combinations of PM applied at 10 t/ha and FF produced the highest fruit yield. However, there was no significant difference in the obtained yield and integration of PM at 7.5 t/ha plus FF. Poultry manure at 7.5 t/ha mixed with foliar fertilizer is therefore recommended.

Acknowledgement The authors wish to acknowledge the Centre for Community Development, Kwara State University, for funding part of the project.

References

Abasi FF, Baloch MA, Zia-ul-hassan Wagan KH, Shah AN, Rajar I (2010) Growth and yield of okra under foliar application of new multinutrients fertilizers products. Pak J Agric Agric Eng Vet Sci 26(2):11–18

Abd El-Kader AA, Shaaban SM, Abd El-Fattah MS (2010) Effect of irrigation levels and organic compost on okra plants (*Abelmoschus esculentus.*) grown in sandy calcareous soil. Agric Biol J N Am 1(3):225–231

Abou El-Magd MA, El-Bassiony M, Fawzy ZF (2006) Effects of organic manure with or without chemical fertilizer on growth, yield and quality of some varieties of Broccoli plants. J Appl Sci Res 2(10):791–798. http://www.insipub.com/jasr/2006/791-798.pdf

Akande MO, Oluwatoyinbo EA, Makinde EA, Adepoju AS, Adepoju IS (2010) Response of okra toorganic and inorganic fertilization. Nat Sci 8(11):261–266

Akinrinde EA, Olubakin DA, Omoloso SO, Ahmed AA (2006) Agric J 1(2):96–103

Bayu W, Rethman NFG, Hammers PS, Alemu G (2006) Effects of farm yard manure and inorganic fertilizers on sorghum growth, yield and nitrogen use in a semi arid area of Ethiopia. J Plant Nutr 29:391–407. http://www.informatworld.com/smpp/content.doi:10.1080/019041605000320962

Dekissa T, Short I, Allen J (2008) In: proceedings of theUCWR/NIWA Annual conference. Intl Water Resources Education; July 22–24, Durhan, NC

ECA (2001) State of the Environment in Africa. Economic Commission of Africa. P. O. Box 3001, Addis Ababa, Ethiopia. ECA/FSSDD/01/06. Htt://www.uneca.org/water/State Environ Afri.pdf

El-Aal FSA, Shaheen AM, Ahmed AA, Mahmoud AR (2010) Effects of foliar fertilization of urea amino acids mixtures as antioxidants on growth, yield and characteristics of squash. Res J Agric Bio Sci 6:583–588

Fageria NK, Filhoa MPB, Moreirab A, Guimarea CM (2009) Foliar fertilization of crop plants. J Plant Nutr 32(6):1044–1064

Food and Agricultural Organization (2006) The world sorghum and millet economies: facts trends and outlook. FAO Document repository W1808/E. Friend et al. Robert SD, Schoenholt SH, Mobley JA, Gerard PO

Garge S, Bahla GS (2008) Bioresour Technol 99(13):5773–5777

Makinde EA, Ayoola AA (2008) Residual influence of early season crop fertilization and cropping systems on growth and yield of cassava. Am J Agric Biol Sci 3(4):715–715. http://www.scipub.org/fulltex/AJAB/34712-715

Paliwal R, Naruka IS, Yadav JK (1999) Effect of Nitrogen and Sulphur on growth and yield of Okra (*Abelmoschus esculentus* L. Moench) cv. Pusa Sawani. Prog Hort 31(1–2):94–97

Philips M (2004) Economic benefits from using micronutrients for farmer and the fertilizer producer. In: IFA, international symposium on micronutrients, 23-25 Feb 2004

Roy M, Kamar S, Debsarcar A, Sen PK, Mukherjee J (2013) Application of rural slaughterhouse waste as an organic fertilizer for pot cultivation of solanaceous vegetables in India. Int J Recycl Org Waste Agric 2:1–11

Roy M, Das R, Debsarcar A, Sen PK (2015) Conversion of rural abattoir wastes to an organic fertilizer and its application the field cultivation of tomato in India. Renew Agric Food Syst. doi:10.1017/s1742170515000289

Satyanarayana A, Thiyagarajan TM, Uphoff N (2007) Opportunities for water saving with higher yield from the system of rice intensification. Irrig Sci 25(2):99–115

Selvi D, Rani P (2000) Effect of integrated nutrient management on yield and economics of okra in inceptisol. Veg Sci 27(2):207–208

Shehu YWS, Alhassan WK, Aliyu MA, Phillips CJC (1997) Tropical greenland 32:139–142

Statistical Analysis System Institute Inc (1989) SAS/STAT User guide. Version 4th edn.Vol.1, Cary, North Carolina

Tu C, Ristaino JB, Hu S (2006) Soil microbial biomass and activity in organic Tomato farming systems: effects of organic inputs and straw mulching. Soil Biol Biochem 38:247–255

Uwah DF, Eneji AS, Eshiet UJ (2011) Organic and mineral fertilizers effects on the performance of sweet maize (*Zea mays L.* saccharata strut.) in south eastern rainforest zone of Nigeria. Int J Agric Sci 3(1):54–61

Zodape ST, Gupta A, Bhandari SC, Rawat US, Chaudhry DR, Eswaran K, Chikara K (2011) Foliar application of seaweed sap as a biostimulant for enhancement of yield and quality of tomato (*Lycopersicum esculentum.* Mill). J Sci Ind Res 70:215–219

7

Vermicompost and vermiwash as supplement to improve seedling, plant growth and yield in *Linum usitassimum L.* for organic agriculture

Cinny Makkar[1] · Jaswinder Singh[2] · Chander Parkash[1]

Abstract

Purpose To study the organic production in two varieties of *Linum* (Linseed) crop using different proportions of vermicomposts and combination of vermiwash as an environment friendly substitute to chemical fertilizers for organic farming of Linseed.

Method Vermicompost prepared from cattle dung was mixed in different proportions, i.e. 0, 20, 40, 60, 80, and 100% vermicompost v/v with soil and commercial potting media (Perlite, Peat and Coconut coir). To study the germination, vegetative growth, reproductive growth and yield in two varieties of *Linum usitassimum L.* LC-54 and LC-2063 were grown in a polyhouse for a 2 years trial. A total of 960 seedlings and 240 plants were studied for assessment of result in 24 treatments. Additional treatments with foliar application of 1:1 v/v vermiwash and recommended dose of inorganic fertilizer were done to determine and compare the role of vermicompost, vermiwash and inorganic treatment.

Results Substituting soil with 60% (v/v) vermicompost in LC-54 and with 40% (v/v) vermicompost in LC-2063 improved the performance of seeds, root morphology and stem growth. Life cycle of crop was shortened in two varieties of Linseed raised under different treatments of vermicompost and vermiwash. Performance of LC-54 was better at 60% and that of LC-2063 was at 40%, which can be attributed to their genotypic differences. Vermicompost proportion above 60% was of no use at any stage of life cycle in both the varieties of *Linum*. Germination and yield were further improved when foliar application of vermiwash was integrated with vermicompost for nutrient management when compared with recommended chemical fertilizer treatment and control.

Conclusion Bio-efficacy of vermicompost and vermiwash for production of organic Linseed crop has been well established with enhanced yield.

Keywords Agriculture · Earthworm · Linseed · Organic farming · Plant growth · Vermicompost

✉ Jaswinder Singh
singhjassi75@yahoo.co.in

Cinny Makkar
cinnymalhotra@gmail.com

Chander Parkash
chander.ptu@gmail.com

1 Department of Applied Sciences, I.K.G Punjab Technical University, Kapurthala, Punjab, India

2 Department of Zoology, Khalsa College, Amritsar, Punjab, India

Introduction

Organic foods have recently gained importance in developed, as well as developing countries with the growing awareness about adverse effects of agricultural chemicals on human health. Demand for organic food has increased worldwide which has led to increased interest in organic agriculture (Follet et al. 1981; Sinha et al. 2009). Crops with a label of 'organically grown' fetch much higher value to farmers in domestic and international market (Lim et al. 2015; Crowder and Reganold 2015). Organic agriculture also promotes ecological conservation due to sustainable use of natural resources (Reganold et al. 1993; Letourneau and Goldstein 2001; Mäder et al. 2002). In cultivation of organically grown crops, chemical usage is excluded at all stages so organic agriculture broadly provides the dual benefit of soil quality improvement and chemical free organic foods (Gomiero et al. 2008). Currently many studies are being done to establish vermicompost as one of the preferred organic substitute to chemical fertilizers (Sinha and Valani 2009; Adhikary 2012). During vermicomposting, earthworms carry out the non thermophilic transformation of organic waste with the accelerated microbial decomposition and humification (Atiyeh et al. 2001). Vermicomposting being a biotechnological extension of composting, produces harmless, simple and stable organic end product, i.e. vermicompost.

There has been tremendous research focussing on the effects of vermicompost on soil condition and plant growth (Atiyeh et al. 2000a, b; Gutiérrez-Miceli et al. 2007; Singh et al. 2008; Roy et al. 2010; Joshi et al. 2013, 2015). Many studies have reflected the great potential of vermicompost in horticultural and agricultural practices in field or as a substitute in medium for plant growth in greenhouses and nurseries. Various studies have confirmed the beneficial effects of vermicompost on seed germination, seedling growth and plant productivity, but the effect of vermicompost has been unique and specific on various plant species and varieties, so it becomes utmost important to investigate the genus specific suitability of the vermicompost before its beneficial effects are predicted for a crop plant (Roy et al. 2010). Even the same rate or ratio of vermicompost application has shown different effects on different varieties of same plant (Zaller 2007; Lazcano and Domínguez 2010).

India is among top five countries producing Linseed, with U.K and Canada being top two countries, and still India imports Linseed. In India, currently Linseed occupies 468.0 thousand ha with productivity of 3490 hectogram's/ha (www.pdkv.ac.in). Linseed is an oilseed crop grown mainly in Punjab (India) in regions of Gurdaspur, Jalandhar, Hoshiarpur and Ropar. Recent statistics show that Linseed has a production of 0.1 thousand tonnes in Punjab state with average yield of 8.0 quintal per hectare. There is a scope to increase the covered area and yield of cultivating Linseed in India (Table 1).

Linseed (*Linum usitassimum L.*) belongs to the family Linaceae. It is an herbaceous plant grown in winter season in India as a Rabi crop and is sown in first fortnight of October. It grows well in drained silt loamy to clayey soil. Crop is ready for harvest in April. Linseed is the oldest oil and fibre yielding crop. Every part of, Linseed plant has commercial value (Jhala and Hall Linda 2010). The characteristics of Linseed plant are given in Table 2. The present study was designed to establish the practices in organic Linseed farming with following objectives (a) to reveal the impact, suitability, efficacy of vermicompost application on the growth and productivity potential of two regional varieties of a commercially important medicinal plant *Linum usitassimum* (b) to optimise the application rate of vermicompost considering genotype differences of each variety of foresaid oilseed crop (c) to compare the effect of vermicompost and inorganic fertilizer on growth and yield of Linseed (d) to study impact of integrating vermiwash with vermicompost as integrated nutrient management (INM) on Ontogenic development, i.e. seed germination and seedling growth in soil mix and in commercial potting media mix (d) to know positive effect on seedling and vegetative growth, if any, is congruently shown on reproductive growth of *Linum* plants in terms of flowering, fruiting and seed setting in various treatments.

Table 1 Recent data on Linseed production in major countries

Country	Years	Area (ha)	Yield (hg/ha[a])	Production (in tonnes)
Canada	2014	620,800	14,054	872,500
	2013	422,100	17,311	730,700
U.S	2014	122,220	13,234	161,750
	2013	69,610	12,247	85,250
India	2014	284,000	4965	141,000
	2013	338,000	4339	147,000
U.K	2014	15,000	26,000	39,000
	2013	34,000	18,235	62,000

Data sourced from http://faostat3.fao.org/download/Q/QC/E

[a] hg/ha—hectograms per hectare

Table 2 Characteristics of the *Linum usitassimum* (Linseed) plant

Characteristics of a Linseed plant	
Plant height	30–120 cm
Root system	Slender, shallow tap root with many lateral roots in top 30 cm of the soil
Leaves	Narrow, small, alternate leaves on stem
Stem	Narrow and thin, branched from base
Leaf shape	Oval with blunt apex
Inflorescence	Terminal panicle that bears numerous flowers
Flower	Perfect with 5 petals, 5 sepals, 5 stamen undergoes self pollination
Fruit	Globular capsule with 5 locules
Seeds	Flat shiny small seed of brown colour

Materials and methods

Procurement of *Linum* seed, vermicompost and vermiwash

Certified seeds of two varieties of *Linum usitatisimum*, i.e. LC-54 (LV_1) and LC-2063 (LV_2) were procured from Punjab Agricultural University (PAU), Ludhiana (Punjab),

Table 3 Physico-chemical properties of vermiwash, vermicompost and soil

Parameter	Vermiwash (mg/L)	Vermicompost	Soil
pH	6.12	7.11	8.6
N[a]	0.5	2.5	0.12
K[a]	1.2	8.0	4.2
P[a]	0.8	8.4	0.9
Ca[b]	130.0	240	149.9
Mn[b]	0.01	0.09	0.008
Cu[b]	0.04	5.0	0.87
Zn[b]	0.9	24.5	1.1

[a] Values are in g/kg

[b] Values are in mg/kg for vermicompost and soil

India. Both are lodging resistant varieties of Linseed. LC-54 (LV_1) is a latest release with brown seeds and white flowers and is recommended in Punjab, Himachal Pradesh, Jammu & Kashmir and Haryana. LC-2063 (LV_2) is high yielding tall variety with golden brown bold seeds, profuse branching, and blue flowers. Both the varieties have average life cycle of 160 days.

Vermicompost prepared from cattle dung was procured from Mahavir Farm, Phillaur, and Punjab. Vermiwash was prepared by earthworm secretion, siphoned as the seepage drain from the worm bed and was kept in earthen pots. 100 ml Vermiwash was diluted with water to prepare (1:1 v/v) foliar spray. Soil was taken from the land which was not cultivated since 5 years.

Physico-chemical properties of vermicompost, vermiwash and soil were analysed for pH, Nitrogen (N), Phosphorus (P), Potassium (K), Calcium (Ca), and heavy metals (Mn, Cu, Zn). pH, was measured using PCSTestr 35 series Digital meter. Nitrogen was estimated by the method of Bremner and Mulvaney (1982). Method of John (1970) was used for measuring phosphorus. Potassium and calcium were analysed by Systronics Flame Photometer-128. Heavy metals analysis was also measured by method of APHA (2012) using AAS Thermofisher iCE 300. Physico-chemical characteristics of vermicompost, vermiwash and soil are given in Table 3.

Experimental site

The experiment was conducted at Department of Horticulture, Jalandhar Cantt, Punjab (India) with Latitude/longitude: 31°38′11″N 74°52′29″E and at altitude 218 m above mean sea level. The mean maximum and minimum temperatures were 45.5 and 4 °C, respectively. The mean annual rainfall is 649 mm, distributed over 120 rainy days. Relative humidity range was 78–96%.

Experimental lay out

Experiment was set up at a polyhouse in germination trays and pots for 2 years trial during Oct–April 2014–15 and 2015–16. Seeds of both the varieties were hand sown in mid October.

Germination trays

To determine the germination indices for twelve replicates of six treatments, vermicompost (v/v) was mixed in different proportions 0% (G_0), 20% (G_{20}), 40% (G_{40}), 60% (G_{60}), 80% (G_{80}), and 100% (G_{100}) with soil. To understand and compare the impact of vermiwash and inorganic fertilizer, two more treatments were set up, i.e. 50% vermicompost (v/v) supplemented with foliar spray of vermiwash (G_{vw}) and inorganic treatment (G_{inor}) to the soil. To compare the effectiveness of vermicompost in substituting commercial potting media, eight more treatments were prepared, i.e. vermicompost was mixed with commercial media in proportion of 0, 20, 40, 60, 80 and 100% in treatments G_{CM0}, G_{CM20}, G_{CM40}, G_{CM60}, G_{CM80} and G_{CM100}. In treatment G_{CMvw}, vermicompost and commercial media mix was supplied with vermiwash and in G_{CMinor}, chemical fertilizer was applied and germination trays were filled with 20 ml of mixed substrate as per treatments in Table 4. Composition of commercial media consisted of 35% limed peat moss, 35% perlite and 30% coconut coir, procured from a local nursery. Totally, 16 treatments with 12 replicates were used for germination studies. Seed germination and seedling growth were studied up to 5 weeks after sowing seeds in germination trays. Plant sampling was done at 21 days after sowing (DAS), for morpho-physiological studies, and five randomly selected seedlings from twelve replicates of all the 16 treatments were uprooted from germination trays having vermicompost and soil mix; vermicompost and commercial potting mix. At 35 DAS, all the seedlings under treatment in cell plug trays were uprooted for study. To obtain the sample, the root was cut from the root collar and watered with gentle deionised water spray using a sieve below to collect any broken fragment. Root was immersed in graduated cylinder to calculate the root volume. Same sample was used to measure length, fresh and dry root weight. Shoots were also measured for shoot length and fresh shoot weight. Samples were then oven dried at 70 °C to determine the respective dry weights.

Potting studies for complete life cycle

Plantation was done in earthen pots of volume 9.51 (15 cm diameter and 10 cm height) and placed in polyhouse in a complete randomized design with eight treatments in three

Table 4 Proportion of vermicompost, soil, and commercial potting media substituted vol/vol in treatments of germination and pot culture to study life cycle of *Linum*

Treatments for germination studies by substituting soil with vermicompost			Treatment for germination studies by substituting commercial potting media with vermicompost			Treatment for vegetative and reproductive growth by substituting soil with vermicompost in pot culture		
Treatment	Proportion of vermicompost (%)	Soil and additive (%)	Treatment	Proportion of vermicompost (%)	Commercial potting media (%)	Treatment	Proportion of vermicompost (%)	Soil and additive (%)
G_0	0	100	G_{CM0}	0	100% commercial potting media	P_0	100	0
G_{20}	20	80	G_{CM20}	20	80% commercial potting media	P_{20}	80	20
G_{40}	40	60	G_{CM40}	40	60% commercial potting media	P_{40}	60	40
G_{60}	60	40	G_{CM60}	60	40% commercial potting media	P_{60}	40	60
G_{80}	80	20	G_{CM80}	80	20% commercial potting media	P_{80}	20	80
G_{100}	100	0	G_{CM100}	100	0% soil + 0% commercial potting media	P_{100}	0	100
G_{vw}	50	50% soil + vermiwash foliar spray	G_{CMvw}	50	50% commercial potting media + vermiwash foliar spray	P_{vw}	50	50% soil + vermiwash foliar spray
G_{inor}	0	100% soil + inorganic fertilizer	G_{CMinor}	0	100% commercial potting media + inorganic treatment	P_{inor}	0	100% soil + inorganic fertilizer

replicates (Table 4). INM (Integrated Nutrient Management), was done in three treatments (G_{vw}, G_{CMvw} and P_{vw}) by combining use of foliar spray of vermiwash with vermicompost application. A comparison of growth parameters was made among plants under inorganic fertilizer treatment, vermicompost treatments, vermiwash treatment and unamended soil. Separate heaps of soil and vermicompost were made, sieved and mixed so as to make vermicompost application dose to be 0, 20, 40, 60, 80 and 100% (Arancon et al. 2004; Zaller 2006) for treatments P_0, P_{20}, P_{40}, P_{60}, P_{80} and P_{100}. In another two treatments, 50% vermicompost application was integrated with vermiwash (P_{vw}) and inorganic fertilizers (P_{inor}). Four seeds per pot were sown at 5 cm depth. After emergence, seedlings were thinned to two plants per pot. The plants were irrigated, according to the plant water requirement, so as to maintain the moisture up to 60%. Manual weeding was done intermittently. 100 ml of Vermiwash mixed with water (1:1 v/v) was applied fortnightly for first 50 days to the seedlings and then monthly to plants in P_{vw}. Five seedlings from three replicates were randomly selected from pots and uprooted at 35 DAS. At 50 DAS no sample was destroyed and only the features above ground were studied. At 160 DAS, i.e. at the end of life cycle all the plants were uprooted and used as sample for the studies. Phenology of the plants was studied to evaluate the effect of treatments on length of life-cycle of Linseed.

Crop was hand harvested twice. First harvest was done on 128 DAS when 90% of the capsules turned brown and capsule number was counted, second harvest was on 150 DAS. Estimation of unripe green capsule, which would be available for second harvest, was also done. Ratio of ripened to unripened capsules was calculated to know early ripening. Yield (%) in first harvest was calculated to find uniformity in maturity timings. Also the ratios of marketable to non-marketable fruits were also calculated (Atiyeh et al. 2000b) so as to know the effect of different treatments on marketable seed yield. In each of the harvest, malformed and non-marketable fruits were sorted out and counted.

Measurement of plant growth index

Seedling growth is an important indicator for survival and growth of plantlings in field (Thompson 1985).

Morpho-physiological parameters

Test plants both from germination trays and potting experiments were studied. Seedling emergence was assessed daily for allometric growth, up to 35 days after sowing (DAS). Morpho-physiological parameters for shoot length, root length, seedling height, shoot weight, root weight, fresh biomass, dry biomass and root collar diameter were assessed. Branching ratio and biomass accumulation were also studied at 21 and 35 days after seed sowing (DAS). At 50 DAS Plant height stem length and basal area, and dried leaf weight was determined. Plant biomass, root: shoot ratio (by weight) and biomass allocation were also calculated. Dry weight of 50 leaves was taken to estimate photosynthetic productivity with each treatment. Fifty leaves per plant were washed in distilled water, put in paper bags and oven dried at 60 °C for 48 h. Shoot height was measured from root collar to the apex at various DAS. Stem diameter was measured as average taken at three points—slightly above the root collar, apex and median with a calliper. Stem length was taken from the soil level to the apical meristem of plant. First day of branching, flowering, fruiting, seedfill and seed maturity were noted. By submerging roots in a measuring cylinder filled with water, root volume (cm^3) was measured. Plant samples were dried in oven at 70 °C for 48 for dry weight of shoot and root. At 160 DAS, Plant height, weight, capsule number per plant, capsule weight, number of seeds per capsule and seed weight per plant was taken. Marketable yield was calculated by rejecting malformed fruits which were without rattling sound. Mean capsule weight was calculated from total fruit weight of each treatment. Fruits were thrashed to obtain seeds. Data was pooled from all replicates to calculate average yield per plant.

Germination indices

Total germination (Final germination percentage-G) and T_{50} It is most widely used index and shows germination capacity of seeds under specific treatment.

$$G = \frac{N_T \times 100}{N},$$

where N_T is proportion of seeds germinated at each treatment for last time; N is total number of seeds used.

$T_{50} = T_{50}$ is the number of days or time when half of the total seeds have germinated (Josep and Maria 2002).

Speed of emergence (S.E) Speed of germination considers the number of seeds germinated between two exposure times

$$\text{S.E} = (N_1 \times 1) + (N_2 - N_1)\frac{1}{2} + (N_3 - N_2)\frac{1}{3} + \ldots\ldots (N_n - N_{n-1})\frac{1}{n};$$

where Proportion of germinated seeds observed at first, second, third......... (n – 1), (n) days or hours (Bradbeer 1988; Wardle et al. 1991).

Germination index (G.I)

$$\text{G.I} = \frac{\text{RSG\%} \times \text{RRG\%}}{100},$$

where RSG is relative seed germination in percentage; RRG is relative root growth in percentage (Zucconi et al. 1981; Tiquia 2010).

$$\text{Relative seed germination (R.S.G) \%} = \frac{N_{\text{GT}} \times 100}{N_{\text{GC}}},$$

where N_{GT} is no. of seeds germinated in treatment; N_{GC} is the no. of seeds germinated in control (Tiquia 2010).

$$\text{Relative root growth (R.R.G) \%} = \frac{\text{Rl}_{\text{T}} \times 100}{\text{Rl}_{\text{C}}},$$

where Rl_{T} is mean root length in treatment; Rl_{C} is mean root length in control.

Seedling vigour index (S.V.I) S.V.I $= \frac{\text{Sl} \times G\%}{100}$ where Sl is seedling length and G is Total seed germination in percentage (Abdul-Baki and Anderson 1973).

Branching ratio (B_R)

It is the ratio of total number of plants and number of plants having branches. Branching ratio of 1:1 shows that all plants in a given treatment have started developing branches at a given time.

$$\text{Branching ratio } (B_{\text{R}}) = \frac{N}{N_{\text{B}}},$$

where N is total no. of plants; N_{B} is no. of plants having branches at a given time.

Yield indices

Fruit weight/100 capsule

$$\frac{W_{\text{TNF}} \times 100}{N_{\text{TF}}},$$

where W_{TNF} is weight of total number of fruits obtained from a treatment; N_{TF} is total number of fruits obtained in same treatment.

Harvest index (H.I) Harvest index (H.I) is the ratio of economic yield and biological yield taken in percentage.

For this H.I was calculated as $= \frac{W_{\text{f}} \times 100}{W_{\text{s}}}$,

where W_{f} is the fruit weight (economic yield) and W_{s} (biological yield) is weight of dry stem (Donald and Hamblin 1976).

Statistical analysis

Experiment was run in triplicate and One-way ANOVA was used to calculate the differences among various treatments. Tukey's test at a level of 5% of probability was applied between different treatments. Student's paired t test was applied to evaluate differences between two varieties of plants. Statistical analysis was done with the help of Assistat 7.7 beta version and Minitab version 14.0 computer software programs.

Results and discussion

Comparison of germination indices in different treatments with vermicompost, commercial potting media (CPM) and vermiwash

All the treatments G_{40}, G_{60}, G_{80}, G_{100}, and G_{vw}, supported the total seed germination when compared with control and G_{inor}. Even the low dose of vermicompost application in soil has improved the total seed germination in G_{20} by 13.8% in LV_1 and 37.8% in LV_2, whereas 13.1 and 15.4% increase was observed in both varieties in G_{CM20} (Fig. 1). There was a significant difference ($p < 0.05$) between two varieties of plants when treated with soil and commercial media mixed with vermicompost. The 20% dosage of vermicompost when mixed in soil reduced half time of germination up to 58% in LV_1 and 37% in LV_2 and further increase in vermicompost dosage showed no improvement in seed germination.

Speed of seedling emergence (S.E) in different treatment showed significant differences ($p < 0.05$) for both the varieties when compared with vermicompost (G_{60}, G_{80}, G_{100}), vermiwash (G_{vw}) and inorganic fertilizer (G_{inor}) but showed no significant difference ($p > 0.05$) within soil treatment of two varieties. Speed of seedling emergence was highest in G_{60}, G_{CM20} in LV_1 (8.8) and G_{CM20}, G_{40} in LV_2 (8.5). Usage of foliar spray of vermiwash in Combination with 50% vermicompost have further shown qualitatively healthy early maturing seedlings, it can be attributed to availability of higher nitrogen content (Chaoui et al. 2003). Relative root growth in both the varieties increased with increasing proportion of vermicompost in soil, as well as Commercial Potting media but not in 100% vermicompost application and inorganic treatment. Relative root growth has been higher in treatments G_{60}, G_{CM60}; G_{80}, G_{CM80}; G_{vw} G_{CMvw}. Highest germination index was achieved with application of 60, 80% vermicompost in soil and 80% vermicompost in commercial potting media in LV_1 and LV_2, respectively. The seedling vigour index

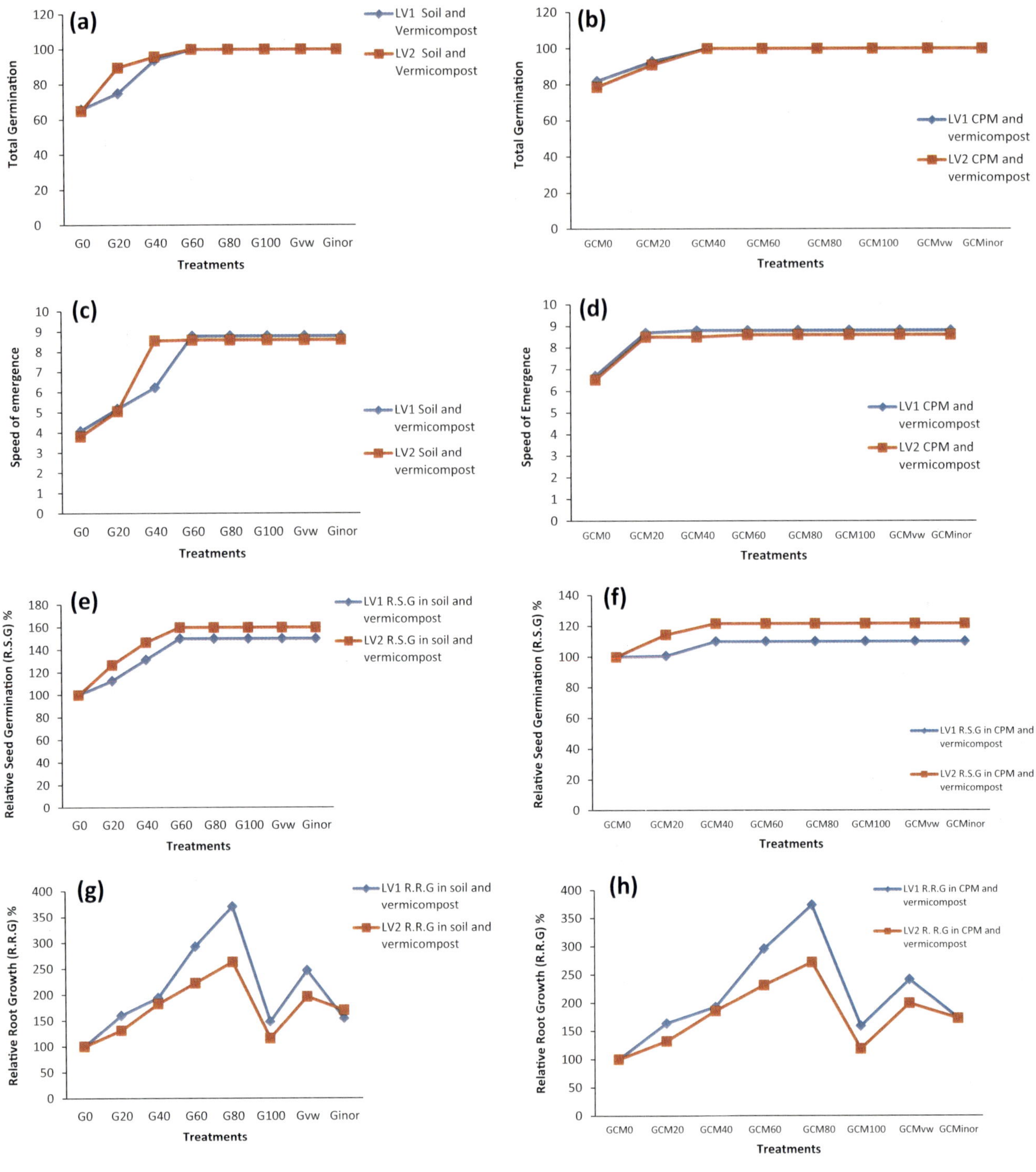

Fig. 1 Analysis of various seedling growth parameters in two varieties Linseed Plants LC- 54 (LV_1) and LC- 2063 (LV_2): **a**, **b** total germination; **c**, **d** speed of emergence; **e**, **f** relative seed germination; **g**, **h** relative root growth under different treatments

(S.V.I) showed significant difference ($p < 0.05$) and was highest with dosage of 60% v/v vermicompost, irrespective of substrate being soil or commercial potting media (Table 5). Substituting soil and commercial potting media with vermicompost show comparable S.V.I and Germination Index (G.I) in treatments G_{20} and G_{CM20}; G_{40} and G_{CM40}; G_{60} and G_{CM60}; G_{vw} and G_{CMvw}. Results indicated that the role of vermiwash and vermicompost as potting media can be no less than commercial potting media.

Scope of vermicompost to be used as an alternative to commercial potting substrates or peat is indicated in present study of germination indices. Vermicompost substitution in commercial potting media and soil showed no negative effect on *Linum* seed germination in all proportions, so vermicompost can be used to replace the expensive commercial potting media partially or whole. Similar results have been put forth by Atiyeh et al. (1999, 2000a, 2001) using pig manure as container media. Replacing vermicompost with commercial potting media would reduce substrate cost in nurseries thus in turn reducing peat harvesting from wetlands. Ievinsh (2011) reported that application of more than 50% vermicompost, inhibited seed germination in radish, cabbage, swedish turnip, beet root, beans and peas. Contrary to this, in present study, germination rate of *Linum* varieties was not negatively affected by increasing the proportion of vermicompost in soil. Even 100% vermicompost did not inhibit the seed germination but seedlings showed lower, relative root growth, S.V.I and biomass at 21 DAS, It can be due to excessive micro and macro nutrients in plant tissues. The concentration of nutrients in plant tissues have been shown to be directly proportional to the concentration of nutrients in medium (Truong and Wang 2015). Atiyeh et al. (2000b) also reported similar results that 100% vermicompost caused tomato seedling to be relatively short, with few leaves and lesser weight.

The morpho-physiological parameters of seedling were compared and analysed in all the treatments at different DAS. Till 21 days after sowing (DAS), shoot length and seedling length increases with increase in vermicompost proportion up to 60% dosage, with highest values of shoot length being (7.23 ± 0.042) mm in LC-54 and (6.48 ± 0.113) mm in LC-2063 (Table 6). Similar pattern has been observed for root length of both varieties. Shoot length in treatment with vermiwash was comparable for LV_1 (6.27 ± 0.156) mm and LV_2 (5.95 ± 0.049) mm. Tall seedlings are believed to be genetically superior (Nienstaedt 1981; Campbell and Sorensen 1984) as height of seedling shows its physiological status. Height and stem diameter of seedling are a measure of field survival and field performance (Mullin and Svaton 1972).

Vegetative growth

Vermicompost application has been proved to be efficient in promoting the root formation, plant height and plant biomass as observed in many horticultural plants (Singh et al. 2008). At 35 DAS root collar diameter was maximum in the plants sprayed with vermiwash (2.15 ± 0.07) in LV_1 and (1.91 ± 0.05) and in LV_2. This indicates the role of foliar spray of vermiwash in promoting root growth. Fresh and dry biomass of the seedling of LV_1 and LV_2 was highest in plants treated with vermiwash, that is, 3.34 ± 0.01 g, 3.27 ± 0.01 g and 0.35 ± 0.05 g, 0.34 ± 0.01 g, respectively, when compared with 60%

Table 5 Seed germination index (G.I) and seedling vigour index (S.V.I) in LV_1 and LV_2 varieties of *Linum* in soil and commercial potting media substituted with vermicompost, vermiwash and inorganic fertilizers

Germination studies using vermicompost in soil					Germination studies using vermicompost in commercial media				
Treatments	Germination index		Seedling vigour index		Treatments	Germination index		Seedling vigour index	
	LV_1	LV_2	LV_1	LV_2		LV_1	LV_2	LV_1	LV_2
G_0	100 ± 0^g	100 ± 0^g	2.16 ± 0.19^e	1.41 ± 0.08^g	G_{CM0}	100 ± 0^f	100 ± 0^g	2.18 ± 0.04^e	1.78 ± 0.04^e
G_{20}	179.6 ± 9.70^f	166.03 ± 2.69^f	3.69 ± 0.43^d	2.13 ± 0.08^f	G_{CM20}	174.12 ± 5.50^e	151.2 ± 16.92^f	3.73 ± 0.08^d	2.32 ± 0.14^d
G_{40}	254.9 ± 3.81^d	267.7 ± 5.69^d	5.45 ± 0.30^{bc}	4.03 ± 0.15^d	G_{CM40}	211.86 ± 3.08^d	225.88 ± 1.83^{cd}	5.58 ± 0.05^b	4.52 ± 0.15^c
G_{60}	438 ± 2.26^b	355.7 ± 3.60^b	7.26 ± 0.75^a	6.56 ± 0.079^a	G_{CM60}	325.27 ± 1.56^b	281.91 ± 2.36^b	7.29 ± 0.17^a	7.02 ± 0.16^a
G_{80}	555.5 ± 5.65^a	420.9 ± 6.87^a	4.8 ± 0.62^{cd}	4.96 ± 0.04^c	G_{CM80}	410.5 ± 1.9^a	331.41 ± 4.21^a	4.32 ± 0.079^{cd}	4.81 ± 0.02^c
G_{100}	221.68 ± 19.85^e	184.83 ± 7.73^e	1.9 ± 0.43^e	1.92 ± 0.23^f	G_{CM100}	220.57 ± 1.55^d	184.45 ± 5.80^e	1.91 ± 0.13^e	1.96 ± 0.08^e
G_{vw}	369.45 ± 3.85^c	313.79 ± 2.10^c	6.38 ± 0.19^{ab}	5.93 ± 0.03^b	G_{CMvw}	265.43 ± 1.49^c	242.54 ± 0.90^c	7.21 ± 0.02^a	6.32 ± 0.072^b
G_{inor}	230.59 ± 7.25^{de}	272 ± 11.55^d	4.1 ± 0.26^d	2.5 ± 0.1^e	G_{CMinor}	189.86 ± 14.86^e	209.78 ± 1.03^d	4.9 ± 0.66^c	2.34 ± 0.05^d

Turkey test at a level of 5% of probability was applied. The averages followed by the same letter in a row do not differ statistically

Table 6 Plant growth parameters studied in two varieties of *Linum* at different time intervals under various treatments

Parameters studied	*Linum* varieties	Treatments subjected to the two varieties of *Linum* plants							
		P_0	P_{20}	P_{40}	P_{60}	P_{80}	P_{100}	P_{vw}	P_{inor}
Above ground characteristics									
At 21 (DAS)									
Shoot length[a]	LV_1	3.11 ± 0.170	4.450 ± 0.212	6.25 ± 0.071	7.23 ± 0.042	4.8 ± 0.000	2.15 ± 0.354	6.27 ± 0.156	2.9 ± 0.141
	LV_2	2.23 ± 0.042	2.6 ± 0.141	4.25 ± 0.212	6.48 ± 0.113	4.68 ± 0.396	2 ± 0.141	5.965 ± 0.049	1.383 ± 0.165
Seedling length	LV_1	5.38 ± 0.396	7.37 ± 0.672	10.41 ± 0.156	13.83 ± 0.099	12.96 ± 0.092	6.39 ± 1.287	11.77 ± 0.106	7.8 ± 1.556
	LV_2	5.69 ± 0.007	7.3 ± 0.141	10.48 ± 0.255	12.94 ± 0.346	13.84 ± 0.205	6.38 ± 0.735	12.73 ± 0.099	7.79 ± 0.993
Fresh shoot weight[b]	LV_1	1.17 ± 0.035	2.01 ± 0.000	2.15 ± 0.049	3.03 ± 0.071	1.51 ± 0.028	1.27 ± 0.021	2.72 ± 0.042	1.59 ± 0.170
	LV_2	1.03 ± 0.247	2.1 ± 0.035	2.09 ± 0.049	1.73 ± 0.339	1.39 ± 0.113	1.12 ± 0.007	2.7 ± 0.170	1.41 ± 0.516
Below ground characteristics									
At 21 (DAS)									
Root length	LV_1	2.365 ± 0.092	3.625 ± 0.106	4.31 ± 0.014	6.5 ± 0.000	8.115 ± 0.163	3.54 ± 0.057	5.535 ± 0.092	3.55 ± 0.212
	LV_2	3.365 ± 0.092	4.5 ± 0.000	6.28 ± 0.028	7.665 ± 0.049	9.015 ± 0.021	3.53 ± 0.608	6.715 ± 0.021	5.915 ± 0.120
Above ground characteristics									
At 35 (DAS)									
Shoot length	LV_1	13.78 ± 0.587	21.50 ± 0.148	35.11 ± 0.445	38.2 ± 0.240	26.75 ± 0.919	28.45 ± 0.071	36.33 ± 0.191	22.71 ± 0.969
	LV_2	14.56 ± 0.092	23.35 ± 0.212	41.08 ± 0.587	35.3 ± 0.849	28.78 ± 0.163	18.25 ± 0.212	37.25 ± 0.071	18.95 ± 0.919
Biomass accumulation	LV_1	0.327 ± 0.036	0.399 ± 0.012	0.408 ± 0.001	0.442 ± 0.004	0.273 ± 0.080	0.344 ± 0.001	0.507 ± 0.001	0.375 ± 0.005
	LV_2	0.356 ± 0.014	0.377 ± 0.037	0.461 ± 0.001	0.440 ± 0.007	0.404 ± 0.023	0.320 ± 0.033	0.449 ± 0.033	0.373 ± 0.006
Dry shoot weight	LV_1	0.175 ± 0.002	0.188 ± 0.001	0.196 ± 0.003	0.221 ± 0.004	0.160 ± 0.003	0.140 ± 0.004	0.235 ± 0.005	0.187 ± 0.006
	LV_2	0.184 ± 0.003	0.214 ± 0.004	0.217 ± 0.003	0.021 ± 0.001	0.190 ± 0.002	0.176 ± 0.005	0.240 ± 0.004	0.197 ± 0.001
Fresh shoot weight	LV_1	1.45 ± 0.450	2.31 ± 0.308	2.61 ± 0.180	3.08 ± 0.274	1.69 ± 0.245	1.85 ± 0.080	2.88 ± 0.046	1.66 ± 0.587
	LV_2	0.963 ± 0.212	2.107 ± 0.287	2.617 ± 0.154	2.060 ± 0.233	1.710 ± 0.170	1.233 ± 0.320	2.883 ± 0.158	1.250 ± 0.066
Root diameter[c]	LV_1	0.67 ± 0.05	0.88 ± 0.05	2.81 ± 0.11	3.06 ± 0.08	1.80 ± 0.03	2.24 ± 0.22	3.23 ± 0.16	2.45 ± 0.21
	LV_2	0.72 ± 0.05	0.87 ± 0.16	1.94 ± 1.68	1.78 ± 1.45	1.39 ± 0.90	1.14 ± 0.55	2.02 ± 1.80	1.79 ± 1.47
Below ground characteristics									
At 35 (DAS)									
Root length	LV_1	4.19 ± 0.127	5.22 ± 0.035	6.43 ± 0.092	7.94 ± 0.064	8.16 ± 0.226	4.71 ± 0.021	6.52 ± 0.537	5.76 ± 0.057
	LV_2	4.64 ± 0.057	5.74 ± 0.205	8.76 ± 0.049	7.91 ± 0.014	10.66 ± 0.233	5.64 ± 0.064	8.15 ± 0.078	5.89 ± 0.233
Root collar diameter	LV_1	1.15 ± 0.354	1.335 ± 0.049	1.475 ± 0.035	1.555 ± 0.205	0.97 ± 0.085	1.335 ± 0.771	2.15 ± 0.071	1.435 ± 0.092
	LV_2	0.855 ± 0.064	1.205 ± 0.007	1.34 ± 0.085	1.315 ± 0.021	0.795 ± 0.177	0.875 ± 0.078	1.91 ± 0.057	1.04 ± 0.042
Fresh Root weight	LV_1	0.063 ± 0	0.123 ± 0.01	0.45 ± 0.03	0.42 ± 0.01	0.673 ± 0.01	0.676 ± 0.07	0.446 ± 0.01	0.566 ± 0.13
	LV_2	0.070 ± 0.026	0.127 ± 0.021	0.407 ± 0.025	0.450 ± 0.036	0.657 ± 0.065	0.667 ± 0.057	0.383 ± 0.006	0.550 ± 0.135
Dry root weight	LV_1	0.057 ± 0.006	0.075 ± 0.003	0.097 ± 0.001	0.08 ± 0.001	0.043 ± 0.012	0.048 ± 0.001	0.119 ± 0.003	0.07 ± 0.001
	LV_2	0.065 ± 0.004	0.080 ± 0.009	0.01 ± 0.001	0.092 ± 0.002	0.037 ± 0.004	0.056 ± 0.004	0.107 ± 0.006	0.073 ± 0.001

Table 6 continued

Parameters studied	*Linum* varieties	Treatments subjected to the two varieties of *Linum* plants							
		P_0	P_{20}	P_{40}	P_{60}	P_{80}	P_{100}	P_{vw}	P_{inor}
Above ground characteristics									
At 50 DAS									
Stem basal area	LV_1	2.365 ± 0.092	3.625 ± 0.106	4.31 ± 0.014	8.115 ± 0.163	6.5 ± 0.000	3.54 ± 0.057	5.535 ± 0.092	3.55 ± 0.212
	LV_2	3.365 ± 0.092	4.5 ± 0.000	9.015 ± 0.021	7.665 ± 0.049	6.28 ± 0.028	3.53 ± 0.608	6.715 ± 0.021	5.915 ± 0.120
Leaf weight	LV_1	45.95 ± 0.495	46.95 ± 0.212	47.25 ± 0.071	48.4 ± 0.424	44.55 ± 1.485	46.25 ± 0.636	53 ± 0.990	47.5 ± 0.424
	LV_2	47.3 ± 0.141	49.1 ± 0.283	52.59 ± 0.70	50.1 ± 0.990	47.2 ± 0.424	47.35 ± 0.778	54.4 ± 0.141	48.95 ± 0.919
Complete plant characteristics									
At harvest									
Shoot length	LV_1	114.65 ± 9.758	130.7 ± 3.111	135.15 ± 0.707	136.4 ± 0.566	125 ± 1.131	126.1 ± 7.354	135.95 ± 0.990	130.75 ± 1.273
	LV_2	122.2 ± 2.546	134.2 ± 7.637	139.4 ± 1.414	138.8 ± 0.283	125.7 ± 5.940	126.25 ± 1.556	138.4 ± 0.566	132.45 ± 2.687
Root length	LV_1	11.2 ± 0.566	14.93 ± 0.587	17.78 ± 0.191	19.08 ± 0.205	10.62 ± 0.863	9.9 ± 0.566	18.17 ± 0.233	15.83 ± 0.643
	LV_2	12.38 ± 1.089	14.93 ± 0.587	17.96 ± 0.078	17.63 ± 0.134	10.35 ± 0.813	11.32 ± 0.042	18.27 ± 0.488	16.8 ± 0.311
Root weight	LV_1	1.175 ± 0.078	1.775 ± 0.064	1.895 ± 0.049	2.255 ± 0.078	1.145 ± 0.092	1.53 ± 0.622	2.99 ± 0.113	1.805 ± 0.021
	LV_2	1.555 ± 0.177	1.885 ± 0.049	2.115 ± 0.134	1.925 ± 0.064	1.71 ± 0.113	1.165 ± 0.205	2.95 ± 0.028	1.925 ± 0.049
Shoot weight	LV_1	14.8 ± 0.170	15.985 ± 0.431	16.825 ± 0.134	17.425 ± 0.148	14.205 ± 0.884	13.865 ± 0.205	19.01 ± 0.057	15.94 ± 0.537
	LV_2	15.5 ± 0.255	17.755 ± 0.219	19.11 ± 0.297	17.96 ± 0.042	14.35 ± 0.806	13.875 ± 0.361	20.025 ± 0.290	16.98 ± 0.354
Plant biomass	LV_1	14.21 ± 1.584	16.995 ± 0.021	18.035 ± 0.134	18.385 ± 0.474	14.38 ± 0.849	13.925 ± 0.573	19.82 ± 0.269	18.275 ± 0.502
	LV_2	16.06 ± 0.212	17.655 ± 0.445	19.54 ± 0.410	18.96 ± 0.099	16.995 ± 0.021	15.685 ± 0.898	20.455 ± 0.559	18.185 ± 0.219
Biomass accumulation	LV_1	0.054 ± 0.040	0.076 ± 0.051	0.079 ± 0.051	0.091 ± 0.057	0.054 ± 0.036	0.063 ± 0.024	0.115 ± 0.067	0.079 ± 0.051
	LV_2	0.070 ± 0.052	0.074 ± 0.041	0.081 ± 0.048	0.075 ± 0.041	0.084 ± 0.065	0.058 ± 0.049	0.107 ± 0.056	0.078 ± 0.045

[a] Length in cm; [b] weight in g and [c] diameter in mm

vermicompost in both the varieties that is 2.9 ± 0.27 g, 2.8 ± 0.15 g in LV_1 and 0.328 ± 0.01 g, 0.113 ± 0.01 g in LV_2.

Plants treated with chemical fertilizer showed thinner stems and less deep root system as compared to vermicompost. Beyond 60% dosage of vermicompost showed no further increase in shoot length, root length, shoot weight, root weight and stem diameter at 50 DAS (Table 6). With the application of greater proportions of vermicompost, the plant growth parameters have not improved, this is in accordance with studies done by Subler et al. (1998) so higher doses of vermicompost had detrimental effects on Linseed plants after 35 days of sowing, which may be attributed to higher ammonium content disrupting the mechanism of nutrient absorption through plant roots (Atiyeh et al. 2000a).

At harvest, plants treated with foliar spray of vermiwash had shorter shoot length with high stem basal area than the plants treated with 60% vermicompost in LV_1 and 40% vermicompost in LV_2. Vermiwash-treated plants had extensive root system with greater root volume and root length, whereas plants treated with chemical fertilisers had weak, thin and poorly branched root system. These results are in corroboration with the studies done by Lazcano et al. (2009). Enhanced root growth parameters can be attributed to the presence of humic acid in vermicompost and vermiwash. Humic acids have been known to enhance root growth (Tallini et al. 1991) and nutrient uptake by increasing the root cell membrane permeability (Valdrighi et al. 1996).

At low to moderate concentrations of vermicompost increased vegetative and reproductive growth is due to appropriate modification of physico-chemical and microbiological status of growing medium. This is in accordance with the studies of Truong and Wang (2015) on tomato seedlings. Optimized dosage of vermicompost was 20% in French marigold seedlings (Bachman and Metzger 2008), up to 60% of cattle manure vermicompost in Petunia (Arancon et al. 2008), 40% of vermicompost in green house peppers(Arancon et al. 2004) and below 50% in tomato (Atiyeh et al. 2000b; Hashemimajd et al. 2006). Though some studies have also shown the best plant performance from 50 to 100% (Lazcano et al. 2009).

There was no significant difference ($p > 0.05$) between biomass allocation and branching ratio in soil and commercial media. Higher branching ratio at 35 days showed the potential of the plantling for higher fruiting and seed setting. Branching in both the varieties is up to third order, i.e. primary, secondary and tertiary being up to 35, 20 and 5 cm long, respectively. There was great variation in branching pattern and number in all treatments (Table 7). The treatment with vermicompost and vermiwash (P_{vw}) showed bushier appearance of plants with branching up to fifth order. Profuse branching is directly correlated to the crop yield in Linseed. Plants treated with 40–60% vermicompost and with vermiwash treatment had started early and uniform branching in all the plants, by vermicompost substitution both in soil or commercial potting media, which clearly signifies the role of vermicompost in promoting vegetative growth. Interestingly, at harvest the highest fruits were also obtained in same treatments, i.e. with 40–60% dosage of vermicompost and vermiwash application.

In both the varieties, plants provided with vermicompost gave better results than that supplemented with the inorganic fertilisers. All the replicates treated with vermicompost and vermiwash had relatively uniform growth pattern than the plants treated with inorganic fertilisers. Although, the nutrients are sufficiently provided in the inorganic treatment but better growth with vermicompost may be attributed to presence of unique mesophylic bacteria, fungi and worm secretions in vermicompost. This signifies that vermicompost is not all about nutritional enrichment but involvement of richer microbial, enzymatic activity of soil and plant growth regulators (Subler et al. 1998; Zaller 2006).

Vermicompost seems to promote nutrient absorption, so the plant with higher doses of vermicompost show relatively poor growth likely due to excessive nutrient absorption and toxicity; reduced aeration and porosity of medium; absorption of heavy metal and excessive phytotoxic substances; and excessive humic acid (Sani 2014). Combination of vermicompost and foliar spray of vermiwash presented better growth till flowering and fruiting stage due to frequent supplementation and foliar absorption of the vermiwash (Khan et al. 2014).Vermiwash has also been proven to replenish the deficiency of potassium and phosphorus in Tomato seedlings (Arthur et al. 2012). Plants treated with foliar spray of vermiwash and vermicompost were consistently been observed to have shorter height and sturdier basal area than the plants treated with 40 or 60% vermicompost till the completion of the study in both the varieties. Sturdier plants proved to be an added advantage to the crop till maturity. Till harvest, shoot length, root length, root weight, shoot weight, plant biomass and biomass accumulation was appreciably high in P_{vw} in both the varieties, in the studies conducted in both years.

Flowering, fruiting and yields

Two varieties respond significantly ($p < 0.05$) during flowering and seedfill stage. Results explain that foliar application of vermiwash has shortened the lifecycle of Linseed plant, thus favouring farmers (Table 8). Vermiwash also caused early ripening of capsules and early browning of stems and leaves. Leaf stages, flowering, fruit

Table 7 Comparison of seedling performance in terms of biomass allocation and branching ratios in two varieties of *Linum* LV_1 and LV_2 in different treatments

Germination studies using vermicompost in soil					Germination studies using vermicompost in commercial media				
Treatments	Biomass allocation		Branching ratio		Treatments	Biomass allocation		Branching ratio	
	LV_1	LV_2	LV_1	LV_2		LV_1	LV_2	LV_1	LV_2
G_0	0.043 ± 0.0^{e}	0.072 ± 0.005^{g}	4:01	03:01	G_{CM0}	0.038 ± 0.007^{e}	0.064 ± 0.005^{de}	1.5:1	1.71:1
G_{20}	0.053 ± 0.004^{e}	0.057 ± 0.004^{g}	1.71:1	1.71:1	G_{CM20}	0.047 ± 0.003^{e}	0.048 ± 0.002^{e}	1.09:1	1.51:1
G_{40}	0.172 ± 0.006^{c}	0.197 ± 0.001^{d}	01:01	01:01	G_{CM40}	0.159 ± 0.010^{c}	0.179 ± 0.002^{c}	01:01	01:01
G_{60}	0.136 ± 0.001^{d}	0.172 ± 0.001^{e}	1.09:1	01:01	G_{CM60}	0.162 ± 0.002^{c}	0.166 ± 0.002^{c}	01:01	01:01
G_{80}	0.39 ± 0.026^{a}	0.383 ± 0.002^{b}	1.71:1	02:01	G_{CM80}	0.236 ± 0.005^{b}	0.363 ± 0.002^{b}	1.33:1	1.71:1
G_{100}	0.295 ± 0.003^{b}	0.31 ± 0.018^{c}	2.4:1	02:01	G_{CM100}	0.297 ± 0.001^{a}	0.301 ± 0.003^{b}	2.4:1	02:01
G_{vw}	0.154 ± 0.081^{cd}	0.132 ± 0.103^{f}	01:01	1.71:1	G_{CMvw}	0.165 ± 0.075^{c}	0.147 ± 0.089^{cd}	01:01	04:01
G_{inor}	0.039 ± 0.003^{e}	0.44 ± 0.007^{a}	02:01	01:01	G_{CMinor}	0.073 ± 0.007^{d}	0.69 ± 0.081^{a}	03:01	01:01

Tukey's test at a level of 5% of probability was applied. The averages followed by the same letter in a column are not differ statistically

maturity was earlier achieved in P_{vw} (15, 106.5 and 130 days) than in control P_0 (18.5, 128, 148 days) in LV_1, whereas in LV_2 earliest stages were achieved in P_{vw} (13.5, 114.5 and 135 days) than in control (17, 121.5 and 144 days). Foliar application of vermiwash during seedfill and seed maturity stage seemed to be non-significant but fruit yield was higher in P_{vw} may be due to the previous applications of vermiwash. Variety of LV_1 had delayed flowering and fruiting than LV_2 which can be attributed to genotypic differences in two varieties. The varied effects of vermicompost among different varieties of same genus indicate the role of genotype in specific response to any organic amendment (Zaller 2007; Lazcano et al. 2010) as some varieties may be more sensitive and better responding at specific dosage.

Plants treated with the combination of 50% vermicompost and a foliar spray of vermiwash (P_{vw}) turned out to be the highest yielding plants with more branches, higher number of capsules, highest plant dry weight and maximum number of seeds (Table 9). Positive effects of vermiwash in this study coincide with results from Gutierrez-miceli et al. (2008) and Tejada et al. (2008).Vermiwash is believed to contain plant growth hormones, enzymes and vitamins from earthworm-associated microbes (Suthar 2010).Vermiwash promotes plant growth by physical amelioration of substrate and influencing nutrient uptake mechanism (Alvarez and Grigera 2005).

The percentage of capsule number in P_{vw} is 49.35, 45.51% higher than P_{inor} and 138.72, 95.5% higher than P_0 in LV_1 and LV_2, respectively (Table 9). During plant growth, initially the root growth is fast but later on, flower, fruit and seed become the sink of plant resources. Leaf weight was consistently high in P_{vw} indicating high photosynthetic efficiency with foliar application of vermiwash. Thus, greater fruit and seed number in P_{VW} is resultant of increased photosynthetic efficiency of the plant with vermiwash treatments. Total capsule number in P_{vw} was 4.04% higher than in P_{40} in LV_1 and 3.8% higher in LV_2 than P_{60}, in both the trials. Maximum fruit yield in numbers was obtained in P_{vw} in first harvest after 128 DAS. In second harvest, fruit yield was very low in treatment with vermiwash in both the varieties. The highest yield of fruit (%) in first harvest of both varieties in P_{vw} (93.97 ± 1.93, 95.27 ± 0.84) shows even and uniform ripening of the fruit when fruit yield in first and second harvest was compared. It suggests the uniform maturation and fruit ripening is achieved with foliar spray of vermiwash.

Average seed number per capsule was highest being 7.58 in first trial and 7.3 in second trial in LV_1 and was 7.25 and 7.05 in LV_2 in treatment with vermiwash. With inorganic treatment, seed number per capsule in P_{inor} was 5.66, 5.5 and 5.88, 5.95 in two trials of both varieties, respectively. Maximum seed yield (in g/plant) was obtained in P_{vw} in LV_1 and LV_2 followed by P_{40} in LV_1 and P_{60} in LV_2. High capsule number and seed yield in treatments P_{vw} can be attributed to extensive and profusely branched root system. Harvest index in LV_1 in P_{vw} is (66.75 ± 0.661) and in LV_2 is (57.56 ± 0.751), which is comparable to P_{40} in LV_1 and P_{60} in LV_2 and is 60.3, 34.97% higher than P_{inor} in both varieties. Even the application of low doses, i.e. 20% vermicompost has significantly enhanced the capsule number per plant (82.4% more in LV_1, 64.1% in LV_2), Fruit ripening (212% in LV_1, 185.5% in LV_2), yield in first harvest (26.8% in LV_1, 34.2% in LV_2), seed yield (56.2% in LV_1, 39.1% in LV_2) and H.I (29.7% in LV_1, 26.6% in LV_2) when compared with control P_0.

Furthermore, the foliar applications of vermiwash reduced the incidence of capsule malformation. This relatively healthier, higher yield and a lesser malformed fruit

Table 8 Effect of vermicompost on the duration of major stages in life cycle of *Linum* plants under different treatments

Treatments	Stage-1 Average number of days to seedling emergence		Stage-2 Average number of days to leaf stages		Stage-3 Average number of days to flowering		Stage-4 Average number of days to seedfill and seed maturity	
	LV_1	LV_2	LV_1	LV_2	LV_1	LV_2	LV_1	LV_2
P_0	14.5 ± 0.50^{a}	16.0 ± 1.14^{a}	18.5 ± 0.43^{a}	17.0 ± 0.50^{a}	128 ± 0.86^{a}	121.5 ± 0.43^{bc}	148 ± 1.32^{a}	144 ± 0.90^{bc}
P_{20}	10.0 ± 0.00^{b}	11.5 ± 0.43^{b}	14.5 ± 0.4^{d}	13.5 ± 0.43^{cd}	109.5 ± 0.66^{cd}	114 ± 0.86^{c}	133 ± 1.8^{bcd}	141.5 ± 3.04^{c}
P_{40}	10.0 ± 0.00^{b}	10.5 ± 0.43^{bc}	16.5 ± 0.50^{bc}	13.0 ± 0.50^{d}	106 ± 0.90^{e}	122.5 ± 1.80^{abc}	129.5 ± 1.29^{d}	142.5 ± 3.01^{c}
P_{60}	9.50 ± 0.50^{b}	10.5 ± 0.43^{bc}	16.5 ± 0.50^{bc}	14.5 ± 0.43^{bcd}	110.5 ± 1.00^{c}	121.5 ± 1.32^{bc}	136 ± 3.96^{bc}	144.5 ± 1.73^{bc}
P_{80}	9.50 ± 0.50^{b}	10.0 ± 0.00^{c}	16.5 ± 0.50^{bc}	15.0 ± 0.89^{bc}	115 ± 1.80^{b}	128 ± 7.79^{ab}	145 ± 3.46^{a}	148.5 ± 0.43^{ab}
P_{100}	9.50 ± 0.50^{b}	10.0 ± 0.00^{c}	17.6 ± 5.7^{ab}	15.5 ± 0.43^{ab}	118 ± 0.86^{b}	131 ± 2.61^{a}	149.5 ± 0.43^{a}	151.5 ± 0.43^{a}
P_{vw}	9.5 ± 0.50^{b}	10.0 ± 0.0^{c}	15.0 ± 0.0^{cd}	13.5 ± 0.44^{cd}	106.5 ± 1.31^{de}	114.5 ± 0.50^{c}	130 ± 1.32^{cd}	135 ± 1.00^{d}
P_{inor}	9.5 ± 0.50^{b}	10.0 ± 0.00^{c}	16.0 ± 0.9^{cd}	15.5 ± 0.44^{ab}	109.5 ± 0.66^{cd}	127.5 ± 0.50^{ab}	137.5 ± 2.16^{b}	148.5 ± 2.17^{ab}

Values are the average number of days after seed sowing (DAS). Tukey's test at a level of 5% of probability was applied. The averages followed by the same letter in a column do not differ statistically

Stage-1: When cotyledons first seen unfolded in 50% of plants

Stage-2: When four leaves unfolded in all seedlings

Stage-3: When at least single flower was observed on all the plant

Stage-4: When at least 10% of seeds have reached final size

Table 9 Effect of various treatments on yield and harvest indices of the two varieties of *Linum*

Parameters studied	Varieties under treatment	Yield and harvest indices of two varieties of *Linum* under different treatments							
		P_0	P_{20}	P_{40}	P_{60}	P_{80}	P_{100}	P_{vw}	P_{inor}
Capsule number per plant	LV_1	97.66 ± 10.37	178.16 ± 5.42	221 ± 11.31	179.83 ± 4.48	153 ± 5.65	167.66 ± 9.42	233 ± 16.97	156 ± 7.07
	LV_2	123.16 ± 5.89	202.16 ± 12.4	204.16 ± 4.47	218.5 ± 6.36	122.16 ± 6.83	143.16 ± 5.89	240.83 ± 11.5	165.5 ± 9.1
Capsule weight/100 capsules (gm)	LV_1	3.865 ± 0.72	3.77 ± 0.14	4.795 ± 0.24	5.86 ± 0.07	3.845 ± 0.10	2.71 ± 0.49	5.55 ± 0.22	3.505 ± 0.44
	LV_2	3.525 ± 0.24	3.84 ± 0.04	4.47 ± 0.21	5.765 ± 0.20	3.87 ± 0.07	3.135 ± 0.10	5.45 ± 0.08	3.81 ± 0.41
Number of seeds per capsule	LV_1	5.36 ± 0.06	5.91 ± 0.82	7.05 ± 0.50	7.22 ± 0.03	6.34 ± 0.33	6.445 ± 0.30	7.44 ± 0.19	5.77 ± 0.15
	LV_2	5.04 ± 0.05	5.81 ± 0.37	6.5 ± 0.00	6.96 ± 0.16	6.66 ± 0.23	6.56 ± 0.26	7.15 ± 0.14	5.725 ± 0.31
Ratio of ripe to unripe fruit	LV_1	1.415 ± 0.17	4.415 ± 0.51	8.345 ± 0.6	6.41 ± 1.28	3.495 ± 0.45	4.1 ± 0.50	11.23 ± 0.6	2.755 ± 0.02
	LV_2	1.73 ± 0.14	4.92 ± 0.89	11.18 ± 1.8	7.955 ± 0.38	3.77 ± 0.84	3.46 ± 0.72	13.02 ± 1.1	3.375 ± 0.19
Yield in 1st harvest (%)	LV_1	65.86 ± 6.99	83.53 ± 2.54	91.09 ± 2.29	88.59 ± 2.20	81.82 ± 3.03	84.98 ± 4.78	93.97 ± 1.93	84.78 ± 3.84
	LV_2	66.24 ± 0.87	88.91 ± 2.43	93.21 ± 2.27	92.00 ± 0.21	82.98 ± 3.19	82.41 ± 2.25	95.27 ± 0.84	81.14 ± 0.56
Seed yield per plant (gm)	LV_1	5.76 ± 0.06	9.00 ± 1.08	12.14 ± 0.76	10.19 ± 0.04	8.37 ± 0.39	8.96 ± 0.37	13.23 ± 0.31	7.60 ± 0.17
	LV_2	6.29 ± 0.45	8.75 ± 0.26	9.36 ± 0.43	10.81 ± 0.05	6.43 ± 0.10	7.48 ± 0.37	11.78 ± 0.48	7.76 ± 0.88
Harvest index	LV_1	40.80 ± 4.95	52.94 ± 6.26	67.28 ± 3.69	55.41 ± 1.19	58.25 ± 0.74	64.35 ± 0.01	66.75 ± 0.66	41.61 ± 2.09
	LV_2	39.14 ± 2.26	49.57 ± 2.73	47.92 ± 3.21	57.03 ± 0.58	37.83 ± 0.60	47.83 ± 5.08	57.56 ± 0.75	42.64 ± 4.32

incidence can be attributed to the presence of certain micro-nutrient as calcium, magnesium, manganese, copper, boron, iron, amino acids and silicic acid in vermiwash (Jarecki et al. 2005).Yield of the Linseed oil from the crop is positively co-related to the seed number and the seed weight per capsule per plant. Thus, high seed and capsule number will increase gains from Linseed. Organically grown Linseed in no way has negative effect on farming economics, as the positive increase in marketable yield is there in organic Linseed production.

In Linseed, the dry weight of stem at harvest is of great importance as linen fibre is commercially obtained from its retted stems. Plant biomass at harvest is highest in P_{vw} (19.82 ± 0.269)g in LV_1 and (20.455 ± 0.559)g in LV_2. High plant biomass (g) is obtained in P_{20} (16.99 ± 0.021) in LV_1 and (17.65 ± 0.445) in LV_2, P_{40} (18.03 ± 0.134) in LV_1 and (19.54 ± 0.410) in LV_2. P_{60} (18.38 ± 0.474) in LV_1,(18.96 ± 0.09) in LV_2 is comparable with that of P_{inor}. Biomass allocation reflects the shift of plant resources which may be due to availability of nutrients in the micro environment. In the present study, in treatment P_{vw}, Biomass accumulation has shifted more to root from 35 DAS (0.507 ± 0.001; 0.449 ± 0.033) to harvest (0.115 ± 0.067; 0.107 ± 0.056) in LV_1 and LV_2 which is likely to cause better nutrient absorption and hence higher harvest indices.

Vermicompost and vermiwash application may have provided certain resistance to plants by indirectly affecting the pathogens (Edwards et al. 2004) or due to bacteria and fungi in vermicompost (Szczech 1999). No fungicide or pesticide was used in the present study rather no need was felt at any point of trial as there was no incidence of disease manifestation in the crop probably due to the pesticidal and nematicidal properties of vermicompost (Edwards et al. 2004, 2006).

Conclusion

The present study indicates that vermicompost is an environmentally friendly substitute in organic agriculture and have shown beneficial effects with 40–60% as optimum proportion in LC-54 and in LC-2063 varieties of *Linum*. Application of vermicompost has positively enhanced growth of both vegetative and reproductive phase of the plant. The yield of Linseed in current pilot study under organic agriculture was more than inorganic treatment, reflecting the scope of organic production of Linseed. Foliar application of the 50% vermicompost and vermiwash had synergistic effect in *Linum* plant. Vermiwash application has shortened the life cycle of *Linum* and improved the growth parameters like stem diameter, root volume, branching ratio, biomass allocation, fruit weight, average seed number per capsule by improving the physico-chemical, biological and microbial properties of the growing medium; by improving the nutrient absorption through roots and foliar absorption of vermiwash. Yield and performance of LC -54 was better than LC-2063 under vermicompost regime. The consistently different response of two varieties of *Linum* at different doses in both the trials showed that LV_1 is more responsive even at relatively lower doses. Thus, improved plant growth parameters and higher yield obtained with vermicompost alone and integrated with foliar application of vermiwash opens new vistas for organic Linseed production.

Acknowledgements We are thankful to Department of Applied Sciences, Punjab Technical University, Kapurthala, Punjab (India) and Department of Zoology, Khalsa College, Amritsar (India) for providing necessary lab facilities.

References

Abdul-Baki AA, Anderson JD (1973) Vigour determination of soybean seed by multiple criteria. Crop Sci 13:630–633

Adhikary S (2012) Vermicompost, the story of organic gold: a review. Agric Sci 3:905–917

Alvarez R, Grigera S (2005) Analysis of soil fertility and management effects on yields of wheat and corn in the rolling Pampa of Argentina. J Agron Crop Sci 191:321–329

APHA (2012) Standard methods for examination of water and wastewater, 22nd edn. American Public Health Association, Washington, DC

Arancon NQ, Edwards CA, Atiyeh R, Metzger JD (2004) Effects of vermicomposts produced from food waste on the growth and yields of greenhouse peppers. Bioresour Technol 93:139–144

Arancon NQ, Edwards CA, Babenko A, Cannon J, Galvis P, Metzger JD (2008) Influences of vermicomposts, produced by earthworms and microorganisms from cattle manure, food waste and paper waste, on the germination, growth and flowering of Petunias in the greenhouse. Appl Soil Ecol 39:91–99

Arthur GD, Aremu OA, Kulkarni MG, Van Staden J (2012) Vermicompost leachate alleviates deficiency of phosphorus and potassium in tomato seedlings. Hortic Sci 47:1304–1307

Atiyeh RM, Subler S, Edwards CA, Metzger JD (1999) Growth of tomato plants in horticultural media amended with vermicompost. Pedobiologia 43:724–728

Atiyeh RM, Subler S, Edwards CA, Bachman G, Metzger JD, Shuster W (2000a) Effects of vermicomposts and compost on plant growth in horticultural container media and soil. Pedobiologia 44:579–590

Atiyeh RM, Arancon NQ, Edwards CA, Metzger JD (2000b) Influence of earthworm-processed pig manure on the growth and yield of greenhouse tomatoes. Bioresour Technol 75:175–180

Atiyeh RM, Edwards CA, Subler S, Metzger JD (2001) Pig manure vermicompost as a component of a horticultural bedding plant medium: effects on physicochemical properties and plant growth. Bioresour Technol 78:11–20

Zaller JG (2007) Vermicompost in seedling potting media can affect germination, biomass allocation, yields and fruit quality of three tomato varieties. Eur J Soil Biol 43:S332–S336

Zucconi F, Forte M, Monaco A, De-Bertoldi M (1981) Biological evaluation of compost maturity. Biocycle 22:27–29

Bachman GR, Metzger JD (2008) Growth of bedding plants in potting substrate amended with vermicompost. Bioresour Technol 99:3155–3161

Bradbeer JW (1988) Seed dormancy and germination. Blackie, Glasgow

Bremner JM, Mulvaney CS (1982) Nitrogen—total. Methods of soil analysis. Part 2. Chem Microbiol Prop:595–624

Campbell RK, Sorensen FC (1984) Genetic implications of nursery practices. Forestry nursery manual: production of bareroot seedlings. Springer, The Netherlands, pp 183–191

Chaoui HI, Zibilske LM, Ohno T (2003) Effects of earthworm casts and compost on soil microbial activity and plant nutrient availability. Soil Biol Biochem 35:295–302

Crowder DW, Reganold JP (2015) Financial competitiveness of organic agriculture on a global scale. Proc Natl Acad Sci 112:7611–7616

Donald CM, Hamblin J (1976) The biological yield and harvest index of cereals as agronomic and plant breeding criteria. Adv Agron 28:361–405

Edwards CA, Domínguez J, Arancon NQ (2004) The influence of vermicomposts on plant growth and pest incidence. In: Shakir SH, Mikhaïl WZA (eds). Soil zoology for sustainable development in the 21st century, Cairo, pp 397–420

Edwards CA, Arancon NQ, Greytak S (2006) Effects of vermicompost teas on plant growth and disease. Biocycle 47:28–31

Follet R, Donahue R, Murphy L (1981) Soil and soil amendments. Prentice hall Inc, New Jersey

Gomiero T, Paoletti MG, Pimentel D (2008) Energy and environmental issues in organic and conventional agriculture. CRC Crit Rev Plant Sci 27:239–254

Gutierrez-Miceli FA, Gracia-Gomez RC, Rincon RR, Abud-Archila M, Maria-Angela OL, Gullin-Cruz MJ, Dendooven L (2008) Formulation of liquid fertilizer for *Sorghum* (*Sorghum bicolour* (L.) *Moench*) using vermicompost leachate. Bioresour Technol 99:6174–6180

Gutiérrez-Miceli FA, Santiago-Boraz J, Molina JAM, Nafate CC, Abdul-Archila M, Llaven MAO, Rincón-Rosales R, Dendooven L (2007) Vermicompost as a soil supplement to improve growth, yield and fruit quality of tomato (*Lycopersicum esculentum*). Bioresour Technol 98:2781–2786

Hashemimajd K, Kalbasi M, Golchin A, Shariatmadari H (2006) Comparison of vermicompost and composts as potting media for growth of tomatoes. J Plant Nutr 27:1107–1123

Ievinsh G (2011) Vermicompost treatment differentially affects seed germination, seedling growth and physiological status of vegetable crop species. Plant Growth Regul 65:169–181

Jarecki MK, Chong C, Voroney RP (2005) Evaluation of compost leachates for plant growth in hydroponic culture. J Plant Nutr 28:651–667

Jhala AJ, Hall Linda M (2010) Flax (*Linum usitatisimum* L.): current uses and future applications. Aust J Basic Appl Sci 4:4304–4312

John MK (1970) Colorimetric determination of phosphorus in soil and plant material with ascorbic acid. Soil Sci 109:214–220

Josep AR, Maria M (2002) Seed germination and reproductive features of *Lysimachia minoricensis* (Primulaceae), a Wild-extinct Plant. Ann Bot 89:559–562

Joshi R, Vig AP, Singh J (2013) Vermicompost as soil supplement to enhance growth, yield and quality of *Triticum aestivum* L.: a field study. Int J Recycl Org Waste Agric 2:2–7

Joshi R, Singh J, Vig AP (2015) Vermicompost as an effective organic fertilizer and biocontrol agent: effect on growth, yield and quality of plants. Rev Environ Sci Biotechnol 14:137–159

Khan MH, Meghvansi MK, Gupta R, Veer V, Singh L, Kalita MC (2014) Foliar spray with vermiwash modifies the arbuscular mycorrhizal dependency and nutrient stoichiometry of Bhut Jolokia (*Capsicum assamicum*). PLoS ONE 9:e92318. doi:10.1371/journal.pone.0092318

Lazcano C, Domínguez J (2010) Effects of vermicompost as a potting amendment of two commercially-grown ornamental plant species. Span J Agric Res 8:1260–1270

Lazcano C, Arnold J, Tato A, Zaller JG, Domínguez J (2009) Compost and vermicompost as nursery pot components: effects on tomato plant growth and morphology. Span J Agric Res 7:944–951

Lazcano C, Sampedro L, Zas R, Domínguez J (2010) Vermicompost enhances germination of the maritime pine (*Pinus pinaster Ait.*). New For 39:387–400

Letourneau DK, Goldstein B (2001) Pest damage and arthropod community structure in organic vs. conventional tomato production in California. J Appl Ecol 38:557–570

Lim SL, Wu TY, Lim PN, Shak KP (2015) The use of vermicompost in organic farming: overview, effects on soil and economics. J Sci Food Agr 95:1143–1156

Mäder P, Fliessbach A, Dubois D, Gunst L, Fried P, Niggli U (2002) Soil fertility and biodiversity in organic farming. Science 296:1694–1697

Mullin RE, Svaton J (1972) A grading study with white spruce nursery stock. Commonw For Rev 51:62–69

Nienstaedt H (1981) "Super" spruce seedlings continue superior growth for 18 years. U.S. Department of Agriculture, Forest Service, North Central Forest Experiment Station

Reganold JP, Palmer AS, Lockhart JC, Macgregor AN (1993) Soil quality and financial performance on biodynamic and conventional farms in New Zealand. Science 260:344–349

Roy S, Arunachalam K, Dutta BK, Arunachalam A (2010) Effect of organic amendments of soil on growth and productivity of three common crops viz. Zea mays, *Phaseolus vulgaris* and *Abelmoschus esculentus*. Appl Soil Ecol 45:76–84

Sani B (2014) Foliar application of humic acid on plant height in Canola. APCBEE Procedia 8:82–86

Singh R, Sharma RR, Kumar S, Gupta RK, Patil RT (2008) Vermicompost substitution influences growth, physiological disorders, fruit yield and quality of strawberry (*Fragaria x ananassa Duch.*). Bioresour Technol 99:8507–8511

Sinha RK, Valani D (2009) Agronomic studies of earthworms and vermicompost vis-à-vis conventional cow dung compost and chemical fertilizers on tomato (Lycopersicum esculentum). Vermiculture Project of School of Engineering (Environment), Griffith University, Brisbane, Australia

Subler S, Edwards C, Metzger J (1998) Comparing composts and vermicomposts. Biocycle 39:63–66

Suthar S (2010) Evidence of plant hormone like substances in vermiwash: an ecologically safe option of synthetic chemicals for sustainable farming. Ecol Eng 36:1089–1092

Szczech MM (1999) Suppressiveness of vermicompost against *Fusarium* wilt of tomato. J Phytopathol 147:155–161

Tallini M, Bertoni LA, Traversim ML (1991) Effect of humic acids on growth and biomass partitioning of container growth olive plants. Acta Hortic 294:75–80

Tejada M, Gonzalez JL, Hernandez MT, Garcia C (2008) Agricultural use of leachates obtained from two different vermicomposting processes. Bioresour Technol 99:6228–6232

Thompson BE (1985) Seedling morphological evaluation-what you can tell by looking. In: Duryea ML (ed) Proceedings: evaluating seedling quality: principles, procedures, and predictive ability of major tests. Workshop held 16-18 October, 1984. Forestry Research Laboratory, Oregon State University, Corvallis, pp 59–71

Tiquia SM (2010) Reduction of compost phytotoxicity during the process of decomposition. Chemosphere 79:506–512

Truong HD, Wang CH (2015) Effects of different combination of vermicompost on growth, yield, and fruit quality of two tomato varieties under greenhouse conditions. J Agric Sci 7(11). doi:10.5539/jas.v7n11p216

Valdrighi MM, Pera A, Agnolucci M, Frassineti S, Lunardi D, Vallini G (1996) Effect of compost-derived humic acids on vegetable biomass production and microbial growth within a plant (*Cichoriumintybus*)—soil system: a comparative study. Agric Ecosyst Environ 58:133–144

Wardle DA, Ahmed M, Nicholson KS (1991) Allelopathic influence of nodding thistle (*Carduus nutans* L.) seeds on germination and radicle growth of pasture plants. N Z J Agric Res 34:185–191

http://faostat3.fao.org/download/Q/QC/E

https://www.pdkv.ac.in/?s=linseed&x=0&y=0#sthash.BQPNc6yb.dpbs

Zaller JG (2006) Foliar spraying of vermicompost extracts: effects on fruit quality and indications for late-blight suppression of field-grown tomatoes. Biol Agric Hortic 24:165–180

Sinha RK, Herat S, Valani D, Chauhan K (2009) Vermiculture and sustainable agriculture. Am Eurasian J Agric Environ Sci 5(S):1–55

8

Enriched cocoa pod composts and their fertilizing effects on hybrid cocoa seedlings

Chris Fidelis[1] · B. K. Rajashekhar Rao[1]

Abstract

Purpose Composting has the potential to recycle wastes as a means of conserving natural resources. The study was aimed at examining feasibility of producing nutrient-enriched composts from pest infested cocoa pods with chemical amendments and using manure composts as a fertilizing material in cocoa seedling nursery.

Methods Cocoa pod waste was composted in static vessels, aerobically, with chemical enrichments (triple super phosphate charged at 0.4% P or urea charged at 0.8% N or poultry manure charged at 22%) along with a control at the Cocoa and Coconut Institute, Papua New Guinea. The reaction (pH) of the composting mixtures (pH) and macro-nutrients dynamics was monitored at periodic intervals. Effect of soil incorporation of cocoa pod manure composts at 10 g kg^{-1} was assessed on the growth and foliar concentration of macro-nutrients in hybrid cocoa seedlings.

Results In the finished manure composts, dry matter loss ranged from 30.6 to 63.3%; greatest in composting mixtures charged with super phosphate and poultry manure. Besides, super phosphate enriched mixture lost small fraction of initial N (6.6%) compared to un-enriched cocoa pod waste (30.2%). Composting mixtures with greater pH values during composting process showed higher losses of N. Super phosphate charged manure compost outperformed the control, in terms of C/N ratio and concentration of macro-nutrients (P, K, Ca, Mg and S). Quality parameters for all the manure composts conformed to the Canadian Compost Guidelines indicating satisfactory standards. Waste cocoa pods enriched with superphosphate did not show any deleterious effects on cocoa seedlings' growth, rather, improved plant height, dry matter production and foliar N concentration.

Conclusion Waste cocoa pods, co-composted with triple super phosphate and poultry manure, produced composts of desirable quality and can be effectively used to fertilize the cocoa seedlings.

Keywords Composting · Infested cocoa pods · C/N ratio · Cocoa seedlings · Compost quality · Macro-nutrients

Introduction

Cocoa (*Theobroma cacao* L., family: Sterculiaceae) is a major cash crop in Papua New Guinea (PNG) and many tropical countries. Global annual production of dry cocoa beans exceeded 3.98 million tons in 2011–2012 (ICCO 2012). Each ton of dry beans produced generates approximately ten tons of cacao pod husks and the global burden of pod husk waste continues to increase well over 40 million tons (Vriesmann et al. 2012). Valorization of cocoa pod husk waste has been attempted by producing animal feed (Aregheore 2002; Alexander et al. 2008), black soap (Taiwo and Osinowo 2001), crude gums and pectins for food and pharmaceutical industries (Samuel 2006; Vriesmann et al. 2012) and as a fertilizer in crop production (Oladokun 1986; Iberemo 2010; Oyewole et al. 2012; Gusli 2013). PNG cocoa is known globally for its fine-flavored cocoa beans and country earns an annual export income of almost $100 million. The crop is grown in

✉ B. K. Rajashekhar Rao
rsraobk@rediffmail.com

[1] Department of Agriculture, PNG University of Technology, Private Mail Bag, Lae, Morobe Province 411, Papua New Guinea

100,000–130,000 ha area and exports over 47,000 t of cocoa (Nelson et al. 2011).

Generally, agro-wastes are converted into composts or farm yard manure before being used in crop production to improve soil fertility. Enrichments in composting improve nutritive value; augment disease suppressive activity and beneficial microbial populations in the composts (Postma et al. 2003; Biswas and Narayanasamy 2006; Pugliese et al. 2011; Diaz et al. 2011; Singh and Amberger, 1995; Nishanth and Biswas 2008; Biswas 2011; Meena and Biswas 2014). Such enriched composts partly substitute mineral fertilizer requirements (Adamtey et al. 2009), improve soil organic matter stocks, augument soil quality and most importantly prevent loss of nutrient N through volatilization (DeLaune et al. 2004; Steiner et al. 2010).

Soil fertility depletion and widespread nutrient deficiencies are matters of concern in tropical regions of the world due to lack of proper plant nutrient management. For example in PNG, N and Fe were deficient in >89% of the cocoa farms and P was deficient in about 25% (Nelson et al. 2011). Nutrient outflow through cocoa beans and husks have created large deficits in nutrient balance sheets. At the average yield of 0.4 t ha^{-1} y^{-1}, a total of 120 kg N, 24 kg P, 78 kg K, 9 kg Ca, and 21 kg Mg would be removed from each hectare in 15 years. These figures could be ten times more if cocoa pod husks are not recycled (Nelson et al. 2011; Fidelis 2014).

Proper waste recycling strategy has become all the more important with the outbreak of cocoa pod borer (*Conopomorpha cramerella* Snellen; family: Gracillariidae) (Yen et al. 2010). The cocoa pod borer (CPB) larva causes damage to cocoa by boring into the pod and disrupting the development of the beans. Although burial of infested pods is the recommended practice for the control of CPB infestation (DAFF 2014; Fidelis 2014), majority of growers find this as a cumbersome activity (Curry et al. 2011). Hence, of cocoa growers abandon CPB-infested pods and opened husks in piles scattered on the farms thus directly contributing to the multiplication of CPB. The husks and pods scattered on the farm, can also harbor and pre-dispose spread of black pod disease (*Phytophthora palmivora*) inoculum (Yen et al. 2010).

The study reported herein envisaged to generate nutrient-enriched composts from the CPB-infested pods. Although a few brief and inconclusive studies have been documented on the production of compost from cocoa husks or direct recycling of husks, no studies ever attempted using CPB-infested cocoa pods as feedstock for compost. The present study, therefore, aimed at evaluating pH and macro-nutrient dynamics during enriched composting of CPB-infested cocoa pods and at testing the fertilizing efficiency of composts on the growth and foliar nutritional status of hybrid cocoa seedlings.

Materials and methods

Study location

The composts were prepared at the Tavilo Research Centre, Papua New Guinea Cocoa and Coconut Institute (PNGCCI) located in Kerevat, East New Britain Province. The average annual rainfall of the area is 2700 mm per annum and daily mean temperature ranges from 30 °C during the day to 24 °C at night. The mean sunshine hour is 5.6 h day^{-1}.

Composting feed-stocks and enrichments

Cocoa pod borer (CPB) infested pods were collected from the field after a phyto-sanitary operation in 2013. The pest CPB attacks both young and mature pods (DAFF 2014) and therefore the pods used comprised of materials of diverse biochemical composition. Before initiating the composting processes, the pods were subjected to phyto-sanitation procedures for preventing possible spread of the CPB pest. The pods were enveloped in thick (500 μ) polythene sacks under sun for seven days. The heating of pods by solarization ensured elimination of CPB eggs, larvae and pupae (Upadhyay et al. 2009). After 7 days of solar treatments, the pods were uncovered and diced into small fragments (1.0–3.0 cm^3) using machetes to serve as the feedstock for composts.

The pods were co-composted with different combinations of mineral N or P sources or poultry manure to ensure improved compost quality and compost recovery. The details of the four treatments used are presented in Table 1. Fertilizer grade urea and triple superphosphate (TSP) were used as mineral N and P source for co-composting, respectively. The poultry manure was collected from layer cages under deep litter system at National Agriculture Research Institute, Kerevat, air-dried and stored in polyethylene sacks. The top soil (0–25 cm depth) served as a starter and was collected from the Tavilo plantation fields, air-dried and sieved to 2 mm.

Experimental set-up and procedure

Composting was done in polythene bins or vessels with diameter of 43 cm, height 41 cm and volume of ~50 L. A total of 30 small vents of ~1.3 mm diameter were made for air circulation at equal distance on the sides, lid and at the base of each bin. The composting vessels were filled with feedstock, enrichments and starter in layers. In brief, bottom of the bins were filled with 2 kg saw dust to absorb any excess moisture and also to act as a bulking agent. Subsequent layer had 5 kg shredded pod husks (1–3 cm^3

Table 1 Details of composting treatments with feedstock composition

Treatments and composition	Quantities of feedstock, starter and enrichments used	
	In a layer	In a vessel
Pod husk + top soil (CC1)	5 kg + 2 kg	15 kg + 6 kg
Pod husk + top soil + super phosphate (CC2)	5 kg + 2 kg + 133 g	15 kg + 6 kg + 400 g
Pod husk + top soil + urea (CC3)	5 kg + 2 kg + 133 g	15 kg + 6 kg + 400 g
Pod husk + top soil + poultry manure (CC4)	5 kg + 2 kg + 1.67 kg	15 kg + 6 kg + 5 kg

Table 2 Chemical properties and macro-nutrients in feed-stock and other enrichments used in composting

Parameters	Topsoil	Cocoa pod	Poultry manure	Sawdust
pH in suspension	7.25 ± 0.85	5.66 ± 0.41	6.29 ± 0.23	3.45 ± 0.78
Total OC (mg g^{-1})	20.4 ± 6.77	409 ± 56.9	425 ± 44.1	561 ± 26.2
Total N (mg g^{-1})	2.41 ± 1.20	15.6 ± 3.51	21.0 ± 9.93	0.80 ± 0.33
C:N ratio	8.50 ± 4.72	26.2 ± 8.30	24.4 ± 11.6	701 ± 43.1
Total P (mg g^{-1})	1.15 ± 0.44	3.24 ± 0.72	10.2 ± 3.14	0.32 ± 0.15
Total K (mg g^{-1})	9.62 ± 3.97	25.2 ± 8.41	10.3 ± 4.55	0.14 ± 0.06
Total Ca (mg g^{-1})	24.2 ± 15.2	1.84 ± 0.57	10.0 ± 3.32	2.47 ± 0.88
Total Mg (mg g^{-1})	7.53 ± 4.61	2.62 ± 0.81	3.57 ± 1.12	0.32 ± 0.20
Total S (mg g^{-1})	0.64 ± 0.33	1.36 ± 0.55	2.48 ± 0.87	0.24 ± 0.17

size) with or without enrichments spread on the layer. The layer of shredded pods along with enrichments was enveloped by 2 kg of top soil layer. The enrichments were 130 g TSP powder and/or urea prills or 1.6 kg poultry manure to each layer of pods according to the treatment description outlined in Table 1. Thus, the pods were charged with TSP or urea or poultry manure at 0.4% P, 0.8% N and 22% mass of materials used for composting, respectively. The important properties of the pod husk, top soil, poultry manure and saw dust used are presented in Table 2. Moisture content of the composting mixtures was adjusted to 65% by weight with the addition of tap water. Composting process was terminated on day 90 by spreading the contents of the vessels on clean polythene sheets for air drying.

Sampling and physico-chemical analysis

Representative sample of pod husk, poultry manure, topsoil and saw dust were drawn in triplicates, oven dried at 65 °C for 48 h and analyzed for moisture content, pH and macronutrients content. Composting mixtures were turned once in ~10 days (on composting days 11, 16, 28, 32, 44, 53, 61, 70, 80 and 90). The turning frequency was chosen according to Nikaeen et al. (2015). Turning of compost materials gives the benefit of forcing aeration, homogenizing materials and adjusting water content. On days 17, 45, 61 and 90 samples were collected, fresh weight of the composting mass was recorded on a balance. Three subsamples (~200 g) were taken from the top, middle, and bottom of each vessel were combined to form one composite sample per composting vessel. Dried samples were ground with a grinding mill and sieved through a 1-mm sieve. Total organic C and total N were analyzed by the dry combustion and micro-Kjeldahl method, respectively. Total elemental composition (P, K, Ca, Mg and S) was estimated in *Aqua Regia* (concentrated HCl-HNO_3 at 3:1 v/v ratio) digested samples in an Inductively Coupled Plasma-Optical Emission Spectrophotometer (Varian 725ES model) (Chen and Ma 2001). The pH was recorded after suspending the feed-stocks and compost sample powders in deionized water (1:10 w/v) for 1 h using a pH meter. The mass loss of composting mixture at termination of experiment was calculated by gravimetric procedure on oven-dry weight basis (Frederick et al. 2004).

Assessment of fertilizing efficacy of composts

The efficacies of pod composts were assessed on hybrid cocoa seedlings were grown in polybags. Composts obtained at day 90 were tested as fertilizer in the nursery growth medium of hybrid cocoa seedlings in Oct–Dec, 2013.

Nursery preparation

The experimental site was situated in the Cocoa Breeding Nursery of Tavilo Research Center, PNG. The seedling bays were constructed with iron pickets and tire wire to hold the polybags in rows. The top surface of the seedling bays was covered with Visqueen® sheet. The seedling bay area was partially shaded with a green shade cloth (50% sunlight absorption property). The topsoil collected from a garden plot at the PNGCCIL's Raulawat plantation served as the growth medium. The topsoil was air-dried for five days under shade, sieved (<2 mm) and filled in standard size polybags (175 cm × 350 mm) at the rate of 2 kg topsoil per bag. A sub-sample of the topsoil analyzed for the physico-chemical properties revealed that top soil used as substrate for the experiment was slightly alkaline soil with a pH 7.29 (1:2.5 soil to water ratio), non-saline (electrical conductivity 0.05 dSm^{-1}) and poor in organic C and nitrogen contents (0.21 and 0.16%, respectively). Polybags with topsoil were arranged in the seedling bays for imposition of treatments.

Treatments, nursery management and monitoring

The topsoil in polybags was either amended with one of the cocoa pod composts at the rate of 20 g $polybag^{-1}$ to correspond a compost application rate of 20 t ha^{-1} (Adejobi et al. 2011). A control was also maintained with soil only. None of the polybags received any mineral fertilizer which is the usual practice by the farmers. Each treatment was replicated six times and was randomly (completely randomized design) arranged in the seedling bays. Disease free hybrid cocoa seeds of KA6-106 × KEE42 parentage (Upper Amazonian × Trinitario) were collected from ripe pods at Tavilo Cocoa Seed Gardens, PNGCCIL. The seeds were then pre-germinated by wrapping in a clean, moist cloth for 6 days and were then sown at three seeds per poly bag (Asare and David 2011). They were watered sufficiently and regularly with stored rain water. 2 weeks after planting seedlings were thinned to retain one healthy and uniform growing seedling per polybag. Seedlings were watered regularly with the stored rainwater. The experiment was terminated 75 days after planting by cutting the plants close to soil. Stem and leaf blades were separated, root portion was carefully removed from soil with a jet of water, rinsed with water and oven dried to record dry weight. Dried leaf blades were ground in an electric blender and assayed for macro-nutrients as explained in "Sampling and physico-chemical analysis".

Statistical analysis

One-way analysis of variance (ANOVA) was used to determine whether the chemical characters and macro-nutrient contents differed among the composts at different sampling intervals. Same test was employed to crop growth parameters upon compost application to the soil. When F-test was significant, means were separated by an LSD test. The samples collected from individual composting vessels were treated as replicates for each sampling time. All statistical analyses were performed with the Statistix 8.0 software.

Results and discussions

The pH values of the major feed-stocks were within the range of 3.45 and 7.25 (Table 2). Topsoil used as a starter had the highest pH (7.25) whilst poultry manure, pods and sawdust had pH of 6.29, 5.66 and 3.45, respectively. Poultry manure contained higher amounts of N, P and S nutrients than topsoil, cocoa pods and sawdust.

Mature composts were harvested on day 90 and the mass loss of wastes in un-enriched mixture (CC1) was significantly ($P < 0.05$) smaller and produced significant mass of compost than the other mixtures (Table 3). The dry matter loss% also differed significantly ranging from 30.6 to 63.3% among the mixtures with greatest losses from P-enriched mixture (CC2) and poultry manure-enriched mixtures (CC4). Mass loss depends on composting methods followed and the biochemical qualities of the feed-stocks used for composting. For example, different methods of composting wheat straw and labile agricultural waste could result in 40–50% mass loss upon composting (Verma et al. 2014); our results were comparable to these findings. Decomposition and composting of cocoa pods depends on biochemical composition of wastes. Certain microbial inhibitors such as phenolic substances occurring in the cocoa pod husks, besides, higher lingo-cellulosic materials could retard the organic matter decomposition rates (Yapo et al. 2013). Besides saw dust with a wide C:N ratio was also used in the study. However, the major feedstock in the study, CPB-infested pods were mostly immature. Narrow C:N ratio of the waste pods, thus, possibly helped in the decomposition process. In co-composting treatments, top soil and sawdust also helped to circumvent the effect of phenolic substances through bulking or dilution.

Co-composting of CPB-infested pods with various additives significantly influenced ($P < 0.05$) evolution of pH (Fig. 1). The pH among the composting mixtures varied between pH 6.93 and pH 8.46. Generally, the pH values of CC1 and CC3 composting mixtures increased through sampling days 17–90. Enriching the pods with poultry manures and super phosphate prevented the upsurge of pH in the compost mixture. This could be due to increased accumulation of mineralized N as NH_4^+–N. The accumulated NH_4^+–N could be volatized as NH_3, incorporated into microbial cells, if sufficient readily degradable C is available or converted into NO_2^-—and then into NO_3^-—by the nitrifying bacteria depending on the stage of composting and chemical characteristics of the composting pile (Gibbs

Table 3 Compost mass, organic matter loss and C/N ratio of the manure composts (day 90.) Dry matter loss and C/N ratio are on dry weight basis

Composting mixture	Compost mass (kg)	Dry matter loss (%)	C/N ratio
Pod husk + top soil (CC1)	13.3[b]	30.6[a]	11.1[b]
Pod husk + top soil + super phosphate (CC2)	6.80[a]	62.2[c]	10.4[b]
Pod husk + top soil + urea (CC3)	8.24[a]	54.4[b]	6.35[a]
Pod husk + top soil + poultry manure (CC4)	8.05[a]	63.3[c]	7.46[a]
P value	0.02	0.04	0.001
LSD (0.05)	2.70	7.30	1.50

Treatment means with different lower case letters in a column indicates significant difference at $P = 0.05$

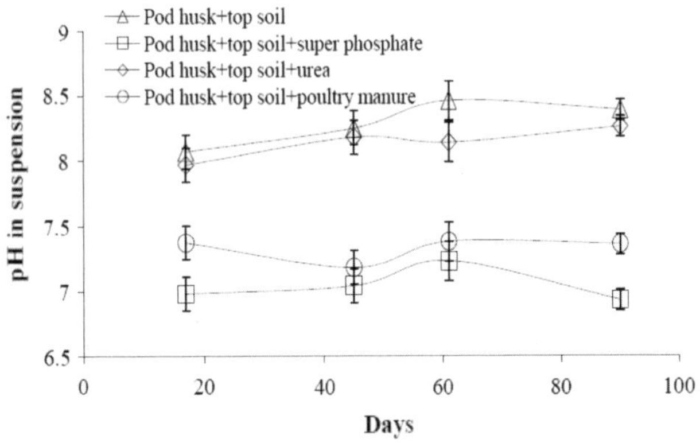

Fig. 1 Profile of pH in composting mixtures during the composting process. *Error bars* indicate LSD values at $P = 0.05$

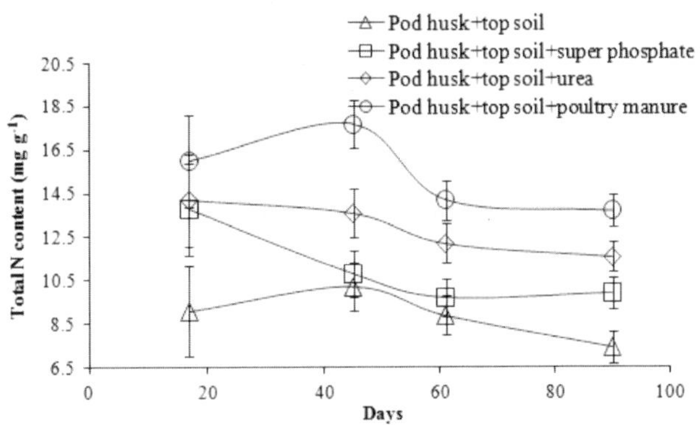

Fig. 2 Changes in total N content of composting mixtures during the composting process (on dry weight basis). *Error bars* indicate LSD values at $P = 0.05$

et al. 2002). Lowering of the pH in N-enriched compost could be due to acidification associated with production of NO_3^-–N during the nitrification process.

The total N concentration of composting mixtures progressively declined during composting (Fig. 2). The NH_4^+–N produced during mineralization could be lost through volatilization process and by leaching (Gibbs et al. 2002; Frederick et al. 2004; Tripetchkul et al. 2012). Gaseous emissions as nitrous oxide through denitrification and nitrogen volatilization have also been common (He et al. 2000; Veeken et al. 2002). The total N loss over 90 day was 30.2, 6.6, 28.8 and 9.7% of the initial N contents for CC1, CC2, CC3 and CC4 composts, respectively. Similar findings were made by Tran et al. (2012) with rice straw compost charged with super phosphate. The extent of volatile N loss was comparable to values reported by Wu et al. (2010). Enrichments improved the N content of the final composts; prominent effects were noticed in CC3 and CC4 mixtures (Fig. 2). Chemical enrichment agents such as phosphogypsum, phosphate rocks, phosphoric acid, alum and simple/single superphosphate have been traditionally used to reduce volatilization losses of N. Findings of the study indicate that triple superphosphate is effective to prevent volatilization loss of N from cocoa pod composting mixtures.

Frequently, the decomposition of composting feed-stocks might experience N or P limitations and enrichment of the feedstock with mineral N and/or P could have improved the decomposition rates as evidenced by maximum mass loss in TSP and poultry manure enrichments. However, in the current study, possibility of N limitation is very remote as the cocoa pods had a C/N ratio of 26, which could be an easily decomposable material. C/N ratio has been used as an indicator of the potential of compost stability. A mature compost should have an C/N ratio ≤ 25 (Sangamithirai et al. 2015). The C/N ratio of the final composts was also lower than 12 (Table 3). When enrichment with mineral N and/or poultry manure was made, final compost products showed C/N ratio ~ 7. One of the consequences of using such composts with low C/N ratio is the possible phytotoxic effects on sensitive crops but might have nematicidal value (Rodriguez-Kabana et al. 1987).

The total P content of the composting mixtures were significantly ($P < 0.05$) different amongst the compost treatments (Fig. 3). As expected, the pod wastes fortified with 0.6% P (CC2) produced composts with higher total P content compared to other composts generated. The P concentration increase in all composting mixtures is attributed to loss of carbon (C) during decomposition and addition of P through co-composting substrates. Poultry manure is a P- rich source and used widely to improve the

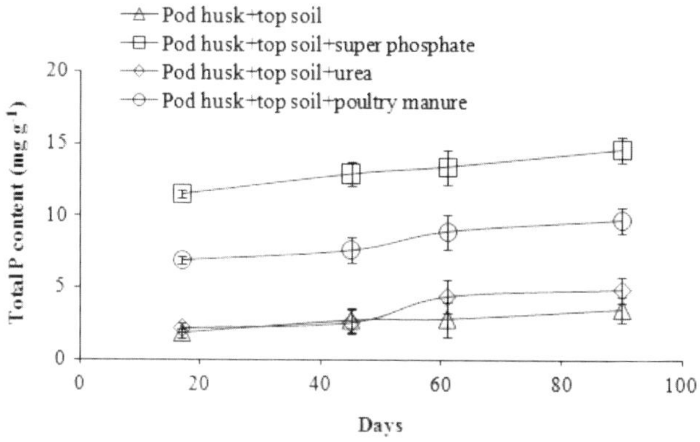

Fig. 3 Changes in total P content of composting mixtures during the composting process. *Error bars* indicate LSD values at $P = 0.05$

P content of composts (Ch'ng et al. 2013). Similar results were observed for other macro-nutrients (K, Ca, Mg and S) (Table 4). Enrichments significantly improved the macro-nutrient concentration of compost mixtures. Among the manure composts, CC2 was superior to control, in terms of K, Ca, Mg and S concentration. The elevated concentrations of Ca and S nutrients were a result of nutrient contributions from triple superphosphate besides the mass loss. Triple superphosphate contains 15% Ca and small amounts of S (~1%) thereby enhanced the nutritive value of the final compost (Mudahar 2013). Although, K content in CC3 and Mg content in CC4 composts were higher than that from CC2, the differences were not statistically significant. All the composts produced from waste pods and used in the study met the quality standards of Canadian

Table 4 Nutrient composition of composting mixtures at the beginning and at maturity

Composting mixture	Total K ($mg\ g^{-1}$)		Total Ca ($mg\ g^{-1}$)		Total Mg ($mg\ g^{-1}$)		Total S ($mg\ g^{-1}$)	
	17 days	90 days	17 days	90 days	17 days	90 days	17 days	90 days
Pod husk + top soil (CC1)	51.7^{c}	62.4^{b}	25.3^{a}	34.9^{a}	8.61^{a}	9.10^{a}	0.40^{a}	0.80^{a}
Pod husk + top soil + super phosphate (CC2)	72.3^{a}	86.8^{a}	88.7^{c}	94.0^{c}	10.6^{a}	14.2^{b}	1.50^{b}	1.90^{c}
Pod husk + top soil + urea (CC3)	63.2^{b}	89.2^{a}	26.3^{a}	43.8^{b}	9.33^{a}	16.3^{b}	0.70^{a}	1.40^{b}
Pod husk + top soil + poultry manure (CC4)	67.4^{ab}	85.1^{a}	36.5^{b}	39.2^{ab}	14.9^{b}	16.7^{b}	0.80^{a}	1.30^{b}
P value	0.004	0.04	0.0001	0.03	0.004	0.001	0.01	0.03
LSD (0.05)	8.50	9.10	6.50	5.07	3.10	4.20	0.60	0.12

Treatment means with different lower case letters in a column indicates significant difference at $P = 0.05$

Table 5 Growth response of hybrid (F1) cocoa seedlings to the soil amendment of different pod manure composts

Fertilization treatments	Plant height (cm)	Total dry matter ($g\ plant^{-1}$)
Control (without compost)	29.4^{ab}	11.9^{b}
Soil + un-enriched compost (CC1)	28.3^{bc}	11.5^{b}
Soil + P-enriched compost (CC2)	28.3^{bc}	12.7^{a}
Soil + N-enriched compost (CC3)	26.3^{c}	11.4^{b}
Soil + Poultry manure-enriched compost (CC4)	30.9^{a}	12.8^{a}
P value	0.01	0.01
LSD (0.05)	2.4	0.51

Treatment means with different lower case letters in a column indicates significant difference at $P = 0.05$

Table 6 Effect of different pod composts on the foliar nutrient content ($mg\ g^{-1}$ dry matter) of hybrid cocoa seedlings

Fertilization treatments	Total N	Total P	Total K	Total Ca	Total Mg	Total S
Control (without compost)	20.6^{a}	1.81	20.1	9.03	3.15	1.90
Soil + un-enriched compost (CC1)	24.9^{bc}	1.50	19.4	9.50	2.90	1.63
Soil + P-enriched compost (CC2)	26.7^{c}	1.82	22.4	11.3	3.44	1.95
Soil + N-enriched compost (CC3)	21.5^{a}	1.74	22.7	9.24	3.02	1.92
Soil + Poultry manure-enriched compost (CC4)	23.0^{ab}	1.70	20.2	10.2	3.16	2.10
P value	0.04	0.62	0.64	0.83	0.98	0.07
LSD (0.05)	3.37	–	–	–	–	–

Treatment means with different lower case letters in a column indicates significant difference at $P = 0.05$

Compost Guidelines (1996)—C/N ratio below 25, pH 5.5–8.5, a minimum concentration of total N, P, K, Ca and Mg at 0.60, 0.25, 0.20, 3.0 and 0.30%, respectively.

The growth response of hybrid cocoa seedlings differed significantly ($P < 0.05$) among the soil incorporation of various enriched manure composts (Table 5). CC3 compost addition to soil suppressed the plant height by 10.5% compared to the soil that did not receive any compost. Crop growth suppression in CC3 compost could be due to phytotoxic consequences of using compost with low C/N ratio (Rodriguez-Kabana et al. 1987). Suppression effects in terms of plant height were not visible in seedlings grown on soils amended with CC4 compost. Suppressive effects on dry matter production were not prominent in composts CC1 and CC3. In summary, two manure compost formulations (CC2 and CC4) encouraged cocoa seedlings dry matter production. These findings were similar to that of observations made by Moyin-Jesu (2015) in cabbage crop supplemented with poultry manure.

Soil incorporation of various manure composts significantly ($P < 0.05$) affected the foliar N concentration in cocoa seedlings (Table 6). Soil amendment with CC2 compost evinced the greatest N concentration in foliage followed by CC1, CC4, CC3 and control. Nutrient concentration of P, Ca and Mg were also greater in CC2 compost amendment. Contrary to our findings, N-enriched compost increased N concentration and uptake in wheat crop (Akhtar et al. 2011). This could be partly due to N dilution in the actively growing biomass. In the current study, the nutrient contents were measured at active growing stage (vegetative) when significant nutrient dilution can occur. Secondly, in the current study no mineral fertilizers were used along with manure composts in accordance with local farmer's practices. Phosphorous-enriched cocoa pod compost improved N concentration in cocoa leaves due to possible improvement in absorption and accumulation of N. The results were in line with previous work done by Kavitha and Subramanian (2007).

Conclusions

The results clearly showed that borer infested cocoa pods can be effectively used to produce nutrient-enriched composts of acceptable quality. Composting without any adjuvant and urea resulted in composts with high pH values with the potential for some N loss through volatilization process. Co-composting cocoa pods with super phosphate and poultry manure prevented the upsurge of pH and conservation of N during composting process. Super phosphate charged cocoa pod compost was also superior in respect of P, S and Ca nutrients.

Nutrient-enriched cocoa pod composts can be successfully used as a nutrient supplement in the growth medium of the cocoa seedlings. Fertilization of cocoa seedlings with enriched cocoa pod composts (charged with superphosphate or poultry manure) improved growth (dry matter) through improved N uptake.

However, before the infected pod composts are widely used, further studies are needed. Results of the study indicate some losses of N through volatilization during cocoa pod composting without any amendments. Cheaper alternatives to superphosphate need to be identified that prevent upsurge of pH and possible losses of N from composting mixture. Such studies must also focus on microbiological parameters to ensure destruction of dormant stages of CPB and pathogens.

Acknowledgements First author (CF) wishes to thank CCIPNG for allowing him to conduct this study and for all the logistic supports.

References

Adamtey N, Cofie O, Ofosu-Budu GK, Danso SKA, Forster D (2009) Production and storage of N-enriched co-compost. Waste Manage 29:2429–2436

Adejobi KB, Famaye AO, Adeniyi DO, Orisajo SB, Adeyemi EA (2011) Effect of cocoa pod husk ash and goat dung on nutrient content and growth performance of cashew (*Anacardium occidentale*). Adv Environ Biol 5(7):1536–1542

Akhtar M, Naeem A, Akhter J, Bokhari SA, Ishaque W (2011) Improvement of nutrient uptake and yield of wheat by combined use of urea and compost. Soil Environ 30(1):45–49

Alexander J, Benford D, Cockburn A, Cravedi JP, Dogliotti E, Dominico AD et al (2008) Theobromine as undesirable substances in animal feed: scientific opinion of the panel on contaminants in the food chain. Eur Food Safety Auth J 725:1–66

Aregheore EM (2002) Chemical evaluation and digestibility of cocoa (*Theobroma cacao*) byproducts fed to goats. Trop Anim Health Prod 34:339–348

Asare R, David S (2011) Good agricultural practices for sustainable cocoa production: a guide for farmer training. Manual no. 1: planting, replanting and tree diversification in cocoa systems, sustainable tree crops programme. International Institute of Tropical Agriculture, Accra

Biswas DR (2011) Nutrient recycling potential of rock phosphate and waste mica enriched compost on crop productivity and changes in soil fertility under potato-soybean cropping sequence in an Inceptisol of Indo-Gangetic Plains of India. Nutr Cycl Agroecosys 89:15–30

Nelson P, Webb M, Berthelsen S, Curry G, Yinil D, Fidelis C, Fisher M, Oberthür T (2011) Nutritional status of cocoa in Papua New Guinea. Better Crops 95:18–20

Yen JD, Waters EK, Hamilton AJ (2010) Cocoa pod borer (*Conopomorpha cramerella* Snellen) in Papua New Guinea:

biosecurity models for New Ireland and the autonomous region of Bougainville. Risk Anal 30(2):293–309

Biswas DR, Narayanasamy G (2006) Rock phosphate enriched compost: an approach to improve low-grade Indian rock phosphate. Bioresour Technol 97:2243–2251

Canadian Compost Guideline (1996) Guidelines for compost quality. CCME documents, PN1340, Manitoba, Canada

Ch'ng HW, Ahmed OH, Kassim S, Majid NMA (2013) Co-composting of pineapple leaves and chicken manure slurry. Int J Recycl Org Waste Agricult 2:23

Chen M, Ma LO (2001) Comparison of three aqua-regia digestion for twenty Florida soils. Soil Sci Soc Am J 65:491–499

Curry G, Lummani J, Omuru E (2011) Socioeconomic impact assessment of cocoa pod borer in East New Britain province, Papua New Guinea. Project no. ASEM/2008/034, Final report no. FR2010-25. Australian Centre for International Agricultural Research (ACIAR), Canberra

DAFF (Queensland Department of Agriculture, Fisheries and Forestry) (2014) Cocoa pod borer. Queensland Department of Agriculture, Fisheries and Forestry (DAFF), Queensland, Australia. http://www.daff.qld.gov.au/plants/health-pests-diseases/a-z-significant/cocoa-pod-borer#. Accessed 02 Nov 2015

DeLaune PB, Moore PA Jr, Daniel TC, Lemunyon JL (2004) Effect of chemical and microbial amendments on ammonia volatilization from composting poultry litter. J Environ Qual 33(2):728–734

Diaz LF, de Bertoldi M, Bidlingmaier W (2011) Compost Science and Technology. Elsevier, The Netherlands

Fidelis C (2014) Development of composting protocols of cocoa (*Theobromae cacao*) pod-based composts and their efficacy. Dissertation, PNG University of Technology, Papua New Guinea

Frederick C, Michel JA Jr, Pecchia JR, Keener HM (2004) Mass and nutrient losses during the composting of dairy manure amended with sawdust or straw. Compost Sci Util 12:323–334

Gibbs PA, Parkinson RJ, Misselbrook TH, Burchett S (2002) Environmental impacts of cattle manure composting. In: Insam H, Riddech N, Klammer S (eds) Microbiology of composting. Springer Verlag, Heidelberg, pp 445–456

Gusli S (2013) Zero-waste integrated cocao production system (ZWICP). In: Rangga I (ed) Coklat. Cocoa Sustainability Partnership, Indonesia, p 12

He Y, Inamori Y, Mizuochi M, Kong H, Iwami N, Sun T (2000) Measurements of N_2O and CH_4 from the aerated composting of food waste. Sci Total Environ 254:65–74

Iberemo OS (2010) Effect of organic fertilizer fortified with phosphate fertilizers and arbuscular mycorrhizal fungi inoculation on the growth of cashew in two ecologies in Nigeria. J Agric Sci 1(2):101–107

ICCO (International Cocoa Organization) (2012) Quarterly bulletin of cocoa statistics, vol 39, London, UK

Kavitha R, Subramanian P (2007) Effect of enriched municipal solid waste compost application on growth, plant nutrient uptake and yield of rice. J Agron 6:586–592

Meena MD, Biswas DR (2014) Changes in biological properties in soil amended with rock phosphate and waste mica enriched compost using biological amendments and chemical fertilizers under wheat-soybean rotation. J Plant Nutr 37:2050–2073

Moyin-Jesu EI (2015) Use of different organic fertilizers on soil fertility improvement, growth and head yield parameters of cabbage (Brassica oleraceae L). Int J Recycl Org Waste Agricult 4:291–298

Mudahar MS (2013) Fertilizer sulfur and food production. Springer Science and Business Media, The Netherlands

Nikaeen M, Nafez AH, Bina B, Nabavi BF, Hassanzadeh A (2015) Respiration and enzymatic activities as indicators of stabilization of sewage sludge composting. Waste Manage 39:104–110

Nishanth D, Biswas DR (2008) Kinetics of phosphorus and potassium release from rock phosphate and waste mica enriched compost and their effect on yield and nutrient uptake by wheat (*Triticum aestivum*). Bioresour Technol 99(9):3342–3353

Oladokun MAO (1986) Use of cocoa pod husk as fertilizer for maize production. Nigerian J Agron 1:103–109

Oyewole OS, Omueti I, Ajayi J, Rotimi RI (2012) Growth of cocoa (*Theobroma cacao L.*) seedlings on old cocoa soils amended with organic and inorganic fertilizers. Afr J Agric Res 7:3604–3608

Postma J, Montanari M, van den Boogert PHJF (2003) Microbial enrichment to enhance the disease suppressive activity of compost. Eur J Soil Biol 39:157–163

Pugliese M, Liu B, Gullino ML, Garibaldi A (2011) Microbial enrichment of compost with biological control agents to enhance suppressiveness to four soil-borne diseases in greenhouse. J Plant Dis Protect 118:45–50

Rodriguez-Kabana R, Morgan-Jones G, Chet I (1987) Biological control of nematodes: soil amendments and microbial antagonists. Plant Soil 100:237–247

Samuel YKC (2006) Crude gum from cocoa of Malaysian origin: part I: rheological properties. Malaysian Cocoa J 2:28–31

Sangamithirai KM, Jayapriya J, Hema J, Manoj R (2015) Evaluation of in-vessel co-composting of yard waste and development of kinetic models for co-composting. Int J Recycl Org Waste Agricult 4:153–165

Singh CP, Amberger A (1995) The effect of rock phosphate-enriched compost on the yield and phosphorus nutrition of rye grass. Am J Altern Agric 10:82–87

Steiner C, Das KC, Melear N, Lakly D (2010) Reducing nitrogen loss during poultry litter composting using biochar. J Environ Qual 39(4):1236–1242

Taiwo OE, Osinowo FAO (2001) Evaluation of various agro-wastes for traditional black soap production. Bioresour Technol 79:95–97

Tran TM, Bui HH, Luxhøi J, Jensen LS (2012) Application rate and composting method affect the immediate and residual manure fertilizer value in a maize–rice–rice–maize cropping sequence on a degraded soil in northern Vietnam. Soil Sci Plant Nutr 58:206–223

Tripetchkul S, Pundee K, Koonsrisuk S, Akeprathumchai S (2012) Co-composting of coir pith and cow manure: initial C/N ratio vs physic-chemical changes. Int J Recycl Org Waste Agricult 1:15

Upadhyay RK, Mukerji KG, Chamola BP, Dubey OP (2009) Integrated pest and disease management. APH Publishing Corporation, New Delhi

Veeken A, de Wilde V, Hamelers B (2002) Passively aerated composting of straw-rich pig manure: effect of compost bed porosity. Compost Sci Util 10(2):114–128

Verma R, Badole WP, Deewan P, Meena VS (2014) Carbon and weight loss during composting of wheat straw by different methods. Ann Biol 30(2):354–357

Vriesmann LC, Teófilo RF, Petkowic CLV (2012) Extraction and characterization of pectin from cacao pod husks (*Theobroma cacao* L.) with citric acid. LWT Food Sci Technol 49:108–116

Wu D, Liu P, Luo Y, Tian G, Mahmood Q (2010) Nitrogen transformations during co-composting of herbal residues, spent mushrooms, and sludge. J Zhejiang Univ Sci B 11(7):497–505

Yapo BM, Besson V, Koubala BB, Koffi KL (2013) Adding value to cocoa pod husks as a potential antioxidant–dietry fibre source. Am J Food Nutr 1(3):38–46

9

In-vessel co-composting of yard waste and food waste: an approach for sustainable waste management in Cameron Highlands, Malaysia

Amirhossein Malakahmad[1] · Natasha Binti Idrus[1] · Motasem S. Abualqumboz[1] · Sara Yavari[1] · Shamsul Rahman M. Kutty[1]

Abstract

Purpose Huge amount of yard waste is produced in cities with excessive agricultural activities like Cameron Highlands, Malaysia where most of the time the yard waste is being managed poorly and big portion of it ends in dump sites. Therefore, this study aims to evaluate the applicability of converting yard waste generated in Cameron Highlands Malaysia into high-quality and fast compost via in-vessel method.

Methods In-vessel composting technique was applied for speedy biotransformation of yard waste. Addition of food waste, effective microorganisms (EM) and Shimamoto Enzyme® (SE) were investigated for improvement of compost quality. Four compositions of feedstock with different yard waste (YW) and food waste (FW) ratios were tested. The compositions were 70%YW + 30%FW, 80%YW + 20%FW, 90%YW + 10%FW and 100%YW. Physicochemical properties of compost including pH, moisture content and C/N ratio were monitored throughout the experiment. Furthermore, quality of compost and its potential for direct application after production were evaluated based on germination index (GI) and nutrient content (NPK).

Results The compost samples had pH ranging from 7 to 9 and moisture content of 15.45–32.13%. Initial C/N ratio of all feedstock was decreased throughout the composting process by more than 50%. Seed germination test showed that only 70%YW + 30%FW feedstock produced immature compost with GI < 80%. The highest GI of 130% was obtained when FW represented in 10% of the feedstock with addition of EM. Average concentrations of nitrogen, phosphorus and potassium were 1.73, 1.21 and 1.66% in case of EM additive and 1.47, 0.56 and 1.74% in case of SE additive.

Conclusion Application of in-vessel composting can improve solid waste management in Cameron Highlands, Malaysia and yield a high-demand product. The approach used in this study can be a good practice for the societies have difficulties in managing their yard waste.

Keywords Aerobic composting · Compost quality · Municipal solid waste · Effective microorganisms

Introduction

Landfilling has been the most adopted option for disposing of growing quantities of municipal solid waste (MSW) in Malaysia as it is the cheapest disposal option (Agamuthu 2013). Besides 114 non-operational landfills, the Malaysian ministry of Housing and Local Government is responsible for other 176 in-operation landfills (Fazeli et al. 2016). Out of the currently operating landfills, only 14 have been classified as sanitary landfills whereas the remaining are unsanitary (Noor et al. 2013). Many illegal dumpsites have been used in Malaysia for disposing of the waste (Malakahmad et al. 2013). Statistics show that more than 90% of collected comingled solid waste in Malaysia is disposed of in unsanitary landfills (Badgie et al. 2012; Johari et al. 2014). Nevertheless, organic fractions of MSW such as yard waste can be transformed into useful products through biological conversion technologies such as composting (Parthan et al. 2012; Rajaie and Tavakoly 2016). Yard waste and food waste collectively represented nearly 50% of waste stream in Malaysia (Dinie et al. 2013;

✉ Amirhossein Malakahmad
amirhossein@utp.edu.my; amalakahmad@gmail.com

[1] Department of Civil and Environmental Engineering, Universiti Teknologi Petronas, 32610 Seri Iskandar, Perak Darul Ridzuan, Malaysia

Periathamby et al. 2009). However, composting of organic wastes has been practiced in Malaysia for only 1% of solid waste from 2000 to 2006 and it is projected to increase to 8% by 2020 (Johari et al. 2014). Mansor et al. (2015) stated that ample quantities of yard waste get generated in Malaysia from areas that have high agriculture activities such as Cameron Highlands. Cameron Highlands is the largest vegetable, flower and tea supplier in Malaysia (Aminu et al. 2015). While considerable portion of generated waste in Cameron Highlands is disposed of in unsanitary dumpsite near Tanah Rata, organic MSW including yard wastes can be converted into useful products via biological transformation methods.

Composting, as a biological waste transformation technique, can be defined as decomposition and stabilization of organic matters, commonly in the presence of oxygen, for production of stable and free-of-pathogen final product that can be beneficially applied to land (Haug 1993) for soil amendment (Ch'ng et al. 2013; Mhindu et al. 2013; Pandey et al. 2016). Generation of renewable energy (Irvine et al. 2010), saving lands by increasing lifetime of landfills (Hoornweg et al. 1999) and reduction of transportation costs (Larney et al. 2006; Mandal et al. 2014) are reported as other benefits of composting. Moreover, less amount of landfill gases (LFG) and leachate would be generated when organic materials are composted rather than landfilled (Lou and Nair 2009; Sudharsan Varma and Kalamdhad 2014).

Composting of organic MSW can be done using different methods such as windrow, aerated static pile, and in-vessel techniques. Selection of composting methods can be considered in respect to factors such as capital and operating costs, land availability, operational complexity and potential of nuisance problems. Although more expensive compared to other methods, in-vessel composing is more controlled, needs smaller spaces and resulted in less bothers than other techniques (Cekmecelioglu et al. 2005). In-vessel composting technique allows the highest degree of temperature control. In addition, it can be used to shorten the time needed for production of compost drastically (An et al. 2012). In Malaysia, in-vessel composting also has been successfully implemented for industrial wastes (Razali et al. 2012; Zahrim et al. 2016). Therefore, in this study composting of yard waste generated in Cameron Highlands was conducted using an in-vessel composter. Quality of compost was assessed via measurement of germination index and nutrient values.

Methodology

Feedstock materials

Yard waste (YW) was gathered from a plant nursery in Cameron Highlands, Malaysia. It was then shredded using mechanical shredder into smaller sizes with 4–5 cm maximum length. Shredding of organic waste into small sizes prior to composting results in faster stabilization (Tognetti et al. 2007). Food waste (FW) was collected from a nearby restaurant and comprised rice, chicken, potatoes, sausages, cooked spaghetti and green vegetables. The initial pH values of YW and FW were 6.17 $\pm$ 0.2 and 4.68 $\pm$ 0.2, respectively. Even though the optimum pH range for composting is 5.5–8, raw feedstock having a pH value within the range of 3–11 can also be composted (Pichtel 2005). YW used in this study had low initial moisture content of 35.63 $\pm$ 0.4% while FW recorded moisture content of 58.24 $\pm$ 0.4%. Optimum initial moisture content of raw feedstock ranges between 50 and 60% (Kalamdhad and Kazmi 2009). The analyses showed that YW and FW contained 17.17 $\pm$ 0.5% and 31.77 $\pm$ 0.5% of carbon, and 0.28 $\pm$ 0.2% and 3.43 $\pm$ 0.2% of nitrogen. The C/N ratios of YW and FW were 61.32 $\pm$ 0.3 and 9.26 $\pm$ 0.2, respectively. Initial carbon to nitrogen ratios between 25 and 50 are optimum for aerobic composting (Kumar et al. 2010; Zhu 2007). High C/N ratio causes slow growth of microorganisms and consequently organic decomposition will be slow (Maheshwari 2014). On the other hand, low C/N ratio contributes to odor problems due to formation of ammonia or nitrous oxides. Blending of YW and FW can provide a near optimum C/N ratio.

In-vessel composter

An in-vessel actual-size rotary drum composter (COMPOSTECH CT200) with the dimensions l: 2.0 $\times$ w: 1.1 $\times$ h: 1.3 m^3 was used in this study. Proper rotation in these kinds of composting machines is an efficient technique which produces uniform end product due to good agitation, aeration and mixing of the compost (Rich and Bharti 2015). The composter was able to handle 200 kg of waste at maximum. It was equipped with three rotary blades for mixing of materials and an air blower to maintain enough oxygen levels inside the chamber. Away from ambient temperatures which were 30 and 23 °C on average for day time and night time, respectively, the temperature inside the composter was automatically controlled at 56 $\pm$ 4 °C using heating functions and sensors equipped in the composter. The temperature inside the composter changed mainly due to heat generation as a result of degradation of organic matters by microorganisms and because of ambient air blown inside the composter for aeration purposes (Arslan Topal et al. 2016). Greatest thermophilic activity would be observed during composting process when the temperature ranges between 52 and 60 °C (Kalamdhad and Kazmi 2009; Mohee and Mudhoo 2005).

Table 1 Combinations of yard waste (YW), food waste (FW) and additives in samples

Compost proportion	Weight of material (kg)		
	YW	FW	Additives
Effective microorganisms (EM)			
100%YW	12.0	0	8.0
90%YW + 10%FW	10.8	1.2	8.0
80%YW + 20%FW	9.6	2.4	8.0
70%YW + 30%FW	8.4	3.6	8.0
Shimamoto Enzyme® (SE)			
100%YW	20.0	0	1.0
90%YW + 10%FW	18.0	2.0	1.0
80%YW + 20%FW	16.0	4.0	1.0
70%YW + 30%FW	14.0	6.0	1.0

Composting process

Four different compositions of feedstock with different YW + FW mixtures were examined (Table 1). Addition of food waste was driven by the fact that at specific C/N ratios for initiating composting, digester operating characteristics and end performance metrics can be improved (Brown and Li 2013). The compositions were (1) 100%YW, (2) 90%YW + 10%FW, (3) 80%YW + 20%FW, and (4) 70%YW + 30%FW. Each composition was composted twice using different additives. 8.0 kg of Effective Microorganisms (EM) and 1.0 kg of Shimamoto Enzyme® (SE) were used in each trial. EM and SE were added to the compost in values recommended by their manufacturing companies to improve and facilitate composting process. Compost additives increase the macro- and micronutrient content and result in more nitrogen (N), phosphorus (P) and potassium (K) nutrient content (NPK) (Jusoh et al. 2013). All prepared feedstock were composted in batches inside the composter for 24 h. Subsequently, approximately 7 kg of the product of each batch was unloaded from the composter and processed further for curing purposes. The curing process lasted for 35 days. The remaining amounts of the product (approximately 13 kg) were left inside the composter and recycled in the next batch.

Compost quality control

Existence of toxic substances is one of the common problems in immature composts (Itävaara et al. 2002). Therefore, phytotoxicity test is an important investigation to check the quality of compost before its application in different crops. In this study, compost extract was prepared initially with 200 mL of distilled water and 20 g of compost. The compost–water mixture was then stirred for 6 h and centrifuged at 8000 rpm for 20 min (Gariglio et al. 2002; Mitelut and Popa 2011). Seed germination was tested for 20 radish seeds in Petri dishes (El Fels et al. 2014). Seeds were soaked with 5 mL of compost extract and left for 5 days in darkness at 22 ± 3 °C. Then, they were preserved in distilled water overnight before germination test. A blank sample filled with distilled water was used as control. At the end of the experiment, the germination index (GI) was obtained based on the following equations explained by Zucconi et al. (1981).

$$\text{Relative seed germination } (\%) = \frac{\text{No. of seeds germinated in compost extract}}{\text{No. of seeds germinates in control}} \times 100\%, \quad (1)$$

$$\text{Relative root growth } (\%) = \frac{\text{Mean root length in compost extract (mm)}}{\text{Mean root length in control (mm)}} \times 100\%, \quad (2)$$

$$\text{Germination index GI } (\%) = \frac{\text{Relative seed germination} \times \text{relative root growth}}{100}. \quad (3)$$

Besides the seed germination test, the NPK content of produced compost was also determined. Total Kjeldahl nitrogen (TKN) test was used for nitrogen content determination using BUCHI Distillation Unit K-314. As for phosphorus and potassium content, collected samples were first digested using dry digestion method. Then, 1 g of dry solid sample was placed in a 10-mL crucible and heated at 550 °C in a muffle furnace for 5 h. Afterwards, total phosphorus and total potassium were determined using PhosVer®3 ascorbic acid and atomic absorption spectroscopy (AAS), respectively, and the results were compared with available standards. In addition, and for benchmarking purpose, quality of samples was compared with a compost produced in windrow system using similar yard waste applied in this study. Nitrogen, phosphorous and potassium content of the compost produced in windrow system were 0.77 ± 0.2, 0.92 ± 0.1 and 1.58 ± 0.1, respectively.

Results and discussion

Composting process

Figure 1 shows initial pH of FW and YW feedstock and final pH of four matured compost compositions. pH varied from 7.01 to 9.17 in investigated compositions. Mature compost usually has a pH range of 6–8 (Cooperband 2002). In compost samples with EM additives, pH values

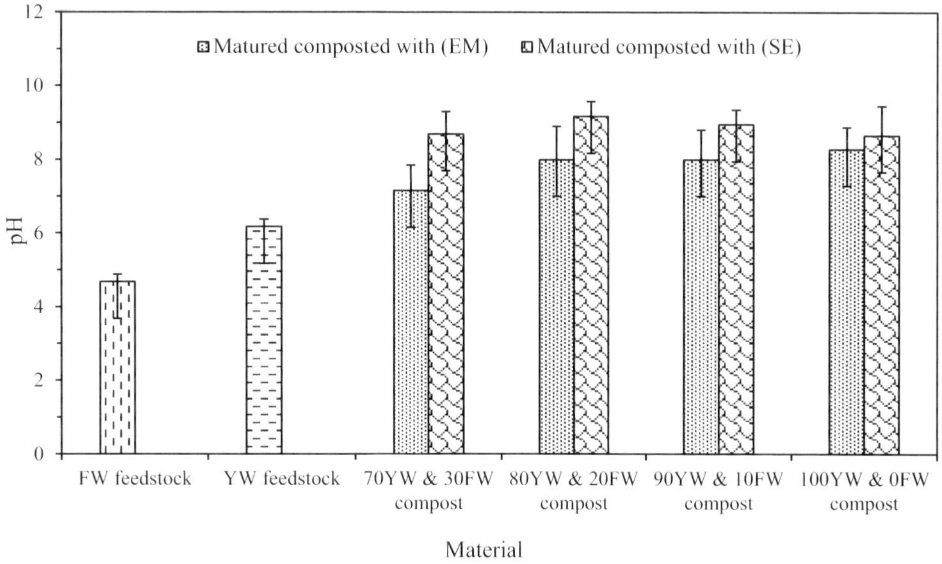

Fig. 1 pH levels of FW and YW feedstock and matured compost samples collected after curing with application of EM and SE additives

decreased and reached an optimum range. However, samples with SE additives had pH higher than optimum range. Effective microorganisms produce organic acids and cause reduction of pH (El-Shafei et al. 2008). High pH values of compost with SE additive could be due to consumption of produced organic acids at the beginning of the process and their conversion to ammonia (Sundberg 2005). Reduction of final pH value of compost can be achieved using different ways, e.g., addition of elemental sulfur (Roig et al. 2004). It was also found that addition of FW could slightly reduce pH of samples. Effect of food waste in lowering pH values was also confirmed by Saad et al. (2013).

Figure 2 shows moisture content values of raw FW and YW feedstock and of composts produced from different compositions of YW and FW. Samples recorded moisture contents ranging from 15.4 to 32.1% which were lower than the optimum levels. Moisture content should be in the range between 50 and 60% during the composting process (Tchobanoglous et al. 1993). Low initial moisture content of YW is believed to be the reason for having low final moisture content in compost samples. Application of yard waste for reducing compost pile moisture due to its high consumption of moisture content has been reported (Rihani et al. 2010). In addition, water contained in the feedstock is consumed during decomposition of organics and consequently the moisture content decreases over time. Therefore, adding enough water into the feedstock prior to composting to maintain an acceptable percentage of moisture content could improve the final moisture content of the compost samples.

C/N ratio of compost ranged from 20 to 27 after 1 day of feedstock processing inside the composter (Fig. 3). Obtained C/N ratios indicated that composting process succeeded in decreasing initial C/N ratio of the feedstock in large percentage. For instance, initial C/N ratio of the 100%YW feedstock was reduced around 56 and 65%, respectively, in case of EM and SE additives' application. Similar reduction percentages in initial C/N ratio were reported by other researchers (Jusoh et al. 2013; Makan and Mountadar 2012; Tumuhairwe et al. 2009). Organic carbon gets converted to CO_2 during composting process and, with minimal nitrogen loss the C/N ratio of compost will eventually decrease (Rasapoor et al. 2016). C/N ratio of finished compost is desired to be close to 10/1. Curing process can reduce further the C/N ratio of compost and resulted in a proper ratio (Shaffer 2010).

Compost quality

Figure 4 shows germinated radish seeds in control and in sample of composts with EM and SE additives. Results of seed germination test are shown in Fig. 5. All samples, except sample of 70%YW + 30%FW with EM additive, could produce satisfactory germination indexes.

Minimum acceptable germination index (GI) of mature compost for different seeds types have been investigated previously. Riffaldi et al. (1986) studied GI of garden cress and reported that compost is certainly not phytotoxic to plants when GI is more than 80%. Similar to garden cress seeds, germination indexes of at least 80% in radish seeds (Mitelut and Popa 2011) and cucumber seeds (Sangamithirai et al. 2015) indicated production of compost which is not phytotoxic. Investigated compost samples in this study recorded germination index higher than those achieved by Kumar et al. (2010) during co-composting of green waste and food waste at low C/N ratio using an in-vessel lab-scale reactor. Generally, the phytotoxicity effect is most likely due to the presence of fatty substance,

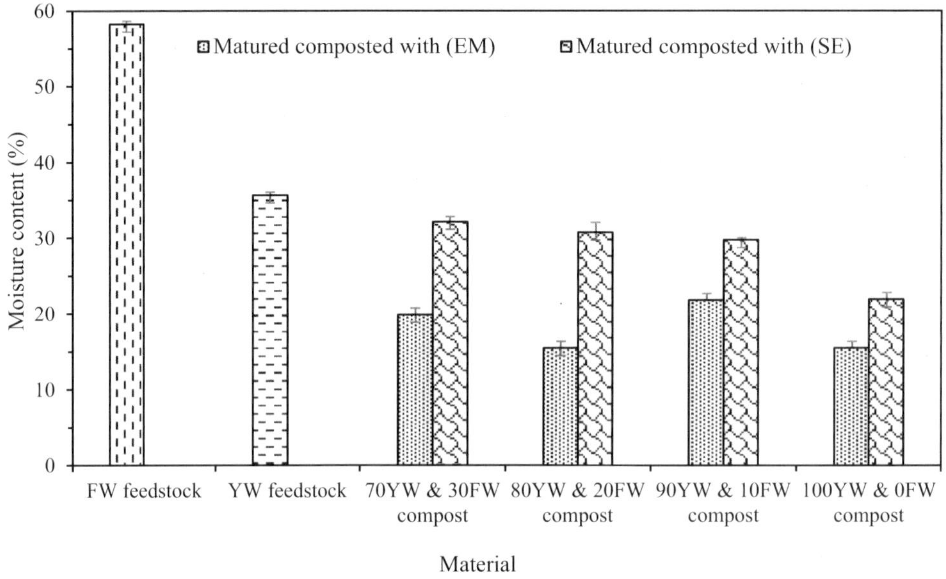

Fig. 2 Moisture content of FW and YW feedstock and matured compost samples collected after curing with application of EM and SE additives

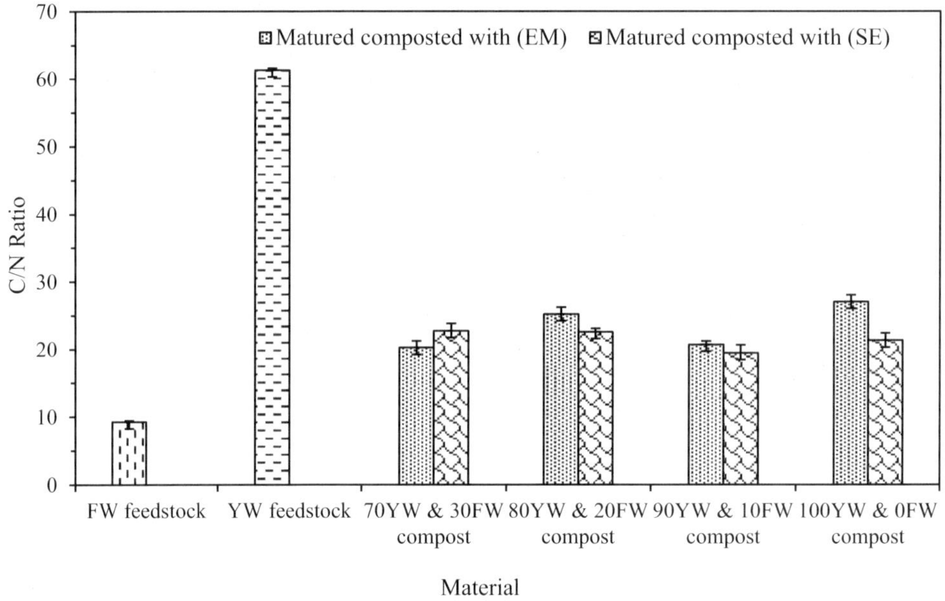

Fig. 3 C/N ratio of FW and YW feedstock and matured compost samples collected after curing with application of EM and SE additives

soluble phenol content as well as organic (Zahrim et al. 2016; Piotrowska-Cyplik et al. 2013).

Figure 6 shows NPK contents of produced compost with EM and SE additives. Nitrogen (N), phosphorus (P) and potassium (K) are among the most demanding nutrients by plants (Maathuis 2009).

Nitrogen mainly gives the plants dark green color and strong growth (Bojović and Marković 2009; Gastal and Lemaire 2002) while phosphorous (P) primarily results in healthy growth of roots and enhances crop quality (Ristvey et al. 2007). Enough potassium (K) content enhances plants' resistance against drought and diseases (Malvi 2011). Hence, NPK content of compost, as a benchmark for compost quality, should be sufficient for plant growth (Rihani et al. 2010). All feedstock produced composts with acceptable levels of NPK. Good quality compost should have 0.4–3.5% nitrogen, 0.3–3.5% phosphorus and 0.5–1.8% potassium (Jaafarzadeh Haghighi Fard et al. 2015; Sadeghi et al. 2015). Addition of EM resulted in production of compost with higher percentages of nitrogen and phosphorus compared to SE. With addition of EM, feedstock with 10% FW recorded the best NPK value. Percentages of phosphorus and potassium decreased as FW percentage in the feedstock increased. In composts with SE additive, supplement of FW has no effect on phosphorus and potassium contents and almost similar percentages of P

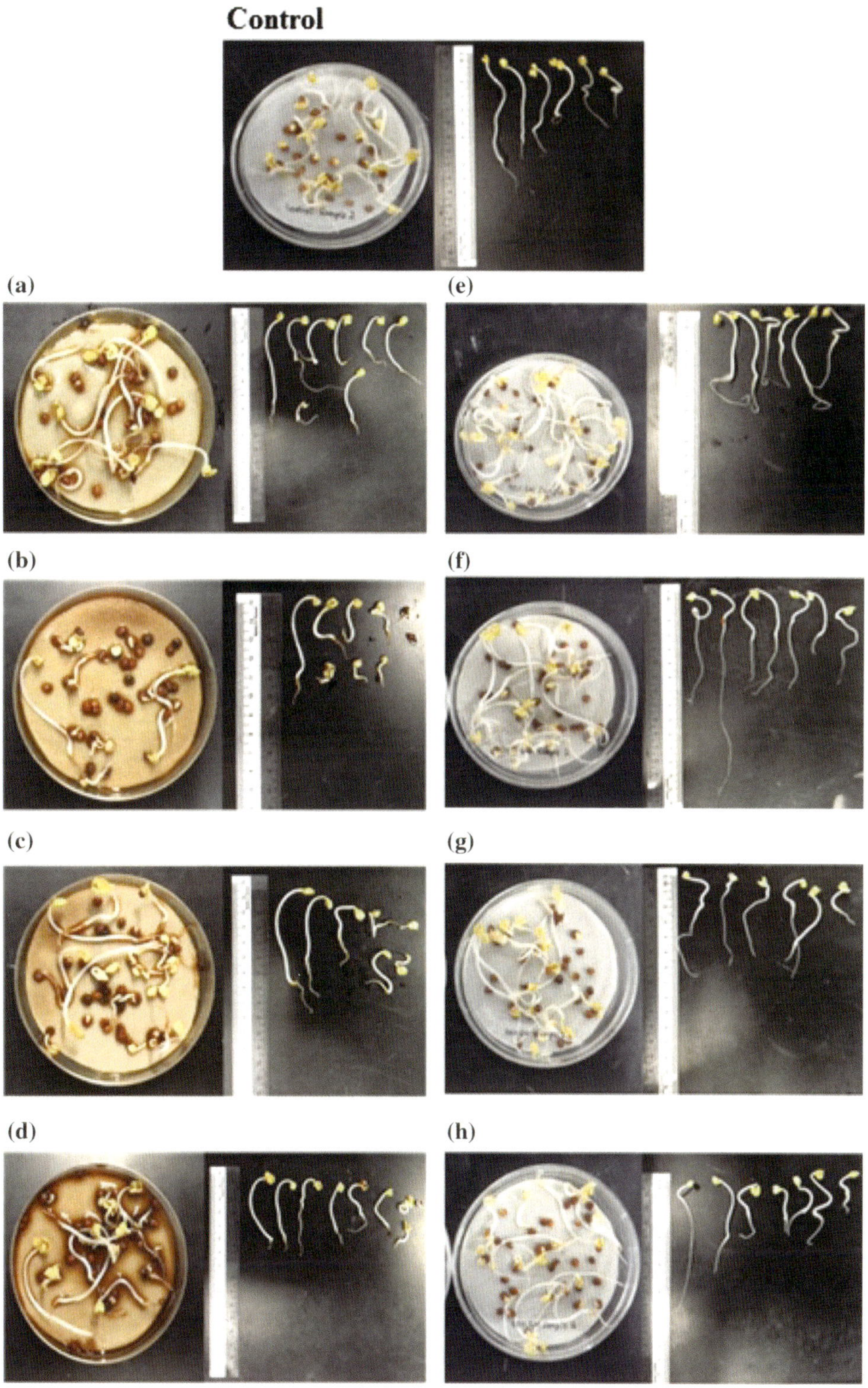

Fig. 4 Seed germination test of control and compost produced using the 100%YW, 90%YW + 10%FW, 80%YW + 20%FW and 70%YW + 30%FW feedstock in case of EM (**a**–**d**) and SE (**e**–**h**) additives

and K were recorded in all compost samples. Nitrogen content, however, varied based on FW percentages. When compared with the compost produced using windrow method, with NPK values of 0.77% ± 0.2, 0.92% ± 0.1 and 1.58% ± 0.1, composts produced in this study had better NPK content.

Conclusion

In-vessel co-composting of yard waste and food waste generated in Cameron Highlands, Malaysia was investigated with addition of effective microorganisms (EM) and Shimamoto Enzyme® (SE) in an actual-size composter.

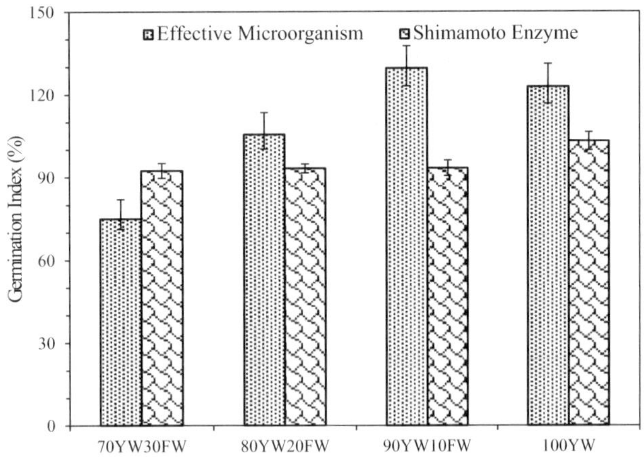

Fig. 5 Germination index results of produced compost with application of EM and SE additives

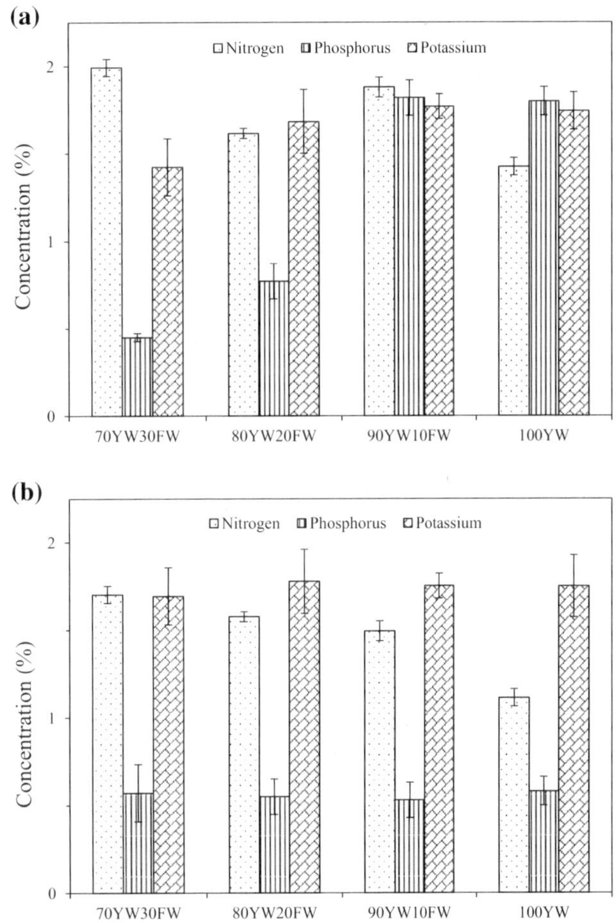

Fig. 6 NPK content of different compost compositions in case of **a** EM and **b** SE additives

Acceptable reduction of pH and C/N ratio was observed, specifically with the addition of effective microorganisms to the feedstock. All samples had insufficient moisture contents due to high percentage of yard waste in feedstock. Maturity tests of germination index and NPK value indicated better quality of compost with EM additives compared to those with SE additive. Composts produced in this study have shown superior quality compared to the one produced in windrow method. Biological transformation of yard waste instead of its landfilling is not only a promising solution to overcome growing problems of solid waste management, but also reduces air and water pollution. In-vessel composting of yard waste instead of its windrow or aerated pile composting in Cameron Highlands, where the lands are in great demand, is an environmentally friendly and sustainable approach for solid waste management.

Acknowledgements This study was supported by the Ministry of Higher Education, Malaysia (MOHE) under classification of MyRA Grant scheme (0153AB-J13). Authors are thankful to Mr. Zaaba Mohammad and Mr. Ali Boualili for their kind assistance in this research. Cooperation from Mr. Leong Hua Qian and Magnificent Associated Nurseries Sdn Bhd for providing yard waste is highly appreciated.

References

Agamuthu P (2013) Landfilling in developing countries. Waste Manag Res 31:1–2. doi:10.1177/0734242X12469169

Aminu M, Matori AN, Yusof KW, Malakahmad A, Zainol RB (2015) A GIS-based water quality model for sustainable tourism planning of Bertam River in Cameron Highlands, Malaysia. Environ Earth Sci 73:6525–6537. doi:10.1007/s12665-014-3873-6

An C-J, Huang G-H, Yao Y, Sun W, An K (2012) Performance of in-vessel composting of food waste in the presence of coal ash and uric acid. J Hazard Mater 203–204:38–45. doi:10.1016/j.jhazmat.2011.11.066

Arslan Topal EI, Ünlü A, Topal M (2016) Effect of aeration rate on elimination of coliforms during composting of vegetable–fruit wastes. Int J Recycl Org Waste Agric 5:243–249. doi:10.1007/s40093-016-0134-6

Badgie D, Samah M, Manaf LA, Muda AB (2012) Assessment of municipal solid waste composition in Malaysia: management, practice and challenges. Pol J Environ Stud 21:539–547. http://www.nswaienvis.nic.in/Waste_Portal/Research_papers/pdf/Assessment_of_MWS_Composition.pdf

Bojović B, Marković A (2009) Correlation between nitrogen and chlorophyll content in wheat (*Triticum aestivum* L.). Kragujevac J Sci 31:69–74. http://www.pmf.kg.ac.rs/KJS/volumes/kjs31/kjs31bojovicmarkovic69.pdf

Brown D, Li Y (2013) Solid state anaerobic co-digestion of yard waste and food waste for biogas production. Bioresour Technol 127:275–280. doi:10.1016/j.biortech.2012.09.081

Cekmecelioglu D, Demirci A, Graves RE, Davitt NH (2005) Applicability of optimised in-vessel food waste composting for windrow systems. Biosyst Eng 91:479–486. doi:10.1016/j.biosystemseng.2005.04.013

Saad M, Fatin N, Nadrah Ma'min N, Zain SM, Basri NEA, Zaini M, Sofia N (2013) Composting of mixed yard and food wastes with effective microbes. J Teknol 65:89–95. doi:10.11113/jt.v65.2196

Zucconi F, Pera A, Forte M, De Bertoldi M (1981) Evaluating toxicity of immature compost. BioCycle 22:54–57. http://agris.fao.org/agris-search/search.do?recordID=US8248707

Ch'ng HY, Ahmed OH, Kassim S, Majid NMA (2013) Co-composting of pineapple leaves and chicken manure slurry. Int J Recycl Org Waste Agric 2:23. doi:10.1186/2251-7715-2-23

Cooperband L (2002) The art and science of composting, a resource for farmers and compost producers. Center for Integrated agricultural systems, University of Wisconsin-Madison, Wisconsin, USA. http://www.cias.wisc.edu/wp-content/uploads/2008/07/artofcompost.pdf

Dinie M, Samsudin M, Don M (2013) Municipal solid waste management in malaysia: current practices, challenges and prospects. J Teknol 62:95–101. doi:10.11113/jt.v62.1293

El Fels L, Zamama M, El Asli A, Hafidi M (2014) Assessment of biotransformation of organic matter during co-composting of sewage sludge-lignocellulosic waste by chemical, FTIR analyses, and phytotoxicity tests. Int Biodeter Biodegr 87:128–137. doi:10.1016/j.ibiod.2013.09.024

El-Shafei A, Yehia M, El-Naqib F (2008) Impact of effective microorganisms compost on soil fertility and rice productivity and quality. Misr J Ag Eng 25:1067–1093. http://www.mjae.eg.net/pdf/2008/july/26.pdf

Fazeli A, Bakhtvar F, Jahanshaloo L, Che Sidik NA, Bayat AE (2016) Malaysia's stand on municipal solid waste conversion to energy: a review. Renew Sust Energ Rev 58:1007–1016. doi:10.1016/j.rser.2015.12.270

Gariglio NF, Buyatti MA, Pilatti RA, Russia DEG, Acosta MR (2002) Use of a germination bioassay to test compost maturity of willow (*Salix* sp.) sawdust. New Zeal J Crop Hort 30:135–139. doi:10.1080/01140671.2002.9514208

Gastal F, Lemaire G (2002) N uptake and distribution in crops: an agronomical and ecophysiological perspective. J Exp Bot 53:789–799. https://www.ncbi.nlm.nih.gov/pubmed/11912222

Haug RT (1993). The practical handbook of compost engineering. CRC Press, Taylor & Francis Group, USA. https://www.crcpress.com/The-Practical-Handbook-of-Compost-Engineering/Haug/p/book/9780873713733

Hoornweg D, Thomas L, Otten L (1999). Composting and its applicability in developing countries. Working paper series No. 8. Urban Development Division, The World Bank, Washington DC, USA. http://www.bvsde.paho.org/bvsacd/cd48/paperseries8.pdf

Irvine G, Lamont E, Antizar-Ladislao B (2010) Energy from waste: reuse of compost heat as a source of renewable energy. Int J Chem Eng 2010:1–10. doi:10.1155/2010/627930

Itävaara M, Venelampi O, Vikman M, Kapanen A (2002) Compost maturity—problems associated with testing. Microbiology of composting 373–382. Springer, Berlin. doi:10.1007/978-3-662-08724-4_31

Jaafarzadeh Haghighi Fard N, Moradi B, Abbasi M, Alivar Babadi R, Bahrani H, Mirzaie A, Zare Javid A, Ravanbakhsh M (2015) Feasibility study of green wastes composting with digested and dewatering sludge from municipal wastewater treatment plant in Iran. Environ Health Eng Manag J 2:149–155. http://ehemj.com/browse.php?a_id=95&sid=1&slc_lang=en

Johari A, Alkali H, Hashim H, Ahmed SI, Mat R (2014) Municipal solid waste management and potential revenue from recycling in Malaysia. Mod Appl Sci 8:37–49. doi:10.5539/mas.v8n4p37

Jusoh MLC, Manaf LA, Latiff PA (2013) Composting of rice straw with effective microorganisms (EM) and its influence on compost quality. Iran J Environ Health Sci Eng 10:1–9. doi:10.1186/1735-2746-10-17

Kalamdhad AS, Kazmi AA (2009) Rotary drum composting of different organic waste mixtures. Waste Manag Res 27:129–137. doi:10.1177/0734242X08091865

Kumar M, Ou Y-L, Lin J-G (2010) Co-composting of green waste and food waste at low C/N ratio. Waste Manag 30:602–609. doi:10.1016/j.wasman.2009.11.023

Larney FJ, Sullivan DM, Buckley KE, Eghball B (2006) The role of composting in recycling manure nutrients. Can J Soil Sci 86:597–611. doi:10.4141/S05-116

Lou XF, Nair J (2009) The impact of landfilling and composting on greenhouse gas emissions—a review. Bioresour Technol 100:3792–3798. doi:10.1016/j.biortech.2008.12.006

Maathuis FJ (2009) Physiological functions of mineral macronutrients. Curr Opin Plant Biol 12:250–258. doi:10.1016/j.pbi.2009.04.003

Maheshwari DK (2014) Composting for sustainable agriculture. Springer International Publishing, Switzerland. doi:10.1007/978-3-319-08004-8

Makan A, Mountadar M (2012) Effect of C/N ratio on the in-vessel composting under air pressure of organic fraction of municipal solid waste in Morocco. J Mater Cycles Waste Manag 14:241–249. doi:10.1007/s10163-012-0062-0

Malakahmad A, Nasrudin S, Basri NEA, Zain SM (2013) Anaerobic transformation of biodegradable waste; simultaneous production of energy and fertilizer. Am J Environ Sci 9:113–119. doi:10.3844/ajessp.2013.113.119

Malvi UR (2011) Interaction of micronutrients with major nutrients with special reference to potassium. Karnataka J Agric Sci 24:106–109. http://www.inflibnet.ac.in/ojs/index.php/KJAS/article/viewFile/865/795

Mandal P, Chaturvedi MK, Bassin JK, Vaidya AN, Gupta RK (2014) Qualitative assessment of municipal solid waste compost by indexing method. Int J Recycl Org WasteAgric 3:133–139. doi:10.1007/s40093-014-0075-x

Mansor N, Mat Rashid K, Mohamad Z, Abdullah Z (2015) Agro tourism potential in Malaysia. Int Acad Res J Bus Technol 1:37–44. http://www.iarjournal.com/wp-content/uploads/IBTC2015-p37-44.pdf

Mhindu RL, Wuta M, Ngorima E (2013) Composting of selected organic wastes from Peri-urban areas of Harare, Zimbabwe. Int J Recycl Org Waste Agric 2:14. doi:10.1186/2251-7715-2-14

Mitelut AC, Popa ME (2011) Seed germination bioassay for toxicity evaluation of different composting biodegradable materials. Rom Biotechnol Lett 16:121–129. http://www.rombio.eu/rbl1vol16Supplement/18%20Amalia%20Mitelut.pdf

Mohee R, Mudhoo A (2005) Analysis of the physical properties of an in-vessel composting matrix. Powder Technol 155:92–99. doi:10.1016/j.powtec.2005.05.051

Noor ZZ, Yusuf RO, Abba AH, Abu Hassan MA, Mohd Din MF (2013) An overview for energy recovery from municipal solid wastes (MSW) in Malaysia scenario. Renew Sustain Energy Rev 20:378–384. doi:10.1016/j.rser.2012.11.050

Pandey PK, Vaddella V, Cao W, Biswas S, Chiu C, Hunter S (2016) In-vessel composting system for converting food and green wastes into pathogen free soil amendment for sustainable agriculture. J Clean Prod 139:407–415. doi:10.1016/j.jclepro.2016.08.034

Parthan SR, Milke MW, Wilson DC, Cocks JH (2012) Cost estimation for solid waste management in industrialising regions—precedents, problems and prospects. Waste Manag 32:584–594. doi:10.1016/j.wasman.2011.11.004

Periathamby A, Hamid FS, Khidzir K (2009) Evolution of solid waste management in Malaysia: impacts and implications of the solid waste bill, 2007. J Mater Cycles Waste Manag 11:96–103. doi:10.1007/s10163-008-0231-3

Pichtel J (2005) Waste management practices: municipal, hazardous, and industrial. CRC Press, Taylor & Francis Group, Boca Raton. https://www.crcpress.com/Waste-Management-Practices-Municipal-Hazardous-and-Industrial-Second/Pichtel/p/book/9781466585188

Piotrowska-Cyplik A et al (2013) Composting of oiled bleaching earth: fatty acids degradation, phytotoxicity and mutagenicity changes. Int Biodeterior Biodegrad 78:49–57. doi:10.1016/j.ibiod.2012.12.007

Rajaie M, Tavakoly AR (2016) Effects of municipal waste compost and nitrogen fertilizer on growth and mineral composition of tomato. Int J Recycl Org Waste Agric 5:339–347. doi:10.1007/s40093-016-0144-4

Rasapoor M, Adl M, Pourazizi B (2016) Comparative evaluation of aeration methods for municipal solid waste composting from the perspective of resource management: a practical case study in Tehran, Iran. J Environ Manag 184:528–534. doi:10.1016/j.jenvman.2016.10.029

Razali WAW, Baharuddin AS, Talib AT, Sulaiman A, Naim MN, Hassan MA, Shirai Y (2012) Degradation of oil palm empty fruit bunches (OPEFB) fibre during composting process using in-vessel composter. BioResources 7:4786–4805. doi:10.1007/s40093-016-0131-9

Rich N, Bharti A (2015) Assessment of different types of in-vessel composters and its effect on stabilization of MSW compost. Int Res J Eng Technol 2:37–42. https://www.irjet.net/archives/V2/i3/Irjet-v2i308.pdf

Riffaldi R, Levi-Minzi R, Pera A, De Bertoidi M (1986) Evaluation of compost maturity by means of chemical and microbial analyses. Waste Manag Res 4:387–396. doi:10.1177/0734242X8600400157

Rihani M, Malamis D, Bihaoui B, Etahiri S, Loizidou M, Assobhei O (2010) In-vessel treatment of urban primary sludge by aerobic composting. Bioresour Technol 101:5988–5995. doi:10.1016/j.biortech.2010.03.007

Roig A, Cayuela ML, Sánchez-Monedero MA (2004) The use of elemental sulphur as organic alternative to control pH during composting of olive mill wastes. Chemosphere 57:1099–1105. doi:10.1016/j.chemosphere.2004.08.024

Ristvey AG, Lea-Cox JD, Ross, DS (2007) Nitrogen and phosphorus uptake efficiency and partitioning of container-grown azalea during spring growth. J Am Soc Hortic Sci 132:563–571. http://journal.ashspublications.org/content/132/4/563.full

Sadeghi S, Dehvari M, Bahmani P, Teymouri P, Fattahi A, Sadeghnia M, Fallahzadeh RA (2015) Physical–chemical analysis and comparison with standards of the compost produced in Sanandaj, Iran. Open Access Libr J 2:1–6. doi:10.4236/oalib.1101855

Sangamithirai KM, Jayapriya J, Hema J, Manoj R (2015) Evaluation of in-vessel co-composting of yard waste and development of kinetic models for co-composting. Int J Recycl Org Waste Agric 4:157–165. doi:10.1007/s40093-015-0095-1

Shaffer B (2010) Curing compost; an antidote for thermal processing. Acres U.S.A. 40. http://www.soilwealth.com.au/imagesDB/news/Nov2010_ShafferCompost.pdf

Sudharsan Varma V, Kalamdhad AS (2014) Stability and microbial community analysis during rotary drum composting of vegetable waste. Int J Recycl Org Waste Agric 3:52. doi:10.1007/s40093-014-0052-4

Sundberg C (2005) Improving compost process efficiency by controlling aeration, temperature and pH (vol. 2005). Swedish University of Agricultural Sciences, Uppsala, Sweden. http://pub.epsilon.slu.se/950/

Tchobanoglous G, Theisen H, Vigil S (1993) Integrated solid waste management: engineering principles and management issues. McGraw-Hill, New York

Tognetti C, Mazzarino MJ, Laos F (2007) Improving the quality of municipal organic waste compost. Bioresour Technol 98:1067–1076. doi:10.1016/j.biortech.2006.04.025

Tumuhairwe JB, Tenywa JS, Otabbong E, Ledin S (2009) Comparison of four low-technology composting methods for market crop wastes. Waste Manag 29:2274–2281. doi:10.1016/j.wasman.2009.03.015

Zahrim A, Leong P, Ayisah S, Janaun J, Chong K, Cooke F, Haywood S (2016) Composting paper and grass clippings with anaerobically treated palm oil mill effluent. Int J Recycl Org Waste Agric 5:221–230

Zhu N (2007) Effect of low initial C/N ratio on aerobic composting of swine manure with rice straw. Bioresour Technol 98:9–13. doi:10.1016/j.biortech.2005.12.003

10

Vermicompost leachate reduces some negative effects of salt stress in pomegranate

Siamak Shirani Bidabadi[1] · Safoora Dehghanipoodeh[2] · Glenn C. Wright[3]

Abstract

Purpose Growth and productivity of plant crops are negatively affected by salt stress. Vermicompost leachate (VCL) has beneficial effects on plant growth in several different ways. This experiment was conducted to investigate the potential of VCL for raising the tolerance of pomegranate against salinity.

Methods A pot experiment was organized to analyze some physiological attributes and antioxidant activities of two Iranian pomegranate cultivars under different concentrations of NaCl with or without VCL.

Results Salinity negatively affected the growth properties of pomegranate, while foliar spray of VCL-induced salt tolerance by reducing the accumulation of Na^+ in the seedlings. Leaf area and shoot and root fresh and dry weight of VCL-treated pomegranate plants showed a significant increase regardless of whether the seedling growth was under non-saline or saline conditions. Chlorophyll loss and reduced photosynthesis efficiency caused by salinity were also ameliorated using VCL. Foliar application of VCL also improved the activity of antioxidant enzymes and caused a reduction in oxidative stress and electrolyte leakage.

Conclusions These results suggested that VCL could alleviate the damage caused by salt stress in pomegranate by limiting the agglomeration of Na^+ and amending the activities of antioxidant enzymes by which the efficiency of plant increases.

Keywords Antioxidant activities · Fruit crops · Organic wastes · Photosynthesis efficiency · Salinity · Seedlings · Sodium accumulation

✉ Siamak Shirani Bidabadi
siamakshirani.upm@gmail.com

Safoora Dehghanipoodeh
safoora.dehghanipoodeh@yahoo.com

Glenn C. Wright
gwright@ag.arizona.edu

1 Department of Horticulture, College of Agriculture, Isfahan University of Technology, IUT, Isfahan 84156-83111, Iran

2 Department of Horticultural Science, College of Agriculture and Natural Resources, University of Tehran, Karaj, Iran

3 The School of Plant Sciences, College of Agriculture and Life Sciences, University of Arizona, Tucson, AZ 85364, USA

Introduction

Growth and productivity of crops are negatively affected by salt stress. The entry of high amounts of salt in the soil and irrigation water disrupts all physiological and biochemical processes in the plant (Hasanuzzamaan et al. 2013). The reduced osmotic potential of soil solution, imbalance in nutritional elements, and the effect of ion toxicity are consequences of salinity which has a detrimental effect on the growth of plants (Ashraf 1994; Parvaiz and Satyawati 2008). During the onset and spread of salt stress in the plant, all physiological and biochemical processes of the plant are affected. (Parvaiz and Satyawati 2008). However, the negative effects of salt stress on plants depend on the severity and exposure time of salinity. Plant genotypes and environmental factors also play a decisive role; therefore, finding a suitable moderator or stress relievers is one of the duties of researchers (Hasanuzzamaan et al. 2013). Due to the high cost and hazardous effects of chemical fertilizers, using organic compound solutions as foliar fertilizer sprays is developing. Moreover, restricted nutrient supply and restriction in chemical fertilizer use in saline soils are considerable problems that should be solved by proper methods and efficient management practices (Astaraei and Ivani 2008; Chinsamy et al. 2013). Organic fertilizers improve soil structure and enhance nutrient availability, thus helping to protect the yield and quality, and are more affordable than synthetic fertilizers (Ayyobi et al. 2014). Use of organic sources of nutrients improves the soil health by sustaining the balance of organic matter, thereby helping to amend physical, chemical, and biological properties of the soil (Mir et al. 2013). Vermicomposts, which are created by crushing organic wastes from earthworm, contain nutrients in a form that is easily absorbed by the plant. Vermicompost leachate (VCL) is a liquid phase that comes from vermicompost and it has a very positive impact on soil improvement and plant growth (Chinsamy et al. 2013). Previous reports have shown that vermicompost modulates the harmful effects of salinity stress on several plant crops (Atiyeh et al. 2000; Oliva et al. 2008; Sallaku et al. 2009; Nandi et al. 2013; Ayyobi et al. 2014). Beneficial effects of vermicompost leachate have also been reported in bulbous plants and some commercially important crops such as banana and tomato (Aremu et al. 2012; Arthur et al. 2012; Aremu et al. 2014). Pomegranate (*Punica granatum*) is a strategic horticultural crop cultivated all over the world (Mirzapour and Khoshgoftarmanesh 2013). It is one of the oldest fruit crops which has been cultivated since ancient times for its economic, ornamental, and medicinal properties (Pirseyedi et al. 2010; Ramezanian et al. 2009). Leaves, fruits, and bark of the pomegranate tree are used in the traditional medicine (Hulya Orak 2009). Fruits are a rich source of minerals and sugars, along with vitamin C and certain phenolic compounds which act as natural antioxidants (Zaouaya et al. 2012). The origin of pomegranate is arid and semi-arid regions of Iran, where in seriously affected by high salinity (Okhovatian-Ardakani et al. 2010). Pomegranate trees are moderately salt tolerant under field conditions (El-Khawaga et al. 2013). Hence, an attempt has been made to examine the interactive effects of salinity and vermicompost leachate on growth indices, physiological parameters, and antioxidative enzyme activities of two commercial cultivars of pomegranate. This work was, therefore, aimed to assess the efficiency of foliar application of VCL to mitigate the negative effects of salinity on pomegranate growth and productivity.

Materials and methods

Plant material and growing conditions

Seed extraction and germination was achieved according to the method recommended by Rawat et al. (2010). Ripe fruits of *Punica granatum* cultivars "Rabab" and "Malas Daneh Ghermez" were harvested in the month of August 2013. Then, their seeds were collected and dried in the shade. Afterwards, the imbibed seeds were placed between layers of moist peat moss and exposed to chilling (5 °C) for up to 30 days. Following germination under greenhouse conditions, 60-day-old seedlings were transplanted into black plastic pots 10 cm in diameter (one plant per pot) filled with soil garden having loamy texture, pH 7.8, with an EC of 1.7 ds cm^{-1}, perlite, and sand with a ratio of 2:1:1, respectively, and grown under glasshouse conditions. Plants were irrigated and fertilized with a complete floral fertilizer 20-20-20 (a product of CIFO S.P.A., Bologna, Italy.), to promote strong growth until 30 July 2014, when salinity and VCL treatments were imposed (seedlings were 30–40 cm in height).

Treatments and experimental design

On 1 August 2014, 36 pomegranate seedlings that were used for the experiment were arranged in a completely randomized factorial design of six treatments with three single-seedling replications along a glasshouse bench under 28 °C daytime and 18 °C nighttime greenhouse temperatures and 250 $\mu molm^{-2}\ S^{-1}$ light intensity. The salinity of the irrigation water was 0, 30, and 60-mM NaCl. Eighteen plants, 12 of each cultivar, received each salinity treatment. NaCl treatments started in 1-year-old seedlings. Plants were irrigated with 1 L of salt solution as needed and leached with tap water, after every fourth irrigation to prevent salt accumulation. Salt treatments were continued for 3 months, and plants were destructively harvested on 1 November 2014. VCL treatment was applied along with saline treatments, beginning on 1 August 2014 using VCL as spray and tap water as a control. A hand sprayer was used to apply VCL to run off. Seedlings that were grown under normal or saline conditions sprayed with VCL and tap water.

VCL preparation and specification

Vermicompost (VC) and vermicompost leachate (VCL) were prepared according to the approach of Ayyobi et al. (2014). Vermicompost was processed by Earthworms (*Eisenia fetida*) in proportion to 25 g of earthworms per 1 kg of cow manure for 2 months. The amount of 50 kg of obtaining vermicompost was squeezed by adding 25 L of water and allowed the leachate to be extracted for 48 h and then gathered in a tank. Table 1 shows the characteristics of VCL derived from cow manure.

Analysis of growth parameters

Leaf area was measured in cm^2 using a leaf area meter (WinArea- UT- 11, Iran). After washing the plants for the separation of mud and other foreign particles sticking to the plants, fresh mass of shoots and roots was recorded. The shoots and the roots were then dried in an oven at 80 °C for 48 h to specify a fixed dry weight. The leaves were collected and frozen in liquid nitrogen and kept at −80 °C until used for biochemical and physiological measurements.

Chlorophyll fluorescence, chlorophyll, and proline measurements

Chlorophyll fluorescence (Fv/Fm) was measured applying a fluorometer (Walz, Effeltrich, Germany). The chlorophyll content was measured with a portable leaf chlorophyll meter (*SPAD 502, Minolta Co.*, Osaka, Japan). The proline content was measured according to the method of Bates et al. (1973).

Determination of endogenous MDA and H_2O_2 concentrations and electrolyte leakage (EL) of pomegranate leaves

Lipid peroxidation indicated by accumulated malondialdehyde (MDA) was assessed according to Wang et al. (2009). After recording the absorbances of the supernatant at 450-, 532-, and 600-nm wavelengths, MDA content was calculated on a fresh weight basis using the following formula:

$$(\text{nmol MDA g}^{-1}\text{FW}) = 6.45(OD_{532} - OD_{600}) - 0.56(OD_{450}) \times 1000.$$

Hydrogen peroxide (H_2O_2) was measured according to method of Velikova and Loreto (2005). Electrolyte leakage (EL) was measured by applying a conductivity meter according to Ozden et al. (2009).

Table 1 Characteristic of vermicompost leachate obtained from cattle manure

Characteristic	Vermicompost leachate
pH	7.56
Electrical conductivity (ds m^{-1})	5.42
N (%)	1.3
P_2O_5 (%)	0.7
K_2O (%)	0.9
Cu (%)	0.05
Zn (%)	0.09
Mn (%)	0.18

These values are subjected to variations depending on the cattle feeding

Determination of sodium (Na^+) concentration

For measurement of Na^+ concentration, the leaf samples were dried in 60 °C for 48 h. Then, 1 g of leaves was ground using an electric grinder and scorched in 560 °C. The obtained ashes were digested in 10 ml of 1 N HCl. Then, the concentration of Na^+ in the digested samples was determined using a flame photometer (Model CL 378, Nanolytik, Germany).

Enzyme extractions and assays

Three months after salinity and VCL treatments, the fresh leaves were collected and enzyme activities were assayed. The samples were prepared as described by Azooz et al. (2012) for CAT (catalase): APX (ascorbate peroxidase) and SOD (superoxide dismutase) activity assay. The activity of SOD was estimated by recording the decrease in absorbance of superoxide-nitroblue tetrazolium complex by the enzyme (Cavalcanti et al. 2004). In addition, then, one SOD unit was defined as the amount of enzyme required to inhibit 50% of the NBT photo-reduction compared to the tubes without the plant extract and expressed as a unit of enzyme activity per mg of protein. Catalase (CAT) activity was measured according to Aebi (1984). The decrease of H_2O_2 was monitored at 240 nm and quantified by its molar extinction coefficient (36 M^{-1} cm^{-1}) and the results expressed as CAT units per mg of leaf fresh weight.

Ascorbate peroxidase (APX) activity was determined by the method of Nakano and Asada (1981). One unit of APX activity was defined as 1-M ascorbate oxidized per minute at 290 nm and the specific activity was expressed as U mg^{-1} protein.

The statistical procedure was performed by SAS 9.1 and means were compared with the LSD multiple range test.

Results and discussion

Effect of foliar application of VCL on leaf area, fresh and dry weight of shoots and roots of pomegranate seedlings under different NaCl levels is presented in Table 2. Leaf area, fresh and dry weight of shoots and roots of both cultivars were significantly reduced by increasing salinity, whereas in cultivar Rabab, leaf area, fresh and dry weight of shoots and roots were reduced by 32.86, 37.48, 50.19, 53.33, and 50.97%, respectively, while in the Malas Daneh Ghermez cultivar, leaf area, fresh and dry weight of shoots and roots were reduced by 46.90, 37.80, 49.34, 50.92, and 52.13%, respectively, at the highest salinity level (60 mM NaCl) in comparison with non-saline controls (Table 2). The negative impacts of salinity on the plants growth have been reported by many researchers (Parvaiz

Table 2 Effects of VCL treatments on shoot fresh and dry weight, root, fresh and dry weight and leaf area of pomegranate seedlings under NaCl-induced salt stress

Cult.	VCL treatment	Salinity (mM)	Shoot fresh weight (g)	Shoot dry weight (g)	Leaf area (cm^2)	Root fresh weight (g)	Root dry weight (g)
"Rabab"	Treated	0	5.84 ± 0.18^{a}	2.84 ± 0.07^{a}	2.84 ± 0.07^{a}	2.93 ± 0.22^{a}	1.62 ± 0.33^{a}
		30	4.90 ± 0.17^{bc}	2.44 ± 0.11^{bc}	2.58 ± 0.11^{ab}	2.15 ± 0.21^{b}	1.47 ± 0.33^{a}
		60	3.80 ± 0.10^{de}	2.01 ± 0.16^{d}	2.30 ± 0.07^{cd}	1.86 ± 0.20^{c}	1.19 ± 0.06^{b}
	Untreated	0	5.63 ± 0.37^{a}	2.65 ± 0.18^{b}	2.80 ± 0.10^{a}	2.85 ± 0.10^{a}	1.55 ± 0.08^{a}
		30	4.06 ± 0.14^{d}	$1.91 \pm .013^{d}$	2.21 ± 0.10^{d}	1.64 ± 0.15^{c}	1.16 ± 0.09^{b}
		60	3.52 ± 0.18^{ef}	1.32 ± 0.08^{f}	1.88 ± 0.08^{e}	1.33 ± 0.12^{d}	0.76 ± 0.06^{cd}
"Malas Daneh Ghermez"	Treated	0	5.17 ± 0.33^{b}	2.37 ± 0.18^{bc}	2.51 ± 0.39^{bc}	2.27 ± 0.14^{b}	0.99 ± 0.19^{bc}
		30	4.67 ± 0.14^{c}	1.99 ± 013^{d}	2.50 ± 0.12^{bc}	1.84 ± 0.09^{c}	0.74 ± 012^{cd}
		60	3.29 ± 0.17^{fg}	1.55 ± 0.13^{e}	2.10 ± 0.20^{de}	1.28 ± 0.15^{de}	0.53 ± 0.08^{de}
	Untreated	0	4.92 ± 0.36^{bc}	2.29 ± 0.16^{c}	2.58 ± 0.16^{ab}	2.18 ± 0.14^{b}	0.94 ± 0.13^{bc}
		30	3.43 ± 0.25^{efg}	1.33 ± 0.09^{ef}	2.22 ± 0.08^{d}	1.24 ± 0.06^{de}	0.60 ± 0.07^{de}
		60	3.06 ± 0.22^{g}	1.16 ± 0.10^{f}	1.37 ± 0.11^{f}	1.07 ± 0.11^{e}	0.45 ± 0.09^{e}
	$LSD_{(0.05)}$		0.40	0.23	0.26	0.25	0.27

Means followed by the same letter are not significantly different by the LSD Multiple Range test at $P \leq 0.05$

and Satyawati 2008; Chinsamy et al. 2013; El-Khawaga et al. 2013 and Hasanuzzamaan et al. 2013). Naeini et al. (2006) reported that pomegranate cuttings tolerate salinity up to 40 Mm. Leaf area, fresh and dry weight of shoots and roots of both cultivars were increased by foliar spray of VCL irrespective to the seedling growth under non-saline or saline conditions. In cultivar Rabab treated with VCL, leaf area, fresh and dry weight of shoots and roots increased by 22.34, 7.95, 52.27, 39.85, and 56.58%, while in VCL-treated Malas Daneh Ghermez cultivar, these attributes were increased 53.28, 7.52, 33.62, 19.62, and 17.78%, respectively, at the highest salinity level (60 mM NaCl) in comparison with their respective controls (Table 2). Similar results in improving morphological attributes of VCL-treated tomato seedlings such as leaf area and overall aboveground growth were observed in tomato seedlings under a high NaCl-tested concentration (100 mM) (Chinsamy et al. 2013). The lack of sensitivity of all morphological attributes to salinity for VCL-treated seedlings was largely caused by a significant reduction in these attributes for non-VCL-treated seedlings under stressed conditions. The prolonged availability of nutrients during the crop growth provided by the application of enriched vermicomposts might have enhanced the plant growth (Nandi et al. 2013). Ability to high water absorption in vermicompost associated with high amount of macro- and micronutrients results in improving photosynthesis (Fathi et al. 2012). Arthur et al. (2012) reported that the presence of VCL relieved the negative effects caused by shortage of P and K during the growth tomato seedlings under nutrient stress. Improvement of plant growth using organic fertilizer such as vermicompost is reported by Ayyobi et al. (2014) in peppermint plants and Nandi et al. (2013) in pomegranate.

For the interaction between the two levels of salinity (30 and 60 mM) and the two pomegranate cultivars, the best results in terms of morphological characteristics were noticed with cultivar Rabab (Table 2). Accordingly, El-Khawaga et al. (2013) concluded that these differences may be attributed to the various responses of some pomegranate cultivars to environmental conditions. Although the application of NaCl decreased the content of relative chlorophyll and photosynthesis efficiency of pomegranate seedlings, the foliar application of VCL limited the decreases. VCL increased photosynthesis efficiency in both saline treatments in comparison with the untreated counterparts for both cultivars (Table 3). A reduced amount of chlorophyll and photosynthesis efficiency has also been observed in several species of plants grown in saline conditions (Okhovatian-Ardakani et al. 2010; Hasanpour et al. 2014). Similarly, leaf chlorophyll increased following the application of VCL to tomato seedlings and some bulbous plants under salinity, temperature, water, and nutrient stresses (Chinsamy et al. 2013; Aremu et al. 2014; Chinsamy et al. 2014). VCL may have an important role in alleviating the salinity-induced damage to the chloroplasts by reducing chlorophyllase activity (Chinsamy et al. 2013). Chlorophyll fluorescence provides useful information about primary reactions of photosynthesis and has been widely used in assessing plant responses to environmental stress (Schnabel et al. 1998). Reducing chlorophyll pigments of leaf in response to

Table 3 Effects of VCL treatments on chlorophyll, photosynthesis efficiency, and proline of pomegranate seedlings under NaCl-induced salt stress

Cult.	VCL treatment	Salinity (mM)	Relative chlorophyll	Photosynthesis efficiency (Fv/Fm)	Proline (μmol g^{-1} FW)
"Rabab"	Treated	0	13.76 ± 1.35^{a}	0.27 ± 0.04^{a}	6.69 ± 0.57^{cde}
		30	11.78 ± 0.56^{c}	0.21 ± 0.02^{b}	6.90 ± 0.46^{bcde}
		60	9.39 ± 0.46^{de}	0.13 ± 0.03^{cde}	8.71 ± 0.15^{a}
	Untreated	0	13.85 ± 0.36^{a}	0.16 ± 0.04^{c}	6.76 ± 0.39^{cde}
		30	10.02 ± 0.67^{d}	0.09 ± 0.02^{efg}	6.49 ± 0.50^{de}
		60	7.87 ± 0.45^{fg}	0.06 ± 0.01^{fg}	6.71 ± 0.38^{cde}
"Ghermez"Malas Daneh	Treated	0	13.26 ± 0.85^{ab}	0.25 ± 0.04^{ab}	6.73 ± 0.65^{cde}
		30	10.30 ± 0.50^{d}	0.14 ± 0.03^{cd}	6.28 ± 0.22^{e}
		60	8.17 ± 0.32^{f}	0.10 ± 0.02^{def}	7.00 ± 0.28^{bcd}
	Untreated	0	12.35 ± 0.44^{bc}	0.23 ± 0.02^{ab}	6.87 ± 0.23^{cde}
		30	8.55 ± 0.82^{ef}	0.11 ± 0.02^{de}	7.17 ± 0.22^{bc}
		60	6.83 ± 0.50^{g}	0.05 ± 0.02^{g}	7.55 ± 0.21^{b}
	$LSD_{(0.05)}$		1.12	0.05	0.65

Means followed by the same letter are not significantly different by the LSD Multiple Range test at $P \leq 0.05$

salinity results in the decline in the values of Fv/Fm ratio (Hasanpour et al. 2014). The results of significant reduction in the photosynthetic efficiency (Fv/Fm) by the higher doses of vermicompost agree with the findings of Lazcano and Dominguez (2010) on the primula plant.

Although VCL-sprayed seedlings of the Malas Dane Ghermez cultivar treated with 60 mM salt had almost similar proline content to untreated counterparts with no significant difference, the highest content of proline was recorded with VCL-treated seedlings of cultivar Rabab under the highest level of NaCl (60 mM)-induced salt stress (Table 3). Among other roles, proline acts as an osmolyte under conditions of water deficit and salinity-induced stress and it is recognized as an osmo-protectant compound, stabilizing proteins, inhibiting lipid peroxidation and scavenging of ROS (Garcial et al. 2012). Chinsamy et al. (2013) showed a significant increase in proline concentration in salt-stressed tomato seedlings by VCL treatment. A similar effect was observed with humic acids of vermicompost as reported by Garcial et al. (2012) in rice seedlings under water stress.

The contents of H_2O_2 and MDA in pomegranate leaves were strongly affected by both salinity and VCL (Table 4). As expected, salinity increased H_2O_2 and MDA contents of the leaves of both cultivars. In Males Dane Ghermez, the content of H_2O_2 and MDA was 16.67 and 6.13% greater than in cultivar Rabab under the highest level of NaCl (60 Mm) (Table 4). The H_2O_2 and MDA contents of VCL-treated seedlings receiving salts were less than those of the untreated seedlings under all levels of NaCl-induced salt stress (Table 4). Clearly, VCL improved the quality of growth, as salt-stressed seedlings showed less foliar necrosis, and less defoliation when treated with VCL. An adaptive mechanism in VCL-treated tomato seedlings reported by Chinsamy et al. (2013) indicates the positive effects of VCL on physiological attributes of plants under salinity. Electrolyte leakage (EL) of leaves in both cultivars increased significantly with increasing levels of NaCl-induced salt stress (Table 4). Malas Daneh Ghermez cultivar showed significantly higher EL (35.33%) resulting from the highest level of NaCl (60 mM)-induced salt stress without VCL treatment. Meanwhile, foliar application of VCL caused a reduction in electrolyte leakage when seedlings subjected to salt stress. According to the results of this investigation, under salinity stress, Malas Daneh Ghermez was more sensitive than Rabab.

The Na^{+} concentration of seedlings receiving no vermicompost leachate increased with increasing salinity, as the highest sodium concentration was recorded at 60-mM NaCl (Table 5). Accumulation of Na^{+} in both cultivars of pomegranate seedlings under salinity and the tendency of Malas Daneh Ghermez to accumulate more Na^{+} than Rabab (Table 5) might partly clarify the apparent greater salt sensitivity of Malas Daneh Ghermez. Naeini et al. (2004) reported that pomegranate accumulates sodium in salt concentrations greater than 40 mM. When subjected to increasing salt levels, the seedlings of both cultivars receiving VCL showed a significant decrease in value of Na^{+} compared to the seedlings receiving no VCL and there was no significant difference between two cultivars at 60-mM NaCl (Table 5). Our present results are consistent with the findings of Maie Mohsen et al. (2016) on

Table 4 Effects of VCL treatments on H_2O_2 content, MDA, and EL of pomegranate seedlings under NaCl-induced salt stress

Cult.	VCL treatment	Salinity (mM)	H_2O_2 content (μmol g^{-1} FW)	MDA (nmol g^{-1} FW)	EL (%)
"Rabab"	Treated	0	20.92 ± 1.42^{g}	7.71 ± 0.63^{f}	20.33 ± 1.53^{fg}
		30	24.90 ± 1.40^{e}	9.74 ± 0.35^{e}	22.67 ± 1.15^{ef}
		60	29.53 ± 0.72^{d}	12.08 ± 0.35^{c}	26.33 ± 0.58^{cd}
	Untreated	0	20.34 ± 0.69^{g}	8.02 ± 0.72^{f}	22.67 ± 1.53^{ef}
		30	30.48 ± 1.18^{d}	11.18 ± 0.35^{d}	28.33 ± 1.53^{c}
		60	35.76 ± 0.89^{b}	13.86 ± 0.31^{ab}	32.67 ± 1.53^{b}
"Malas Daneh Ghermez"	Treated	0	21.31 ± 1.36^{fg}	8.11 ± 0.81^{f}	19.33 ± 1.53^{g}
		30	29.42 ± 0.73^{d}	9.67 ± 0.34^{e}	24.67 ± 0.58^{de}
		60	33.10 ± 0.92^{c}	13.14 ± 0.33^{b}	28.67 ± 1.53^{c}
	Untreated	0	23.23 ± 0.47^{ef}	9.28 ± 0.48^{e}	21.00 ± 2.00^{fg}
		30	37.38 ± 1.43^{b}	13.06 ± 0.73^{b}	28.33 ± 1.16^{c}
		60	41.72 ± 1.81^{a}	14.71 ± 0.60^{a}	35.33 ± 1.53^{a}
	$LSD_{(0.05)}$		1.94	0.89	2.37

Means followed by the same letter are not significantly different by the LSD Multiple Range test at $P \leq 0.05$

Table 5 Effects of VCL treatments on Na^+ content of pomegranate leaf samples under NaCl-induced salt stress

Cult.	VCL treatment	Salinity (mM)	Na^+ (%)
"Rabab"	Treated	0	0.055 ± 0.004^{f}
		30	0.099 ± 0.013^{e}
		60	0.174 ± 0.005^{c}
	Untreated	0	0.055 ± 0.002^{f}
		30	0.104 ± 0.011^{e}
		60	0.187 ± 0.004^{b}
"Malas Daneh Ghermez"	Treated	0	0.062 ± 0.005^{f}
		30	0.117 ± 0.003^{d}
		60	0.180 ± 0.002^{bc}
	Untreated	0	0.064 ± 0.004^{f}
		30	0.122 ± 0.008^{d}
		60	0.205 ± 0.008^{a}
	$LSD_{(0.05)}$		0.012

Means followed by the same letter are not significantly different by the LSD Multiple Range test at $P \leq 0.05$

Majorana hortensis L. They reported that vermicompost reduced Na^+ content of the shoots.

CAT, APX, and SOD activity data presented in Table 6 revealed that a significant effect was observed in enzyme activity. CAT activity showed significant negative correlation with hydrogen peroxide content, in the presence of increasing salinity (Table 7). Following application of a VCL treatment, the highest activity of CAT (0.51) was found in Rabab under 0 mM NaCl (control conditions) (Table 6). The VCL-treated seedlings of cultivar Rababs had greater CAT activity under 60-mM NaCl-induced salt stress, but there was no significant difference in both cultivars. Results in Table 6 showed that different salinity levels exhibited variable effects on increasing APX activity and spraying VCL caused no significant changes of APX activity in response to the varying levels of NaCl-induced salinity (Table 6). Applying the VCL-induced noticeable changes in SOD activity among both cultivars under different levels of salinity (Table 6). The VCL-treated plants of both cultivars had significantly greater SOD activity compared with non-VCL-treated plants under the highest level (60 mM NaCl) of salinity (Table 6). Garcia et al. (2012) reported that vermicompost enhanced the activity of POX resulted in a reduction of H_2O_2 content and greater conservation of membrane permeability in rice seedlings under water stress. Aremu et al. (2014) showed that increased antioxidant activities of some bulbous plants detected when VCL was applied under nutrient stress.

Table 6 Effects of VCL treatments on the activity of CAT, APX, and SOD of pomegranate seedlings under NaCl-induced salt stress

Cult.	VCL treatment	Salinity (mM)	Activity of CAT (U/mg protein)	Activity of APX (U/mg protein)	Activity of SOD (U/mg protein)
"Rabab"	Treated	0	0.51 ± 0.08^{a}	3.19 ± 0.39^{b}	230.90 ± 24.27^{c}
		30	0.31 ± 0.07^{c}	3.71 ± 0.43^{ab}	256.45 ± 9.47^{ab}
		60	0.19 ± 0.02^{de}	3.77 ± 0.41^{ab}	204.43 ± 6.27^{def}
	Untreated	0	0.42 ± 0.05^{b}	3.19 ± 0.51^{b}	224.32 ± 18.06^{cd}
		30	0.24 ± 0.03^{cd}	3.89 ± 0.35^{a}	185.00 ± 6.00^{fg}
		60	0.13 ± 0.03^{ef}	2.20 ± 0.27^{c}	176.45 ± 10.29^{g}
"Malas Daneh Ghermez"	Treated	0	0.30 ± 0.02^{c}	3.18 ± 0.25^{b}	237.23 ± 9.58^{bc}
		30	0.13 ± 0.02^{ef}	3.89 ± 0.23^{a}	262.62 ± 12.71^{a}
		60	0.11 ± 0.03^{f}	3.57 ± 0.51^{ab}	271.29 ± 10.79^{a}
	Untreated	0	0.46 ± 0.05^{ab}	3.20 ± 0.30^{b}	221.03 ± 8.75^{cde}
		30	0.19 ± 0.02^{de}	3.93 ± 0.11^{a}	240.51 ± 16.04^{bc}
		60	0.11 ± 0.03^{f}	3.38 ± 0.41^{ab}	202.21 ± 12.83^{ef}
	$LSD_{(0.05)}$		0.08	0.61	22.05

Means followed by the same letter are not significantly different by the LSD Multiple Range test at $P \leq 0.05$

Table 7 Correlation between measured traits of pomegranate seedlings

	1	2	3	4	5	6	7	8	9	10	11	12
1. Shoot dry weight	1	0.76^{**}	0.86^{**}	0.90^{**}	0.81^{**}	-0.17^{ns}	-0.94^{**}	-0.92^{**}	-0.87^{**}	0.90^{**}	0.02^{ns}	0.21^{ns}
2. Root dry weight		1	0.71^{**}	0.74^{**}	0.55^{**}	-0.001^{ns}	-0.75^{**}	-0.70^{**}	-0.61^{**}	0.58^{**}	-0.004^{ns}	-0.07^{ns}
3. Leaf Area			1	0.84^{**}	0.75^{**}	-0.22^{ns}	-0.85^{**}	-0.87^{**}	-0.88^{**}	0.77^{**}	0.05^{ns}	0.35^{*}
4. Chlorophyll content				1	0.82^{**}	-0.30^{*}	-0.95^{**}	-0.94^{**}	-0.90^{**}	0.90^{**}	-0.10^{ns}	0.22^{ns}
5. chlorophyll fluorescence					1	-0.16^{ns}	-0.81^{**}	-0.83^{**}	-0.83^{**}	0.86^{**}	-0.04^{ns}	0.35^{*}
6. Proline						1	0.27^{ns}	0.38^{*}	0.21^{ns}	-0.38^{*}	0.98^{ns}	-0.25^{ns}
7. H_2O_2							1	0.94^{**}	0.91^{**}	-0.88^{**}	0.06^{ns}	-0.24^{ns}
8. MDA								1	0.90^{**}	-0.93^{**}	-0.03^{ns}	-0.31^{*}
9-EL									1	-0.86^{**}	-0.03^{ns}	-0.37^{*}
10. Activity of CAT										1	-0.05^{ns}	0.22^{ns}
11. Activity of APX											1	0.36^{*}
12. Activity of SOD												1

*, ** and ns: significant at 0.05 and 0.01 probability levels and no significant, respectively

These results confirm the effectiveness of VCL in stimulating antioxidant mechanisms in plants.

Conclusion

The results of the present research strongly proved that foliar application of VCL improved growth parameters and physiological factors, enabling VCL-treated pomegranate seedlings to perform better. Therefore, application of VCL is recommended under salt-stress conditions.

Acknowledgements We thank all people who helped us to complete this study and wish to thank the college of agriculture, Isfahan University of Technology for allowing to conduct this study.

References

Aebi H (1984) Catalase in vitro. Methods Enzymol 105:121–126

Aremu AO, Kulkarni MG, Bairu MW, Finnie JF, Van Staden J (2012) Growth stimulation effects of smoke-water and vermicompost leachate on greenhouse grown-tissue-cultured 'Williams' bananas. Plant Growth Regul 66:111–118. doi:10.1007/s10725-011-9634-6

Aremu AO, Masondo NA, Staden JV (2014) Physiological and phytochemical responses of three nutrient-stressed bulbous plants subjected to vermicompost leachate treatment. Acta Physiol Plant 36(3):721–731. doi:10.1007/s11738-013-1450-3

Arthur GD, Aremu AO, Kulkarni MG, Staden JV (2012) Vermicompost leachate alleviates deficiency of phosphorus and potassium in tomato seedlings. HortScience 47(9):1304–1307

Ashraf M (1994) Organic substances responsible for salt tolerance in Eruca sativa. Biol Plantarum 36:255–259. doi:10.1007/BF02921095

Astaraei AR, Ivani R (2008) Effect of organic sources as foliar spray and nutrition of cowpea plant. Am Eurasian J Agric Environ Sci 3:352–356

Atiyeh RM, Subler S, Edwards CA, Bachman G, Metzger JD, Shuster W (2000) Effect of vermicomposts and composts on plant growth in horticultural container and soil. Pedobiologia 44(5):579–590. doi:10.1078/S0031-4056(04)70073-6

Ayyobi H, Olfati JA, Peyvast GA (2014) The effects of cow manure vermicompost and municipal solid waste compost on peppermint (*Mentha piperita* L.) in Torbat-e-Jam and Rasht regions of Iran. Int J Recycl Org Waste Agric 3:147–153. doi:10.1007/s4009

Azooz MM, Abou-Elhamd MF, Al-Fredan MA (2012) Biphasic effect of copper on growth, proline, lipid peroxidation and antioxidant enzyme activities of wheat (*Triticum aestivum* cv. Hasaawi) at early growing stage. Aust J Crop Sci 6:688–694

Bates LS, Waldern RP, Teare ID (1973) Rapid determination of free proline from water stress studies. Plant Soil 39:205–207. doi:10.1007/BF00018060

Cavalcanti FR, Oliveira JT, Martins-miranda AAS, Viegas RA, Silveira JAG (2004) Superoxide dismutase, catalase and peroxidase activities do not confer protection against oxidative damage in salt-stressed cowpea leaves. New Phytol 163:563–571. doi:10.1111/j.1469-8137.2004.01139.x

Chinsamy M, Kulkarni MG, Staden JV (2013) Garden-waste-vermicompost leachate alleviates salinity stress in tomato seedlings by mobilizing salt tolerance mechanisms. Plant Growth Regul 71:41–47. doi:10.1007/s10725-013-9807-6

Chinsamy M, Kulkarni MG, Staden JV (2014) Vermicompost leachate reduces temperature and water stress effects in tomato seedlings. HortScience 49(9):1183–1187

El- Khawaga AS, Zaeneldeen EMA, Yossef MA (2013) Response of three pomegranate cultivars (*Punica granatum* L.) to salinity stress. Middle East J Agric Res 1(1):64–75

Fathi T, Golchin A, Safikhani F (2012) Effect of drought stress and vermicompost on clary sage. Ann Biol Res 3(7):3346–3349

Garcial AC, Berbara RLL, Farias LP, Izquierdo FG, Hemandez OL, Campos RH, Castro RN (2012) Humic acids of vermicompost as an ecological pathway to increase resistance of rice seedlings to water stress. Afr J Biotechnol 11(13):3125–3134. doi:10.5897/AJB11.1960

Hasanpour Z, Karimi HR, Mirdehghan SH (2014) Effects of salinity and water stress on echophysiological parameters and micronutrients concentration of pomegranate (*Punica granatum* L.). J Plant Nutr 38(5):795–807. doi:10.1080/01904167.2014.944711

Hasanuzzamaan M, Nahar K, Fujita M (2013) Plant response to salt stress and role of exogenous protectants to mitigate salt-induced damages. In: Ahmad P, Azooz MM, Prasad MNV (eds) Ecophysiology and responses of plants under salt stress. Chapter 2, Springer, New York, pp 25–87. doi:10.1007/978-1-4614-4747-4_2. ISBN-13: 9781461447467

Hulya Orak H (2009) Evaluation of antioxidant activity, colour and some nutritional characteristics of pomegranate (*Punica granatum* L.) juice and its sour concentrate processed by conventional evaporation. Int J Food Sci Nutr 60(1):1–11. doi:10.1080/09637480701523306

Lazcano C, Dominguez J (2010) Effects of vermicompost as a potting amendment of two commercially-grown ornamental plant species. Span J Agric Res 8(4):1260–1270

Maie Mohsen MA, Abo- Kora HA, Abeer Kassem HM (2016) Effect of vermicompost and calcium silicate to reduce the soil salinity on growth and oil determinations of Marjoram plant. Int J Chemtech Res 9(5):235–262

Mir M, Hassan GI, Mir A, Hassan A, Sulaimani M (2013) Effects of bio-organics and chemical fertilizers on nutrient availability and biological properties of pomegranate orchard soil. Afr J Agric Res 8(37):4623–4627

Mirzapour MH, Khoshgoftarmanesh AH (2013) Effect of soil and foliar application of iron and zinc on quantitative and qualitative yield of pomegranate. J Plant Nutr 36(1):55–66. doi:10.1080/01904167.2012.733049

Naeini MR, Khoshgoftarmanesh AH, Lessani H, Fallahi E (2004) Effects of sodium chloride induced salinity onmineral nutrients and soluble sugar in three commercial cultivars on pomegranate. J Plant Nutr 27:1319–1326. doi:10.1081/PLN-200025832

Naeini MR, Khoshgoftarmanesh AH, Fallahi E (2006) Partitioning of chlorine, sodium, and potassium and shoot growth of three pomegranate cultivars under different levels of salinity. J Plant Nutr 29:1835–1843. doi:10.1080/01904160600899352

Nakano Y, Asada YK (1981) Hydrogen peroxide is scavenged by ascorbate specific peroxidase in spinach chloroplasts. Photochem Photobiol 37:679–690. doi:10.1093/oxfordjournals.pcp.a076232

Nandi B, Bhandari SC, Meena RH, Meena RR (2013) Effect of vermicompost on plant growth, fruit yield and quality of pomegranate cv Ganesh. Environ Ecol 31(1A):322–324

Okhovatian-Ardakani A, Mehrabanian RM, Dehghani F, Akbarzadeh A (2010) Salt tolerance evaluation and relative comparison in cuttings of different pomegranate cultivars. Plant Soil Environ 56(4):176–185

Oliva MA, Ricon R, Zenteno E, Pinto A, Dendooven L, Gutierrez F (2008) Vermicompost rol against sodium chloride stress in the growth and photosynthesis in tamarind plantlets (*Tamarindus indica* L.). Gayana Bot 65(1):23–31

Ozden M, Demirel U, Kahraman A (2009) Effects of proline on antioxidant system in leaves of grapevine (*Vitis vinifera* L.) exposed to oxidative stress by H_2O_2. Sci Hortic 119:163–168. doi:10.1016/j.scienta.2008.07.031

Parvaiz A, Satyawati S (2008) Salt stress and phyto-biochemical responses of plants—a review. Plant Soil Environ 54:88–99

Pirseyedi S, Valizadehghan MS, Mardi M, Ghaffari MR, Mamhmoodi P, Zahravi M, Zeinalabadini M, Nekoui SMK (2010) Isolation and characterization of novel microsatellite markers in pomegranate (*Punica granatum* L.). Int J Mol Sci 11:2010–2016. doi:10.3390/ijms11052010

Ramezanian A, Rahemi M, Vazifehshenas MR (2009) Effects of foliar application of calcium chloride and urea on quantitative and qualitative characteristics of pomegranate fruits. Sci Hortic 121:171–175. doi:10.1016/j.scienta.2009.01.039

Rawat JMS, Tomar YK, Rawat V (2010) Effect of stratification on seed germination and seedling performance of wild pomegranate. J Am Sci 6(5):97–99

Sallaku G, Babaj I, Kaciu S, Balliu A (2009) The influence of vermicompost on plant growth characteristics of cucumber (*Cucumis sativus* L.) seedlings under saline conditions. J Food Agric Environ 7(3 and 4):869–872

Schnabel G, Strittmatter G, Noga G (1998) Changes in photosynthetic electron transport in potato cultivars with different field resistance after infection with *Phytophthora infestans*. J Phytopathology 146(5–6):205–210. doi:10.1111/j.1439-0434.1998.tb04681.x

Velikova V, Loreto F (2005) On the relationship between isoprene emission and thermo tolerance in *Phragmites ausrralis* leaves exposed to high temperatures and during the recovery from a heat stress. Plant Cell Environ 28:318–327. doi:10.1111/j.1365-3040.2004.01314.x

Wang F, Zeng B, Sun Z, Zhu C (2009) Relationship between proline and Hg^{+2}-induced oxidative stress in tolerant rice mutant. Arch Environ Contam Toxicol 56:723–731. doi:10.1007/s00244-008-9226-2

Zaouaya F, Menab P, Garcia-Viguerab C, Marsa M (2012) Antioxidant activity and physico-chemical properties of Tunisian grown pomegranate (*Punica granatum* L.) cultivars. Ind Crops Prod 40:81–89. doi:10.1016/j.indcrop.2012.02.045

Soil chemical properties and growth response of *Moringa oleifera* to different sources and rates of organic and NPK fertilizers

A. G. Adebayo[1] · H. A. Akintoye[1] · A. O. Shokalu[1] · M. T. Olatunji[1]

Abstract

Purpose The need for increasing production of *Moringa oleifera* in Nigeria can be achieved through adequate fertilization. This study investigated the effects of sources and rates of NPK (15:15:15) and compost on soil properties and productivity of Moringa at National Horticultural Research Institute, Ibadan, Nigeria.

Methods NPK was applied at 30, 60, and 90 kg N/ ha, cow dung (CD), poultry manure (PM) and organomineral (OM) were applied at 10, 20 and 30 tons/ha. The experimental design was a randomized complete block design with three replicates. Organic fertilizers were incorporated into the soil 2 weeks before sowing; NPK was split applied at 2 and 5 weeks after sowing. Seeds were sown at 75 cm × 75 cm spacing. Data were taken on plant height (cm), stem girth (cm), number of leaves, leaf biomass, stems weight and post-planting soil properties.

Results Growth values for NPK and compost treatments were higher than the control. PM applied at 30 tons ha^{-1} resulted in highest growth values: plant height (65.91 cm), stem girth (1.51 cm) and number of leaves (14.20). PM applied at 30 tons ha^{-1} gave higher stem weight (2249.9 g) and leaf biomass (3610.5 g). Post-planting soil analysis indicated that nitrogen, phosphorus, calcium, sodium, manganese, iron and zinc contents were higher in plots with organic and inorganic fertilizers except for potassium and magnesium.

Conclusion PM proved more superior to CD manure and others because it produced better growth attributes such as shoot height, stem girth and number of leaves and leaf biomass than its counterparts produced.

Keywords Soil amendments · *Moringa oleifera* · Vegetative growth · Leaf biomass · Soil nutrient

Introduction

Moringa oleifera, also known as horseradish tree, is a medium-sized tree belonging to the family Moringaceae; it is adapted to a wide range of soil types, but grows best in well-drained loam to clay loam, neutral to slightly acidic soils, but cannot withstand prolonged water logging. It grows where temperature ranges from 26 to 40 °C and annual rainfall totals at least 500 mm.

According to Fuglie (1999), the uses of Moringa included, but not limited to alley cropping (biomass production), animal forage (leaves and treated seed-cake), biogas (leaves), domestic cleaning agent (crushed leaves), blue dye (wood), fencing (living trees), fertilizer (seed-cake), foliar nutrient (juice expressed from the leaves), green manure (leaves), gum (tree trunks), honey and sugar cane juice-clarifier (powdered seeds), medicine (all plant parts), ornamental plantings, biopesticide (soil incorporation of leaves to prevent seedling damping off), pulp (wood), rope (bark), tannin for tanning hides (bark and gum), water purification (powdered seeds) and Moringa seed oil (yield 30–40% by weight). It is a good source of vitamin A, vitamins B and C, essential minerals, and the sulphur-containing amino acids, methionine and cysteine (Bosch et al. 2004). The composition of the amino acids in

✉ A. G. Adebayo
ajisafeag@yahoo.com

[1] Department of Vegetable and Floriculture, National Horticultural Research Institute, Ibadan, Nigeria

the leaf protein is well balanced (Foidl et al. 2001). The available data (Yazzie et al. 1994; Nordeide et al. 1996) show that leaves contain 13–15% protein, 60–70% carbohydrate, 4–10% fat, around 11% fiber, 16% ash and the energy value varies from 1180 to 1900 kJ/100 g of which 80% are metabolisable energy.

Interest in Moringa in recent times is skewed towards its medicinal properties. Hence, researchers focus on its value as medicine. Consequently, the demand for the plant products has been on ascendancy (Imoro et al. 2012). However, not much research has been done on its cultivation and fertilization in Nigeria with respect to its growth and productivity using different types of inorganic and organic manures commonly used by local farmers and more so, until recently *M. oleifera* was not popular in southwestern Nigeria despite its acclaimed economic and health importance (Odeyinka et al. 2007); hence, very little research has been done on the species, although it is widely used by the rural poor as a food resource.

Note This research was carried out in the southwestern part of Nigeria, where before now *M. oleifera* was not a popular field crop.

Nigeria is among countries with low to medium productivity soils (Eswaram et al. 1997) and are classified as being generally infertile (Aduayi et al. 2002), thus, necessitating the addition of fertilizers for optimum yield. Amelioration of soil by the use of inorganic or organic fertilizers is considered as a component of intervention for increased crop production around the world (Chand et al. 2006). The positive effects of organic manures on soil structure, aggregate stability, water-holding capacity and crop yield were reported in several studies (Togun and Akanbi 2003; Jedidi et al. 2004; Odlare et al. 2008). Low cost and ready availability of materials for their preparation, gradual release of plant nutrients, aeration, the ability of the soil to hold nutrients and being environmental friendly have made organic fertilizers popular among farmers.

This experiment was carried out to investigate the effects of different sources and rates of organic and inorganic fertilizers on early growth of *M. oleifera* and soil characteristics.

Materials and methods

Field experiments were carried out at the National Horticultural Research Institute (3°54′E, 7°30′N, and 213 m above sea level), Ibadan, to determine the optimum rate at which different rates and sources of fertilizers can be applied to soil to increase its productivity and growth of *M. oleifera*. The experiment comprises thirteen treatments: NPK (15:15:15) applied at 30, 60 and 90 kg N ha^{-1}, cow dung (10, 20 and 30 tons ha^{-1}), poultry manure (10, 20 and 30 tons ha^{-1}) and organomineral (10, 20 and 30 tons ha^{-1}) along with the control. The experimental design was a randomized complete block design (RCBD) with three replicates. Organic fertilizers (poultry manure and cow dung) were incorporated into the soil 2 weeks before planting to allow for their mineralization, while NPK fertilizer was split applied at 2 and 5 weeks after planting while organominerals were applied at 2 weeks after planting to enable the seedlings absorb the mineral nutrients. Three seeds were directly sown per hole at 75 cm × 75 cm spacing, on plots of size 9 m^2. The seedlings were thinned to one plant per stand at 2 weeks after sowing (2WAS). Data on growth parameters (plant height, stem girth and number of leaves) were taken fortnightly from 4WAS for ten (10) weeks. Also leaf biomass was determined by cutting four stands per plot at 1 m above ground level; the pruned shoot were separated into leaves and stems, air dried at room temperature until uniform weight was achieved. This process was repeated every fortnight and the total sum of the leaf weight over time was recorded as leaf biomass. Samples of soil from each plot were taken before and 14WAS for laboratory analysis to determine the effects of treatments on soil properties. The soil samples were air dried, crushed and sieved through 2 mm mesh for the determination of the physical and chemical properties of the soil. Total nitrogen (N) was determined by Kjeldhal method after digestion of material (A.O.A.C. 1999). Available P was determined using Bray 1 method (Murphy and Rikey 1962). Exchangeable bases were done by extraction with 1 m NH_4OAC. Calcium (Ca) and magnesium (Mg) in the extract were determined by atomic absorption spectrophotometer (AAS). Effective CEC was the summation of exchangeable bases (Ca, Mg, K, and Na); potassium (K) was determined by flame photometer, and pH (H_2O) using a pH meter. Percentage organic carbon was estimated by chromic acid digestion. Data were subjected to analysis of variance and the mean data were compared using the least significant difference (LSD) at $p > 0.05$.

Results

Pre-field soil analysis

The results of the physical and chemical analysis of the potted soil before cropping are presented in Table 1. The soil was mostly sandy loam in texture with pH of 7.7. Total N content was 0.9 mg g^{-1}, available P was 1.95 mg g^{-1}, K 0.21 c mol kg^{-1}, Fe 6.15 mg g^{-1}, Zn 1.18 mg g^{-1} and Mn value was 45.45 mg kg^{-1} (Table 1).

Table 1 Pre-cropping physical and chemical properties of soils

Soil property	
pH	7.7
Sand (mg g^{-1})	770
Clay (mg g^{-1})	56
Silt (mg g^{-1})	174
Ca (cmol kg^{-1})	1.80
Mg (cmol kg^{-1})	0.74
Na (cmol kg^{-1})	0.35
K (cmol kg^{-1})	0.21
Exch. acidity	0.06
ECEC me/100 g	3.16
C (mg g^{-1})	11.1
N (mg g^{-1})	0.9
Av. P (mg kg^{-1})	1.95
Cu (mg g^{-1})	0.34
Zn (mg g^{-1})	1.18
Fe (mg g^{-1})	6.15
Mg (mg g^{-1})	45.45

Analysis of organic material

The nutrient content of poultry manure, cow dung and organomineral used in the experiment is represented in Table 2. It indicated that there were differences in nutrient composition of organic components. N, P and K contents of poultry manure were higher in value than cow dung and organomineral.

Effects of soil amendments on vegetative growth of *M. oleifera*

Plant height

There were significant ($p > 0.05$) differences in the plant height between those treated with fertilizers and the control. Results indicated that at all growth stages, plant height observed for all the fertilizer types and rates were all higher than the control, where the average values for control were 13.0, 26.1, 47.5 and 74.7 cm for 5, 7, 9 and 11WAS; plant height for 20 tons ha^{-1} of poultry manure (PM) was 27.2, 61.3, 111.6 and 157.4 cm, respectively. At 11WAS, 30 tons ha^{-1} PM had the highest plant height. This performance was followed in ranking by 20 tons ha^{-1} PM (157.4 cm) and 10 tons ha^{-1} PM (Table 3).

Stem girth

Stem girth values obtained for all organic and NPK rates were higher than the control at all stages of data measurements, where the values obtained for control were 0.3, 0.5, 0.9 and 1.4 cm; the highest values of 0.8, 2.0, 2.2 and 2.5 cm were obtained from 20 tons ha^{-1} PM. At 11WAS, 30 tons ha^{-1} PM application resulted in the highest value of stem girth obtained in the experiment. Poultry manure at all levels of application in the study resulted in stronger stem. Performance of PM was followed by application of 20 and 30 tons ha^{-1} cow dung (Table 3).

Number of leaves

The number of leaves produced by *M. oleifera* was highly influenced by application of fertilizers (Table 3). Pots without any fertilizer application had the least number of leaves at 5, 7, 9 and 11WAS (8.1, 9.3, 12.8 and 17.8), respectively, while pots with 20 tons ha^{-1} of PM had the highest number of leaves at 5WAS but pots with 30 tons ha^{-1} application had the highest number of leaves from 7WAS (15.6, 24.8, 32.3). Generally, PM at all rates produced higher number of leaves than all rates of OM, CD and NPK. The performance of seedlings with 30 tons ha^{-1} PM application was followed by 20 tons ha^{-1} OM and 30 tons ha^{-1} of CD in descending order. The higher the rate of poultry manure, the higher the number of *M. oleifera* leaves produced.

Effects of soil amendments on stem and leaf biomass

All treatments with organic manure had higher stem weight and leaf biomass than the NPK rates and the control, but among the organic manures, *M. oleifera* treated with 30 tons ha^{-1} poultry manure (30 tons ha^{-1} PM) had the maximum stem weight (2249.9 g) and leaf biomass (3610.5 g). This performance was followed in descending order 30 tons ha^{-1} OM (3335.7 g) > 20 tons ha^{-1} OM (3276.2 g) > 20 tons ha^{-1} PM (3254.5 g). So also for stem weight, it was 30 tons ha^{-1} PM (2249.9 g) > 20 tons ha^{-1} OM (2181.1 g) > 20 tons ha^{-1} PM (1992.3 g) (Table 4).

Table 2 Chemical properties of the different types of organic materials used

Treatment	N (%)	P (%)	K (%)	Ca (%)	Mg (%)	Fe (%)	C (%)
Cow dung	0.85	0.12	1.49	1.57	0.51	0.09	30
Poultry manure	1.18	0.99	1.71	2.87	0.61	0.78	17
Org. mineral	0.45	0.55	0.89	1.44	0.74	1.41	40

Table 3 Growth response of *Moringa oleifera* to different sources and rates of organic and inorganic fertilizers

	Weeks after sowing											
	Plant height (cm)				Stem girth (cm)				Number of leaves			
TRT	5	7	9	11	5	7	9	11	5	7	9	11
PM10	28.4	56.6	105.4	154.2	0.7	1.1	1.8	2.2	13.2	15.6	22.0	27.3
PM20	27.7	61.3	111.6	157.4	0.8	2.1	2.0	2.5	13.8	15.1	20.6	24.6
PM30	27.9	57.2	108.8	161.6	0.7	1.3	1.8	2.6	13.7	15.6	24.8	32.3
OM10	18.4	39.7	77.8	126.9	0.5	1.8	1.5	2.0	10.2	13.5	19.8	23.3
OM20	22.0	46.2	88.4	136.4	0.6	1.0	1.5	1.9	10.3	14.1	21.4	30.1
OM30	17.9	34.5	63.1	98.5	0.5	0.8	1.1	1.4	9.2	11.3	16.2	22.6
CD10	20.2	39.3	75.2	104.4	0.5	0.8	1.3	1.7	10.8	12.4	19.4	22.5
CD20	19.5	39.6	75.4	121.2	0.5	0.8	1.5	2.0	11.3	13.7	19.7	22.3
CD30	25.6	53.2	93.8	131.4	0.6	1.0	1.5	2.0	13.0	14.6	21.8	25.8
NPK30 (kg N ha^{-1})	15.0	28.4	58.4	98.6	0.4	0.7	1.1	1.5	9.5	10.8	17.2	19.7
NPK60 (kg N ha^{-1})	24.0	47.0	87.9	131.2	0.6	1.0	1.5	2.0	11.8	13.8	21.0	24.0
NPK90 (kg N ha^{-1})	18.8	39.2	76.8	123.2	0.5	0.8	1.3	1.8	11.0	12.7	20.2	22.8
CT	13.0	26.1	47.5	74.7	0.3	0.5	0.9	1.4	8.1	9.3	17.8	17.8
LSD	8.4	20.3	35.92	47.5	1.0	1.1	0.6	0.7	3.1	4.0	4.5	6.6

Table 4 Effects of different sources and rates of NPK and organic fertilizers on leaf and stem biomass of *Moringa oleifera*

Treatment	Stem biomass (g)	Leaf biomass (g)
PM10	1541.5	2641.5
PM20	1992.3	3254.5
PM30	2249.9	3610.5
OM10	1813.3	2712.4
OM20	2181.1	3276.2
OM30	1912.8	3335.7
CD10	1091.5	2240.5
CD20	1256.5	2071.1
CD30	1825.2	2437.2
NPK30 (kg N ha^{-1})	1095.5	1871.3
NPK60 (kg N ha^{-1})	1129.2	1808.7
NPK90 (kg N ha^{-1})	1130.1	2147.0
CT	888.5	1271.3
LSD	738.8	532.9

Effects of different sources and rates of nutrients on post-field soil properties

Soil nitrogen (N), phosphorus (P) and potassium (K) were significantly influenced by different rates and sources of fertilizers. For post-planting soil N and P, addition of both organic and NPK fertilizers resulted in the higher N and P soil contents than the control plots except for K (Fig. 1). Organomineral (OM) applied at 10 tons ha^{-1} had higher residual soil N content; this was followed by poultry manure applied at 30 tons $ha^{-1.}$ For K and P, it was also OM applied at 30 tons ha^{-1}; this performance was followed by poultry manure. Other soil properties were determined which were significantly influenced by different rates and sources of fertilizers include: Ca, Na, Mn, Fe and Zn. Application of 30 kg ha^{-1} of NPK resulted in the higher residual soil content of Ca; this was followed by 30 tons ha^{-1} application of OM. 20 tons ha^{-1} application of CD on *M. oleifera* resulted in the higher residual content of Mg and Mn in the soil, while for Fe, it was 10 tons ha^{-1} application of CD; this was followed by PM applied at 30 and 20 tons ha^{-1} application, respectively (Fig. 2).

Discussion

The pre-planting major nutrient content of the soil was inadequate, where the minimum critical level required for crop growth in the area is 0.6–1.0 mg kg^{-1} for N, 3–7 mg kg^{-1} for P, and 0.21–0.3 cmol kg^{-1} for K (Adeoye and Agboola 1985). This nutrient status of the soil agreed with the results of Aduayi et al. (2002) who reported that most Nigerian soils are deficient in major nutrient, thus emphasis the need for use of sustainable soil amendments. Also nutrient analysis of organic manure (Table 2) indicated that poultry manure is higher in content in major nutrients required for optimum plant growth, such as N, P and K. This may have been responsible for significance performance by Moringa plants treated with higher rates of poultry manure. Addition of fertilizers either organic or inorganic improves Moringa growth and productivity of soils, but their performance depends on the crop type, the environment, season, variety, fertilization, and irrigation

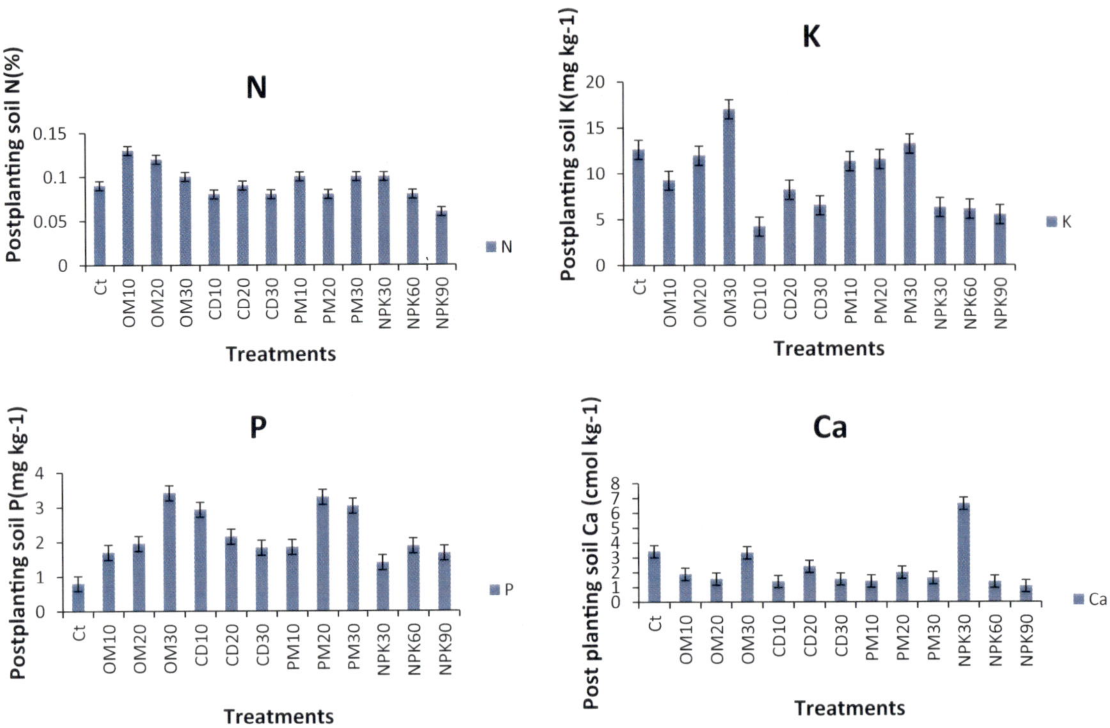

Fig. 1 Effect of different sources and rates of organic and inorganic fertilizers on post-planting soil properties of *Moringa oleifera*

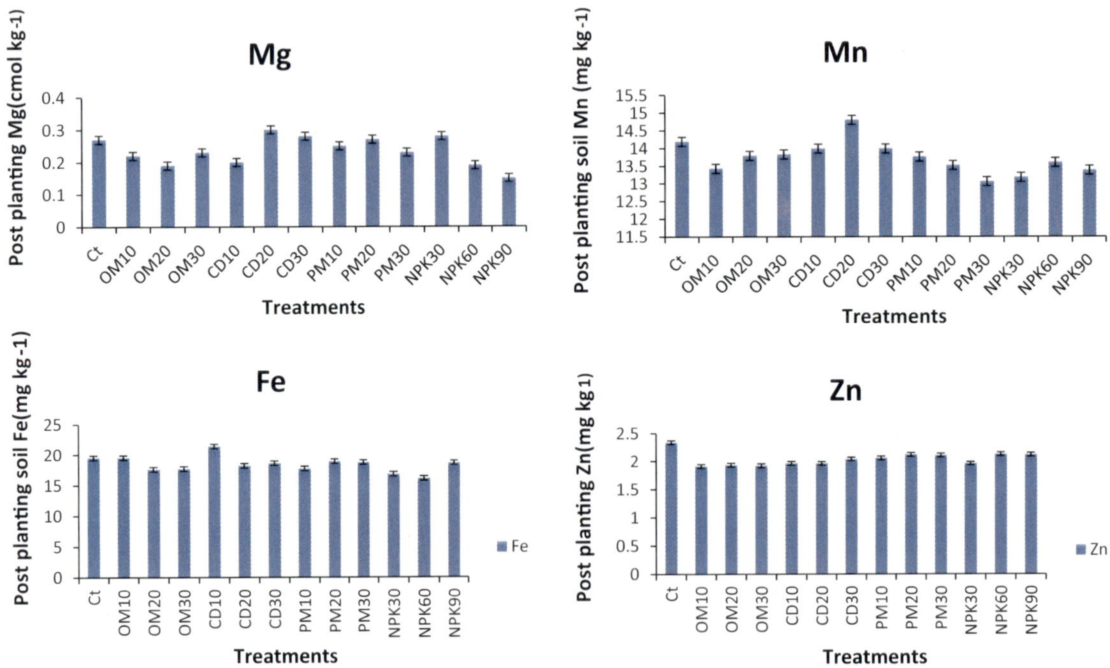

Fig. 2 Effect of different sources and rates of NPK and organic fertilizers on post-planting soil properties of *Moringa oleifera*

regime. Moringa yields best under warm, dry conditions, with some supplemental fertilizer and irrigation (Radovich and Paul 2008). In this experiment, poultry manure applied at 20 tons ha^{-1} treated seedlings showed a significant ($p > 0.05$) difference over other amendments and the control in terms of height, girth, number of leaves and leaf biomass; this observation was similar to Usman (2015) report, when he compared the performance of tomato crop treated with 20 tons ha^{-1} of cow dung, goat droppings and poultry manure, respectively; application of poultry manure (PM) resulted in the higher growth parameters and fruit yield of tomato plants. The increasing effects of PM on moringa development and soil properties in this study also corroborated the observation of Gupta et al. (1997), Adeniyan and Ojeniyi (2005), Akanbi et al. (2005), Adenawoola and Adejoro (2005) and Agyenim-Boateng et al. (2006) that poultry manure is very rich animal manure that releases considerable soil organic matter, available P and exchangeable cations when applied to soil. Poultry manure contains all the essential plant nutrients that are used by plants; these include nitrogen (N), phosphorus (P), potassium (K), calcium (Ca), magnesium (Mg), sulphur (S), manganese (Mn), copper (Cu), zinc (Zn), chlorine (Cl), boron (B), iron (Fe) and molybdenum (Mo) (Amanullah et al. 2010). It increases the moisture holding capacity of the soil and improves lateral water movement, thus improving irrigation efficiency and decreasing the general doughtiness of sandy soils. Poultry manure application improves soil retention and uptake of plant nutrients (Amanullah et al. 2010). It increases the number and diversity of soil microorganisms, particularly in sandy conditions. This effect enhances crop health by increasing water and nutrient availability, as well as suppressing harmful levels of plant parasitic nematodes, fungi and bacteria. The cumulative effect supports the observation that poultry manure added organic matter and nutrients to the soil. This may probably be due to the fact that poultry manure contains concentrated nutrients and hence led to enhanced plant growth in those seedlings treated with poultry manure. The nutrient quality in poultry manure might surpass the ones in cow dung manure leading to more enhanced plant growth in those treated with poultry manure. This observation was between week 5 and 7 after planting. This generally implies that the plant is a fast growing species and this conforms to the earlier findings by Odee (1998) and Imoro et al. (2012) that *Moringa* is a fast growing plant and grows between 6 to 7 m per annum even in areas receiving less than 400 mm of rainfall. The significance influence of organomineral (OM) on residual soil content of N, P and K may also be because, OM was fortified with mineral N, P and K. The bulk of cow dung manure might probably be materials that do not significantly enhance plant growth as compared to those found in poultry manure. Other authors found significant improved height in *M. oleifera* using poultry manure (Agyenim-Boateng et al. 2006; Imoro et al. 2012). The organic matter component of poultry manure were higher, and when decomposed in soil, nutrients were released to soil. Hence the findings that poultry manure increased soil N, P, K, Ca, and Mg significantly.

Conclusion

Application of poultry manure, cow dung manure, organomineral and NPK had significant effects on the growth of *M. oleifera* over controls. However, poultry manure treatments had the best growth attributes for optimal yield. Therefore, the treatment with 20 tons ha^{-1} is recommended for sustainable leaf dry matter production of *M. oleifera* in South–West, Nigeria.

Acknowledgements Funding was provided by the National Horticultural Research Institute.

References

Adenawoola AR, Adejoro SA (2005) Residual effects of poultry manure and NPK fertilizer residues on soil nutrient and performance of jute (*Corchorus olitorius* L.). Niger J Soil Sci 15:133–135

Adeniyan ON, Ojeniyi SO (2005) Effect of poultry manure, NPK (15:15:15) and combination of their reduced levels on maize growth and soil chemical properties. Niger J Soil Sci 15:34–41

Adeoye GO, Agboola AA (1985) Critical levels for soil pH, available phosphorus (P), potassium (K), zinc (Zn) and manganese (Mn) and maize ear leaf content of boron (B), cupper (Cu) and manganese (Mn) in sedimentary soils of south western, Nigeria. Fert Res 6:65–71

Aduayi EA, Chude VO, Adebusuyi BA, Olayiwola SO (2002) Fertilizer use and management in Nigeria. Produced by Federal Ministry of Agriculture Rural Development, Abuja, p 188

Agyenim-Boateng S, Zickerman J, Kornahrens M (2006) Poultry manure effect on growth and yield of maize. West Afr J Appl Ecol 9:61–70

Akanbi WB, Akande MO, Adediran JA (2005) Suitability of composted maize straw and mineral nitrogen fertilizer for tomato production. J Veg Sci 11(1):57–65. doi:10.1300/484v11n01-06

Amanullah MM, Sekar S, Muthukrishnan P (2010) Prospects and potential of poultry manure. Asian J Plant Sci 9:172–182. doi:10.3923/ajps.2010.172.182

Association of Official Analytical Chemists (A.O.A.C) (1999) Official methods of analysis, 11 edn. Washington, D.C. https://www.ajol.info/index.php/wajae/article/viewFile/45682/29162

Bosch CH, Sié K, Asafa BA (2004) *Adansonia digitata* L. Record from protabase. In: Grubben GJH, Denton OA (eds) PROTA (Plant resources of tropical Africa). Wageningen, Netherlands

Chand S, Anwar M, Patra DD (2006) Influence of long-term application of organic and inorganic fertilizer to build up soil fertility and nutrient uptake in mint-mustard cropping sequence. Commun Soil Sci Plant Anal 37:63–76. http://www.tandfonline.com/doi/abs/10.1080/00103620500408704

Eswaram H, Almaraz R, van den Berg E, Reich P (1997) An assessment of the soil resources of Africa in relation to productivity. Geoderma 77:1–18. http://www.sciencedirect.com/science/article/pii/S0016706197000074

Nordeide MB, Hatloy A, Folling M, Lied E, Osbaug A (1996) Nutrient composition and nutritional importance of green leaves and wild food resources in an agricultural district, Koutiala, in southern Mali. Int J Food Scand Nutr 47:455–468. https://www.ncbi.nlm.nih.gov/pubmed/8933199

Foidl N, Makkar HPS, Becker K (2001) The potential of *Moringa oleifera* for agricultural and industrial uses. In: Fuglie LJ (ed) The miracle tree/the multiple attributes of Moringa. CTA, USA

Fuglie LJ (1999) The miracle tree: *Moringa oleifera*: natural nutrition for the tropics. Dakar, Church World Service, p 68 **(revised in 2001 and published as The Miracle Tree)**

Gupta G, Borowiee J, Okoh J (1997) Toxicity identification of poultry litter aqueous leachate. Poult Sci 76:1364–1367. doi:10.1093/ps/76.10.1364

Imoro WM, Sackey I, Abubakar AH (2012) Preliminary study on the effects of two different sources of organic manure on the growth performance of *Moringa oleifera* seedling. J Biol Agric Healthc 2(10):147–158

Jedidi N, Hassen A, van Cleemput O, M'Hiri A (2004) Microbial biomass in a soil amended with different types of organic wastes. Waste Manag Res 22:93–99. journals.sagepub.com/doi/pdf/10.1177/0734242X04043930

Murphy J, Rikey JP (1962) A modified single solution method for the determination of phosphate in natural water. Anal Chem Acts 27:31–36

Odee D (1998) Forest biotechnology research in drylands of Kenya: the development of *Moringa* species. Dryland Biodivers 2:7–12

Odeyinka SM, Torimiro DO, Oyedele JO, Asaolu VO (2007) Farmer's awareness and knowledge of *Moringa oleifera* in Southwestern Nigeria: a perceptional analysis. Asian J Plant Sci 6:320–325. doi:10.3923/ajps.2007.320.325

Odlare M, Pell M, Swenson K (2008) Changes in soil chemical and microbiological properties during 4 years of application of various organic residues. Waste Manag 28:1246–1253. doi:10.1016/j.wasman.2007.06.005

Radovich TJK, Paul R (2008) Early growth, leaf yield, protein content and pod yield of four Moringa accessions in Hawaii. Hortic Sci 43:1135

Togun AO, Akanbi WB (2003) Comparative effectiveness of organic-based fertilizers to mineral fertilizers on tomato growth and fruit yield. Compost Sci Util 11:337–42. http://www.tandfonline.com/doi/pdf/10.1080/1065657X.2003.10702143

Usman M (2015) Cow dung, goat and poultry manure and their effects on the average yields and growth parameters of tomato crop. J Biol Agric Healthc 5(5). http://www.iiste.org. 04/01/2017. http://www.iiste.org/Journals/index.php/JBAH/article/viewFile/20621/21555

Yazzie D, Vanderjagt DJ, Pastuszyn A, Okolo A, Glew RH (1994) The amino acid and mineral contents of baobab (*Adansonia digitata* L.) leaves. J Food Compos Anal. doi:10.1006/jfca.1994.1018

12

Accumulation pattern of trace metals in *Spinacia oleracea* harvested from soil treated with urine in comparison with other soil amendments in Pretoria, South Africa

L. L. Mugivhisa[1] · J. O. Olowoyo[1]

Abstract

Purpose The study assessed trace metal levels in the leaves, stalks and roots of *Spinacia oleracea* harvested from soil treated with urine in comparison to chemical fertilizers and biosolids.

Methods *Spinacia oleracea* seedlings were planted on soils pretreated with urine, chemical fertilizers and biosolids.

Results The soil treated with chemical fertilizers resulted in an increase of Cr, Pb, Cd, Cu, Ni and Sb while there was an increase in the concentrations of Zn and Hg in the soil treated with urine. The soil with no amendments recorded higher mean values of As and Mn, whereas the biosolids treatment did not show any increases of the trace metals in the soil. The concentration of Mn, Pb and Ni in the leaves and stalks of *S. oleracea* harvested from soil treated with urine were below the recommended limits for trace metals in edible plants as set by WHO even though the urine treatment recorded the highest concentration of Cd in the roots, stalks and leaves. The *S. oleracea* harvested from the soil treated with chemical fertilizers showed an accumulation Cu and Mn in the stalks and leaves while those harvested from soil treated with biosolids showed an accumulation of Cr, As, Zn and Ni in the stalks and Cr, Pb and Sb in the leaves and all trace metals in the roots except Cd and As. *S. oleracea* harvested from the soil with no amendments showed an accumulation of Cr, As, Zn and Ni in the stalks and Cr, Pb, Zn and Sb in the leaves. The transfer factor showed that Cd, Zn, Mn and Sn were translocated from the soil to the leaves even though the concentrations were below acceptable limits for human consumption.

Conclusions The study demonstrated that the use of urine as a soil amendment may not facilitate or increase the bioavailability of trace metals in the plant tissues.

Keywords Urine · Organic waste · Trace metals · Agriculture

Introduction

Agricultural lands in developing countries especially in sub-Saharan Africa are less productive as a result of recurrent droughts (Cofie et al. 2010). The soils in areas which are poor in plant nutrients and also over-utilized result in millions of people being exposed to shortage of food and diseases due to malnutrition (Egigu et al. 2014). Soils need to be fertilized and the demand for the synthetic chemical fertilizers has been escalating in order for the food demand to be addressed so that the high populations in developing countries can be supported. However, high prices of chemical fertilizers result in the reduction of production and yields mainly in developing countries where there are limitations as with regards to tools and inaccessibility of fertilizers (Sene et al. 2013). There has been a mission to explore new alternatives such as recycling of wastes like human urine to improve the fertility of the soil (Mnkeni et al. 2008).

The use of human urine as a valuable resource has been practised to improve the growth of mainly leafy vegetables in countries such as Germany, Sweden, USA, Mexico, Zimbabwe and Denmark even though there is a limitation of information in South Africa (Mnkeni et al. 2008). There

✉ L. L. Mugivhisa
Liziwe.Mugivhisa@smu.ac.za

[1] Department of Biology, Sefako Makgatho Health Sciences University, Medunsa Campus, P.O Box 139, Pretoria 0204, South Africa

has been broad research carried out to compare the effectiveness of urine in fertilisation of crop in comparison of commercial fertilizers on a wide variety of cereals and vegetables such as cucumber (Heinonen-Tanski et al. 2007), pumpkin (Pradhan et al. 2009a), tomato (Pradhan et al. 2009b), red beet (Pradhan et al. 2010a), radish, potato, cauliflower (Pradhan et al. 2010b), cabbage (Mnkeni et al. 2005), spinach (Mnkeni et al. 2005), carrot (Mnkeni et al. 2008), beetroot (Mnkeni et al. 2008) and maize (Guzha et al. 2005). From all the research human urine was found to have compared equally favourable with the chemical fertilizers when both amendments were utilized at equal doses (Mnkeni et al. 2008).

The direct utilisation of human urine as a fertilizer for agricultural purposes is controversial and associated with problems with respect to hygiene, transport, storage and spreading (Lindt et al. 2000). Some of the harmful substances which are present in urine include pathogens, salts and pharmaceuticals (Sene et al. 2013). However, the urine of a healthy individual is considered sterile in the bladder even though as it passes through different types of dermal tissues, bacteria are added to it resulting in less than ten thousand bacteria per ml being excreted with urine (Hoglund et al. 2002). Urine is also considered sterile unless it has been cross-contaminated by faeces (Mnkeni et al. 2008). The majority of the agents such as bacteria, protozoa, viruses and helminthes in excreta which cause diseases are usually shed in faeces rather than in urine (Hoglund et al. 2002). As a result, the health risks which are associated with the consumption of urine for agricultural purposes have been found to be insignificant (Upreti et al. 2011). The pathogens which cause venereal diseases can also be excreted through urine but there is no indication of whether they have a potential to thrive outside the human body and if this would be of any health significance (Hoglund et al. 2002). The drying of soil which results from evaporation from the plants can result in the decreased chances of survival of these pathogenic agents in the soil (Heinonen-Tanski and Wijk-Sijbesma, Heinonen-Tanski and Van Wijk-Sijbesma 2005). The risk of contamination by the microbes is reduced since microbes which are pathogenic are unable to survive in the soil for a long period of time (Sene et al. 2013).

Storage of human urine for a couple of weeks results in a great reduction of the number of enteric microbes and this renders urine a safer fertilizer compared with animal manure which requires more than 6 months for decomposition (Egigu et al. 2014). The microorganisms die during the hygienization step of simple urine storage even though in Nordic countries the storage of urine might need to be longer and be for a period of more than 6 months (Sene et al. 2013). There has also been an introduction of a set of guidelines by the Vietnamese government for effective and proper compositing of excreta before its consumption so as to limit and reduce the risk of infections by the helminths (Jensen et al. 2007).

The levels of heavy metals in urine have been found to be low and to be 50–1000 times lower than in soils or composts produced in gardens. However, the trace metals levels have been found to be 100–500 times more abundant in urine than in rainwater (Kirchmann and Petterson 1995). According to a study by Paschal et al. (1998), from the 13 trace metals which were measured in the urine of United States residents eight of the analytes were present at detectable concentrations. Trace metals which are of great concern are those which exist almost everywhere in the environment such as Cr, Cu, Zn, Fe, Mn and Pb. The heavy metals receive significant interest throughout the world because of their effects which are toxic, teratogenic and mutagenic even at low concentration levels (Mohod 2015).

Anthropogenic activities such as the utilisation of soil amendments and fertilizers in high-production agriculture contribute to the accumulation of trace metals in the soils (Kidd et al. 2007). The avenue of the entry of trace elements into the human food chain is mainly through the uptake of these trace elements from the soil especially by leafy vegetables (Dingkwoet et al. 2013). The essential nutrients like Fe, Co, Ni, Mn and Co have low permissible limits in living organisms (Salawu et al. 2015); however, the deficiency or excess of these trace elements could cause several disorders (Uwah et al. 2011).

The increasing levels of heavy metals in the environment have been attributed to agricultural practices such as organic and inorganic fertilizer supplements (Agrawal et al. 2007). Heavy metals in the soil can be translocated to different parts of the plants through the uptake by the roots. The uptake and accumulation of the trace metals by the different parts of the plants is dependent on the concentrations and the forms of the heavy metals which are available in the soil. The accumulation of the heavy metals can result in the alteration of the physico-chemical properties of the soil and also result in the plants becoming toxic and the food chain being contaminated (Agrawal et al. 2007). Leafy vegetables like *Spinacia oleracea* have been found to result in increased translocation of trace metals to the parts of the plant which are above the ground resulting in higher levels of heavy metals in the leafy parts (Agrawal et al. 2007).

Spinacia oleracea is one of the most common and popular green leafy vegetable crops used in South Africa as a source of iron (Olowoyo et al. 2011). It grows throughout the whole year with a maturation period of about six weeks and above, hence producing a large mass of fresh leaves in a minimum period of time (Zeka et al. 2014). The study assessed the accumulation of trace metal levels in *S. oleracea* harvested from soil treated with urine in comparison to the other commonly used amendments (chemical fertilizers and biosolids).

Materials and methods

Experimental design

The study was conducted at the production unit of the Sefako Makgatho Health Sciences University. The size of the plot was 5.0 m by 8.45 m. It was separated into four beds measuring 3.45 m by 0.55 m each. The four beds representing different treatments were separated with an interspace of 0.6 m. The plot was chosen to emulate conditions which occur in the real world with regards to agricultural practices.

Soil amendments collection

For urine collection, consent was sought from the male students. A 25 L plastic container was placed in the male students' toilet and students were asked to urinate in it. The collected urine was then stored for a period of 3 months in a sealed plastic container in order for the possible losses of ammonia to be minimised (Kutu et al. 2010). Male students were chosen so as to exclude the interferences of hormones and contraceptives which could be present in the urine samples of female students. The dry biosolids were collected from a sewage treatment area on campus. Chemical fertilizers (NPK) were purchased from a registered trading nursery.

Treatments

Equal and comparable amounts of soil amendments [human urine, chemical fertilizers (2:3:2 NPK fertilizers) and biosolids] were added to each of the four beds representing each group of the amendments and thoroughly mixed with the soil. The treatments/amendments were 5.0 L of urine, 5.0 kg chemical fertilizers (NPK) and 5.0 kg biosolids and each amendment was applied into separate beds. The soil with no amendment did not receive any treatment. From the literature, urine was used in liquid form and fertilizers were used in solid forms (Ranasinghe et al. 2016; Akpan-Idiok et al. 2012). The idea is to check whether urine will perform as a cheap alternative source of nutrients to soil. From all the studies conducted and that are available to us, litres of urine were varied and compared to commercially available fertilizers (Ranasinghe et al. 2016; Akpan-Idiok et al. 2012). The majority of the plant nutrients which are present in human excreta are found in urine. An estimate from the data from five countries (India, Haiti, Uganda and South Africa) which was made was that each person produces about 5 kg of elemental NPK in excreta in a year with about 4 kg being in the urine and 1 kg in faeces (Winblad and Simpson-Hébert 2004). The application of all the amendments was done only once before sowing the seedlings throughout the study. An equal total number of 40 seedlings of *S. oleracea* purchased from a registered marketer were transplanted into each bed a week after the pre-treatment with the soil amendments. All plants were watered in the morning and late in the afternoon on a daily basis. The plants were harvested eight weeks after planting.

Soil and plant analysis

Spinacia oleracea harvested from soil treated with the amendments was washed and separated into roots, stalks and leaves. The separated plant samples were oven dried at 70 °C for 48 h and then grinded with an aid of a mortar and a pestle to achieve homogeneity of the samples. The soil samples which were treated with the amendments were collected after harvesting of *S. oleracea*. All soil samples were air dried, grinded with pestle and mortar and sieved with a 2.0 mm sieve (Germer et al. 2011).

The pH of the soil was determined in distilled water using the 1:12 soil water suspension and in calcium chloride ($CaCl_2$) using 1:12 soil $CaCl_2$ suspension and measured with an aid of a pH metre (Germer et al. 2011). About 0.5 g of the ground soil samples was added with 2.0 ml of $HClO_4$, 2.0 ml of HCl, 8.0 ml $HHNO_3$ and 2.0 ml HF. The solutions resulting from a mixture of the soil and the acids were then analysed for trace metals with an aid of the Inductively Coupled Plasma Mass Spectrometry (ICP–MS) for the determinations of the trace metals concentrations in the soil samples. From the ground plant samples, 0.2 g of each of the different parts was digested with the acids using 2.0 ml HCl, 2.0 ml HF, 5.0 ml HNO_3 and 1.0 ml $HClO_4$. The solutions which resulted from the solutions of the plant samples and the acids were then analysed for levels of trace metals using ICP-MS.

All the apparatuses which were used for homogenization of the soil and plant samples were cleaned with distilled water so that cross contamination could be avoided. For quality assurance purposes, the blanks for the soil and plant samples were separately prepared and the Certified Reference Materials (CRM) for plant and soil samples were done (Lion and Olowoyo 2013). The analyses were also carried out in triplicate (Olowoyo et al. 2015). The CRM 482 was used along with the samples for the purpose of quality assurance.

Statistical analysis

Analysis of variance (ANOVA) and Student's *t* test were used to determine whether the differences obtained from the mean concentrations of trace metals measured in *S. oleracea* harvested from the soil treated with different amendments were significant.

Transfer factor (TF)

The transfer factor (TF) was calculated as the concentration ratio of the trace metals in the stalks and leaves of *S. oleracea* to the concentration of the trace metals in the soil. The TF index was calculated with an aid of the relationship as described in Sithole et al. (2016).

In the stalks, $TF = C_{stalk}/C_{soil}$ and in the leaves, $TF = C_{leaves}/C_{soil}$ with TF representing the transfer factor of *S. oleracea*, C_{stalk} and C_{leaf} representing the concentration of the trace metal in the stalks and leaves of *S. oleracea*, respectively, and C_{soil} representing the concentration of the trace metal in the soil before the cultivation of *S. oleracea*. TF > 1 indicates that *S. oleracea* is enriched with the elements from the soil (accumulator), whereas TF < 1 means that *S. oleracea* is an excluder of the trace metals from the soil.

Results and discussion

From all the trace metals examined, Zinc (Zn) showed the highest mean concentration in the roots, stalks and leaves followed by Manganese (Mn) and then Copper (Cu) (Tables 1, 2, 3). The concentration of Zn in the roots, stalks and leaves ranged from 135.60 ± 8.45 to 2423.00 ± 222.09 mg/kg. The leaves of *S. oleracea* harvested from soil treated with chemical fertilizers showed the highest concentration of Zn while the stalks harvested from soil treated with urine showed the least concentration with no significant difference ($p < 0.05$). The order of the Zn concentration in *S. oleracea* was leaves > roots > stalks for all the amendments and this is in agreement with the study conducted by Farooq et al. (2008) where *S. oleracea* stalks recorded the least concentration of Zn. Even though the concentration of Zn in the stalks harvested from soil treated with urine was below the recommended value of 200 μg/g (Alia et al. 2015; Jung 2008) the leaves harvested from the soil treated with urine was above this recommended value.

Zn is important for the normal growth of the plant and for the development and growth in human beings. Zn is essentially required by plants in minute quantities (25–150 μg/g) in dry tissue (Jung 2008) and the level of Zn in agricultural products should be below 200 μg/g and the

Table 1 Trace metals in the roots of *Spinacia oleracea* harvested from soil treated with different soil amendments

Treatments	Cr (mg/kg)*	Pb (mg/kg)*	Cd (mg/kg)**	As (mg/kg)*	Zn (mg/kg)*	Cu (mg/kg)*	Mn (mg/kg)*	Ni (mg/kg)*	Sb (mg/kg)*	Hg (mg/kg)*
Urine	2.48 ± 0.30	0.95 ± 0.080	1.14 ± 0.12	0.57 ± 0.108	328.70 ± 12.80	17.22 ± 0.98	111.60 ± 11.086	9.12 ± 0.901	0.058 ± 0.009	0.32 ± 0.018
Chemical fertilizers	1.82 ± 0.21	0.83 ± 0.076	0.44 ± 0.034	0.095 ± 0.013	450.10 ± 45.38	15,43 ± 0.954	146.00 ± 13.106	7.155 ± 0.742	0.059 ± 0.00252	0.29 ± 0.017
Biosolids	4.54 ± 0.12	3.78 ± 0.117	0.21 ± 0.013	0.35 ± 0.094	573.7 ± 20.353	30.9 ± 0.988	259.0 ± 8.085	14.82 ± 0.200	0.24 ± 0.009	1.13 ± 0.017
Soil with no amendment	1.36 ± 0.071	0.87 ± 0.016	0.10 ± 0.006	0.12 ± 0.114	321.80 ± 4.652	9.62 ± 0.0225	64.56 ± 1.015	3.36 ± 0.17	0.031 ± 0.006	0.154 ± 0.024

Values: mean ± SD

* Not significantly different

** Significantly different

Table 2 Trace metals in the stalks of *Spinacia oleracea* harvested from soil treated with different soil amendments

Treatments	Cr (mg/kg)*	Pb (mg/kg)*	Cd (mg/kg)**	As (mg/kg)*	Zn (mg/kg)*	Cu (mg/kg)*	Mn (mg/kg)*	Ni (mg/kg)*	Sb (mg/kg)*	Hg (mg/kg)*
Urine	0.65 ± 0.031	0.47 ± 0.030	0.68 ± 0.088	0.059 ± 0.051	135.60 ± 8.448	5.043 ± 0.216	37.74 ± 2.293	1.50 ± 0.026	0.017 ± 0.0027	0.054 ± 0.013
Chemical fertilizers	1.56 ± 0.128	0.42 ± 0.031	0.23 ± 0.025	0.061 ± 0.003	366.5 ± 19.695	7.44 ± 0.444	80.58 ± 4.078	3.09 ± 0.215	0.013 ± 0.005	0.048 ± 0.017
Biosolids	1.46 ± 0.051	0.31 ± 0.020	0.094 ± 0.014	0.014 ± 0.037	195.8 ± 7.805	6.63 ± 0.201	42.64 ± 1.638	2.53 ± 0.759	0.047 ± 0.006	0.033 ± 0.011
Soil with no amendment	1.80 ± 0.053	0.38 ± 0.013	0.15 ± 0.005	0.13 ± 0.023	885.10 ± 18.762	6.44 ± 0.167	30.09 ± 0.570	3.35 ± 0.090	0.019 ± 0.006	0.038 ± 0.14

Values: mean ± SD

* Not significantly different

** Significantly different

Table 3 Trace metals in the leaves of *Spinacia oleracea* harvested from soil treated with different soil amendments

Treatments	Cr (mg/kg)*	Pb (mg/kg)*	Cd (mg/kg)**	As (mg/kg)*	Zn (mg/kg)*	Cu (mg/kg)*	Mn (mg/kg)*	Ni (mg/kg)*	Sb (mg/kg)*	Hg (mg/kg)*
Urine	1.59 ± 0.082	0.44 ± 0.011	2.42 ± 0.098	0.052 ± 0.008	764.90 ± 10.151	16.90 ± 0.14	330.20 ± 4.038	7.73 ± 0.26	0.019 ± 0.006	0.097 ± 0.018
Chemical fertilizers	1.64 ± 0.20	1.00 ± 0.076	0.77 ± 0.067	0.096 ± 0.008	2423.00 ± 222.086	18.40 ± 1.69	446.0 ± 46.147	5.62 ± 0.420	0.023 ± 0.003	0.091 ± 0.018
Biosolids	1.42 ± 0.18	0.50 ± 0.00058	0.18 ± 0.008	0.098 ± 0.032	590.50 ± 15.849	12.63 ± 0.422	87.46 ± 3.484	4.92 ± 0.113	0.018 ± 0.002	0.069 ± 0.013
Soil with no amendment	1.98 ± 0.056	6.52 ± 0.32	0.60 ± 0.053	0.043 ± 0.023	1072.00 ± 58.501	14.20 ± 0.607	81.67 ± 4.175	3.58 ± 0.238	0.031 ± 0.006	0.058 ± 0.016

Values: mean ± SD

* Not significantly different

** Significantly different

dietary intake of Zn should not be above 150 μg/g (Alia et al. 2015; Jung 2008). In humans, excess amounts of Zn may cause metal poisoning and retardation of growth especially in young children (Sithole et al. 2016) and in plants it may inhibit plant growth, cause damage and chlorosis in leaves resulting in a reduction in chlorophyll (Alia et al. 2015).

Mn concentration varied between 30.09 ± 0.57 and 446.00 ± 46.15 mg/kg. The leaves of *S. oleracea* harvested from the soil treated with chemical fertilizers showed the highest concentrations of Mn while the least concentration was in the roots harvested from the soil with no amendment. The concentration of Mn in *S. oleracea* harvested from the soil treated with urine and other amendments was lower than the permissible limit of 500 mg/kg as recommended by WHO (2001). According to Dingkwoet et al. (2013), *S. oleracea* has been shown to be an excellent source of Mn in food. In plants, Mn plays a role in the splitting of water molecules which are essential for photosynthesis (Alia et al. 2015). In the human body, Mn is important for the manufacturing of enzymes even though in excess it has a negative impact on the enzyme activity, translocation, absorption and consumption of other mineral elements resulting in oxidative stress (Lion and Olowoyo 2013).

The concentration of Copper (Cu) in the roots, stalks and leaves was 5.04 ± 0.22–30.9 ± 0.99 mg/kg. The roots of *S. oleracea* harvested from soil treated with biosolids showed the highest concentration of Cu while stalks of *S. oleracea* harvested from soil treated with urine showed the least concentration of Cu. There were no significant differences in the concentration of Cu ($p < 0.05$). The decreasing trend of the concentration of Cu for all amendments was leaves > roots > stalks. The concentration of Cu in the stalks of *S. oleracea* harvested from the soil treated with urine was below the critical limit of 10.00 μg/g as set by WHO (Sani et al. 2011; Iqbal et al. 2011), whereas the concentration of Cu in the leaves harvested from soil treated with urine was above the recommended value. The result of Cu concentration in the edible parts (stalks and leaves) in the present study are much lower compared to the level of 37.62 mg/kg in *S. oleracea* harvested from soil under agricultural activities reported in Agrawal et al. (2007). Cu is an essential cofactor for most metalloproteins and plays a role in the manufacturing of enzymes which work as antioxidants and excess amounts of Cu may result in the malfunctioning of enzymes (Singh and Taneja 2010).

The concentration of Cr ranged from 0.65 ± 0.03 to 4.54 ± 0.12 mg/kg (Tables 1, 2, 3). The highest concentration of Cr was present in the roots of *S. oleracea* harvested from soil treated with biosolids, whereas the least concentration of Cr was found in the stalks harvested from soil treated with urine. The concentration of Cr in the leaves harvested from soil treated with urine was above the recommended level of 1.30 mg/kg by WHO (2001) while the concentration in the stalks harvested from soil treated with urine was lower than the recommended value. In humans and animals, Cr is considered to be important for the glucose and protein metabolic processes and it also prevents the increase of the triglycerides and cholesterol. In plants, it promotes growth and in low concentrations it increases plant yield (Peralta-Videa et al. 2009). The translocation and absorption of Cr is mainly altered by the chelating agents, pH and OM of the soil (Peralta-Videa et al. 2009).

The concentration of lead (Pb) in the roots, stalks and leaves of *S. oleracea* ranged from 0.31 ± 0.02 to 6.52 ± 0.32 mg/kg (Tables 1, 2, 3). The highest concentration of Pb was recorded from the leaves of *S. oleracea* harvested from the soil with no amendment while the least concentration was recorded from the stalks of *S. oleracea* harvested from soil treated with biosolids. There was no significant difference in the concentration of Pb ($p < 0.05$). The concentration of Pb in the roots, stalks and leaves of *S. oleracea* harvested from soil treated with urine was below the recommended level of 2.00 μg/g for edible portions of vegetables as set by WHO (Sani et al. 2011; Iqbal et al. 2011). Among the trace metals, Pb is the most highly toxic trace metal to people through the food chain since it is persistent and does not have any biological value (Tang et al. 2015).

Cadmium (Cd) concentration in the leaves, stalks and roots varied between 0.09 ± 0.014 and 2.42 ± 0.098 mg/kg. The leaves of *S. oleracea* harvested from soil treated with urine showed the significantly highest concentration of Cd ($p < 0.05$) which was also higher than and the permissible limit of 0.02 mg/kg in plants (Iqbal et al. 2011) while the least concentration of Cd was in the stalks of *S. oleracea* harvested from soil treated with biosolids. The concentration range of Cd in the present study exceeded the recommended concentration of Cd in edible vegetables which should range between 0.05 and 0.9 μg/g (Jung 2008). Exposure to high concentrations of Cd result in decreased rate of growth, hypertension, anaemia, poor mineralization of bones and damage to the renal tubules (Iqbal et al. 2011). In the present study, the result of the range of the concentration of Cd was also slightly higher than the concentration range of 0.02 ± 0.00–0.28 ± 0.02 μg/g in Lion and Olowoyo (2013) where *S. oleracea* was harvested from waste dump sites.

The concentration of Nickel (Ni) ranged from 1.50 ± 0.026 to 14.82 ± 0.20 mg/kg. The highest concentration of Ni was present in the roots harvested from soil treated with biosolids while the least concentration was in the stalks harvested from soil treated with urine. The concentration of Ni in the roots, stalks and leaves harvested from soil treated with urine were below the recommended limit of 10.0 mg/kg. Ni is a trace metal which occurs only at very low concentrations in the environment and it is essential in small quantities even though it can be dangerous when the maximum tolerable amounts are exceeded (Wuana and Okieimen 2011) resulting in blindness, increased level of sugar in the blood and cholesterol in the serum (Iqbal et al. 2011).

The pH values of the soil treated with different amendments measured in water were slightly neutral to alkaline and ranged between 6.92 ± 0.067 and 7.31 ± 0.023 with a mean value of 7.16, whereas those measured in $CaCl_2$ were slightly acidic to sub-alkaline and ranged between 6.57 ± 0.038 and 7.03 ± 0.025 with a mean value of 6.80. From both methods of measurements, the soil with no amendment had the highest pH followed by the soil amended with biosolids and then the urine treatment soil with the least pH recorded for the soil treated with chemical fertilizers. According to Tang et al. (2015), pH is one of the most crucial factors which affect the availability of trace metals in the soil.

Table 4 shows the concentration of the trace metals in the soil treated with different amendments and used for cultivating *S. oleracea*. The concentrations of Cd, As and Hg were higher in the soil without the amendment compared to the soil which received the amendments and this could have been due to the nature of the soil and the *S. oleracea* harvested from the soil with no amendments might have not absorbed the trace metals from the soil.

The transfer factors (TF) of the stalks and leaves which were calculated from the concentration of the trace metals in the soil are shown in Tables 5 and 6, respectively. The leaves of *S. oleracea* harvested from soil treated with urine bioaccumulated Cd, Zn, Mn and Sb as the TF values of these trace metals were greater than 1. However, the stalks harvested from soil treated with urine did not bioaccumulate any of the trace metals indicating that the trace metals which were bioaccumulated in the leaves were absorbed from the soil by the roots and translocated to the leaves.

Conclusion

The results from the study indicated that urine performed favourably well as a soil amendment in comparison with the other conventional soil amendments used in terms of the concentration of the trace metals in *S. oleracea*. The concentration of Mn, Pb and Ni in the leaves and stalks of *S. oleracea* harvested from soil treated with urine were below the recommended limits for edible vegetables as set out by WHO. The stalks harvested from soil

Table 4 Trace metals of soil samples treated with different soil amendments for cultivation of *Spinacia oleracea*

Treatments	Cr (mg/kg)**	Pb (mg/kg)**	Cd (mg/kg)**	As (mg/kg)**	Zn (mg/kg)**	Cu (mg/kg)**	Mn (mg/kg)**	Ni (mg/kg)**	Sb (mg/kg)*	Hg (mg/kg)**
Urine	114.83 ± 2.58	13.41 ± 0.05	0.068 ± 0.012	3.53 ± 0.05	102.43 ± 0.29	33.47 ± 0.29	209.13 ± 3.15	49.02 ± 0.72	0.0077 ± 0.0015	0.286 ± 0.013
Chemical fertilizers	123.67 ± 1.93	14.77 ± 3.96	0.074 ± 0.008	3.96 ± 0.07	92.71 ± 1.46	37.21 ± 0.32	177.4 ± 2.10	55.39 ± 0.88	0.0097 ± 0.0012	0.236 ± 0.024
Biosolids	113.97 ± 0.70	14.04 ± 0.18	0.074 ± 0.007	3.73 ± 0.33	84.83 ± 0.66	34.48 ± 0.60	215.20 ± 3.71	52.34 ± 0.56	0.0087 ± 0.0015	0.169 ± 0.037
Soil with no amendment	91.27 ± 1.41	11.60 ± 0.09	0.08 ± 0.010	4.60 ± 0.42	70.85 ± 1.50	27.93 ± 0.12	218.33 ± 1.19	47.8 ± 0.29	0.0073 ± 0.0060	0.174 ± 0.012

Values: mean ± SD

* Not significantly different

** Significantly different

Table 5 Translocation factors (TF) of *Spinacia oleracea* stalks harvested from soil treated with different soil amendments

Treatments	Cr	Pb	Cd	As	Zn	Cu	Mn	Ni	Sb	Hg
Urine	0.06	0.50	0.60	0.10	0.41	0.29	0.34	0.16	0.29	0.17
Chemical fertilizers	0.86	0.51	0.54	0.64	0.81	0.48	0.55	0.43	0.22	0.17
Biosolids	0.32	0.08	0.45	0.04	0.34	0.21	0.16	0.17	0.19	0.03
Soil with no amendment	1.35	0.43	1.42	1.04	2.75	0.67	0.47	1.00	0.61	0.25

Table 6 Translocation factors (TF) of *Spinacia oleracea* leaves harvested from soil treated with different soil amendments

Treatments	Cr	Pb	Cd	As	Zn	Cu	Mn	Ni	Sb	Hg
Urine	0.64	0.47	2.12	0.09	2.33	0.98	2.96	0.85	0.33	0.30
Chemical fertilizers	0.90	1.20	1.81	1.01	5.38	1.19	3.05	0.79	0.39	0.31
Biosolids	0.31	0.13	0.85	0.28	1.03	0.41	0.34	0.33	0.07	0.06
Soil with no amendment	1.46	7.47	5.62	0.35	3.33	1.48	1.27	1.07	1.00	0.38

treated with urine showed the lowest concentrations of Cr, Zn, Cu and Ni while the leaves harvested from urine showed the least concentration of Pb compared to the stalks and leaves harvested from soil treated with other amendments. Even though the biosolids performed slightly better than urine and showed the least concentrations of Cr, Cd, Zn, Cu in the leaves and Pb, Cd and Hg in the stalks, the biosolids are not as easily available and affordable as human urine which is readily available at no cost in all the communities. It is hence concluded that the urine amendment can be recommended for the growing of *S. oleracea.* It is further recommended that the health effects of the Pb and Cd in the stalks and Cd, Ni and Hg in the leaves harvested from soil treated with urine which were present in highest concentrations compared to the stalks and leaves harvested from other amendments be further investigated.

Acknowledgements The authors would like to thank the staff, students and management at the Sefako Makgatho Health Sciences University and the Agriculture Research Council (Institute for Soil, Climate and Water).

References

Agrawal SB, Singh A, Sharma RK, Agrawal M (2007) Bioaccumulation of heavy metals in vegetables: a threat to human health. Terrestrial and Aquatic Environmental Toxicology @ 2007 Global Science Books

Akpan-Idiok AU, Udo IA, Braide EI (2012) The use of human urine as an organic fertilizer in the production of okra (*Abelmoschus esculentus*) in South Eastern Nigeria. Resour Conserv Recy 62:14–20

Pradhan SK, Piya RC, Heinonen-Tanski H (2010b) Eco-Sanitation and its benefits: an experimental demonstration program to raise awareness in central Nepal. Environ Dev Sustain. doi:10.1007/s10668-010-9273-5

Alia N, Sardar K, Said M, Salma K, Sadia A, Sadaf S, Toqeer A, Miklas S (2015) Toxicity and bioaccumulation of heavy metals in spinach (*Spinacia oleracea*) grown in a controlled environment. Int J Environ Res Public Health 12:7400–7416

Cofie O, Adeoti A, Nkansah-Boadu F, Awuah E (2010) Farmers' perception and economic benefits of excreta use in Southern Ghana. Resour Conserv Recy 55:161–166

Dingkwoet DJ, Danladi SM, Gabriel MS (2013) Comparative study of some heavy and trace metals in selected vegetables from four local government areas of Plateau State, Nigeria. J Environ Sci Toxicol Food Technol 6:86–93

Egigu MC, Melak B, Kebede A, Muthuswamy M (2014) Use of human urine as fertilizer for vegetation cultivation. Int J Agric Innov Res (IJAIR) 3(1):254–258

Farooq M, Anwar F, Rashid U (2008) Appraisal of heavy metal contents in different vegetables grown in the vicinity of an industrial area. Pak J Bot 40:2099–2106

Germer J, Addai S, Sauerborn J (2011) Response of grain sorghum to fertilization with human urine. Field Crop Res 122:234–241

Guzha E, Nhape I, Rockstrom J (2005) An assessment of the effects of human faeces and urine on maize production and water productivity. Phys Chem Earth 30:840–845

Heinonen-Tanski H, Van Wijk-Sijbesma C (2005) Human excreta for plant production. Bioresour Technol 36:403–411

Heinonen-Tanski H, Sjoblom A, Fabritus H, Karinen P (2007) Pure human urine is a good fertilizer for cucumbers. Bioresour Technol 98:214–217

Hoglund C, Ashbolt N, Stenstrom TA, Svensson L (2002) Viral persistence in source-separated human urine. Adv Environ Res 6:265–275

Iqbal MA, Chaudhary MN, Zaib S, Imran M, Ali K, Iqbal A (2011) Accumulation of heavy metals (Ni, Cu, Cd, Cr, Pb) in agricultural soils and spring seasonal plants, irrigated by industrial waste water. J Environ Tech Manage 2(1):1–9

Jensen PKM, Phuc PD, Knudsen LG, Dalsgaard A, Konradsen F (2007) Hygiene versus fertilizer: the use of human excreta in agriculture-a Vietnamese example. Int J Hyg Environ Health 211:432–439

Jung MC (2008) Heavy metal concentration in soils and factors affecting metal uptake by plants in the vicinity of a Korean Cu-W Mine. Sensors (Basel) 8:2413–2423

Kidd PS, Domìnguez-Rodrìguez MJ, Monterroso C (2007) Bioavailability and plant accumulation of heavy metals and phosphorus in agricultural soils amended by long-term application of sewage sludge. Chemosphere 66:1458–1467

Kirchmann H, Petterson S (1995) Human urine—chemical composition and fertiliser use efficiency. Fert Res 40:149–154

Kutu FR, Muchaonyerwa P, Mnkeni PNS (2010) Complementary nutrient effects of separately collected human faeces and urine on the yield and nutrient uptake of (*Spinacia oleracea*). Waste Manag Res 29:532–539

Lindt B, Ban Z, Byde'n S (2000) Nutrient recovery from human urine by struvite crystallization with ammonia adsorption on zeolite and Wollastonite. Bioresour Technol 73:169–174

Lion GN, Olowoyo JO (2013) Population health risk due to dietary intake of toxic heavy metals from *Spinacia oleracea* harvested from soils collected in and around Tshwane, South Africa. S Afr J Bot 88:178–182

Mnkeni PNS, Austin A, Kutu FR (2005) Preliminary studies on the evaluation of human urine as a source of nutrients for vegetables in the Eastern Cape Province, South Africa. In: Ecological Sanitation: A Sustainable Integrated Solution. Proceedings of the 3rd International Ecological Sanitation Conference, Durban, South Africa, 23–26 May 2005, pp 418–426. http://conference2005.ecosan.org/papes/mnkeni.etal.pdf

Mnkeni PNS, Kutu FR, Muchaonyerwa P (2008) Evaluation of human urine as a source of nutrients for selected vegetables and maize under tunnel house conditions in the Eastern Cape, South Africa. Waste Manag Res 26:132–139

Mohod CV (2015) A review on the concentration of the heavy metals in vegetable samples like spinach and tomato grown near the area of Amba Nalla of Amravati City. Int J Innov Res Sci, Eng Technol. doi:10.15680/IJIRSET.2015.0405019

Olowoyo JO, Okedeyi OO, Mkolo NM, Lion GN, Mdakane STR (2011) Uptake and translocation of heavy metals by medicinal plants around a waste dump site in Pretoria, South Africa. S Afr J Bot 78:116–121

Olowoyo JO, Mugivhisa LL, Busa NG (2015) Trace metals in soil and plants around a cement factory in Pretoria, South Africa. Pol J Environ Stud 24(5):2087–2093

Paschal DC, Ting BG, Morrow JC (1998) Trace metals in urine of United States residents: reference range concentrations. Environ Res 76:53–59

Peralta-Videa JR, Lopez ML, Narayan M, Saupe G, Gardea-Torresdey J (2009) The biochemistry of environmental heavy metal uptake by plants: implications for the food chain. Int J of Biochem Cell Biol 41:1665–1677

Pradhan SK, Pitkanen S, Heinonen-Tanski H (2009a) Fertilizer value of urine in pumpkin (*Cucurbita maxima* L.) cultivation. J Agric Food Sci 18:57–68

Pradhan SK, Holopainen JK, Weisell J, Heinonen-Tanski H (2009b) Stored human urine supplemented with wood ash as fertilizer in tomato (*Solanum lycopersicum*) cultivation and its impacts on fruit yield and quality. J Agric Food Chem 57:7612–7617

Pradhan SK, Holopainen JK, Weisell J, Heinonen-Tanski H (2010a) Human urine and wood ash as plant nutrients for red beet (*Beta vulgaris*) cultivation: impacts on yield quality. J Agric Food Chem 58:2034–2039

Ranasinghe ESS, Karunarathne CLSM, Beneragama CK, Wijesooriya BGG (2016) Human urine as a low cost and effective nitrogen fertilizer for bean production. Procedia Food Sci 6:279–282

Salawu K, Barau MM, Mohammed D, Mikailu DA, Abdullahi BH, Uroko RI (2015) Determination of some selected heavy metals in spinach and irrigated water from Samaru area within Gusau Metropolis in Zamfara State, Nigeria. J Toxicol Environ Health Sci 7(8):76–80

Sani HA, Tsafe AI, Bagudo BU, Itodo AU (2011) Toxic metals uptake by Spinach (*Spinacia oleracea*) cultivated in Sokoto: a comparative study. Pak J Nutr (PJN) 10(6):572–576

Sene M, Ken NH, Ushijima K, Funamizu N (2013) Effects of extra human urine volume application in plant and soil. Glob J Pests Dis Crop Prot (GJPDCP) 1:051–060

Singh KB, Taneja SK (2010) Concentration of Zn, Cu and Mn in vegetables and meat foodstuff commonly available in Manipur: a North Eastern state of India. Electron J Environl Agric Food Chem 3:610–616

Sithole SC, Mugivhisa LL, Olowoyo JO (2016) Pattern and concentrations of trace metals in mushrooms harvested from trace metal-polluted soils in Pretoria, South Africa. S Afr J Bot. doi:10.1016/j.sajb.2016.08.010

Tang X, Li X, Liu MZ, Hashmi J, Xu, Brookes PC (2015) Effects of inorganic and organic amendments on the uptake of lead and trace elements by *Brassica chinensis* grown in acidic red soil. Chemosphere 119:177–183

Upreti HK, Shrestha P, Paudel P (2011) Effect of human urine as fertilizer on crop production. Agron J Nepal 2:168–172

Uwah EL, Ndahi NP, Abdulrahman FI, Ogugbuaja VO (2011) Heavy metal levels in Spinach (*Amaranthus caudatus*) and Lettuce (*Lactuca sativa*) grown in Maiduguri, Nigeria. J Environ Chem Ecotoxicol 3(10):71–78

Winblad U, Simpson-Hébert M (2004) Ecological Sanitation—revised and enlarged edition. Stockholm, Sweden: Stockholm Environmental Institute, pp. 14–22. Available at: http://docs.zetatalk.com/Sanitation/Ecological-Sanitation-2004.pdf. Accessed January 2008

World Health Organization (WHO) (2001) Toxicological Evaluation of Certain Food Additives. Joint FAO/WHO Expert Committee on Food Additives. Food Additive Series No. 683. World Health Organization, Geneva

Wuana RA, Okieimen FE (2011) Heavy metals in contaminated soils: a review of sources, chemistry, risks and best available strategies for remediation. International Scholarly Research Network (ISRN). Ecology. doi:10.5402/2011/402647

Zeka N, Mero F, Skenderasi B, Gjanci S (2014) Effects of nitrogen sources and levels on yield and nutritive values of spinach (*Spinacia oleracea* L.). J Int Acad Res Multidiscip (IJARM) 2(2):327–337

13

Growth and proximate composition of *Amaranthus cruentus* L. on poor soil amended with compost and arbuscular mycorrhiza fungi

Oyeyemi A. Dada[2] · Francis Imade[1] · E. M. Anifowose[2]

Abstract

Purpose The study was carried out to examine growth, shoot yield, dry matter and proximate composition of *Amaranthus cruentus* on poor soil augmented with compost or AMF either singly or in combination.

Methods The experiment was arranged in completely randomized designed in six replicates with four treatments. Four treatments: comprised control and three other amendment types derived from the application of compost made from cattle dung and maize stover, arbuscular mycorrhiza fungi singly or in combination with compost. The treatments were applied a week before sowing to allow for proper mineralization. Growth characteristics, chlorophyll content, ascorbic acid content and proximate composition were assessed.

Results The results revealed that the compost supplied sufficient plant nutrients needed for improving biological and economic yields of *Amaranthus cruentus* . Application of compost significantly ($P \leq 0.05$) influenced growth, dry matter and fresh shoot yield of *A. cruentus*. Applying of combination AMF and compost to nutrient limiting soil had no significant ($P \geq 0.05$) effect on yield and yield components of *A. cruentus*. Proximate composition of *A. cruentus* was significantly enhanced in pots augmented with compost better than pots amended with the combination of AMF and compost.

Conclusions Application of compost to nutrient deficient soil promoted growth, fresh shoot and dry matter yield of *A. cruentus*. Similarly, proximate composition of the crop was appreciably influenced by compost application.

Keywords *Amaranthus cruentus* · AMF · Compost · Proximate composition

✉ Francis Imade
imadefrancis@gmail.com;
imadefrancis@aauekpoma.edu.ng

Oyeyemi A. Dada
oa.dada@mail.ui.edu.ng; oadada247@yahoo.com

[1] Department of Botany, Faculty of Life Sciences, Ambrose Alli University, Ekpoma, Edo State, Nigeria

[2] Department of Botany, University of Ibadan, Ibadan, Oyo State, Nigeria

Introduction

Amaranthus collectively known as amaranth is a leafy vegetable of high dietary value produced and consumed most parts of sub Sahara Africa, particularly in Nigeria (Tindall 1983). Amaranthus has diverse health advantages such as therapeutic value on cardiovascular diseases (Martirosyan et al. 2007), rich in phytosterols which reduce the cholesterol levels and also prevent cancer (Su et al. 2002).

The genus *Amaranthus* has received considerable research attention in many countries because of high nutritional value of some species that are important source of food, either as vegetable or grain. The leaves contain 17.5–38.3% dry matter protein of which 5% is lysine (Oliveira and de Carvallo 1975). Vitamin A and C are also found in significant level. Muloskozi et al. (2004) reported that a total of 100 g of vegetable material cooked without oil can contribute 45% of daily vitamin A requirement. Despite its nutrients advantages to consumers, many factors still limit its growth, particularly climatic conditions and soil characteristics. Growth of *Amaranthus cruentus* L. is often affected by low or unreliable rainfall (Cunningham et al. 1992), low level of soil nutrient, among other abiotic constraints.

Nutrients availability in soil is greatly affected by low pH, coarse texture, soils and by leaching thereby limiting base cations such as Ca^{+}, $Mg^{+}$and K^{+}. Improving and maintaining soil structure is known to aid nutrient retention and soil biological health. Although, nitrogen is the most common limiting nutrient in most soil (Craine and Jackson 2010). Manure or organic fertilizers are valuable fertilizer for any farming operation due to their ability to supply needed nutrients for crop growth. The use of manure has, however, declined generally on many farms over the past fifty years (Kato and Place, 2011). This is due to several factors, chiefly among them are farm specialization with increasing separation of crop and livestock production and cost of transporting bulky manure. Despite these limitations, manure is beneficial to soil and crop grown on it (Carl and Bierman 2005). Organic fertilizer including compost do not only supply many nutrients for crop production including micronutrients but they are also valuable sources of organic matter. Application of organic fertilizers helps increasing water holding capacity of coarse-texture sandy soils, ensures slow release of nutrients and also enhances growth of earthworms and other beneficial soil organisms. Most leafy vegetables return small amount of crop residue to soil, thus application of compost and other organic amendments help to maintain soil organic levels (Carl and Bierman 2005; Oworu and Dada 2009; Ojo et al. 2014).

Arbuscular mycorrhizal Fungi (AMF) play important role in acquisition of mineral nutrients, especially the slowly mobile ions such as phosphorus (Gazey et al. 2004). It has also been shown that plants grown on AMF augmented soil are more resistant to biotic and abiotic stresses in nature (Karaki and Al Radded 1997; Silva-Sánchez et al. 2008). AMF constitutes an important group of symbionts associated with agricultural crops. This symbiosis may enhance the root systems ability to absorb and carry phosphorus and other soil elements of low mobility by means of a network of mycelia, thus promoting optimum plant growth (Raiman et al. 2007).

Utilization of AMF in augmenting poor soil is gaining wide research attention (Ibijbijen et al. 2006). Many researchers have observed different effects of inoculation of free living diazotrophs and AMF on different species of cultivated plants (Pérez-Montãno et al. 2014). Results have indicated that the interaction between these two groups of microorganisms, in many instances, result in improved plant growth and fungal infectivity or stimulate the effectiveness of the free-living diazotroph (Barea et al. 2002). There is, however, sparse information on effect combination of AMF and compost in improving fertility status of nutrient deficient soil. This study, therefore, conducted to investigate growth, dry matter and yield of *Amaranthus cruentus* L. grown on poor soil augmented with *Arbuscular mycorrhiza* Fungus (*Glomus intraradices*) and cattle dung—maize residue based compost (Maize Stover + Cattle Dung) used singly or in combination.

Materials and methods

Study site

The study was a pot experiment carried out in the greenhouse of the Department of Botany, Faculty of Science, University of Ibadan, Nigeria. The geographical area of the site is located at 7°26′N and 3°54′E at the elevation 215 m of the sea level. Polythene pots with a diameter of 14 and 18.5 cm in depth was used for the study. The pots were each filled with 5 kg sterilized soil.

Composting materials and inoculum source

The materials used for composting were fresh cattle dung collected from Faculty of Veterinary Medicine while maize stover were obtained from the University Teaching and Research Farm both in University of Ibadan, Nigeria. Composts were prepared from cattle dung, a well cured-poultry manure and maize Stover. The proportion of the maize stover to the animal wastes (cattle dung) was 3:1 on dry weight basis. Each of the organic waste was laid out in layers, a layer made up of 30 kg maize stover and 10 kg of either cattle dung. Ten of such layers made up a heap of about $3.1 \times 1.0 \times 1.0\ m^3$ for the respective compost. These were prepared using concrete described by Dada and Aminu (2013). *Glomus intraradices* an Arbuscular Mycorizal Fungus (AMF) used for the study was obtained from the Department of Botany, University of Ibadan, Ibadan.

Soil and compost analysis

The soil was collected (0–15 cm depth) using a soil auger from Practical Year Training Programe (PYTP) field of the Faculty of Agriculture and Forestry, University of Ibadan, Nigeria. The PYTP field is continuously cultivated with maize and leafy vegetables used for training agricultural students in their penultimate year. The soil and matured compost samples were air-dried under shade for 2 weeks before taken for analysis at Department of Agronomy, University of Ibadan. Both samples were analysed for pH as described by Mckeague (1978) total nitrogen using Kjeldahl procedure followed by distillation described by Bremner and Mulvaney (1982); Organic carbon using the method described by Walkley (1947); Available phosphorus and extractable cations such as Ca^{2+}, Mg^{2+} and Fe^{2+} by atomic absorption spectrophotometer. Also physical properties such as sand, silt, and clay and bulk density of soil sample were determined using procedures described by International Institute of Tropical Agriculture (IITA 2002).

Plant material and sowing

Seeds of *Amaranthus cruentus* L. were obtained from Department of Agronomy. The seeds were sown using broadcast method. A week after sowing, it was thinned to ten plants per pot. Weeds were rouged regularly and the pots were watered as needed.

Treatments and experimental design

There were four treatments comprising of a control, application of compost, AMF singly or in combination with compost. The four treatments were: control (No amendment), 2.5 g AMF + 2.5 g compost ($\equiv 4$ t ha^{-1}), 5 g of AMF and 5 g compost ($\equiv 8$ t ha^{-1}). The collected soil was sterilized at 1.05 kg/cm^2 (121 °C) for 15 min using an autoclave. The treatments were applied a week before sowing to allow for proper mineralization. The four treatments were arranged in a Completely Randomized Design (CRD) with six replicates.

Data collection

Data collection commenced 2 weeks after sowing (WAS) and continued until final harvest (6WAS). Data were collected from two tagged plants at the middle of each pots on number of leaves and plant height (cm), fresh shoot yield (g) and dry matter (g) components. At 4, 5 and 6 WAS one plant was uprooted per treatment to determine the relative growth rate. At final harvest, the tagged plants were uprooted to determine the weight of fresh shoot. The plant parts were kept in an envelope and oven dried to a constant weight at 80 °C after which they were weighed on a sensitive balance to evaluate the dry matter yield.

Chlorophyll content determination

Chlorophyll contents were determined using the method of Poora (2002). Leaf samples from each treatment were randomly harvested and weighed. Leaf sample was cut into smaller chips and placed in a mortar. 1 g of fresh plant sample was crushed in acetone with pestle and mortar. The mixture was made up to 100 ml volume in a volumetric flask. Absorbance of each sample was read using Atomic Absorption Spectrophotometer (AAS Model SP9) at the desired wavelength of chlorophyll *a*, chlorophyll *b* and total chlorophyll was determined using the formulae below.

$$\text{Chlorophyll } a\,(\mu g/ml) = 12.25(A663.6) - 2.55(A646.6)$$
$$\text{Chlorophyll } b\,(\mu g/ml) = 20.31(A646.6) - 20.31(A663.6)$$
$$\text{Total chlorophyll }(\mu g/ml) = 17.76(A646.6) + 7.34(A663.6).$$

Determination of abscorbic acid

10 g of the slurry sample was weighed into a 100 ml volumetric flask and diluted to 100 ml with 3% meta phosphoric acid solution (0.0033 M EDTA). The diluted samples were filtered using a Whatman filter paper No. 3. The filtrate (10 ml) was pipette into a small flask and titrated immediately with a standardized solution of 2.6 dichlorophenol-in-dephenol to a faint pink end point. The ascorbic acid content of the fruit was calculated from the relationship below.

$$\frac{V \times T}{W \times 100}(\text{mg ascorbic acid per 100g sample}),$$

where: $V =$ ml dye used for titration of aliquot of diluted sample. $T =$ ascorbic acid equivalent of dye solution expressed as mg per ml of dye. $W =$ gram of sample in aliquot titrated.

Proximate composition

The proximate constituents of the crop were determined according to AOAC (1998). All analyses were carried out in triplicate.

Data analysis

Data were analysed with analysis of variance (ANOVA) using a General linear model (GLM) of statistical analysis system (SAS 2002). Differences in means were compared using Least Significant Difference (LSD) at $P \leq 0.05$.

Results

The results of the pre-trial physico-chemical properties of the soil and compost revealed that the compost was rich in both macro and micro elements. The total N (4.6 g/kg), available P (15.3 mg/kg), OC (46.1 g/kg) and K (0.6 cmol/kg) in the compost were higher than those in the soil. The mineral constituents N (1.9 g/kg), K (0.4 Cmol/kg) in the soil were relatively below the critical limit for growing *Amaranthus cruentus* Table 1.

Influence of various treatments on number of leaves produced by *A. cruentus* on soil deficient in nutrients

Table 1 Chemical characteristics of compost and soil (0–15 cm) used for the study Source: Dada and Aminu (2013)

	pH (H_2O)	OC	N	P	Na	Ca	Mg	K	Fe	Silt	Clay	Sand	Bulk density
Soil	6.5	18.4	1.9	13.0	0.4	1.2	0.7	0.4	58	6.1	10.8	83.1	1.7
Compost[a]	6.5	46.1	4.6	15.3	0.4	1.4	0.8	0.6	167	nd	nd	nd	nd

nd not determined

[a] Dry weight basi

augmented with various soil amendments either singly or in combination is shown in Fig. 1a. The results showed that there was no significant ($P > 0.05$) difference in number of leaves of *Amaranthus* grown on nutrient deficient soil treated with different soil amendments. However, pots treated with AMF+ Compost had highest number of leaves (9.67).

Height of *Amaranthus* was influenced by various types of soil amendments (Fig. 1b). Pots treated with compost produced the tallest (16.42 cm) plants while shortest plants (14.07 cm) were observed in pots treated with AMF although there was no significant difference in the response of the crop to different types of amendments with respect to height.

There was significant ($P \leq 0.05$) difference in the fresh weight of *A. cruentus* on poor soil treated with different amendments (Fig. 2a). Pots augmented with 8 t ha^{-1} of compost (CDMS) produced highest (5.59 g) fresh weight whereas, least fresh weight (3.19 g) was produced in pot augmented with AMF. Application of 8 t ha^{-1} compost promoted better dry matter production compared to other amendment treatments (Fig. 2b).

The results obtained for the relative growth rate was not significantly ($P > 0.05$) different among the different amendments. Application of different amendments to poor soil had no significant effect on relative growth rate of *A. cruentus* grown on such soil. However, the crop had better relative growth rate (0.26 g day^{-1}) in pots augmented with compost than pot amended with AMF which had the least (0.17 g day^{-1}) rate (Fig. 3a). Application of different organic amendment on poor soil had significant influence on chlorophyll content of *A. cruentus*. Highest chlorophyll content (0.89 µ/100 g) was recorded in pots treated with 8 t ha^{-1} of CDMS while pots treated with the combination of AMF and CDMS had least (0.76 µ/100 g) chlorophyll content (Fig. 3b).

Generally, the proximate composition of *A. cruentus* grown on poor was significantly influenced by soil amendments. The results obtained for the determination of crude protein content showed significant ($P < 0.05$) difference among the different amendments. *Amaranthus* treated with 8 t ha^{-1} compost had highest crude protein (16.33%) while least crude protein content was recorded in AMF treated pots (Table 2).

Crude fibre content of *A. cruentus* on poor soil was significantly influenced by the amendments (Table 2). Application of different organic amendments to poor soil had significant influence on fibre content of *A. cruentus* whereas application of combination of AMF+ compost treated pots had the highest (16.85%) crude fibre content while least fibre (15.31%) was observed in AMF treated pots.

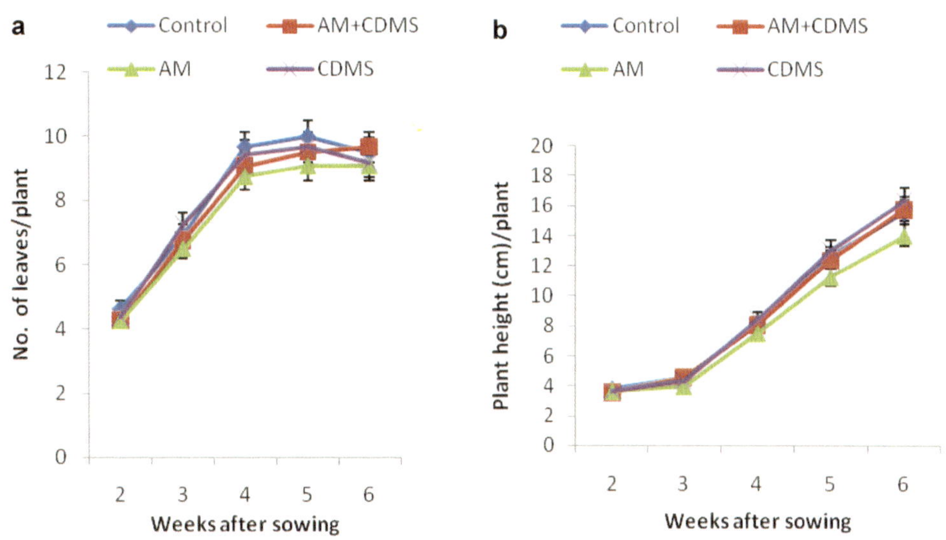

Fig. 1 **a** Influence of various soil amendments on number of leaves of *A. cruentus*. **b** Influence of various soil amendements on height (cm) of *A. cruentus*

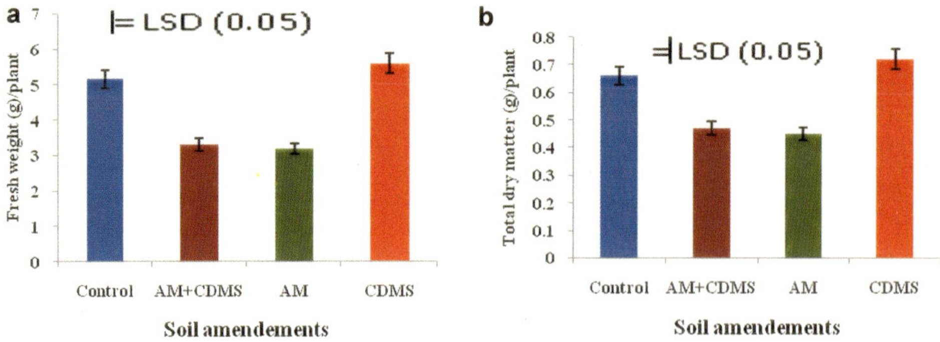

Fig. 2 **a** Influence of various soil amendements on weight of fresh shoot (g) of *Amaranthus cruentus*. **b** Influence of various soil amendements on biomass yield (g) of *A. cruentus*

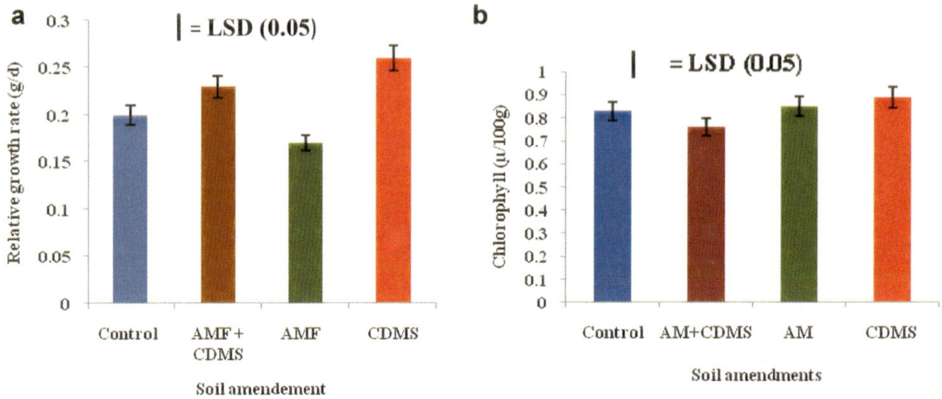

Fig. 3 **a** Influence of different amendments on relative growth rate of *A. cruentus*. **b** Influence of different amendments on chlorophyll contents of *A. cruentus*

Table 2 Effect of application of different organic amendments on proximate compositions of *Amaranthus cruentus* L. grown on nutrient deficient soil

Organic amendments ($t\ ha^{-1}$)	Percent (%)					Ascorbic acid (mg/g)
	Crude protein	Crude fibre	Fat	Ash	Moisture content	
CDMS	16.33	16.70	5.80	12.36	13.41	14.61
AM + CDMS	15.93	16.85	4.97	11.69	12.9	12.41
AM	14.88	15.31	4.78	11.25	11.94	13.93
Control	15.74	16.77	5.08	12.19	12.94	14.49
Mean	15.72	16.41	5.16	11.87	12.80	13.86
CV	0.17	2.22	0.23	0.29	0.25	0.14
LSD ($P \leq 0.05$)	0.09	1.16	0.04	0.11	0.10	0.06

CDMS compost, *AM* arbuscular mycorrhiza

There was significant difference in fat content of poor soil treated with different organic amendments. Application of 8 $t\ ha^{-1}$ of compost significantly enhanced fat content of *A. cruentus* on poor soil. Least fat (4.78%) content was observed in pots treated with the combination of AM + CDMS. There was significant difference in the ash content of *A. cruentus* on poor soil treated with different organic amendments (Table 2). Pots augmented with 8 $t\ ha^{-1}$ of CDMS produced highest (12.36%) ash content. The results of moisture content was significantly ($P < 0.05$) different among the different amendments. Highest (13.41%) moisture content was recorded in pots treated with 8 $t\ ha^{-1}$ of CDMS. AMF treated pot had least (11.94%) moisture content. The ascorbic acid contents was highest (14.61 mg/100 g) in CDMS treated pots whereas least (12.41 mg/100 g) was observed in pots treated with combination of AMF + CDMS (Table 2).

Discussion

Application of compost singly or in combination with AMF enhanced leaf formation in *Amaranthus cruentus* on poor soil. Compost and AMF had been reported to enhance plant growth, plant water stress tolerance (Cruz et al. 2000), nutrient cycling and soil quality (Ishii et al. 2004). This possibly explains why organic amendments promoted better growth of *A. cruentus* than in the control pots. The superior improvement in plant height on compost treated pot suggests that compost is likely to possess growth promoting substances which enhanced stem elongation and ultimately the plant height. Miezah et al. (2008) reported that growth substances such as IAA, GA3 and cytokinins were isolated from compost and co-compost. Carl and Bierman (2005) had reported that application of compost supply 70% organic nitrogen, phosphorus and potassium uptake require for improving plant growth. This simply explains reasons why compost augmented pots had better growth performance than other amendments.

Augmentation of poor soil with compost, promoted higher fresh leaf yield better than other amendments. This means that compost could serve as an alternative to expensive and scarce inorganic fertilizers (AdeOluwa and Akinyemi 2014) that could be used by poor resource leavy vegetable growers. The poor fresh shoot yield response of *Amaranthus cruentus* to AMF augmentation either singly or in combination with compost presumes a likelihood of antagonistic effects between AMF and compost.

Compost application to poor soil significantly promoted better production of fresh shoot weight which is the main economic yield of *A. cruentus*. Manios and Kepetanois (1992) had reported that nutrient supplied by farmyard manure affects and increases crop biomass yield. Our observation in this study implies, that the applied compost probably supplied sufficient nutrient which was taken up to synthesis vegetative and morphological development better than when AMF was applied singly or in combination with compost.

The impressive performance of *A. cruentus* with respect to total biomass accumulation in pots augmented with compost over other treatments showed that the applied compost supplied the needed nutrient as required by the crop for its metabolic functions. Several authors have reported that pattern of nutrient release by organic amendments is a vital factor which determines the plant growth and dry matter accumulation. (Akanbi and Togun 2002; Odeleye et al. 2007; Eifediyi et al. 2010).

Amaranthus cruentus had better relative growth rate in compost-augmented pots than in other treatments which perhaps might be due to steady release of nutrients as required by the crop. This invariably tends to enhance photosynthetic activities and hence performance of the crop better than other amendments. Similar observation had been reported on horticultural plants (Abad et al. (1997), on pineapple (Liu et al. 2013) and on tomato (Mashavira et al. 2015). The minimal relative growth rate observed in other treatments apart from compost treated pots might be due to utilization of available P and C for structural development by AMF at the expense of the plant, this agrees with the observation of Graham and Abbott (2000).

Amaranthus cruentus treated with compost had better proximate compositions possibly due to the availability of sufficient nutrients in the applied compost which were translocated into appropriate sources where they were partitioned into respective sinks. Akhtar et al. (2010) had reported linear relationship between quality of constituents of compost and vegetative development and nutrient uptake in lettuce and amaranthus. Similarly, superior chlorophyll content recorded in compost augmented pots suggests abundance of chlorophyll synthesis by the leaves and other tissues resulting from adequate nutrient supply and other precursors of chlorophyll. Bittenbender et al. (1998) and Naikwade (2014) in separates reports concluded that compost application to nutrient limiting soil improved quality and quantity of both biomass and proximate constituents of maize.

The poor performance of combination of AMF + compost suggests that the compost probably provided a good substrate for the AMF to thrive being an obligate saprophyte/symbionts. Consequently, the nutrients that could have been released for plant growth might have been used up by the organism. Since *A. cruentus* is a short duration crop, the time required for the AMF to release the nutrient for uptake might be behind the crop growth range. Hence, the crop might have been harvested before the nutrients were released thus not benefiting the crop. Similar observations had been made on cherry fruit by Stefano et al. (2004), on *Linum usitatissimum* by Cavagnaro et al. (2005) and on onions by Khang et al. (2011). Besides, it is not unlikely that the organism (*Glomus intraradices*) was a poor symbiont with *Amaranthus cruentus* whereby the organism used large amount of carbon from its host. Report of Burleigh et al. (2002) explained similar observations where they found that *Gi. rosea* was a very poor symbiont with *Lycopersicon esculentum* and *Medicago truncatula* compared with the other AMF tested.

Conclusion

Application of compost to nutrient deficient soil promoted growth, fresh shoot and dry matter yield of *A. cruentus*. Similarly, proximate composition of the crop was appreciably influenced by compost application. Applying combination of compost and AMF to augment poor soil may not be a good option in amaranthus production as the outcome in this study showed that amarathus performed poorly under such treatments. Therefore, this study recommends utilization of 8 t ha^{-1} cattle dung based compost in augmenting poor soil to enhance growth and quality of *Amaranthus cruentus* L.

References

Abad M, Berjon MD, Climent M, Camarero SA (1997) The influence of solid urban waste compost and nitrogen-mineral fertilizer on growth and productivity in potatoes. Commun Soil Sci Plant Anal 28(17–18):1653–1661

AdeOluwa OO and Akinyemi O (2014) Amaranths (*Amaranthus viridis*) dry matter and soil qualities: organic vs inorganic fertilizers. In: Rahmann G and Aksoy U (eds) In: Proceedings of the 4th ISOFAR Scientific Conference. 'Building Organic Bridges', at the Organic World Congress 2014, 13–15 October, Istanbul, Turkey, p 879–882

Akanbi WB, Togun AO (2002) The influence of maize- stover compost and nitrogen fertilizer on growth yield and nutrient uptake of amaranth. Sci Hortic 93:1–8

Akhtar JM, Young I, Irvine RJ, Sturrock C (2010) Assessing nitrogen supply potential and influence on growth of lettuce and *amaranthus* of different aged composts. Pak J Bot 42(1):527–536

AOAC (2005) Official method of analysis. In: Association of Analytical Chemists. 15th edn. Washington, DC, pp 11–14

Barea JM, Azcon R, Azcon-Aguiliar C (2002) Mycorrhizosphere interactions to improve plant fitness and soil quality. Antone Van Leeuwenhock 81:343–351

Bittenbender HC, Hue NV, Fleming K, Brown H (1998) Sustainability of organic fertilization of macadamia with macadamia husk-manure compost. Commun Soil Sci Plant Anal 29:409–419

Bremner JM, Mulvaney CS (1982) Nitrogen-Total. In: Page AL, Miller RH (eds) Methods of soil analysis, Part 2, 2nd edn. Agron. Monogr. 9. ASA and SSSA, Madison, pp 595–624

Burleigh SH, Cavagnaro TR, Jakobsen I (2002) Functional diversity of arbuscular mycorrhizae extends to expression of plant genes involved in P nutrition. J Exp Bot 53:1–9

Carl JR and Bierman PM (2005) Nutrient management for fruit and vegetable crop production *Bulletin of University of Minnesota Extension service* University of Minnesota. p 1–10. http://www.swac.umn.edu/People/Faculty/CarlRosen/. Accessed 12 April 2016

Cavagnaro TR, Smith FA, Smith SE, Jakobsen I (2005) Functional diversity in arbuscular mycorrhizas: exploitation of soil patches with different phosphate enrichment differs among fungal species *Plant*. Cell Environ 28:642–650

Craine JM, Jackson RD (2010) Plant nitrogen and phosphorus limitation in 98 North American grassland soils. Plant Soil 334(73–84):1007. doi:10.1007/s11104-009-0237-1

Cruz AF, Ishii T, Kadoya K (2000) Effects of arbuscular mycorrhizal fungi on tree growth, leaf water potential, and levels of 1-aminocyclopropane-1-carboxylic acid and ethylene in the roots of papaya under water-stress conditions. Mycorrhiza 10(3):121–123

Cunningham AB, de Jager PJ and Hansen LCB (1992) The indigenous plant use programme foundation for research development, Pretoria. http://www.bioversityinternational.org/fileadmin/bioversity/publications/Web_version/500/ch24.htm. Accessed 12 April 2016

Su Q, Rowley KG, Itsiopoulos C, O'Dea K (2002) Identification and quantitation of major carotenoids in selected components of the Mediterranean diet: green leafy vegetables, figs and olive oil. Eur J Clin Nutr 56(11):1149–1154

Tindall HD (1983) Vegetables in the tropics, 1st edn. Macmillian Education Ltd. Houndmills, Hampshire, p 553

Walkley A (1947) A critical examination of a rapid method for determining organic carbon m soils: effect of variations in digestion conditions and of inorganic soil constituents. Soil Sci 63:251–263

Dada OA, Aminu JA (2013) The performance of lowland rice (*Oryza sativa* L.) cultivars on iron toxic soil augmented with compost. J Stress Physiol Biochem 9(4):207–218

Eifediyi EK, Remison SU, Okaka VB (2010) Effects of farmyard manure on the dry matter components of two cucumber varieties. Nat Sci 8(5):16–22

Gazey C, Abbott LK, Robson AD (2004) Indigenous and introduced arbuscular mycorrhizal fungi contribute to plant growth in two agricultural soils from south-western Australia. Mycorrhiza 14(6):355–362

Graham JH, Abbott LK (2000) Wheat responses to aggressive and non-aggressive arbuscular mycorrhizal fungi. Plant Soil 220:207–218

Ibijbijen J, Urquiaga S, Ismaili M (2006) Effect of arbuscular mycorrhizal fungi on growth, mineral nutrition and nitrogen fixation of three varieties of common beans (*Phaseolus vulgaris*). N Phytol 134:353–360

International Institute of Tropical Agriculture (I.I.T.A.) (2002) Standard methods of analyzing nutrient in the laboratory. In: Manual Series, p 55

Ishii T, Kitabayashi H, Nakano M and Motosugi H (2004) Growth stimulatory substances for mycorrhizal fungi in composted organic matter. In: Proc. XXVI IHC—sustainability of horticultural systems. Bertschinger L and Anderson JD (eds). Acta Hort 638: 289–292

Karaki Al, Al Radded A (1997) Effects of arbuscular mycorrhizal fungi and drought stress on growth and nutrient uptake of two weeks genotype differing in drought resistance. Mycorrhiza 7:83–88

Kato E and Place FM (2011) Heterogeneous treatment effects of integrated soil fertility management on crop productivity: evidence from Nigeria. IFPRI Discussion Paper 01089, p 20

Khang VuT, Patil HM, Gudedhe NN (2011) Effect of integrated nutrient management on onion yield and soil properties in soybean—onion cropping sequence. Omonrice 18:112–120

Liu CH, Liu Y, Fan C, Kuang SZ (2013) The effects of composted pineapple residue return on soil properties and the growth and yield of pineapple. J Soil Sci Plant Nutr 13(2):433–444

Manios VI, Kapetanios A (1992) Effects of town refuse compost as soil amendment on greenhouse tomato crp. Acta Hort 302:193–201

Martirosyan DM, Miroshnichenko LA, Kulakova SN, Pogojeva AV, Zoloedov VI (2007) Amaranth oil application for coronary heart disease and hypertension. Lipids Health Dis 6:1. doi:10.1186/1476-511X-6-1

Mashavira M, Chitata T, Mhindu RL, Muzemu S, Kapenzi A, Manjeru P (2015) The effect of water hyacinth (*Eichhornia crassipes*) compost on tomato (*Lycopersicon esculentum*) growth attributes, yield potential and heavy metal levels. Am J Plant Sci 6:545–553. doi:10.4236/ajps.2015.64059

McKeague JA (1978) Manual on soil sampling and methods of soil analyses, 2nd edn. Canadian Society of Soil Science, Ottawa, p 212

Miezah K, Ofosu-Anim J, Budu GKO, Enu-Kwesi L, Cofie O (2008) Isolation and identification of some plant growth promoting substances in compost and co-compost. Int J Virol 4(2):30–40

Montãno F, Alías-Villegas C, Bellogín RA, del Cerro P, Espuny MR, Jiménez-Guerrero I, López-Baena FJ, Ollero FJ, Cubo T (2014) Plant growth promotion in cereal and leguminous agricultural important plants: from microorganism capacities to crop production. Microbiol Res 169:325–336

Muloskozi G, Hedren E, Svanberg U (2004) *In vitro* accessibility and intake of β-carotene from cooked green leafy vegetables and their estimated contribution to vitamin a requirements. Plant Food Hum Nutr 59:1–9

Naikwade P (2014) Effect of litter compost on yield and nutrient content of *Zea Mays* L. Sci Res Rep 4(1):79–84

Odeleye FO, Odeleye OMO, Dada OA (2007) The performance of soybean (*Glycine max* (L.) Merrill) under varying weeding regimes in south western Nigeria. Not Bot Hort Agrobot Cluj 35:127–136

Ojo JA, Olowoake AA, Obembe A (2014) Efficacy of organomineral fertilizer and un-amended compost on the growth and yield of watermelon (*Citrullus lanatus* Thumb) in Ilorin Southern Guinea Savanna zone of Nigeria. Int J Recycl Org Waste Agric 3(4):121–125. doi:10.1007/s40093-014-0073-z

Oliveira JS, de Carvallo MF (1975) Nutritional value of some edible leaves used in Mozambique. Econ Bot 29:255

Oworu OO, Dada OA (2009) Influence of compost on growth, nutrient uptake and dry matter accumulation of grain amaranths (*Amaranthus hypochondriacus*. l.) on Oxic Paleulstalf. Niger J Sci 43:7–18

Porra RJ (2002) The chequered history of the development and use of simultaneous equations for the accurate determination of chlorophylls *a* and *b*. Photosynth Res 73:149–156

Statistical Analysis System (SAS) (2002) SAS/STAT User's Guide. Version 8, 6th edn. SAS Institute, Cary. NC

Silva-Sánchez C, de la Rosa AP, León-Galván MF, de Lumen BO, de León-Rodríguez A, de Mejía EG (2008) Bioactive peptides in amaranth (*Amaranthus hypochondriacus*) seed. J Agric Food Chem 56(4):1233–1240. doi:10.1021/jf072911z

Stefano P, Dris R, Rapparini F (2004) Influence of growing conditions on yield and quality of cherry. II: fruit quality. J Food Agric Environ 2:307–309

Raiman MP, Albino U, Cruz MF, Lovato GM, Spago F, Ferracin TP, Lima DS, Goulart T, Bernardi CM, Miyauchi M, Nogueira MA, Andrade G (2007) Interaction among free-living N-fixing bacteria isolated from *Drosera villosa* var. *villosa* and AM fungi (*Glomus clarum*) in rice (*Oryza sativa*). Appl Soil Ecol 35:25–34

Nutrient uptake and growth of spinach as affected by cow manure co-composted with poplar leaf litter

Zobia Anwar[1] · Muhammad Irshad[1] · Qaisar Mahmood[1] · Farhan Hafeez[1] · Muhammad Bilal[1]

Abstract

Purpose Wastes were composted and applied as the soils amendment to improve soil fertility and crop productivity. The study aimed at assessing the nutrient uptake and growth of spinach (*Spinacia oleracea* L.) grown in soils amended with cow manure after a co-composting process.

Methods Sandy loam and silt loam soils were amended with cow manure after co-composting with poplar leaf litter at 1:0, 1:1, 1:2 and 1:3. The compost was applied to soil at the rate of 20 t ha^{-1}. Spinach was grown for 8 weeks and then harvested to measure plant shoot biomass. Spinach shoot samples were digested and nutrient contents in the shoots were determined.

Results Co-composted manure significantly improved the growth and nutrients availability to the spinach. Dry biomass, P and K contents in spinach shoot varied among manure: leaf litter ratios: 1:0 < 1:1 < 1:2 < 1:3. Conversely, N, Zn, Fe, Cu and Cd contents in spinach shoot reduced with the manure amendment with increasing amount of leaf litter. Water extractable micro-elements in the post-harvest soils were found in the order of Zn > Fe > Cu > Cd. Co-compost amendments increased the P and K availability except N, NO_3 and NH_4 in the post-harvest soils. Trace elements in the post-harvest soils reduced with leaf additives in the compost.

Conclusions Co-composted cow manure with leaf litter proved to be superior in terms of bioavailability of plant nutrients over the composted manure without leaf litter. This may also assist in mitigating the environmental contamination of heavy metals in the farm lands.

Keywords Spinach growth · Nutrient content · Cow manure · Leaf litter · Co-composting

Introduction

In developing countries, continuous application of inorganic fertilizer in conjunction with paucity of technical prowess among farmers comprises nutrient imbalance in soil and environmental pollution (Azza et al. 2007). Escalating costs of inorganic fertilizers have directed the attention of farmers towards organic sources which enhance soil fertility possibly by improving physico-chemical properties of soils. These may involve nutrient availability and uptake, soil texture, water holding capacity, cation exchange capacity, electrical conductivity, pH, microbial population and soil organic matter (Agbede et al. 2008; Muhammad and Khattak 2009). Likewise, increasing urbanization is intensifying inappropriate management and disposal of organic wastes from different sources, thereby deteriorating the environmental quality and subsequent food chain contamination through bio-magnification (Mrabet et al. 2012; Benjawan et al. 2015). Organic residues constitute valuable plant nutrients which can be transformed into high yielding soil amendments via composting (Silva et al. 2007). Simultaneously, this potentially resourceful soil amendment is being wasted, demonstrating a reduction in the progressive productivity. Intensive cultivation has triggered decline in the soil organic matter (Giannakis et al. 2014).

The application of compost is a common practice for heavy metals immobilization and soil amelioration. It was

✉ Muhammad Irshad
mirshad@ciit.net.pk

1 Department of Environmental Sciences, COMSATS Institute of Information Technology, Abbottabad, Pakistan

reported that organic amendments could reduce the bioavailability of heavy metals in contaminated soils (Gul et al. 2016). Composting was suggested as the sustainable soil fertility management option for teeming population, as well as the strategy for recycling a variety of organic residues, resulting in the reclamation of degraded soils (Li et al. 2008). The reuse of organic wastes in agriculture is an appropriate method of environment management (Torkashvand et al. 2015). Livestock industry produces a great amount of solid waste in Pakistan which pollutes the environment. Moreover, large amount of plant residues especially from different trees including leaf litter is also generated daily which is a potential environmental hazard via land-filling and incineration of these carbon pools (Chaudhry et al. 1999). Therefore, their effective utilization through co-composting of organic wastes to amend nutrient deficient soil and to increase bio-available fraction for plant uptake is yet to be explored. Therefore, an attempt was made to study the nutrient uptake efficiency, growth and extractable plant nutrients in soils as influenced by cow manure co-composted with poplar leaf litter.

Materials and methods

Composting process

The raw materials for composting were cow manure and leaf litter of poplar (*Populus tremula* L.). These materials were collected from Abbottabad area (34.1558°N, 73.2194°E), Pakistan. Cow manure was co-composted with poplar leaf litter in plastic composting bins (15 L) at the cow manure to leaf litter ratios of 1:0, 1:1, 1:2 and 1:3 (mass basis) for 120 days. Each treatment was in triplicate. During the composting process, manure samples were occasionally aerated and moistened to about 30% at room temperature on weekly basis. Samples were collected after 120 days of composting, air dried and passed through a 2 mm sieve for the determination of relevant parameters (Table 1).

Greenhouse study

Two types of soils (sandy loam and silt loam) were sampled at a depth of 0–20 cm from nearby agricultural fields of Abbottabad city. The samples were air-dried for two days and screened through a 2 mm sieve. The physicochemical properties of soils were determined using the methodologies given in Table 2. A pot experiment was carried out in plastic pots. Each pot was filled with 10 kg soil. Co-composted manures were thoroughly mixed with these soils at the application rate of 20 t ha^{-1} (based on 2 million kg soil per plow layer per ha). Co-composted amendments were applied to two types of soils and were compared with the soil applied with manure composted without plant litter (1:0). This treatment was considered as a control. The experiment was a (4 × 2) factorial (ratio of cow manure co-composted with plant residues × soil types) resulting in eight experimental units which were arranged in a completely randomized block design. Each experimental unit was installed in triplicate. Irrigation was applied twice the daily pan evaporation. Spinach (*Spinacia oleracea* L.) was grown for 8 weeks as test crop and then harvested for plant shoot biomass. Subsequently, plants were thoroughly rinsed with distilled water and oven dried for 48 h at 60 °C. Plants' shoots were oven dried for 48 h at 60 °C. The dry weight of spinach was calculated. The samples were milled and digested as given below.

Analytical methods

Dry combustion method was used to determine total carbon in the composted manure (Nelson and Sommers 1982). Total organic matter was determined by multiplying the total carbon values by 1.72 (Nelson and Sommers 1982). Total content of nitrogen (N) was determined by Kjeldahl method. Soil, milled composted manure and spinach shoot samples were digested in a mixture (1:3) of perchloric ($HClO_4$) and nitric (HNO_3) acids to determine the total elemental composition of phosphorous (P), potassium (K), copper (Cu), iron (Fe), zinc (Zn) and cadmium (Cd) in the extracts (Miller 1998). Spectrophotometer (Stalwart UV 900 USA) determined the P contents in all samples (soil, manure and plant) through phosphomolybdate blue method. Absorbance was determined at a wavelength of 710 nm (Olsen and Sommers 1982). Cationic elements in the soil, manure and plant samples were determined using atomic absorption spectrometer (AAS) (Model AAnalyst 700, Perkin-Elmer, Waltham, MA, USA). The pH and electrical conductivity (EC) of soil and organic manures samples were measured in water-manure and water-soil suspensions of 1:5 ratio with pH (Janway 3505) and EC (Janway 470) meters. Nitrate-N and NH_4-N were extracted from soil and manure by shaking 2 g of sample in 2 M KCl solutions for 1 h in 25 mL polypropylene centrifuge tubes on an end-over-end shaker, then centrifuged at 10,000 rpm for 10 min and supernatant was analyzed. Ammonium-N was determined following the indophenol blue method (Keeney and Nelson 1982), whereas the NO_3-N was determined according to Anderson and Ingram (1993) and read from the spectrophotometer at 635 and 410 nm, respectively. Particle size distribution of the soil was measured by the pipette method (Gee and Bauder 1986). The soils used during the study were classified as sandy loam on the basis of 57% sand, 24% silt and 19% clay and loamy sand on the basis of 70% sand, 17% silt and 13% clay.

Table 1 Chemical composition of manure co-composted with poplar leaf litter at different ratios

Parameters	Unit	1:0	1:1	1:2	1:3
Total C	g kg^{-1}	366.8 ± 14	374.4 ± 16	378.3 ± 16	397.5 ± 15
Total N	g kg^{-1}	18.0 ± 2	16.7 ± 2	15.6 ± 2	15.0 ± 2
NO_3-N	mg kg^{-1}	91.6 ± 10	86.8 ± 11	74.5 ± 8	67.9 ± 7
NH_4-N	mg kg^{-1}	59.6 ± 6	60.3 ± 7	57.6 ± 6	34.8 ± 3
Total P	mg kg^{-1}	169.7 ± 8	180.9 ± 9	196.4 ± 8	206.3 ± 10
Total Ca	mg kg^{-1}	167.2 ± 11	178.6 ± 10	188.2 ± 9	203.6 ± 12
Total Mg	mg kg^{-1}	132.6 ± 9	143.9 ± 9	165.2 ± 6	172.6 ± 10
Total K	mg kg^{-1}	400.4 ± 15	467.9 ± 17	489.0 ± 17	511.8 ± 21
Total Na	mg kg^{-1}	41.0 ± 3	33.4 ± 2	31.2 ± 2	25.6 ± 1.8
Total Cu	mg kg^{-1}	93.1 ± 6	63.5 ± 3	66.2 ± 4	56.1 ± 2
Total Fe	mg kg^{-1}	184.9 ± 13	166.4 ± 10	167.1 ± 7	154.4 ± 9
Total Mn	mg kg^{-1}	124.8 ± 8	105.2 ± 5	98.5 ± 7	82.7 ± 6
Total Zn	mg kg^{-1}	140.6 ± 14	115.2 ± 10	100.2 ± 10	87.5 ± 7
Total Ni	mg kg^{-1}	2.7 ± 0.2	2.3 ± 0.2	1.8 ± 0.2	1.9 ± 0.2
Total Cd	mg kg^{-1}	1.5 ± 0.1	1.1 ± 0.1	1.0 ± 0.1	0.9 ± 0.1
EC (1:5)	dS m^{-1}	5.0 ± 0.8	6.1 ± 0.8	6.3 ± 1	6.8 ± 0.9
pH (1: 5)		8.7 ± 1	8.3 ± 1	8.1 ± 1	7.7 ± 0.6

± values indicate standard deviation

Table 2 Chemical composition of soils used for the experiment

Parameters	Unit	Sandy loam soil	Silt loam soil
Total C	g kg^{-1}	25.6 ± 2	28.3 ± 2
Total N	mg kg^{-1}	35.8 ± 3	56.6 ± 4
NO_3-N	mg kg^{-1}	6.6 ± 1	15.3 ± 2
NH_4-N	mg kg^{-1}	8.2 ± 1	10.8 ± 1
Total P	mg kg^{-1}	45.7 ± 2	67.8 ± 3
Total Ca	mg kg^{-1}	218.7 ± 8	267.8 ± 12
Total Mg	mg kg^{-1}	128.6 ± 6	131.6 ± 6
Total K	mg kg^{-1}	147.8 ± 7	158.8 ± 8
Total Na	mg kg^{-1}	23.4 ± 3	34.6 ± 4
Total Cu	mg kg^{-1}	34.5 ± 4	56.4 ± 3
Total Fe	mg kg^{-1}	67.8 ± 5	78.6 ± 6
Total Mn	mg kg^{-1}	47.5 ± 4	67.5 ± 6
Total Zn	mg kg^{-1}	56.7 ± 4	76.9 ± 5
Total Ni	mg kg^{-1}	2.5 ± 0.2	3.0 ± 0.2
Total Cd	mg kg^{-1}	0.3 ± 0.1	1.9 ± 0.1
EC (1:5)	dS m^{-1}	0.5 ± 0.1	0.3 ± 0.02
pH (1: 5)		7.6 ± 0.7	7.8 ± 0.7

± values indicate standard deviation

Statistical analysis

The data on each parameter were subjected to analyses of variance using Stat-view software. The least significant difference test at $P \leq 0.05$ was used to calculate significant differences between means of treatments. Regression plot was calculated between macro- and micro-elements in the composted manure with their concentrations in the spinach shoot.

Results and discussion

The relevant characteristics of soils and composted manure were given in Tables 1, 2.

Spinach biomass

Spinach biomass significantly increased after the application of co-composted manure as compared to the manure without plant litter (Fig. 1). Among the treatments, the highest plant biomass observed with increasing manure: leaf litter ratios while the lower amount of biomass was obtained when manure was applied without tree litter (1:0). The plant biomass was higher in sandy loam soil as compared to the silt loam irrespective of the manure treatments. Spinach biomass enhanced by 62, 105 and 142% in manure co-composted with leaf litter at 1:1, 1:2 and 1:3, respectively, in comparison to the composted manure of 1:0 in sandy loam soil. In case of silt loam soil, biomass increased by 39, 70 and 109% for manure co-composted with the leaf litter at 1:1, 1:2 and 1:3, respectively, as compared to the manure composted at 1:0. The higher biomass yield at greater manure: leaf litter ratio may be due to the optimum availability of the essential plant nutrients. Incremental plant growth in the soils amended with manure indicated that the application of manure after co-composting might

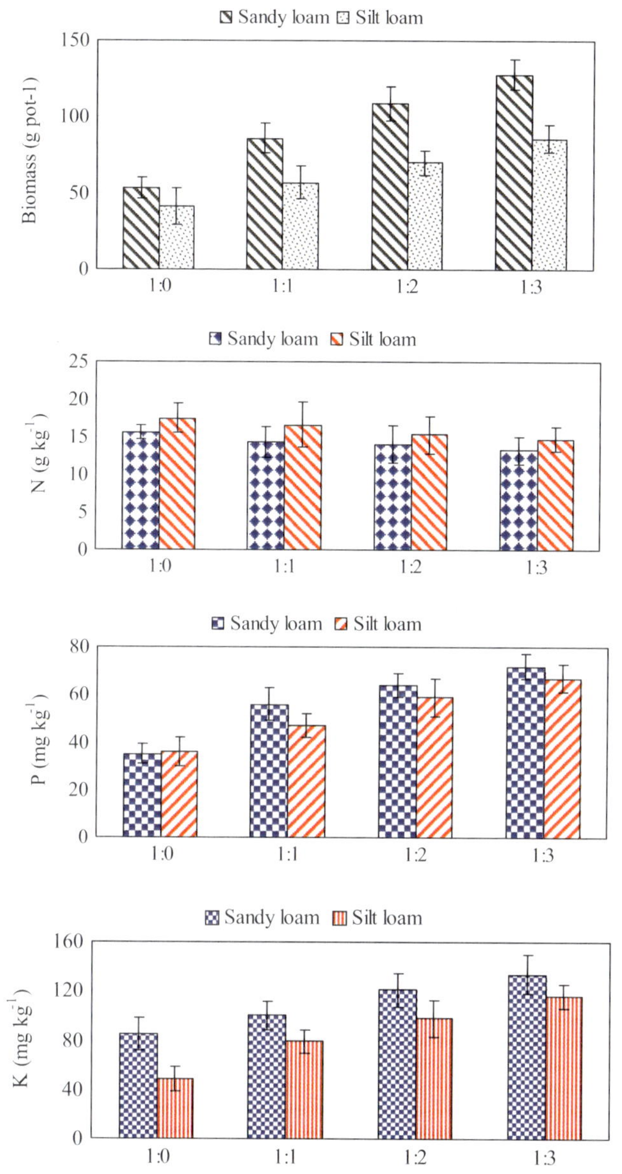

Fig. 1 Spinach biomass and nutrient contents in spinach shoot after application of composted and co-composted manure

be more complimentary to the plant development due to the nutrients availability and conversion of organic fractions into inorganic forms. The application of composted manure with tree residues proved to be a viable resource of essential plant nutrients. Irshad et al. (2002) reported higher biomass yield and promoted mineral concentrations in the maize grown in soil amended with composted manure. These results advocated enhancement of soil fertility subsequent to the application of composted manure. Several researchers have reported the significance of compost in triggering and enhancing soil fertility via supplying nutrients for plant use (Gul et al. 2016; Irshad et al. 2014). D'Hose et al. (2012) also confirmed that composted animal manure considerably increased dry matter yield of potatoes, fodder beets, forage maize and Brussels sprouts.

Nutrient contents

The application of co-composted manure significantly affected N content in spinach (Fig. 1). Increasing amount of leaf litter in the co-compost slightly reduced N content in the spinach as compared to the application of composted manure without leaf litter additives. Nitrogen contents in spinach shoot reduced by 15 and 13% in the mixture of manure and leaf litter of 1:3 as compared to 1:0 in sandy loam and silt loam soils, respectively. This might be related to the changes in the N composition of soil after adding co-composted manure. Spinach exhibited higher N contents in the plant shoot grown in silt loam soil as compared to the sandy loam soil. Zai et al. (2008) demonstrated differential effects of composts in relation to the nutrient status for plant growth and development. Increased N bioavailability was reported after the application of composted manure to alfalfa and the increased nutrients availability from soil was associated with the improvement of soil properties (Malhi 2012).

Phosphorus contents in spinach significantly increased with the addition of leaf litter with the cow manure (Fig. 1). Phosphorus level in the shoot of spinach was lower, i.e., 35 and 36 mg kg^{-1} when grown in sandy loam and silt loam soil, respectively, treated with manure amendment of 1:0. Phosphorus contents in the spinach tissues increased by 60, 82 and 105% in composted manure treatments of 1:1, 1:2 and 1:3, respectively, in the sandy loam soil. Phosphorus content increased by 30, 63 and 86%, respectively, in silt loam soil. The variation of P content among spinach plants after manure amendments might be associated with the altered mineral composition of soils. Higher P content in plants with the addition of manure was reported by Huang et al. (2001). Irshad et al. (2002) reported that increasing composted manure application profoundly increased the nutrients uptake by plants for better growth. Moreover, poultry litter amendment in soil improved the elemental contents in maize (Faridullah et al. 2009). Mustafa et al. (2016) reported that muskmelon plants grown in media-containing guar or jantar composts had greater tissue nutrient concentrations as compared to other compost treatments.

Potassium content in plant tissue significantly increased with the application of co-composts as compared to the plants applied with the manure without leaf litter (Fig. 1). The K contents observed in the spinach shoot were 100, 121 and 134 mg kg^{-1} in sandy loam soil after amendment with manure: leaf litter ratios of 1:1, 1:2 and 1:3, respectively. In silt loam soil, the K contents were 47, 59 and

67 mg kg^{-1}, respectively. Potassium increased in the plant shoot after applying co-compost mixed with greater amount of leaf litter. This indicated high quality of organic matter which might have affected the bioavailability of K. Increased K content has been attributed to the elemental contents of the composts, which improved the root growth and increased the nutrient uptake by plants (Aziz et al. 2006). Malhi (2012) reported that the application of organic manures favored the uptake of N, K and S in wheat.

The application of co-composted manures decreased shoot contents of mico-elements. The contents of these elements in the spinach shoot varied in the order of Zn > Fe > Cu > Cd (Fig. 2). When compared with the manure of 1:0, the manure amendment at 1:3 reduced the Zn contents in spinach from 45.1 to 23.1 mg kg^{-1} in sandy loam and from 48.3 to 25.4 mg kg^{-1} in silt loam soil. Iron concentrations in the spinach shoots were 27.3 and 37.5 mg kg^{-1} and Cu concentrations were 4.5 and 5.1 mg kg^{-1} and Cd concentrations were 1.4 and 1.8 mg kg^{-1} in sandy loam and silt loam soil, respectively, after amendment with composted manure (1:0). The Zn contents reduced to 23.1 and 25.4 mg kg^{-1}, Fe contents were 9.6 and 17.8 mg kg^{-1}, Cu contents were 3 and 3.4 mg kg^{-1}, Cd contents were 0.8 and 0.6 mg kg^{-1} in plants grown in sandy loam and silt loam soil, respectively, after amendment with co-composted manure at 1:3. Low availability of trace elements after the application of co-composted manure with increasing amounts of leaf litter may be responsible for the lower trace elements contents in the spinach. Generally water-extractable fractions of trace elements are easily available to the plants and associated with the noxious and integral component of composts (Singh and Kalamdhad 2014). The reduction in trace elements composition in spinach plants could also be related to the amount and nature of the organic matter. Organic fraction of soil affected the interactions and availability of trace elements in soil by the formation of varyingly stable organometallic complexes called chelates by binding and holding insoluble metal ions and compounds with the –OH and –COOH groups. Such compounds are enriched by cattle manure which consequently may limit metal solubility and bioavailability in soils (Liu et al. 2009; Singh et al. 2015). Heavy metals in soil are held and bound by organic compounds and are not available for plant uptake if these complexes are in soluble, but they are slowly released occurring through microbial decomposition (Guan et al. 2011). Trace elements contents in spinach shoot were higher in silt loam soil than sandy loam soil, irrespective of the type of compost applied. Silveira et al. (2003) reported that the application of potentially mineralized organic matter after the composting resulted in the formation of strong binding sites in soil which diminished heavy metal uptake by plants. Likewise, Singh et al. (2015) reported reduced bioavailability of heavy metals to plants after application of agitated pile co-composted cattle manure with the rice husk. Composting may reduce the mobility of heavy metals and thus their transport to the environment in contrast to the fresh manure. The utilization of composted manure was advantageous and worthwhile in adulterated soils for alleviating heavy metal contamination (Gul et al. 2016).

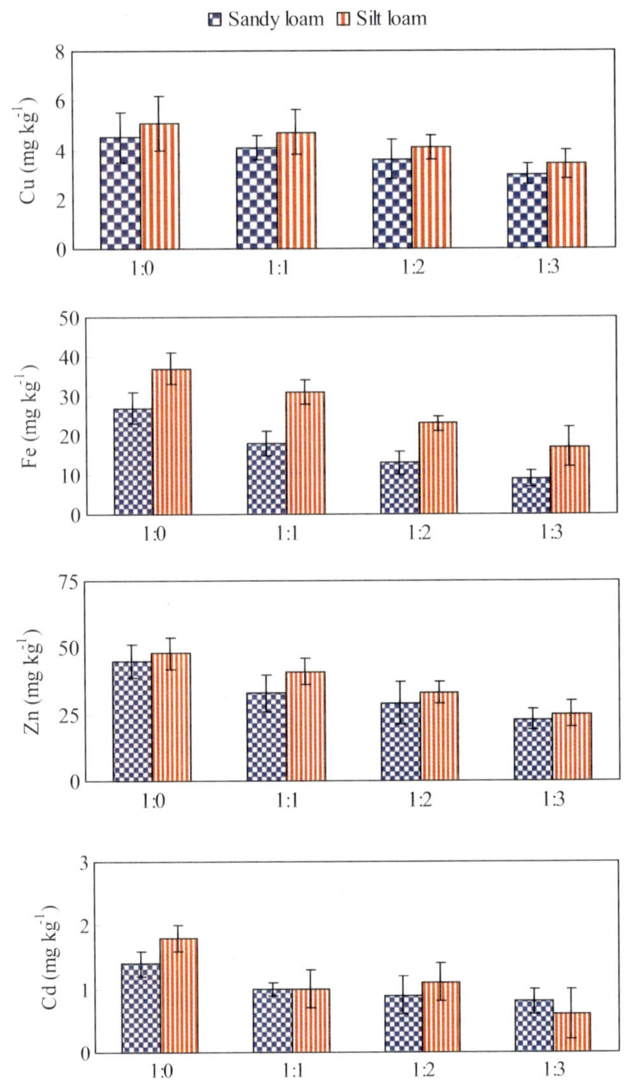

Fig. 2 Micro-elements concentrations in spinach shoot after application of co-composted manure

Regression equations and determinant (R^2) explaining the relationships of nutrient contents in spinach shoot (Y) to their contents in co-composted manure (X) are given in Fig. 3. The higher R^2 values for P, K and Zn showed that the contents of these nutrients elements in spinach shoot were highly dependent on their contents in the co-compost applied to the soil. The other elements were also found positively associated with the composition of co-compost.

Nutrients extractability from post-harvest soil

The application of manure compost in soil significantly affected the extractability of essential plant nutrients. The application of composted manure with plant litter significantly improved soil P contents and its uptake by spinach (Table 3). The soil amendment of higher quantity of poplar leaf litter with manure at the ratios of 1:1, 1:2 and 1:3 enhanced soil P by 22, 35 and 51% in sandy loam and 14, 39 and 53% in silt loam soil, respectively. Eichler-Löbermann et al. (2007) reported increased soil P content after application of manure compost. In the present work, P contents were noticed as 31 and 28 mg kg^{-1} in sandy loam and silt loam soil, respectively, after amendment with compost without leaf litter.

Conversely, the application of co-compost decreased total N content particularly with increasing amount of leaf litter irrespectively of the soil type (Table 3). Nitrogen content of the soil could vary due to the nutrient status of the co-composts of cow manure. The N-enriched manure source after the incorporation of increased amount of leaf litter reduced the N level of soils. The N content in sandy loam soil was found more than that of silt loam soil. Wolka and Melaku (2015) reported significantly lower N and organic carbon after manure addition in soil. Giannakis et al. (2014) demonstrated that compost application gradually reduced soil N and organic C. This declining trend of N might also be due to leaching and/or nitrification etc. Contradictory results of increased N after different levels of compost application have been reported by Nguyen and Shindo (2011). Likewise, significant increase in the N and P was noticed subsequent to the application of chicken manure (Dikinya and Mufwanzala 2010).

Nitrate and NH_4-N fluctuated in accordance with the soil-type and mixing ratios of leaf litter and manure. The contents had inverse relation with the proportion of leaf additives in manure. The greater amount of leaf litter in the manure lowered the NO_3-N and NH_4-N contents in both soils. For instance, sandy loam soil showed significantly declined NO_3-N content (8.2, 7.6, 6.4 and 5.7 mg kg^{-1})

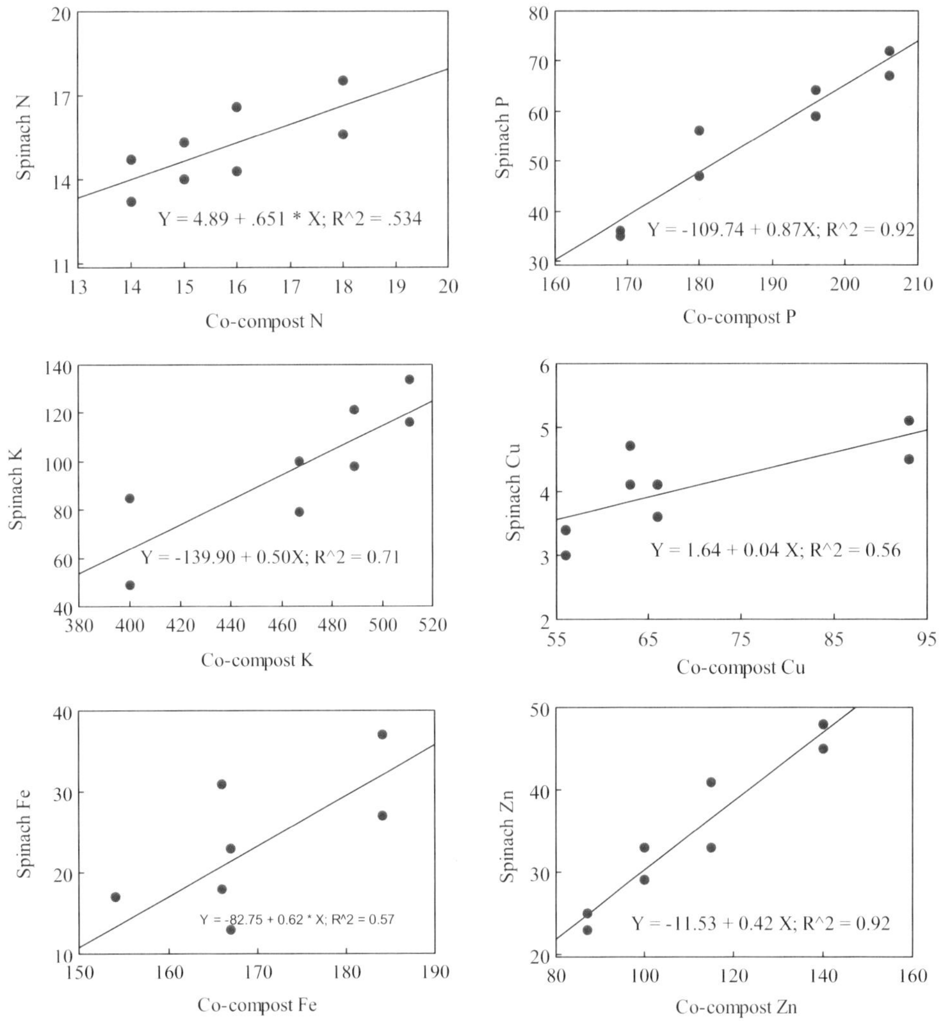

Fig. 3 Regression between macro- and micro-elements in co-compost applied versus these elements in spinach shoot. The contents of all elements in spinach shoot and co-compost are expressed in mg kg^{-1} except N values, i.e., g kg^{-1}

Table 3 Changes in plant macro- and micro-elements (mg kg^{-1}) in post-harvest soil applied with co-composted cow manure (CM) with plant litter (PL)

Soil	CM:PL	N g kg^{-1}	NO_3-N	NH_4-N mg kg^{-1}	P	K	Cu	Zn	Fe	Cd
Sandy loam	1:0	2.3	8.2	6.8	31	61	3.6	36.7	14.1	0.8
	1:1	2.1	7.6	5.9	38	88	3.0	28.4	10.1	0.9
	1:2	1.8	6.4	5.0	42	103	2.7	24.8	8.7	0.6
	1:3	1.5	5.7	4.3	47	120	2.4	20.7	6.5	0.6
Silt loam	1:0	2.1	9.1	7.2	28	32	4.1	58.2	20.1	0.9
	1:1	1.9	7.8	6.6	32	46	3.5	51.9	15.6	0.8
	1:2	1.6	7.1	5.8	39	65	3.0	45.3	13.5	0.7
	1:3	1.3	6.1	4.7	43	89	2.5	41.5	10.4	0.6
LSD (0.05)		0.1	0.7	0.8	1.3	3.4	0.6	1.7	0.8	0.1

after compost amendment of 1:0, 1:1, 1:2 and 1:3. Similarly, the lower NO_3-N content of 6.1 mg kg^{-1} was observed in manure leaf ratio of 1:3 in silt loam soil as compared to 9.1 mg kg^{-1} in compost amendment of 1:0. The reduced NO_3-N level in the soils applied with co-composts of considerable leaf additives could be related to the immobilization and/or denitrification of the cumulated NO_3-N. Furthermore, irrespective of the soil type, NH_4-N was profoundly reduced after co-composts applications. Among compost amendments, the NH_4-N contents in the soil varied in the order of 1:0 > 1:1 > 1:2 > 1:3. Co-composts treated soils better reduced the NH_4 release as compared to the soils amended with compost without tree litter (1:0) which stressed on the negative relationship of soil-NH_4 to the increased ratios of leaf additives in the co-compost. Similarly, declined NO_3 was reported in compost amended turf-grass soil (Wright et al. 2007).

Soils amended with co-compost increased the available K content, particularly with increasing ratio of poplar leaf litter (Table 3). This indicated the effectiveness of organic matter for the K availability in the soil. The maximum K content of 120 mg kg^{-1} was noted in sandy loam soil and 89 mg kg^{-1} in silt loam soil after the manure application of leaf litter at 1:3. The K release dynamics could be attributed to the elemental properties of the manure composted at different ratios along leaf litter. Huang et al. (2001) reported leaf litter as a potential source of nutrients for soil fertility and crop production if co-composted with N-enriched source of cattle manure (Table 4). The application of manure amendment of 1:0 to sandy loam and silt loam soils resulted in K contents of 61 and 32 mg kg^{-1}, respectively. The reduced availability of K through the manure application without leaf litter may be due to the dilution effect. Among amended soils, K differed in the order sandy loam > silt loam. This alteration in the K extraction could be attributed to the modified properties of soil after manure amendments. Phosphorus and K availability in soil and their uptake by plants were enhanced by organic manure (Aziz et al. 2010; Cavigelli and Thien 2003).

Table 4 Changes in pH and EC in post-harvest soil applied with co-composted cow manure (CM) with plant litter (PL)

Soil	CM:PL	pH	EC dS m^{-1}
Sandy loam	1:0	7.1	1.8
	1:1	7.4	2.2
	1:2	7.5	2.2
	1:3	7.6	2.6
Silt loam	1:0	7.0	1.7
	1:1	7.2	1.9
	1:2	7.4	2.0
	1:3	7.5	2.2
LSD (0.05)		0.4	0.3

The micro-element contents in both soils altered in the order of Zn > Fe > Cu > Cd regardless of the co-composts applied. Incorporation of manure amendment to the soils reduced the availability of trace elements and consequently reduced their uptake by spinach. Zinc contents of sandy loam and silt loam soils declined by 44.9 and 29.3% with the application manure: leaf litter of 1:3 as compared to the manure treatment without leaf litter (1:0). Zinc content in soil was within the threshold range of 14–125 mg kg^{-1}. Zinc concentrations in the soils varied with the co-composted manure in the order of 1:0 > 1:1 > 1:2 > 1:3. This reduction might possibly be related to the effective mineralization of the organic manure with the leaf litter additives that could efficiently adsorbed inorganic Zn. Similar finding of reduced metals availability in soil mixed with composted animal manure in contrast to soil amended with the fresh manures was reported earlier (Irshad et al. 2014). The utilization of composted manure instead of fresh manure in metal contaminated soils could be more advantageous in alleviating heavy metal pollution (Gul et al. 2016).

The co-composts reduced the extractable Fe in soil. The Fe content in the sandy loam soil was reduced by 53% with the application of manure: leaf litter at 1:3 ratio as compared to the manure minus leaf litter (1:0). The efficient reduction with the application of co-compost could be attributed to the potential source of essential nutrients and the drop of heavy metals in the soils. The fluctuations observed in the release of Fe could be related to the elemental status of the co-composts applied to the soil. The effectiveness of the manure product could be associated with the formation of organometallic complexes that hold more Fe and reduced the Fe uptake by spinach plants. Furthermore, basic pH of soil helps to adsorb cationic metals on organic matter sites. The pH is evaluated as the controlling factor of ion exchange, redox reactions, adsorption and complexation reactions (Walter et al. 2006). The effectiveness of heavy metal solubility after manure compost application was determined by the extent of organic matter humification and the successive ramification on soil pH (Gupta and Sinha 2007).

Copper content of the soils demonstrated an inverse relationship with the application of compost. Lower Cu content was found in sandy loam soil (2.4 mg kg^{-1}) applied with 1:3 manure amendment as compared to the soil amended with manure without leaf litter having Cu content of 3.6 mg kg^{-1}. Similarly, silt loam soil amended with manure amendments of 1:0, 1:1, 1:2 and 1:3 exhibited Cu contents of 4.1, 3.5, 3.0 and 2.5 mg kg^{-1}. Reduced Cu release from compost amended soil might be associated with the progressive humification of organic manure applied and the formation of stabilized Cu complexes with humic substances (McLaren and Clucas 2001). Compost and poultry manure amendments were recommended for the optimization of Cu in contaminated soil (Thomas and Dauda 2015). A marked reduction in extractable Cd was recorded in soils after co-composts application. Cd level fluctuations in soils depended on the composting ratios and varied as 1:3 < 1:2 < 1:1 < 1:0. For instance, extractable Cd decreased by 25% in sandy loam soil and 32% in silt loam amended with manure: leaf litter ratio of 1:3 as compared to the manure without leaf litter. The maximum permissible limit of Cd in soil was reported as 5 mg kg^{-1} (Dudka et al. 1994). Therefore, the Cd contents in soil samples were found within permissible limits during this study.

EC and pH of post-harvest soil

The respective EC of sandy loam and silt loam soil was 1.8 and 1.7 dS m^{-1} applied with composted manure without leaf additives in contrast to the soils amended with 1:3 of co-composts where the EC values were 3.1 and 2.9 dS m^{-1}, respectively. Comparatively, silt loam soil amended with 1:1, 1:2 and 1:3 composted manure with leaf litter exhibited EC values of 1.9, 2.0 and 2.2 dS m^{-1}, respectively. This increase may be attributed to the greater pool of soluble cations in the co-composts and their release in the respective soil. It was found that co-composts application increased the soluble salts contents in the post-harvest soils that directly measure the salinity after decomposition and the mobilization of organic matter. Dikinya and Mufwanzala (2010) reported increased EC values with the increasing application rates of manure. Results of higher EC values after manure applications in soils have also been reported by different authors (Loper et al. 2010; Wolka and Melaku 2015).

Soil pH slightly differed with the leaf litter application in the manure. The compost demonstrated slightly alkaline pH (7.1–7.6) in sandy loam soil and 7.0–7.5 in silt loam soil. With the application of 1:3 co-composted manure the pH value increased from 7.1 to 7.6 in sandy soil and 7.0–7.5 in silt loam soil. High soil pH for composted manure was also reported by Scharenbroch (2009). On the other hand, a low soil pH after compost amendment was reported by Loper et al. (2010).

Conclusions

Significant impact of co-compost on the bioavailability and extractability of plant nutrients in soils was evident. Dry biomass, P and K contents in spinach shoot enhanced with the application of manure co-composted at different ratio of leaf litter. Conversely N, Zn, Fe, Cu and Cd contents in spinach declined with the application of manure amendment co-composted with increasing amount of leaf litter. Chemical composition of co-compost significantly altered the availability of trace elements in the order of Cd < Cu < Fe < Zn. Micro-elements contents in plants and post-harvest soils were inversely related to the increased amount of leaf additives in the composted manure. Nitrogen content in the post-harvest soil decreased in the sequence of 1:1 > 1:2 > 1:3 of the co-composts. The lower availability of micro-elements in manure may recommend it for the safe and productive utilization in soils.

Acknowledgements We acknowledge Life Sciences Department, COMSATS Institute of Information Technology, Abbottabad, for provision of the lab facilities.

References

Agbede TM, Ojeniyi SO, Adeyemo AJ (2008) Effect of poultry manure on soil physical and chemical properties, growth and grain yield of sorghum in southwest, Nigeria. Am Eur J Sustain Agric 2:72–77

Anderson JM, Ingram JSI (1993) Tropical soil biology and fertility, a handbook of methods, 2nd edn. CABI, Wallingford

Aziz T, Rahmatullah MA, Maqsood MA, Tahir IA, Cheema MA (2006) Phosphorus utilization by six Brassica cultivars (*Brassica juncea* L.) from tri-calcium phosphate; a relatively insoluble P compound. Pak J Bot 38:1529–1538

Aziz T, Samiullah SA, Nasim M, Farooq M, Khan MM (2010) Nutrient availability and maize growth in soil amended with organic manure. Int J Agric Biol 12(4):621–624. doi:10.1016/S0960.8524(02)00042-1

Azza E, Hideto U, Abdel G, Naomi A (2007) Uptake of carbon and nitrogen through rice root from 13C and 15N dual labelled maize residue compost. Int J Biol Chem 1:75–83. doi:10.1007/s11104-008-9543-2

Benjawan L, Sihawong S, Chayaprasert W, Liamlaem W (2015) Composting of biodegradable organic waste from Thai household in a semi-continuous composter. Compost Sci Util 23:11–17. doi:10.1080/1065657X.2014.963742

Cavigelli MA, Thien SJ (2003) Phosphorus bioavailability following incorporation of green manure crops. Soil Sci Soc Am J 67:1186–1194

Chaudhry MG, Ahmad MS, Chaudhry GM (1999) Growth of livestock production in Pakistan: an analysis. Pak Dev Rev 38(4):605–614

D'Hose T, Cougnon M, De Vliegher A, Willekens K, Van Bockstaele E, Rehe D (2012) Farm compost application: effects on crop performance. Compost Sci Util 20:49–56. doi:10.1080/1065657X.2012.10737022

Dikinya O, Mufwanzala N (2010) Chicken manure enhanced soil fertility and productivity: effects of application rates. J Soil Sci Environ Manag 1(3):46–54

Dudka S, Piotrowska M, Chlopecka A (1994) Effect of elevated concentrations of Cd and Zn in soil on spring wheat yield and the metal contents of the plants. Water Air Soil Pollut 76:333–341

Eichler-Löbermann B, Köhne S, Köppen D (2007) Effect of organic, inorganic, and combined organic and inorganic P fertilization on plant P uptake and soil P pools. J Plant Nutr Soil Sci 170:623–628. doi:10.1002/jpln.200620645

Faridullah IM, Yamamoto S, Eneji AE, Uchiyama T, Honna T (2009) Recycling of chicken and duck litter ash as a nutrient source for Japanese mustard spinach. J Plant Nutr 32:1082–1091. doi:10.1080/01904160902943122

Gee GW, Bauder JW (1986) Particle-size analysis. In: Klute A (ed) Method of soil analysis. Part 1. Agronomy series No. 9. ASA, SSSA, Madison

Giannakis GV, Kourgialas NN, Paranychianakis NV, Nikolaidis NP, Kalogerakis N (2014) Effects of municipal solid waste compost on soil properties and vegetables growth. Compost Sci Util 22:116–131. doi:10.1080/1065657X.2014.899938

Guan TX, He HB, Zhang XD, Bai Z (2011) Cu fractions, mobility and bioavailability in soil–wheat system after Cu-enriched livestock manure applications. Chemosphere 82:215–222. doi:10.1016/chemosphere.2010.10.018

Gul S, Naz A, Khan A, Nisa S, Irshad M (2016) Phytoavailability and leachability of heavy metals from contaminated soil treated with composted livestock manure. Soil Sediment Contam 25:181–194. doi:10.1515/pjct-2015-0044

Gupta AK, Sinha S (2007) Phytoextraction capacity of the plants growing on tannery sludge dumping sites. Bioresour Technol 98:1788–1794. doi:10.1016/j.biortech.2006.06.028

Huang GF, Fang M, Wu QT, Zhou LX, Liao XD, Wong JWC (2001) Co-composting of pig manure with leaves. Environ Technol 22:1203–1212. doi:10.1080/09593332208618207

Irshad M, Yamamoto S, Eneji AE, Honna T, Endo T (2002) Influence of composted manure and salinity on growth and nutrient content of maize tissue. Sand Dune Res 49:1–10

Irshad M, Gul S, Eneji AE, Anwar Z, Ashraf M (2014) Extraction of heavy metals from manure and their bioavailability to spinach after composting. J Plant Nutr 37:1661–1675. doi:10.1080/01904167.2014.888748

Keeney DR, Nelson DW (1982) Nitrogen-Inorganic forms. In: page AL, Miller RH, Keeney DR (eds) Methods of Soil Analysis part 2: Chemical and microbiological properties, Agronomy 9/2, 2nd edn. American society of Agronomy, Madison, WI, pp 643–698

Li X, Zhang R, Pang Y (2008) Characteristics of dairy manure composting with rice straw. Biores Technol 99:359–367. doi:10.1016/j.biortech.2006.12.009

Liu L, Hansong C, Peng C, Wei L, Qiaoyun H (2009) Immobilization and phytotoxicity of Cd in contaminated soil amended with chicken manure compost. J Hazard Mater 163:563–567. doi:10.1016/j.jhazmat.2008.07.004

Loper S, Shober AL, Wiese C, Denny GC, Stanley CD (2010) Organic soil amendment and tillage affect soil quality and plant performance in stimulated residential landscape. Hort Sci 45:1522–1528

Malhi SS (2012) Relative effectiveness of various amendments in improving yield and nutrient uptake under organic crop production. Open J Soil Sci 2:299–311. doi:10.4236/ojss.2012.23036

McLaren RG, Clucas LM (2001) Fractionation of copper, nickel, and zinc in metal-spiked sewage sludge. J Environ Qual 30:1968–1975

Miller RO (1998) Nitric-perchloric acid wet digestion in an open vessel. In: Kalra Y (ed) Handbook of reference methods for plant analysis. CRC Press, Washington

Mrabet L, Belghyti D, Loukili A, Attarassi B (2012) Effect of household waste compost on the productivity of maize and lettuce. Agric Sci Res J 2(8):462–469

Muhammad D, Khattak RA (2009) Growth and nutrient concentration of maize in pressmud treated saline-sodic soils. Soil Environ 28:145–155

Mustafa G, Ali MA, Smith D, Schwinghamer T, Lamont JR, Ahmed N, Hussain S, Arshad M (2016) Guar, jantar, wheat straw, and rice hull composts as replacements for peat in muskmelon transplant production. Int J Recycl Org Waste Agric 5:323–332. doi:10.1007/s40093-016-0142-6

Nelson DW, Sommers LE (1982) Total carbon, organic carbon and organic matter. In: Page AL, Miller RH, Keeney DR (eds) Methods of soil analysis. Part 2. Chemical and microbiological properties. ASA, SSSA, Madison, pp 539–580

Nguyen TH, Shindo H (2011) Effects of different levels of compost application on amounts and distribution of organic nitrogen forms in soil particle size fractions subjected mainly to double cropping. Agric Sci 2(3):213–219. doi:10.4236/as.2011.23030

Olsen SR, Sommers LE (1982) Phosphorus. In: Page AL et al (eds) Methods of soil analysis, Part 2, 2nd edn, Agronomy Monograph 9. ASA and ASSA, Madison, pp 403–430

Scharenbroch BC (2009) A meta-analysis of studies published in arboriculture and urban forestry relating to organic materials and impacts on soil, tree, and environmental properties. Arboric Urban For 35:221–231

Silva BMT, Menduíña AM, CendónSeijo Y, Viqueira DF (2007) Assessment of municipal solid waste compost quality using standardized methods before preparation of plant growth media. Waste Manag Res 25:99–108

Silveira MLA, Alleoni LRF, Guilherme LRG (2003) Biosolids and heavy metals in soils. Sci Agric 60:793–806. doi:10.1590/S0103-90162003000400029

Singh J, Kalamdhad AS (2014) Effect of lime on speciation of heavy metals during agitated pile composting of water hyacinth. Front Environ Sci Eng (FESE) 10:10–22. doi:10.1007/s11783-014-0704-7

Singh WR, Pankaj SK, Kalamdhad AS (2015) Reduction of bioavailability and leachability of heavy metals during agitated pile composting of Salvinianatans weed of Loktak Lake. Int J Recycl Org Waste Agric 4:143–156. doi:10.1007/s40093-015-0094-2

Thomas EY, Dauda SO (2015) Comparative effects of compost and poultry manure on bioavailability of Pb and Cu and their uptake by maize. New York Sci J 8:72–81

Torkashvand AM, Alidoust M, Khomami AM (2015) The reuse of peanut organic wastes as a growth medium for ornamental plants. Int J Recycl Org Waste Agric 4(2):85–94. doi:10.1007/s40093-015-0088-0

Walter I, Martinez F, Cala V (2006) Heavy metal speciation and phytotoxic effects of three representative sewage sludge for agricultural uses. Environ Pollut 139:507–514. doi:10.1016/j.envpol.2005.05.020

Wolka K, Melaku B (2015) Exploring selected plant nutrient in compost prepared from food waste and cattle manure and its effects on soil properties and maize yield in Wondo Genet, Ethopia. Environ Syst Res 4:1–7. doi:10.1186/s40068-015-0035-0

Wright AL, Provin TL, Hons FM, Zuberer DA, White RH (2007) Nutrient accumulation and availability in compost amended turfgrass soil. HortSci 42:1473–1477

Zai AKE, Takatsugu H, Tsutomu M (2008) Effects of compost and green manure of pea and their combinations with chicken manure and rapeseed residue on soil fertility and nutrient uptake in wheat-rice cropping system. Afr J Agric Res 3:633–639

15

Changes in microbial pathogen dynamics during vermicomposting mixture of cow manure–organic solid waste and cow manure–sewage sludge

Hossein Karimi[1] · Mehdi Mokhtari[1] · Faezeh Salehi[1] · Sajed Sojoudi[2] · Aliasghar Ebrahimi[1]

Abstract

Purpose Microbial pathogens can lead to health problems and disease transmission. Present study aimed to evaluate the microbial quality and quantity of vermicompost production and to compare to the current Iranian standards to protect public health and environmental concerns.

Method This is a pilot-scale experimental study conducted in the Public Health laboratory of Shahid Sadoughi University of Medical Sciences. The samples included organic municipal solid waste, cow manure and wastewater treatment plant sludge which are used for vermicompost production. The samples are mixed as cow manure–organic waste and cow manure–sewage sludge in two reactors. Microbial tests such as *fecal coliforms* (FC) and parasite egg (*Ascaris*) were carried out during start, processing and curing time with duplication analysis in 56 days. Totally, a number of 128 samples was analyzed. Analyses were conducted according to standard methods. Data analysis was conducted through one-way ANOVA and Duncan tests.

Results The results showed a significant reduction in number of FCs in cow manure–organic waste so that the number of 350,000 MPN/g in the raw sample decreased to 800 MPN/g within 8-week period, also FC in the case of cow manure–sewage sludge was achieved to 2400 from 6,500,000 MPN/g. In two cocomposting cases, the parasite eggs were completely removed in the second week.

Conclusion The results showed vermicomposting as a feasible method to convert waste into fertilizer humus in agriculture which also enables to achieve Iranian class A compost standard.

Keywords Vermicompost · Cow manure · Sewage sludge · Solid waste · Parasite egg · Fecal coliform

Introduction

Due to various agricultural, industrial and restaurant activities in different sectors, lots of organic wastes such as animal manure, sewage sludge, food waste and industrial organic waste are produced. These wastes are disposed by scattering on land surfaces, burning or burial which have the potential to increase global water and soil pollution. Approximately, 50–60% of the buried disposed waste is composed of organic waste material (Edwards et al. 2010). The continuous increase of waste production has led to the innovation of new approaches which are environmental friendly. Also, conversion of waste into useful materials, compost and vermicomposting can be noted among these methods (Mohammadi et al. 2012).

The nurturing of earth worms in organic waste and its processing to produce fertilizer from waste is called vermicomposting. As a result of organic waste by earth worms, vermicompost comes to successfully convert such wastes as wastewater treatment sludge, dairy plant waste, food waste, municipal waste, paper waste and animal waste into fertilizer. Vermicompost is the result of waste decomposition through the interaction of earthworms and microorganisms. Even though microorganisms are mainly

✉ Aliasghar Ebrahimi
ebrahimi20007@gmail.com; ebrahimi20007@ssu.ac.ir

[1] Environmental Science and Technology Research Center, Department of Environmental Health Engineering, Shahid Sadoughi University of Medical Sciences, Yazd, Iran

[2] Department of Occupational Health Engineering, Shahid Sadoughi University of Medical Sciences, Yazd, Iran

responsible for decomposition, earth worms increase microbial activity and changes in biological activity by increasing the contact area (Domínguez 2004).

Vermicomposting is not widely used, due to lack of information about reducing human pathogens in this process. Composting process is exploited in the thermophilic phase, but vermicomposting is exploited in the mesophilic phase in which the temperatures should be kept lower than 35 °C to prevent the worms dying.

According to the standards set by the Environmental Protection Agency (EPA) to reduce pathogen up to a safe level for use, the compost pile should be exposed to the temperatures of 55–70 °C for 72 h. So the question is arisen that at the temperatures lower than the standard temperature, like vermicomposting, does the pathogen reduce to a safe level? (Edwards et al. 2010). Several reports have shown the reduction of pathogen to a safe level in the vermicomposting process under specific processing conditions. The first reduction of pathogens was reported by USEPA (1999) where pathogens are reduced by passing biological solid waste through the worm intestine.

Furthermore, Eastman et al. (2001) report a significant reduction of FC and salmonella, studying vermicomposting municipal wastewater solids. Also, Contreras-Ramos et al. (2005) show decrease of pathogen in the vermicompost of cow manure. Monroy et al. (2009) report a reduction of 85–98% for total coliforms in the process of pig manure vermicomposts. Moreover, Liu et al. (2009) report a decrease of *Escherichia coli* O_{157}:H_7 bacteria in artificial soil vermicomposts. Rodríguez-Canché et al. (2010), find a decrease of pathogen in the process of septic tank sludge vermicompost. Aira et al. (2011) also report a significant reduction in pathogen through the process of cow manure vermicomposting. Hill and Baldwin (2012) show a decrease of coliforms in toilet vermicomposting. Hill et al. (2013) report the effects of vermicomposting on reducing the number of Ascaris eggs. In a study that was conducted in 2015 by María and Gómez-Brandón (2016), entitled "effects of digestate on soil chemical and microbiological properties: a comparative study with compost and vermicompost", the results show that FC and *E. Coli* cannot be identified in 60 days.

Organic fertilizers are beneficial as they maintain and improve the structure of soil nutrients, however, if the microbial pathogen and pathogenic factors are present, they may lead to health problems and disease transmission. For this reason, evaluating the number of pathogens is necessary prior to using vermicompost in soil. Thus, this study aimed to evaluate the microbial quality of the produced vermicomposts and to compare them with existing Iranian standards to use the results for maintaining public health and also to ensure a step towards the use of vermicompost.

Materials and methods

This was a pilot-scale experimental study conducted in the laboratory of Yazd school of health to produce vermicompost from: (1) perishable organic waste, such as food waste and fruit and vegetable leftovers, (2) cow manure, (3) wastewater treatment plant sludge, the combination of cow manure with organic waste and municipal sewage sludge in equal proportions in two separate reactors with dimensions of 15 × 30 × 50 cm.

The terms of operation on the pilot included a period of 8 weeks, aerobic conditions and an ambient temperature of (30 ± 5 °C). Also moisture content range was 75–85% controlled by addition of water or aeration.

The *Eisenia fetida* worms were used to produce this vermicompost; a number of 1000 worms was available into each reactor. Determination of the microbial quality was carried out in early stages, during the process and on the final product. These tests included determining the most probable number (MPN) of FC bacteria and parasite eggs (Ascaris) of which the A_1 Medium for FC and zinc sulfate were used for identifying parasite eggs, respectively (Association et al. 1915; Forbes et al. 2005).

After preparing the desired culture environments, 5 g of the sample was inoculated on a peptone water nonselective liquid enrichment broth and then was placed in an incubator shaker for 5 min at a temperature of 25 °C and then in an incubator at 37 °C for 16–20 h. The nine-test tube fermentation method and A_1 culture were used for identification of the fecal coliforms; the test was also carried out using the multiple test tube fermentation method with three dilutions including 1, 0.1 and 0.01 under bain-marie conditions and serology with 41.5 °C temperature for 20–24 h.

For the test of parasite eggs detection, initially, half a test tube was filled with zinc sulfate solution. After transferring 1 g of the sample into the test tube, the suspension was prepared. The above suspension was passed through a two-layered filtrate and the filtered solution was returned to the test tube. Zinc sulfate solution was added to the test tube as long as a distance of 2–3 mm from the solution to the surface of the test tube opening was achieved. After centrifugation at a speed of 3000 rpm for 1 min, a loop from the zinc solution was obtained and added onto a slide containing a legal solution so that the identification process could be carried out under the microscope.

To evaluate the effects of treatment on the number of coliforms from the analysis test, one-way ANOVA and Duncan test were used, and a value of 0.05 was considered as sig.

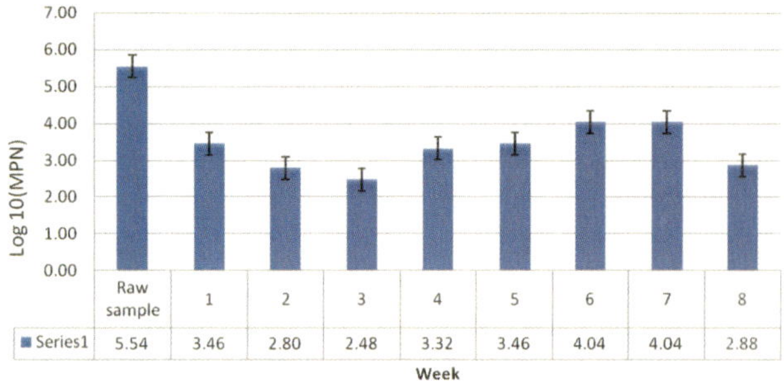

Fig. 1 The trend for the number of FC cow manure–organic waste during the 8 weeks (log 3 is shown, the FC standard in class A compost of Iran)

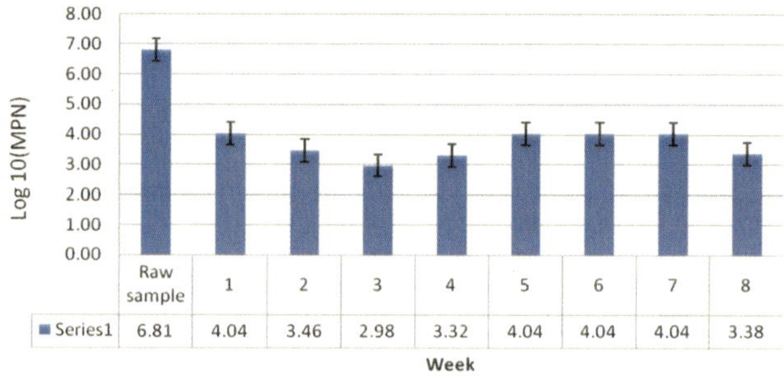

Fig. 2 The trend for the number of FC cow manure–sludge during the 8 weeks (log 3 is shown, the FC standard in class A compost of Iran)

Results and discussion

Given the results, there was a significant decrease in the number of FC in cow manure–organic waste so that the number of FC reached 800 from 350,000 MPN/g in the raw sample within 8 weeks (Fig. 1). In case of the raw samples, FC of cow manure–sludge was also significantly reduced pursuant to the data obtained. In this regard, the number of FC reached 2400 from 6,500,000 MPN/g within 8 weeks (Fig. 2).

In accordance with the Figs. 3 and 4, the mixture of cow manure–sludge and cow manure–organic waste had the highest parasite eggs (22/g) in the raw samples; this amount was completely removed in the second week via the vermicomposting process.

The physico-chemical properties of the raw samples are shown in Tables 1 and 2 including moisture, nitrogen, organic carbon (OC), pH and electrical conductivity (EC). The moisture content of organic waste had a noticeable difference to the cow manure and sewage sludge; however, other parameters showed no significant difference.

Removal of pathogens from such wastes as cow manure and sludge of wastewater treatment plant may be used for soil amendment which is very important in preventing the spread and transmission of disease. Vermicompost is able to remove significant amounts of pathogens. The passing of different treatments through the earthworm gut affects the microbial population in different treatments. Reduction of the number of pathogens in the vermicomposting process depends on different factors such as worm gut enzymes, coelomic fluid secretion (which has antibacterial properties) as well as the competition between different groups of microorganisms (Edwards et al. 2010; Hait and Tare 2011).

In the study of Monroy et al. (2008) on reduction of the density of nematodes, protozoa and FC by passing through earthworm intestine, results show a significant reduction of nematodes and FC and these results are consistent with those of present study. Reduction in the number of nematodes may be due to the fact that they are a part of the earthworms' food; it may also be due to proteolytic enzyme activity to eliminate nematode and pathogens (Monroy et al. 2008). In this study, the highest coliforms removal in the cow manure–organic waste and cow manure–sludge, occurred in the week 3; however, the same result was seen in the week 2 by Monroy et al. (2009), showing the greatest reduction of coliforms in pig manure by passing through earthworm. This result is inconsistent with ours maybe due to differences in type of treatments or conditions of

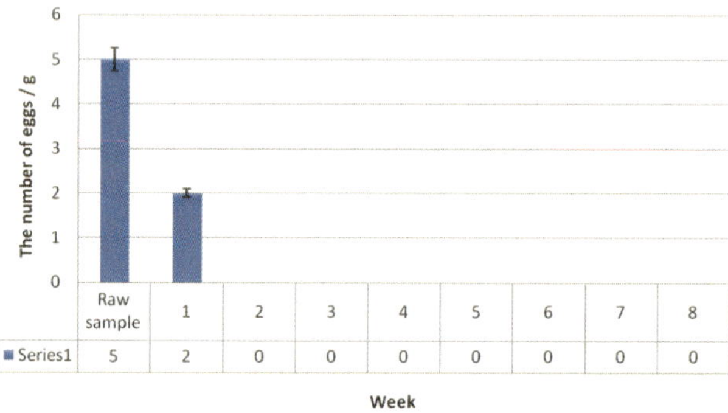

Fig. 3 The trend for the number of parasite eggs in vermicompost produced of a mixture of cow manure–organic waste within 8 weeks (*number 1* in the *figure* shows the standard rate of parasite eggs in class A compost of Iran)

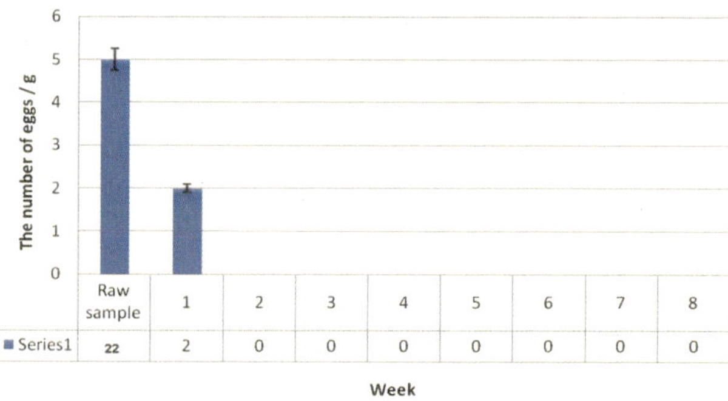

Fig. 4 The trend for the number of parasite eggs in vermicompost produced of a mixture of cow manure–sludge within 8 weeks (*number 1* in the *figure* shows the standard rate of parasite eggs in class A compost of Iran)

Table 1 Physico-chemical parameters of wastewater treatment plant sludge and cow manure in the raw samples

Parameters	Cow manure			Sludge		
	Max	Minimum	Mean	Max	Minimum	Mean
Moisture (%)	40	30	35.75 ± 4.34	20	15	17.25 ± 2.21
EC (mm hose/cm)	1.9	1.2	1.9 ± 0.09	2.9	1.3	2.17 ± 0.69
Organic carbon (%)	75.13	50.21	64.4 ± 11.42	79.8	30.24	70.26 ± 8.06
N (%)	2.26	1.91	2.11 ± 0.15	7.9	2.1	5.41 ± 0.65
pH	7.5	6.2	7.1 ± 0.75	8.11	5.86	7.06 ± 1.09

Table 2 Physico-chemical parameters of the raw domestic waste

	N (%)	Organic carbon (%)	EC (mm hose/cm)	Moisture (%)	pH
Minimum	1.68	53.88	1.45	70	6.2
Max	1.7	54.16	1.46	73.7	6.4

vermicomposting. Also results show an increase of coliforms after the week 4 which may be due to lack of proper food for the earthworm as well as moisture content (Monroy et al. 2009). Another study with results consistent to those of present work is Lui (1982) where a reduction of 98% of pathogens was seen in 17 days while the vermicomposting process (Edwards et al. 2010). Also, Rodríguez-Canché et al. (2010) report reducing septic tank pathogens by vermicomposting, showing a significant reduction in the number of pathogens that is in line with results of present study.

Bowman et al. (2006) studied the impact of vermicomposting on the removal of parasite eggs, indicating low capability of vermicomposting in parasite egg elimination. This result is inconsistent with that of present study which may be due to differences in the conditions of the vermicompost process or laboratory conditions.

One-way ANOVA showed the average FC at least for one of the weeks. In doing so, Duncan test was used to assess the details. The Duncan test results showed that there are significant differences between the raw sample and the product of week 8 ($p < 0.001$). According to the statistics of the Log number of FC for raw sample in different weeks, the mean log FC was found in the week 8 with a difference of 95%. Therefore, at the end of the eighth week, FC reached the standard level.

In a study carried out by Aira et al. (2011) to reduce pathogens in cow manure, results showed that the reduction of FC does not meet EPA standards which are inconsistent with results of present study probably due to differences in the vermicompost system or laboratory conditions. Accordingly, Subrataitail reported that in the final product of vermicompost derived from primary wastewater sludge, pathogens significantly reduced and were lower than the standards set by the USEPA which is consistent with those of present study (Hait and Tare 2011). In another study by Contreras-Ramos et al. (2005), no parasite egg was

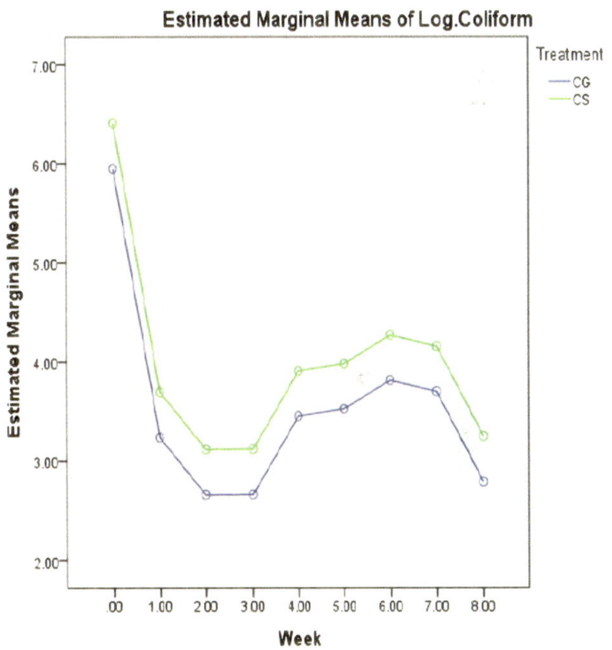

Fig. 5 Changing, the number of FC according to the week, and type of treatment. *CG* cow manure–organic waste, *CS* cow manure–wastewater treatment sludge

detected and the number of FC significantly decreased after 60 days; this result is consistent with that of present work. In a study conducted by Edwards et al. (2010), results showed that the number of FC in the vermicompost process significantly reduced during 7 days; these results were parallel with those of present study. Furthermore, their results show significant decrease of parasitic worms after 3 days of vermicomposting (Edwards et al. 2010); these results are consistent with those of present study.

Eastman et al. (2001) indicated a significant reduction of FC after 7 days of vermicomposting, which is consistent with present study. In addition, the number of parasitic worms significantly reduced after 3 days which was also consistent with present study (Eastman et al. 2001). Finally, estimated marginal means of Log coliform was shown in Fig. 5 which presents changed number of FC according to the week and type of treatment.

Conclusion

The results of this study showed that vermicomposting can be used as a suitable method to transform perishable waste into fertilizer which is usable in agriculture. Earthworms have a great ability to reduce pathogens without the need of increased temperatures and the final product in terms of microbiological quality is placed as class A compost.

Acknowledgements The support President of the School of Public Health and laboratory experts which provide laboratory the possibility to carry out the research to be provided and their valuable guidance in this study helped us to appreciate.

Conflict of interest We have no competing interests.

References

Aira M, Gómez-Brandón M et al (2011) Selective reduction of the pathogenic load of cow manure in an industrial-scale continuous-feeding vermireactor. Bioresour Technol 102(20):9633–9637

Association APH, Association AWW et al (1915) Standard methods for the examination of water and wastewater. American Public Health Association, Washington, DC

Bowman DD, Liotta JL et al (2006) Ascaris suum Egg inactivation and destruction by the vermicomposting worm, *Eisenia foetida*. Proc Water Environ Fed 2006(2):11–18

Contreras-Ramos S, Escamilla-Silva E et al (2005) Vermicomposting of biosolids with cow manure and oat straw. Biol Fertil Soils 41(3):190–198

Domínguez, J. (2004). State-of-the-art and new perspectives on vermicomposting research. In: Earthworm ecology, O· 8-193· 18 1 9· XI04. CRC Press LLC

Eastman BR, Kane PN et al (2001) The effectiveness of vermiculture in human pathogen reduction for USEPA biosolids stabilization. Compost Sci Util 9(1):38–49

Edwards CA, Arancon NQ et al (2010) Vermiculture technology: earthworms, organic wastes, and environmental management. CRC Press, Boca Raton

Forbes BA, Sahm D et al (2005) Diagnostic microbiology. Mosby, St. Louis

Hait S, Tare V (2011) Vermistabilization of primary sewage sludge. Bioresour Technol 102(3):2812–2820

Hill GB, Baldwin SA (2012) Vermicomposting toilets, an alternative to latrine style microbial composting toilets, prove far superior in mass reduction, pathogen destruction, compost quality, and operational cost. Waste Manag 32(10):1811–1820

Hill GB, Lalander C et al (2013) The effectiveness and safety of vermi-versus conventional composting of human feces with *Ascaris suum* ova as model helminthic parasites. J Sustain Dev 6(4):p1

Liu X, Sun Z et al (2009) Growth and stress responses of the earthworm *Eisenia fetida* to *Escherichia coli* O157: H7 in an artificial soil. Microb Pathog 46(5):266–272

Lui TA (1982) Earthworms, producers of biologically active substances. Zh Obshch Biol 24:149–154

Mohammadi A, Ebrahimi A et al (2012) Feasibility energy recovery potential of municipal solid waste in Northwest of Iran. Int J Environ Health Eng 1(1):14

Monroy F, Aira M et al (2008) Changes in density of nematodes, protozoa and total coliforms after transit through the gut of four epigeic earthworms (Oligochaeta). Appl Soil Ecol 39(2):127–132

Monroy F, Aira M et al (2009) Reduction of total coliform numbers during vermicomposting is caused by short-term direct effects of earthworms on microorganisms and depends on the dose of application of pig slurry. Sci Total Environ 407(20):5411–5416

Rodríguez-Canché L, Vigueros LC et al (2010) Pathogen reduction in septic tank sludge through vermicomposting using *Eisenia fetida*. Bioresour Technol 101(10):3548–3553

Usepa E (1999) Control of pathogens and vector attraction in sewage sludge. Environmental Regulations and Technology, USEPA, Cincinnati

16

Organic and inorganic fertilizer application enhances the effect of *Bradyrhizobium* on nodulation and yield of peanut (*Arachis hypogea* L.) in nutrient depleted and sandy soils of Ethiopia

Anteneh Argaw[1]

Abstract
Purpose An investigation was carried out to evaluate the effect of the integrated application of organic and inorganic fertilizer effect on *Bradyrhizobium* effectiveness on nodulation and yield of peanut at the major growing areas of Eastern Ethiopia, Babillae and Fedis sites.
Methods Systemic combination of compost, manure, *Bradyrhizobium* inoculation and NP application was laid out in Randomized Complete Block Design with three replications.
Results The result showed that *Bradyrhizobium* integrated with organic inputs significantly improved the nodule number at Babillae while *Bradyrhizobium* when applied with DAP resulted in a significant increase of nodulation at Fedis site. The highest total biomass and total pods weight at both sites were found to record when *Bradyrhizobium* integrated with manure and compost. Integration of *Bradyrhizobium*, manure and compost at Fedis and *Bradyrhizobium* with manure at Babillae was found to increase the kernel yield by 44 and 66.6% over the control check, respectively. Integration of *Bradyrhizobium*, manure and compost at Babillae and *Bradyrhizobium* with starter N at Fedis significantly increased plant N accumulation. The effect of organic and inorganic application on soil N and organic carbon content was not significant at Fedis, but the slight increase was observed in Babillae site. A significant increase in the soil available P by organic and/or DAP application was found in either of the experimental sites.
Conclusion Organic fertilizer when integrated with Starter N and DAP is better in improving the effectiveness of *Bradyrhizobium*, nodulation and yield of peanut in either of the sites.

Keywords Available P · Compost · Manure · Soil properties · Soil N · Soil organic carbon

Introduction

Most of the research system in Africa emphasizes on managing the three macronutrients nitrogen (N), phosphorus (P), and potassium (K) (Kang and Balasubramanian 1990; Smaling et al. 1993; Smaling and Braun 1996). In this region, crop production is critically dependent on sub-optimum nutrient application, especially N and P which are very low in amount with less than 8 kg ha/year (Crawford and Jayne 2010; Morris et al. 2007; Smaling 2006). Such agricultural practice accelerates the depletion of other macronutrients and micronutrients (Cobo et al. 2010; Sanchez 2002; Smaling et al. 1997) and causes negative soil nutrient balances. In Ethiopia, the nutrient depletion has been 41, 6, and 26 kg/ha/year for N, P, and K, respectively (Stoorvogel and Smaling 1990). This problem is aggravated by the inherent poor fertility in most tropical soils (Okalebo et al. 2003). These practices, consequently, lead to the less responsive or non-responsive soil (Foli 2012; Vanlauwe et al. 2014). When farmers applied fertilizers on this soil, they did not get benefit by increasing crop productivity. Due to this reason, farmers became reluctant to apply inputs, besides the cost of fertilizer is increasing.

✉ Anteneh Argaw
anteneh.argaw@haramaya.edu.et;
antenehargaw@gmail.com

[1] School of Natural Resources Management and Environmental Sciences, College of Agriculture and Environmental Sciences, Haramaya University, Haramaya, Ethiopia

A study conducted across Ethiopia showed that most sites are low in soil organic carbon (SOC), available N, and available P, K and S contents and some micronutrients (Hailu et al. 2015; Laekemariam et al. 2016). These nutrient deficiencies can affect not only plants but also soil microbes including rhizobia populations (O'Hara 2001) and its activities such as N_2 fixation (Giller 2001). To lessen these problems, the use of chemical fertilizer has been limited mainly because of little accessibility and lack of buying ability of the poor farmers in the region (Morris et al. 2007; Giller et al.1998; Mugwira and Murwira 1997; Rufino et al. 2010). For instance, mineral fertilizers in Africa cost at the farm gate, two to six times as much as in Europe, North America, or Asia (Sanchez 2002). Therefore, organic inputs are a viable alternative source of plant nutrients for resource-poor farmers. Application of organic input usually leads to increased crop yields (Ogundare et al. 2012; Kimetu et al. 2004; Mugwira 1984; Vanlauwe et al. 2001a, b). However, yield reduction by organic fertilizer has also been reported (Mugwira and Murwira 1997; Nhamo 2002). The negative effect of low-quality organic fertilizer on the productivity of peanut has been resolved when it has been applied it's with chemical fertilizer and/or fly ash (Burgos et al. 2006).

As residue quality influences the rate of decomposition, the amount of nutrients immobilization–mineralization and timing of nutrient release (Heal et al. 1997). Vanlauwe et al. (2005) found that Class I residues caused net N mineralization, Class II residues had no effect on mineral N, and Class III and IV residues resulted in net N immobilization. The addition of fertilizers with intermediate quality organic inputs may increase organic matter decomposition (Sakala et al. 2000; Zingore et al. 2003) and, thus, enhances the crop production. Animal manures and composts have shown in several trials to increase nutrient availability and to partly substitute mineral fertilizers (Goyal et al. 1999). However, availability of manure in smallholder farms was generally lower in Ethiopia (Lupwayi et al. 2000). In this region, some crop by-products such as Khat leftover are available for use as soil amendments because of fewer alternate uses.

Keeping the above in view, the present investigation was undertaken with a major objective to make the influence towards growing peanut production by promoting the use of integrated nutrient management that is capable of increasing soil fertility status in the small plot based farming systems in the Eastern Ethiopia using locally available resources. The specific objectives of the study were to compare the effect of combined application of different organic input and inorganic fertilizer with Bradyrhizobium inoculation on nodulation and yield of peanut and to assess how these inputs influence selected soil properties.

Materials and methods

Description of the study sites

Field experiments were initiated during the rainy season of 2014 cropping season at Fedis (09°06.941′N and 042°04.835′E at an altitude of 5476 ft above sea level [asl]) and Babillae (09°13.234′N and 042°19.407′E at 5478 ft asl) experimental sites, Eastern Hararghe, Ethiopia under Haramaya University. The study sites comprise lowland and midland agroecological zones that obtain 400–800 mm of annual rainfall in a bimodal pattern. In normal years, 60–70% of the rainfall occurs during the main rains season, between May and October, while the rest falls during the short rains between March and May. At Fedis site, the mean maximum and minimum temperatures are 27.8 and 8.8 °C, respectively, with annual mean rainfall of 714.5 mm. The climate of the Babillae is semi-arid tropical, with the mean annual air maximum and minimum temperature for the duration from 2000 to 2014 of 20 and 9 °C, respectively, and the mean annual precipitation of about 566 mm.

The topography of Babillae is rolling with low SOC concentration (0.56%) and total N (0.06%), available P (2.22 mg/kg), cation exchangeable capacity of 6.26 cmol(+)/kg, exchangeable Ca^{+2}, Mg^{+2}, Na^{+1} and K^{+1} of 4.18, 3.5, 0.15 and 0.34 cmol(+)/kg and sand, silt, clay contents of 76, 6 and 18%, respectively (Tekalign 1991). The soil pH and electric conductivity are 6.66 and 0.04 mS/cm, respectively.

The surface soil in the plot area in Fedis site before commencing the experiment is a silty clay loam, containing 36% sand, 45% silt, and 19% clay. The soil (0–20 cm depth) had a pH of 7.76, electric conductivity of 0.06 mS/cm, ammonium acetate-extractable K of 1.09 cmol(+)/kg, Mg of 12.87 cmol(+)/kg, Na of 0.12 cmol(+)/kg, Ca of 23.12 cmol(+)/kg with cation exchangeable capacity of 32.22 cmol(+)/kg. The soil had low organic carbon (1.32) and Olsen extractable P (1.78 mg/kg) and medium total N (0.12%) (Tekalign 1991).

Source of the test variety

Groundnut variety "*Baha Jidu*" which has been recently approved as high yielder in this region was obtained from groundnut improvement project, Haramaya University, Ethiopia. BaHa-jidu variety is the runner type and medium seeded (Kebede and Bushra 2012).

Organic fertilizer sampling and analyses

The farmyard manure was prepared mainly from cow dung and hay, which is normally used as a bedding material in the cow shed. Compost was prepared following the conventional method using locally available khat leftover organic waste. The following properties were analyzed according to the recommended testing methods: pH by potentiometric; Ammonium acetate extraction, flame photometry for available K determination; organic C by loss of weight on ignition; Olsen (available P); Kjeldahl (total N); DTPA extraction(Zn); Azomethine-H method (available B); KCl extract (NO_3^-) and KCl extract-Magnesium oxide distillation (NH_4^+). The average nutrient composition of FYM and compost applied in the experiment during this period are given in Table 1.

Treatments and crop management

The experiment consisted of eleven treatments: (1) 2 ton manure/ha + 4 ton compost/ha + no inoculation; (2) 2 ton manure/ha + 4 ton compost/ha + *Bradyrhizobium* inoculation; (3) 2 ton manure/ha + no inoculation; (4) 2 ton manure/ha + *Bradyrhizobium* inoculation; (5) 4 ton compost/ha + no inoculation; (6) 4 ton compost/ha + *Bradyrhizobium* inoculation; (7) DAP—(46 P_2O_5 kg/ha + 19 kg N/ha); (8) (46 P_2O_5 kg/ha + 19 kg N/ha) + *Bradyrhizobium* inoculation; (9) 20 kg N/ha; (10) 20 kg N/ha + *Bradyrhizobium* inoculation and (11) the control check. The rates of organic inputs were developed based on its inorganic N (NO_3^- and NH_4^+) concentration. The treatments were replicated three times, and laid out according to a Randomized Complete Block (RBD) design, with plot dimension of 3 m × 3 m.

The experimental plots preparation involved one plowing immediately after getting the first rainfall in June followed by blade harrowing. The organic inputs are surface applied and incorporated minimally with a hoe to a depth of approximately 10 cm. Inorganic N and P fertilizers were applied in the form of urea and triphosphate, respectively. The entire dose of inorganic and organic fertilizers was applied as basal at the beginning of growing season as per the treatment.

Soil sampling and analysis

After the peanut harvest in November 2014, soil samples were taken from 0 to 20 cm soil layers from each plot of two experimental sites. In each plot, the soil samples were collected from four points and were mixed to get a composite sample. Soil samples were air dried, gently ground and passed through a 2-mm sieve. The major chemical composition of the manure and compost was then analyzed using standard laboratory methods of soil and plant analysis.

Agronomic data collection

At the R_2 stage of peanut, five plants from the central three rows were uprooted. The nodulation status (nodule number dry eight) and shoot dry weight were recorded. At harvest, the pod weight, the total biomass yield, and the kernel weight were measured. The yield was calculated by harvesting central three rows of peanut (3.6 m^2).

Plant samples and analysis

At late flowering stage (R_2), three plant samples were uprooted for plant N tissue analysis. The oven dried plant sub-samples were then ground and analyzed for N by the micro-Kjeldahl procedure.

Statistical analysis

Statistical analyses of the data were done using the SAS version 9.2 to analyze variance and to determine the statistical significance of the treatment effects. Analysis of variance (ANOVA) was performed on a fully randomized

Table 1 The selected chemical properties of khat leftover and manure compost

Parameters	Khat leftover compost	Manure compost
pHH_2O (1:2.5)	6.82	7.21
EC (mS/cm)	7.46	7.50
Total N (%)	2.48	1.30
Organic carbon (%)	33.02	18.91
NH_4–N (mg/kg)	43.22	57.59
NO_3–N (mg/kg)	9860.79	5557.60
Available K (cmol(+)/kg soil)	12.36	25.59
Available P (mg/kg)	524.13	763.14
Zn (mg/kg)	26.81	4.49
B (mg/kg)	7.91	3.49

plot design to test for significance of treatments and means were compared by least significance difference (LSD) at the 5% level (SAS, 1996).

Results

The nodulation and yield of peanut showed a significant response to organic (compost and manure) and inorganic fertilizer (Urea and DAP) integrated with *Bradyrhizobium* inoculation (Tables 2, 3, 4). At Fedis, the nodule number was found to be increased by 141, 142.8 and 143.6% due to the applied compost (C) + *Bradyrhizobium*, Manure (M) + *Bradyrhizobium*, and DAP + *Bradyrhizobium*, respectively, compared to the control check (Table 2). Only DAP + *Bradyrhizobium* application resulted in a significant increase in the nodule number of peanut at Babillae. However, the treatments did not affect the pooled NN peanut. Likewise, the nodule dry weight of peanut at Babillae was enhanced significantly by M + C + *Bradyrhizobium* and C + *Bradyrhizobium* application (Table 2). At Fedis, an increase in nodule dry weight of peanut by starter N (20 kg N/ha) application was found. Application of M + C + *Bradyrhizobium* and C + *Bradyrhizobium* resulted in a significant increase in the pooled nodule dry weight.

Excluding C + M, the other organic and inorganic combination of fertilizer improved the effectiveness of *Bradyrhizobium* inoculation on nodulation at Babillae site (Figs. 1a, 2a). At Fedis site, manure and compost application enhanced the effectiveness inoculation on nodule number and dry weight at Fedis site (Figs. 1b, 2b). However, starter N and DAP application did not improve the effectiveness of *Bradyrhizobium* inoculation at Fedis site.

There was no significant effect of organic and inorganic fertilizers' combination with *Bradyrhizobium* on shoot dry weight measured at late flowering and a number of pods per plant at Babillae site (Table 2). In contrary, M, C, and DAP applied with *Bradyrhizobium* caused a significant positive influence on the shoot dry weight at Fedis site. The pooled shoot dry weight significantly increased by DAP applied with *Bradyrhizobium* inoculation. The plants receiving DAP + *Bradyrhizobium* and M + *Bradyrhizobium* were found to record significantly higher number of seed per plant than the control at Babillae and Fedis sites,

Table 2 Nodule number, nodule dry weight and shoot dry weight of peanut as affected by combined application of organic and inorganic fertilizer application at Babillae and Fedis sites, eastern Ethiopia

Treatments	Nodule number			Nodule dry weight			Shoot dry weight		
	Babillae	Fedis	Mean	Babillae	Fedis	Mean	Babillae	Fedis	Mean
M + C	70.0abc	21.0c	5.1a	0.080bc	0.060abc	0.070abc	63.4a	46.0d	54.7bc
M + C + B	63.3abc	35.7abc	49.5a	0.188a	0.082abc	0.135a	53.8a	49.4 cd	51.6c
M	69.3abc	47.3abc	58.3a	0.028c	0.048bc	0.038bc	52.9a	64.6bcd	58.8bc
M + B	94.7a	50.0abc	72.3a	0.056c	0.094ab	0.075abc	61.4a	77.1abc	69.2abc
C	52.7bc	37.7abc	45.2a	0.636c	0.043bc	0bc.053	63.2a	51.7 cd	57.4bc
C + B	94.0a	47.7abc	70.8a	0.142ab	0.053bc	0.0ab98	79.1a	81.9ab	80.5ab
DAP	42.3c	50.3c	46.3a	0.048c	0.080abc	0.064bc	81.4a	65.9bcd	73.6abc
DAP + B	95.0a	31.7a	63.3a	0.097bc	0.062abc	0.079abc	89.9a	95.7a	92.8a
Urea	45.0bc	64.0ab	54.5a	0.039c	0.111a	0.075abc	53.6a	54.8bcd	54.2c
Urea + B	83.3ab	59.3bc	71.3a	0.087c	0.078abc	0.083abc	63.2a	58.2bcd	60.7bc
No amendment	39.0c	30.3	34.7a	0.023c	0.033c	0.028c	72.1a	45.1d	58.6bc
Mean	68.1	43.2	55.6	0.077	0.068	0.073	66.7	62.8	64.7
Significance	***	**	ns	***	***	***	ns	***	***
LSD ($P < 0.05$)	39.0	31.6	42.3	0.078	0.053	0.068	44.3	29.4	15.2
CV (%)	19.63	25.04	39.20	34.34	26.92	48.18	22.72	16.07	20.94

M + C—2 ton manure/ha + 4 ton compost/ha + no inoculation; M + C + B—2 ton manure/ha + 4 ton compost/ha + *Bradyrhizobium* inoculation; M—2 ton manure/ha + no inoculation; M + B—2 ton manure/ha + *Bradyrhizobium* inoculation; C—4 ton compost/ha + no inoculation; C + B—4 ton compost/ha + *Bradyrhizobium* inoculation; DAP—(46 P_2O_5 kg/ha + 19 kg N/ha); DAP + B—(46 P_2O_5 kg/ha + 19 kg N/ha) + *Bradyrhizobium* inoculation; Urea—20 kg N/ha; Urea + B—20 kg N/ha + *Bradyrhizobium* inoculation

Means within the same factor and column followed by the same letter are not significantly different at 5% level of significance

ns non significant

** Significant at 0.01

*** Significant at 0.001

Table 3 Number of pegs per plant, hundred seed weight and total biomass yield of peanut as affected by combined application of organic and inorganic fertilizer application at Babillae and Fedis sites, eastern Ethiopia

Treatment	Number of peg per plant			Hundred seeds weight			Total biomass yield (kg/ha)		
	Babillae	Fedis	Mean	Babillae	Fedis	Mean	Babillae	Fedis	Mean
M + C	10.7a	23.2ab	16.9a	61.0ab	41.9ab	51.5a	6296.3b	5759.3b	6027.8c
M + C + B	11.2a	28.4a	19.8a	59.5ab	41.5ab	50.5a	8833.3a	8933.3a	8883.3a
M	11.9a	26.1ab	19.0a	63.7ab	41.0b	52.3a	6666.7b	5703.7b	6185.2bc
M + B	9.8a	23.4ab	16.6a	61.8ab	54.6a	58.2a	7092.6ab	8600.0a	7846.3ab
C	11.0a	22.1ab	16.6a	54.6b	40.0b	47.3a	5796.6b	5648.1b	5722.4c
C + B	9.2a	27.2a	18.2a	60.2ab	42.8ab	51.5a	5592.6b	7394.4ab	6493.5bc
DAP	10.6a	17.4b	14.0a	64.2ab	41.5ab	52.8a	6666.7b	5277.8b	5972.2c
DAP + B	9.8a	24.8ab	17.3a	69.9a	41.2b	55.6a	6925.9ab	6903.7ab	6914.8bc
Urea	10.2a	22.6ab	16.4a	61.2ab	43.5ab	52.4a	5870.4b	5869.2b	5869.8c
Urea + B	11.0a	26.3ab	18.7a	60.7ab	41.0b	50.9a	6851.9ab	5666.7b	6259.3bc
No amendment	9.0a	20.0ab	14.5a	58.3ab	41.1b	49.7a	6037.0b	6611.1ab	6324.1bc
Mean	10.4	23.8	17.1	61.4	42.7	52.1a	6602.72	6578.85	6560.78
Significance	ns	*	ns	*	*	ns	**	***	***
LSD ($P < 0.05$)	4.0	9.3	26.3	12.8	13.3	22.0	2108.9	2570.7	1737.7
CV (%)	13.13	13.35	45.8	7.15	10.68	21.75	10.94	13.39	13.60

M + C—2 ton manure/ha + 4 ton compost/ha + no inoculation; M +C + B—2 ton manure/ha + 4 ton compost/ha + *Bradyrhizobium* inoculation; M—2 ton manure/ha + no inoculation; M + B—2 ton manure/ha + *Bradyrhizobium* inoculation; C—4 ton compost/ha + no inoculation; C + B—4 ton compost/ha + *Bradyrhizobium* inoculation; DAP—(46 P_2O_5 kg/ha + 19 kg N/ha); DAP + B—(46 P_2O_5 kg/ha + 19 kg N/ha) + *Bradyrhizobium* inoculation; Urea—20 kg N/ha; Urea + B—20 kg N/ha + *Bradyrhizobium*

Means within the same factor and column followed by the same letter are not significantly different at 5% level of significance

ns non significant

* Significant at 0.05

** Significant at 0.01

*** Significant at 0.001

respectively. The pooled number of seeds per pod was not affected by the treatments.

The highest total biomass yield (kg/ha) of peanut at Babillae and Fedis was recorded by combined application of manure, compost and *Bradyrhizobium* (Table 3). This treatment increased the total biomass yield by 46.3 and 35.1% over the control check at Babillae and Fedis site, respectively. The total pod's weight (kg/ha) under application of M + C + *Bradyrhizobium* and Urea + *Bradyrhizobium* was significantly ($P < 0.05$) superior to that of control check at Babillae site (Table 4). Similarly, the total pod's weight at Fedis and pooled total pods weight were found to be significantly increased due to combined application of manure, compost and *Bradyrhizobium*.

The shelling % was not influenced by the treatment at Babillae site (Table 4). However, most of the organic and inorganic fertilizers applied with *Bradyrhizobium* inoculation encouraged the pooled shelling % at Fedis site. A significant increase in plant N accumulation at Babillae site was found in combined application of manure, compost and *Bradyrhizobium*. Most of the organic and inorganic fertilizer combination increased the accumulation of N in plant tissue at Fedis site and the pooled plant N accumulation (Table 4).

At Babillae site, manure and compost applied with *Bradyrhizobium* inoculation significantly increased the plant N accumulation compared to sole application of manure and compost (Fig. 3a). *Bradyrhizobium* inoculation did not increase the plant N accumulation when integrated with manure and compost, and manure at Fedis site (Fig. 3b), but inoculation improved the plant N accumulation when integrated with DAP and Urea.

The kernel yield of peanut was significantly influenced by *Bradyrhizobium* inoculation integrated with organic and inorganic fertilizer at Babillae site (Table 4). At this site, the highest kernel yield was recorded at *Bradyrhizobium* integrated with manure, followed by manure applied with compost. However, Kernel yield at Fedis site and pooled kernel yield were not significantly affected by the treatments. The highest kernel yield (1562.1 and 1818.8 kg/ha) of peanut at Babillae and Fedis site was 66.6 and 44.1% increase over unamended control of respective sites, respectively.

Table 4 Total weight of pods per hectare, shelling percentage, grain yield and plant tissue N accumulation of peanut as affected by combined application of organic and inorganic fertilizer application at Babillae and Fedis sites, eastern Ethiopia

Treatment	Pod (kg/ha)			Shelling %			Kernel yield (kg/ha)			Plant tissue N accumulation		
	Babillae	Fedis	Mean	Babillae	Fedis	Mean	Babillae	Fedis	Mean	Babillae	Fedis	Overall
M + C	2588.6abcd	2189.2b	2388.9abcd	53.6abc	67.4a	60.5a	1388.1ab	1466.8ab	1427.4ab	2.717bc	2.990ab	2.853ab
M + C + B	3420.9ab	3222.2a	3321.6a	35.3d	56.8ab	46.0ab	1182.0abc	1818.8a	1500.4a	3.050a	2.937abc	2.993a
M	2586.8abcd	1691.8b	2139.3bcd	48.0abcd	68.9a	58.5a	1217.2abc	1166.1abc	1191.7abc	2.657bc	2.743bcd	2.700bcde
M + B	3342.1abc	2244.7b	2793.4ab	47.2abcd	62.1ab	54.6ab	1562.1a	1397.5ab	1479.8a	2.607bc	2.567de	2.587cde
C	2061.9bcd	2027.0b	2044.4bcd	41.2 cd	58.3ab	49.7ab	854.0c	1170.2abc	1012.1bc	2.523c	2.610de	2.567de
C + B	1966.0bcd	2143.5b	2054.7bcd	59.3ab	60.3ab	59.8a	1160.9abc	1288.0abc	1224.4abc	2.787abc	2.757bcd	2.772abcd
DAP	1881.4 cd	1546.4b	1713.9d	61.9ab	64.3ab	63.1a	1173.0abc	656.9c	914.9c	2.683bc	2.700 cd	2.692bcde
DAP + B	2747.6abcd	2118.2b	2432.9abcd	45.4bcd	52.8ab	49.1ab	1247.9abc	1120.8bc	1184.3abc	2.737abc	2.817bcd	2.777abcd
Urea	1598.4d	2076.1b	1837.3 cd	63.2a	56.3ab	59.7a	997.1bc	1169.8abc	1083.4abc	2.750abc	2.897abc	2.823abc
Urea + B	3839.5a	2307.7b	3073.6ab	30.6d	48.5b	39.5b	1118.7abc	1090.7bc	1104.7abc	2.873ab	3.143a	3.008a
No amendment	1982.2bcd	2068.0b	2025.1 cd	47.5abcd	61.2ab	54.3ab	937.7bc	1261.9abc	1099.8abc	2.653bc	2.387e	2.520e
Mean	2546.9	2148.6	2347.7	48.0	59.7	54.1	1167.15	1237.04	1202.09	2.731	2.777	2.754
Significance	**	***	***	***	*		**	***	***	***	***	***
LSD ($P < 0.05$)	1528.8	780.28	1032.4	17.6	18.8	18.6	470.56	658.56	464.31	0.315	0.264	0.241
CV (%)	20.57	12.44	22.68	12.4	10.8	17.7	13.81	18.24	19.93	3.95	3.25	4.52

M + C—2 ton manure/ha + 4 ton compost/ha + no inoculation; M + C + B—2 ton manure/ha + 4 ton compost/ha + *Bradyrhizobium* inoculation; M—2 ton manure/ha + no inoculation; M + B—2 ton manure/ha + *Bradyrhizobium* inoculation; C—4 ton compost/ha + no inoculation; C + B—4 ton compost/ha + *Bradyrhizobium* inoculation; DAP—(46 P_2O_5 kg/ha + 19 kg N/ha); DAP + B—(46 P_2O_5 kg/ha + 19 kg N/ha) + *Bradyrhizobium* inoculation; Urea—20 kg N/ha; Urea + B—20 kg N/ha + *Bradyrhizobium* inoculation

Means within the same factor and column followed by the same letter are not significantly different at 5% level of significance

Fig. 1 Effect of *Bradyrhizobium* inoculation on nodule number of peanut at **a** Babillae and **b** Fedis experimental sites

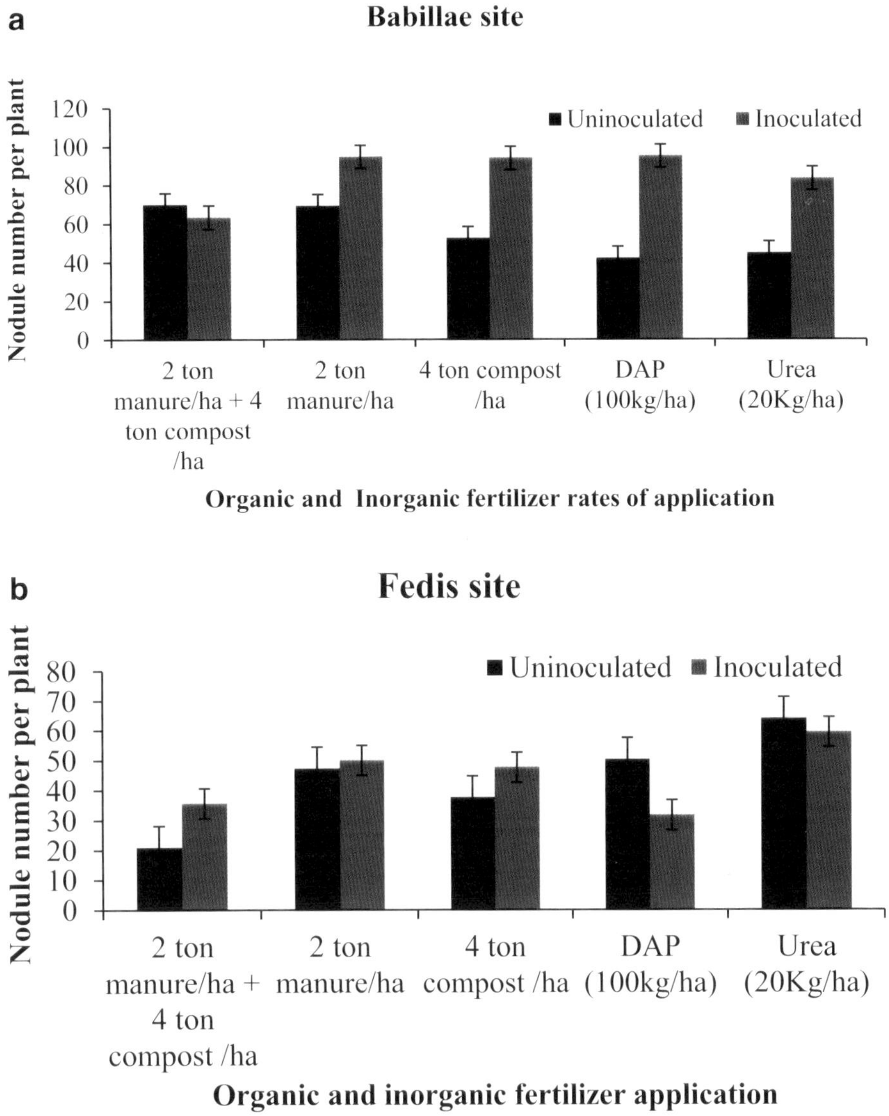

Inoculating *Bradyrhizobium* integrated with compost and manure recorded the higher kernel yield than those organic fertilizers without inoculation at Babillae (Fig. 4a). However, compost and manure application together did not enhance the effect of *Bradyrhizobium* on kernel yield. At Fedis site, *Bradyrhizobium* in combination with organic and DAP application produced higher kernel yield than those obtained from fertilizer without inoculation (Fig. 4b).

The effect of organic and inorganic fertilizers on soil N and organic carbon was not significant at Fedis site (Table 5). The treatments did not also significantly influence the pooled soil N and organic C content. Slight increase in soil N and organic C by the organic and inorganic application was found at Babilale site. However, organic and inorganic application significantly improved the soil available P in both experimental sites. The pooled available P significantly increased when applied DAP alone and in combination with *Bradyrhizobium* over the control check.

Discussion

The result from this experiment showed that *Bradyrhizobium* inoculation in conjunction with organic (Compost and manure) and inorganic fertilizer (NP) significantly increased the nodulation and yield of peanut at both experimental sites. Inoculating *Bradyrhizobium* integrated with organic fertilizer and DAP at both sites were found to increase significantly the nodulation when compared to the corresponding sites control check. This finding is similar to those reported by Panda et al. (2012) who found that poultry manure boosts the effectiveness of *Rhizobium* in cowpea. With the present study, all fertilizer

Fig. 2 Effect of *Bradyrhizobium* inoculation on nodule dry weight of peanut at **a** Babillae and **b** Fedis experimental sites

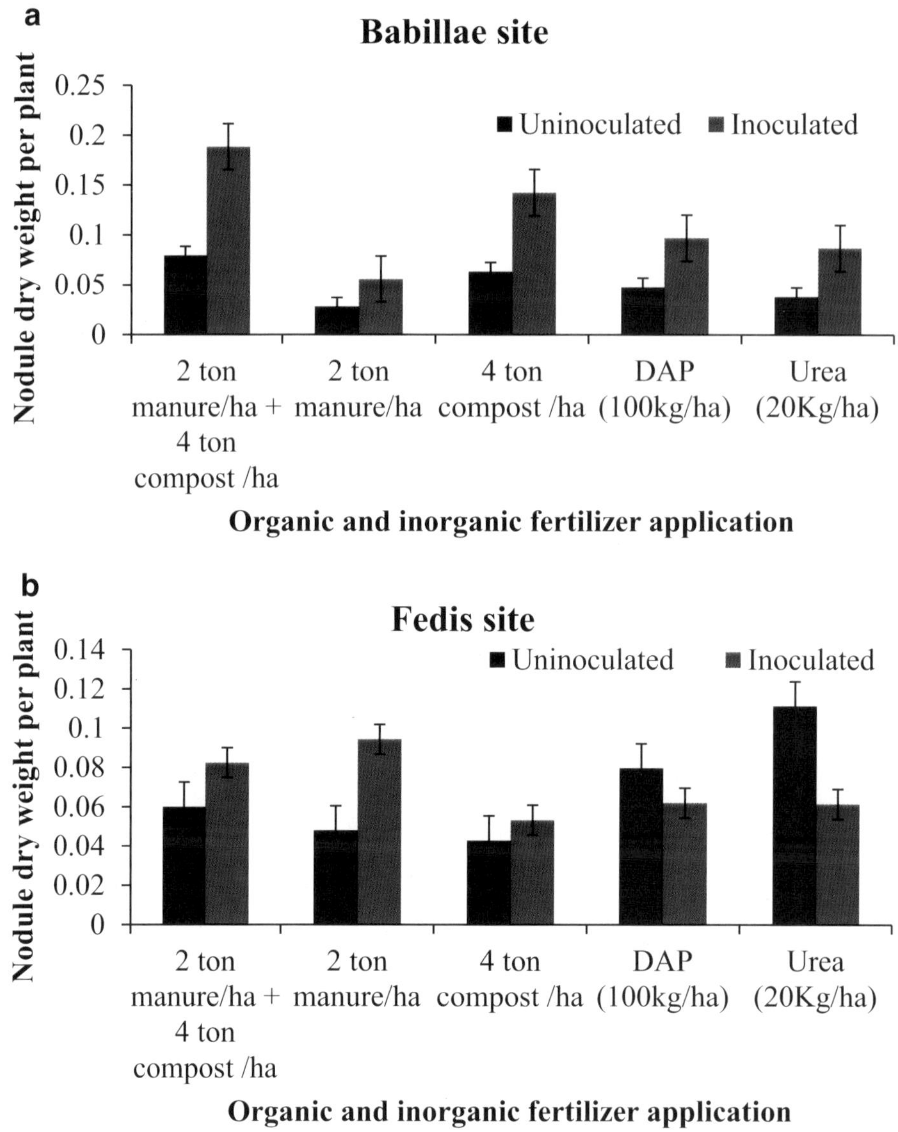

combination improved the effectiveness of *Bradyrhizobium* inoculation on the NN and NDW at Babillae site. However, an increase in NN and NDW due to inoculation was recorded only with the sole and combined application of manure and compost at Fedis. These difference results could have been attributed to the correction of the deficiencies of essential macronutrients and micronutrients on top of NP in Babillae soil. Previous work on organic fertilization effect on the soil has found to buildup of essential plant nutrients (i.e., N, P, K, S, Ca, Mg, Zn, Fe, Mn, and B) in the soil (Dotaniya et al. 2016; Ogundare et al. 2012; Rezig et al. 2012). Organic manure application enhanced the native rhizobia population nodulating cowpea by 23% above control (Kimiti and Odee 2010). This positive influence leads to enhance the root growth and the uptake of nutrients (Ibrahim et al. 2011) and thus improve the nodulation (Basu et al. 2007; Mohammadi et al. 2011; Tsai et al. 1993). This observation is in agreement with other studies where organic matter has been shown to increase the viable number of rhizobia and nodulation of peanut (Basu et al. 2008). These authors found a significant increase in native rhizobia population and nodule formation.

The application of organic fertilizer rich in nitrate did not suppress nodulation in both experimental sites though this N-rich material with C: N ratio less than 17:1 enhanced mineralization by microorganisms. However, the negative effect of N on nodulation did not observe. Lack of inhibition effect of N might be because of the attenuated effect of other essential nutrients found in the organic input (Burgos et al. 2006). Wu and Arima (1992) reported that N applied with other nutrients had increased nodulation whereas N applied alone reduced the nodule formation. This result suggests that application of medium quality

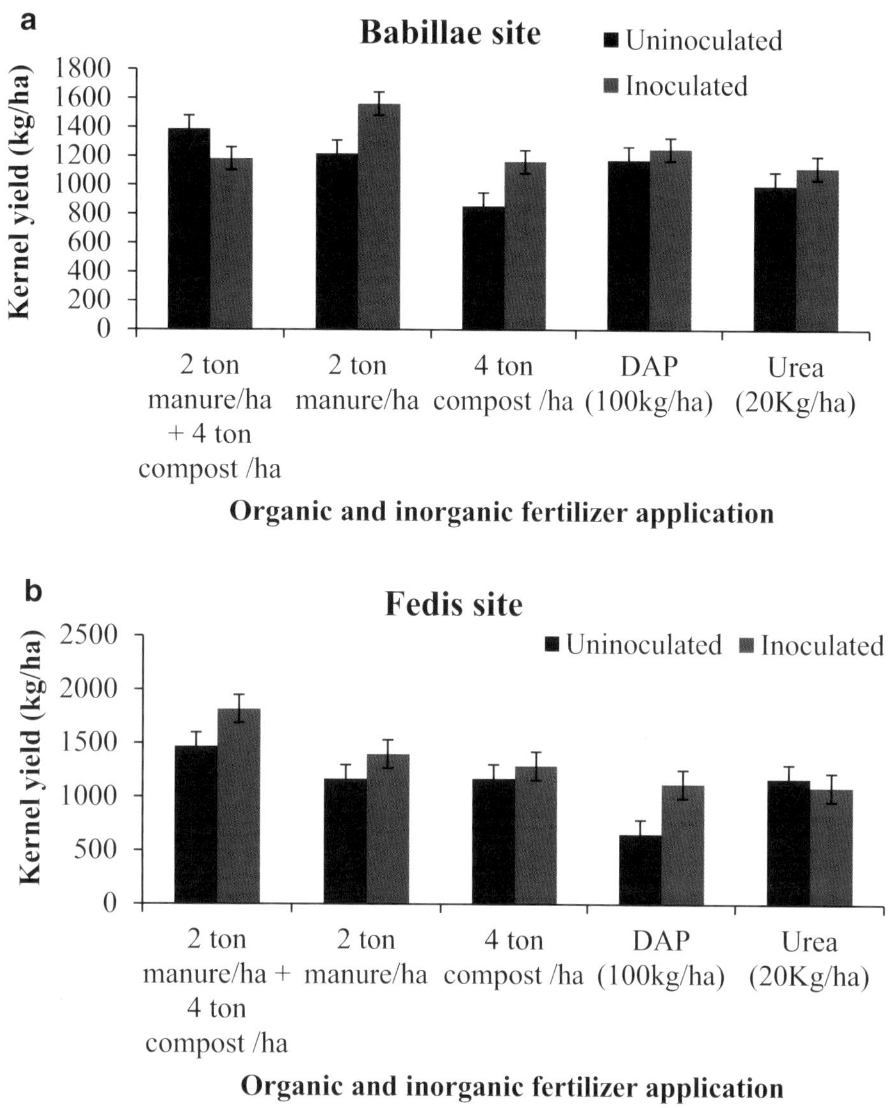

Fig. 3 Effect of *Bradyrhizobium* inoculation on grain yield of peanut at **a** Babillae and **b** Fedis experimental sites

organic fertilizer is essential to enhance nodulation in degraded and low fertile sand soils of Ethiopia.

Organic and inorganic fertilizer application did not increase significantly the shoot dry weight measured at the late flowering stage at Babillae site but it was increased significantly at Fedis site. The effect of organic fertilizer with *Bradyrhizobium* inoculation at Babillae site did not improve the number of pod per plant and hundred seeds weight. However, the treatment where *Bradyrhizobium* was combined with DAP significantly increased the hundred-seed weight at Babillae site. Insufficient nutrient supply at flowering stage of the plants due to immobilization (Petersen et al. 2005) might have been the reason, which limited the remarkable effect of organic inputs. Limitation in plant growth due to nutrients immobilization particularly N and P had also been stated by Herencia et al. (2011).

In contrary, there was an increase in a number of pods per plant, hundred seeds weight, and shoot dry weight of peanut by *Bradyrhizobium* when integrated with organic fertilizer (manure and/or compost) at Fedis site. This result indicates the need of organic fertilizer application at Fedis site beside *Bradyrhizobium* inoculation. This increase in pod number might be through reducing bulk density besides supplying adequate nutrients (Mittra et al. 2005), thereby increasing pegging and pods formation.

The highest total biomass yield, kernel yield and plant N accumulation of peanut at either of the experimental sites were found where *Bradyrhizobium* in conjunction with organic fertilizer was applied. This shows the importance of organic fertilizer application in increasing yield of peanut. This pronounced effect of organic inputs on final yield of peanut could be associated with the mineralization and releasing of nutrients from the organic inputs at a late stage of the plants. It has been known that organic fertilizer is the major source of mineral nutrients (Eghball et al. 2002). These responses of grain yields to organic fertilizer

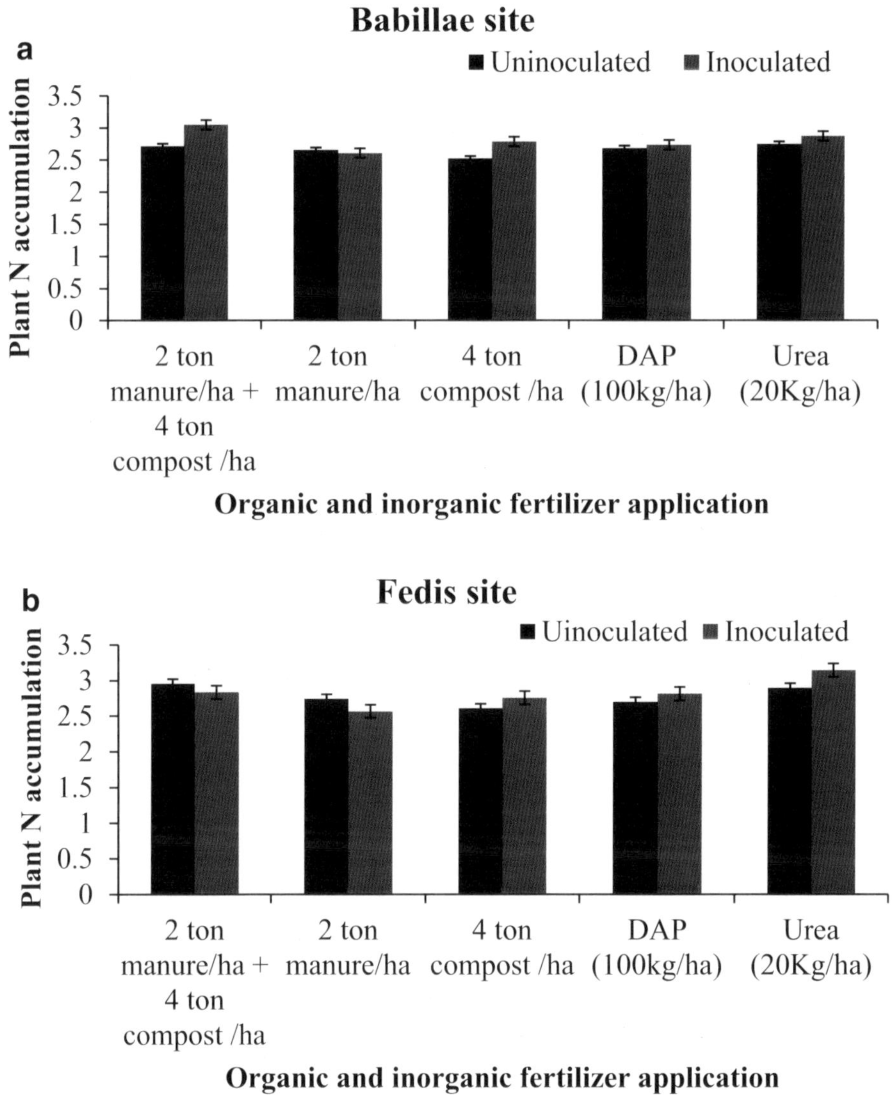

Fig. 4 Effect of *Bradyrhizobium* inoculation on plant N accumulation of peanut at **a** Babillae and **b** Fedis experimental sites

are consistent with other studies on peanut such that there was a significant increase in peanut production by organic input enriched by fly ash (Burgos et al. 2006). The benefits of organic fertilizer and inorganic fertilizer application in combination with rhizobia on the agronomic productivity of food legumes have been previously reported in chickpea (Namvar et al. 2013; Shahzad et al. 2013). The present results are also in accordance with the findings of Shahzad et al. (2013) who demonstrated a significant increase in soil nutrients including available P by organic matter application.

The effect of organic and inorganic fertilizer on soil N and organic C was not significant at both sites indicating that one year application organic fertilizer did not affect the soil N and soil organic matter. However, the results showed that DAP and organic inputs increased the available P in both soils. The increase in P due to organic fertilizer application could be organic acids production as a result of microbial decomposition of organic matter, which can solubilize the native unavailable inorganic P beside serve as source of inorganic P (Ramesh et al. 2009).

Conclusion

In general, organic fertilizer applications are relevant to boost the effectiveness of *Bradyrhizobium* on nodulation and yield of peanut in degraded and sand soil of eastern Ethiopia. Compost and farmyard manure prepared from locally available materials and inoculation of elite *Bradyrhizobium* isolate are needed to increase the productivity of peanut sustainably in sandy and degraded soil of eastern Ethiopia. Although this study finds a handful of beneficial effects of organic amendments with *Bradyrhizobium* on the yield of peanut, its effect on selected soil properties has been negligible. Therefore, further study on

Table 5 Total N, organic matter and available P of Babillae and fedis soils as affected by combined application of organic and inorganic fertilizer application

Treatment	Soil total N			Soil organic carbon			Available P concentration		
	Babillae	Fedis	Mean	Babillae	Fedis	Mean	Babillae	Fedis	Mean
M + C	0.037ab	0.125a	0.081a	0.433ab	1.304a	0.868a	3.78bc	3.95ab	3.87c
M + C + B	0.042ab	0.124a	0.083a	0.487ab	1.270a	0.879a	7.31b	5.25ab	6.28bc
M	0.047ab	0.098a	0.072a	0.541ab	1.137a	0.839a	1.53c	5.20ab	3.37c
M + B	0.028b	0.084a	0.056a	0.325b	1.141a	0.733a	7.19b	6.28a	6.73bc
C	0.047ab	0.093a	0.070a	0.541ab	1.083a	0.812a	5.76bc	2.87b	4.32c
C + B	0.051ab	0.098a	0.075a	0.596ab	1.137a	0.866a	5.11bc	3.58ab	4.35c
DAP	0.700a	0.087a	0.078a	0.812a	1.083a	0.947a	20.46a	5.91ab	13.18a
DAP + B	0.042ab	0.095a	0.068a	0.487ab	1.137a	0.812a	15.42a	6.23a	10.83ab
Urea	0.037ab	0.096a	0.067a	0.433ab	1.124a	0.779a	2.69bc	3.10b	2.89c
Urea + B	0.053ab	0.098a	0.076a	0.575ab	1.104a	0.839a	4.77bc	3.58ab	4.18c
No amendment	0.029b	0.094a	0.062a	0.346b	1.179a	0.762a	3.40bc	3.01b	3.20c
Mean	0.044	0.099	0.072	0.507	1.154	0.83	7.04	4.45	5.74
Significance	*	ns	ns	*	ns	ns	*	**	***
LSD ($P < 0.05$)	0.037	0.046	0.066	0.437	0.479	0.76	5.60	3.09	6.38
CV (%)	28.46	15.78	47.82	29.55	14.21	47.25	27.24	23.76	57.31

M + C—2 ton manure/ha + 4 ton compost/ha + no inoculation; M + C + B—2 ton manure/ha + 4 ton compost/ha + *Bradyrhizobium* inoculation; M—2 ton manure/ha + no inoculation; M + B—2 ton manure/ha + *Bradyrhizobium* inoculation; C—4 ton compost/ha + no inoculation; C + B—4 ton compost/ha + *Bradyrhizobium* inoculation; DAP—(46 P_2O_5 kg/ha + 19 kg N/ha); DAP + B—(46 P_2O_5 kg/ha + 19 kg N/ha) + *Bradyrhizobium* inoculation; Urea—20 kg N/ha; Urea + B—20 kg N/ha + *Bradyrhizobium* inoculation

Means within the same factor and column followed by the same letter are not significantly different at 5% level of significance

the long-term application of organic inputs effect on soil fertility and peanut production would be suggested.

Acknowledgements The author is grateful to Mr. Berhanu Mengistu and Girmay Mekonnen for their assistance in the field and laboratory experiments. My appreciation also goes to Ms. Rahel Berhanu for providing laboratory facilities.

References

Basu M, Bhadoria PBS, Mahapatra SC (2007) Comparative effectiveness of different organic and industrial wastes on peanut: plant growth, yield, oil content, protein content, mineral composition and hydration coefficient of kernels. Arch Agron Soil Sci 53(6):645–658

Basu M, Bhadoria PBS, Mahapatra SC (2008) Growth, nitrogen fixation, yield and kernel quality of peanut in response to lime,

Wu J, Arima Y (1992) Effect of *Rhizobium* inoculation and application of N, P, K fertilizer on the growth and nitrogen fixation of field-grown Chinese milk vetch. Soil Sci Plant Nutr 38(1):75–84

Zingore S, Mafongoya PL, Nyamugafata P, Giller KE (2003) Nitrogen mineralization and maize yields following application of tree pruning on a sandy soil in Zimbabwe. Agrofor Syst 57:199–211

organic and inorganic fertilizer levels. Bioresour Technol 99:4675–4683

Burgos P, Madejon E, Cabrera F (2006) Nitrogen mineralization and nitrate leaching of a sandy soil amended with different organic wastes. Waste Manage Res 24:175–218

Cobo JG, Dercon G, Cadisch G (2010) Nutrient balances in African land use systems across different spatial scales: a review of approaches, challenges, and progress. Agric Ecosyst Environ 136:1–15

Crawford E, Jayne T (2010) The World bank alternative approaches for promoting fertilizer use in Africa. Agriculture and rural development discussion paper 22, Washington DC. pp 69

Dotaniya ML, Datta SC, Biswas DR, Dotaniya CK, Meena BL, Rajendiran S, Regar KL, Lata M (2016) Use of sugarcane industrial by-products for improving sugarcane productivity and soil health. Int J Recycl Org Waste Agric 5:185–194

Eghball B, Wienhold BJ, Gilley JE, Eigenberg RA (2002) Mineralization of manure nutrients. Int J Soil Water Conserv 57(6):470–473

Foli SK (2012) Qualitative and quantitative diagnosis of macro and micronutrient deficiencies in soils across three agro-ecological environments of northern Nigeria using the double-pot technique. MSc internship report, plant production systems group, course code: PPS-70424. Wageningen: Wageningen University

Giller KE (2001) Nitrogen fixation in tropical cropping systems. CAB International, Walliford

Giller KE, Cadisch G, Mugwira LM (1998) Potential benefits from interactions between mineral and organic nutrient sources. In: Waddington SR et al (eds) Soil fertility research for maize-based farming systems in Malawi and Zimbabwe. Soil Fertility Network and CIMMYT-Zimbabwe, Harare, pp 155–158

Goyal S, Chander K, Mundra MC, Kapoor KK (1999) Influence of inorganic fertilizers and organic amendments on soil organic matter and soil microbial properties under tropical conditions. Biol Fertil Soils 29:196–200

Hailu Hillette, Mamo Tekalign, Keskinen Riikka, Karltun Erik, Gebrekidan Heluf, Bekele Taye (2015) Soil fertility status and wheat nutrient content in Vertisol cropping systems of central highlands of Ethiopia. Agric Food Secur 4:19

Heal OW, Anderson JM, Swift MJ (1997) Plant litter quality and decomposition: a historical overview. In: Cadisch G, Giller KE (eds) Driven by nature: plant litter quality and decomposition. CAB International, Wallingford, pp 3–30

Herencia JF, García-Galavísa PA, Ruiz Dorado JA, Maqueda C (2011) Comparison of nutritional quality of the crops grown in an organic and conventional fertilized soil. Sci Hortic 129:882–888

Ibrahim M, Yamin M, Sarwar G, Anayat A, Habib F, Ullah S, Rehman S (2011) Tillage and farm manure affect root growth and nutrient uptake of wheat and rice under semi-arid conditions. Appl Geochem 26:194–197

Kang BT, Balasubramanian V (1990) Long-term fertilizer trials on Alfisols in West Africa. In Trans 14th International Congress of Soil Science, Kyoto, Japan, vol. IV. pp 20–25

Kebede A, Bushra F (2012) Registration of BaHa-jidu and BaHa-gudo groundnut (*Arachis hypogaea* L.) varieties. East Afr J Sci 6(1):79–80

Kimetu JM, Mugendi DN, Palm CA, Mutuo PK, Gachengo CN, Bationo A, Nandwa S, Kungu JB (2004) Nitrogen fertilizer equivalencies or organics of differing quality and optimum combination with inorganic nitrogen source in Central Kenya. Nutr Cycl Agroecosyst 68:127–135

Kimiti JM, Odee DW (2010) Integrated soil fertility management enhances population and effectiveness of indigenous cowpea rhizobia in semi-arid eastern Kenya. Appl Soil Ecol 45:304–309

Laekemariam Fanuel, Kibret Kibebew, Mamo Tekalign, Gebrekidan Heluf (2016) Soil–plant nutrient status and their relations in Maize-growing fields of Wolaita zone, Southern Ethiopia. Commun Soil Sci Plant Anal 47(11):1343–1356

Lupwayi NZ, Girma M, Haque I (2000) Plant nutrient content of cattle manures from smallscale farms and experimental stations in the Ethiopian highlands. Agric Ecosyst Environ 78:57–63

Mittra BN, Karmakar S, Swain DK, Ghosh BC (2005) Fly ash—a potential source of soil amendment and a component of integrated plant nutrient supply system. Fuel 84:1447–1451

Mohammadi K, Ghalavand A, Aghaalikhani M (2011) Effect of different soil fertility strategies on absorption metabolism and molecular nitrogen fixation in chickpea (*Cicer arietinum*). Iran J Pazhuhesh Sazandegi 91:78–89

Morris M, Kelly VA, Kopicki RJ, Byerlee D (2007) Fertilizer use in African agriculture. Lessons learned and good practice guidelines. Directions in development: agriculture and rural development publication 39037 World Bank, Washington DC. pp 162

Mugwira LM (1984) Relative effectiveness of fertilizer and communal area manures as plant nutrient sources. Zimb Agric J 81:85–90

Mugwira LM, Murwira HK (1997) Use of cattle manure to improve soil fertility in Zimbabwe: past and current research and future research needs. In: Soil fertility network research results working paper no. 2

Namvar A, Seyed Sharifi R, Khandan T, Jafari Moghadam M (2013) Seed inoculation and inorganic nitrogen fertilization effects on some physiological and agronomical traits of Chickpea (*Cicer arietinum* L.) in irrigated condition. J Central Eur Agric 14(3):28–40

Nhamo N (2002) An evaluation of the efficacy of organic and inorganic fertilizer combinations in supplying nitrogen to crops. Master of Philosophy thesis, University of Zimbabwe

Ogundare K, Agele S, Aiyelari P (2012) Organic amendment of an ultisol: effects on soil properties, growth, and yield of maize in Southern Guinea savanna zone of Nigeria. J Recycl Org Waste Agric 1:11

O'Hara GW (2001) Nutritional constraints on root nodule bacteria affecting symbiotic nitrogen fixation: a review. Aust J Exp Agric 41(3):417–433

Okalebo JR, Palm CA, Lekasi JK, Nandwa SM, Othieno CO, Waigwa M, Ndungu KW (2003) Use of organic and inorganic resources to increase maize yields in some Kenyan soils: a five year experience. In: Bationo A, Swift MJ (eds) Proceedings of the 8th meeting of the African network of tropical (AfNET) of soil biology and fertility research. Nairobi, Kenya

Panda PK, Nandi A, Swain PK, Patnaik SK, Patnaik M (2012) Soil amendment on growth, nodulation, yield, soil health, and economics of Cowpea. Int J Veg Sci 18:284–297

Petersen BM, Jensen LS, Hansen S, Pedersen AS, Henriksen TM, Sørensen P, Trinsoutrot Gattin I, Berntsen J (2005) CN-SIM: a model for the turnover of soil organic matter, II: short-term carbon and nitrogen development. Soil Biol Biochem 37:375–393

Ramesh P, Panwar NR, Singh AB, Ramana S, Rao AS (2009) Impact of organic-manure combinations on the productivity and soil quality in different cropping systems in central India. J Plant Nutr Soil Sci 172(4):577–585

Rezig AMR, Elhadi EA, Mubarak AR (2012) Effect of incorporation of some wastes on a wheat–guar rotation system on soil physical and chemical properties. J Recycl Org Waste Agric 1:1

Rufino M, Dury J, Tittonell P, van Wijk M, Herrero M, Zingore S, Mapfumo P, Giller K (2010) Competing use of organic resources, village-level interactions between farm types and climate variability in a communal area of NE Zimbabwe. Agric Syst. doi:10.1016/j.agsy.2010.01.001

Sakala W, Cadisch G, Giller KE (2000) Interactions between residues of maize and pigeon pea and mineral N fertilizers during decomposition and N mineralization. Soil Biol Biochem 32:679–688

Sanchez PA (2002) Soil fertility and hunger in Africa. Science 295:2019–2020

Shahzad SM, Khalid A, Arif MS, Riaz M, Ashraf M, Iqbal Z, Yasmeen T (2013) Co-inoculation integrated with P-enriched compost improved nodulation and growth of Chickpea (*Cicer arietinum* L.) under irrigated and rainfed farming systems. Biol Fertil Soils. doi:10.1007/s00374-013-0826-2

Smaling EMA (2006) Fertilizer use and the environment in Africa: friends or foes? Background paper: African fertilizer summit: nourish the soil, feed the continent, 9–13 June 2006 Abuja, Nigeria. pp 25

Smaling EMA, Braun AR (1996) Soil fertility research in sub-Saharan Africa: new dimensions, new challenges. Commun Soil Sci Plant Anal 27(34):365–386

Smaling EMA, Stoorvogel JJ, Windmeijer PN (1993) Calculating soil nutrient balances in Africa at different scales. II: district scale. Fertil Res 35:237–250

Smaling EMA, Nandwa S, Janssen BH (1997) Soil fertility in Africa is at stake. In: Buresh RJ, Sanchez PA, Calhoun F (eds) Replenishing soil fertility in Africa. Soil Science Society of America (SSSA), SSSA Special Publication 51, Madison. pp 47–61

Stoorvogel JJ, Smaling EMA (1990) Assessment of soil nutrient depletion in sub-Saharan Africa: 1983–2000, vols 1–4. Wageningen, Winand Staring Centre

Tekalign T (1991) Soil, plant, water, fertilizer, animal manure and compost analysis. Working Document No. 13. International Livestock Research Center for Africa, Addis Ababa

Tsai SM, Bonetti R, Agbala SM, Rossetto R (1993) Minimizing the effect t of mineral nitrogen on biological nitrogen fixation in common bean by increasing nutrient levels. Plant Soil 15(2):131–138

Vanlauwe B, Aihou K, Houngnandam P, Diels J, Sanginga N, Merckx R (2001a) Nitrogen management inadequate' input maize-based agriculture in the derived savanna benchmark zone of the Benin Republic. Plant Soil 228:61–71

Vanlauwe B, Wendt J, Diels J (2001b) Combined application of organic matter and fertilizer. In: Tian G, Ishida F, Keatinge JDH (eds) Sustaining soil fertility in West Africa. Soil Science Society of America and American Society of Agronomy, Madison, pp 247–279

Vanlauwe B, Gachengo C, Shepherd K, Barrios E, Cadisch G, Palm CA (2005) Laboratory validation of a resource quality-based conceptual framework for organic matter management. Soil Sci Soc Am J 69:1135–1145

Vanlauwe B, Wendt J, Giller KE, Corbeels M, Gerar B, Nolte C (2014) A fourth principle is required to define conservation agriculture in sub-Saharan Africa: the appropriate use of fertilizer to enhance crop productivity. Field Crops Res 155:10–13

17

Effects of co-composting of faecal sludge and agricultural wastes on tomato transplant and growth

Eric Gbenatey Nartey[1] · Philip Amoah[1] · Godfred Kwabena Ofosu-Budu[2] · Ashley Muspratt[3] · Surendra kumar Pradhan[1]

Abstract

Purpose Faecal sludge (FS) has been co-composted with many organic solid wastes globally. Agricultural wastes, such as oil palm empty fruit bunches (EFB) and cocoa pod husks (CPH), have received very little research attention as far as combining with FS is concerned. This study aimed at co-composting these wastes at different ratios to produce safe compost for use as soilless medium for raising tomato transplants.

Methods Dewatered FS (DFS) was mixed with shredded EFB and CPH at five different ratios: 1DFS:1EFB, 1DFS:1CPH, and DFS:EFB:CPH in ratios of 1:1:1, 2:1:1, and 2:2:1 and composted for 3 months. Select physico-chemical parameters and pathogens were monitored every fortnightly and 3 weeks, respectively.

Results Maximum temperatures obtained ranged 46.8–54.5 °C. Though these temperatures were lower than sanitizing temperatures prescribed by USEPA, no *E. coli* was found in any of the piles at the end of composting. The ratio 2DFS:2EFB:1CPH was found to be the safest formulation and hence was used to grow tomato under greenhouse conditions. Tomato seeds were sown in three different growing media: 100% FS-based compost, 100% rice husk biochar, and 50% FS-based compost–50% rice husk biochar mix.

Conclusion Results showed that FS-based compost was a suitable growing medium for tomato. Further studies into the optimal rate and frequency of application of compost teas on tomato are recommended.

Keywords Faecal sludge · Co-composting · Cocoa pod husks · Transplant · Compost tea · Tomato

✉ Eric Gbenatey Nartey
e.nartey@cgiar.org; flerigo@yahoo.com

Philip Amoah
p.amoah@cgiar.org

Godfred Kwabena Ofosu-Budu
obuduster@gmail.com

Ashley Muspratt
ashley@waste-enterprisers.com

Surendra kumar Pradhan
s.k.pradhan@cgiar.org

1 International Water Management Institute (IWMI), PMB CT, 112, Cantonments, Accra, Ghana

2 College of Agricultural and Consumer Sciences, University of Ghana Forest and Horticultural Crops Research Centre-Kade, Legon, Ghana

3 Pivot Ltd, P.O. Box 98394, Mombasa, Kenya

Introduction

Insufficient waste management has led to growing interests in approaches to safe emptying, transport, and disposal of faecal sludge (FS) (Strauss and Montangero 2002). Numerous technical developments have attempted to address the problems of emptying, transporting, and disposal of FS from urban centres (Boot and Scott 2008). Currently, the focus of research is shifting from FS disposal to reuse, with fertilizer and soil conditioner being among the most common reuse options. FS, like animal manure, is a good soil conditioner and a renewable source of plant nutrients, such as nitrogen (N), phosphorus (P) and potassium (K) (Cofie et al. 2004), and organic matter. These are lacking in arable soils found in the tropics. There is, therefore, need to transform this valuable waste into useful soil amendments for crop production. This will represent an important option for both sustainable waste management and sustainable agriculture (Chang and Davila 2007; Dick and McCoy 1993). Although FS contains these beneficial resources that can be applied to arable soils, the raw sludge contains high pathogen concentrations and direct use without prior treatment puts both farmers and consumers at risks. Composting is one way to sanitize dewatered FS (DFS). On its own, there is too little organic content to achieve thermophilic conditions (i.e., temperatures above 55 °C) during composting, which is necessary for complete pathogen destruction. Co-composting with carbon-rich materials, such as oil palm empty fruit bunches (EFB) and cocoa pod husks (CPH), can impart two key benefits: (a) enabling of thermophilic temperatures and (b) prevention of nitrogen volatilization (Jeong and Kim 2001). Previous composting studies in Ghana experimented with co-composting DFS with municipal solid wastes and sawdust (Kone et al. 2007; Cofie et al. 2009). The use of other agricultural wastes, such as EFB and CPH, has received very little research attention in terms of co-composting with DFS. It is estimated that approximately 400,000 tonnes of cocoa beans are produced annually (Ofosu-Budu et al. 2001, personal communication), and based on this annual production, an estimated 550,750 tonnes of dry cocoa husks are produced (Adamtey 2005). Annual production of EFB in Ghana can be estimated based on annual production of approximately 3,135,000 tonnes of fresh oil palm fruit bunches (calculation based on average yield of 11 tonnes per hectare) (Toledano et al. 2004). Agro-industrial wastes are posing serious challenges for the agro-industries (Abu Qdais and Al-Widyan 2016). Similarly, an estimated amount of about 3.5 million tonnes of human excreta is produced in Ghana based on an average production of 400 g cap^{-1} day^{-1} (Mann 1976). This leaves a lot of human and agricultural wastes in the environment, which are traditionally directly incinerated or deposited in landfills resulting in the production of significant amounts of greenhouse gases (Zhang et al. 2013) or sometimes reapplied to land as a conditioner and could pose health and environmental risks. Many studies have reported that organic wastes composts, such as from sewage sludge (Perez-Murcia et al. 2006), municipal solid waste (Ostos et al. 2008), animal manure (Atiyeh et al. 2001; Eklind et al. 2001), green waste (Grigatti et al. 2007; Ribeiro et al. 2007), and agro-industrial waste (Papafotiou et al. 2004; Bustamante et al. 2008) can be used with very good results as growth medium instead of peat (Kala et al. 2012). These composts and other solid manures are great, but they have one major disadvantage of taking a longer time to breakdown in the soil to release nutrients. Therefore, for successful growing of vegetables and other horticultural crops, such as tomato, which require rapid nitrogen sources, the use of compost tea could play a significant role. Compost tea has evolved from historical horticultural practices, such as steeping manure or plants in water, with the liquid applied to crops for nutritional and plant health reasons. In recent decades, the majority of published studies examining compost tea have focussed on foliar plant disease suppression using liquid materials produced by steeping compost in water for several days to weeks (Scheuerell 2004). It is increasingly being used as an alternative plant disease control measure in commercial horticulture (Diver 2001; Scheuerell and Mahaffee 2002). Compared to the numerous reports on the use of compost tea for managing foliar diseases, little work has been done to assess the nutritional benefits of compost tea on plant growth (Pant 2009). Studies into the nutritional effect of compost tea on horticultural plants are, therefore, necessary. As a liquid fertilizer, its high nutrient value and rapid availability make it a great tonic for the plants. This study, therefore, aims at (i) co-composting DFS, EFB, and CPH at different formulations to determine the appropriate formulation that will produce safe compost for use in soilless medium for tomato transplant production and (ii) to assess the nutritional effect of compost tea brewed from the FS-based compost on the vegetative growth of tomato.

Materials and methods

The study was carried out at the University of Ghana, Forest and Horticultural Crops Research Centre (FOHCREC) (6°05′N; 0°05′W). Climate of the area is humid tropical and monthly average temperature reaches a maximum of 28–29 °C in February and March, and a minimum of 25–26 °C in July/August. Rainfall pattern is bimodal with peaks (150–200 mm) in May/June and September/October (Adamtey 2005). Raw FS coming from household and public toilets at a ratio of 2:1 (*v*/*v*) was dewatered on sand drying beds to obtain DFS. The DFS was characterized and mixed at different formulations with shredded (size 2–3 cm) EFB and CPH, obtained from nearby oil palm processing industries and cocoa farms, respectively. The formulations were as follows: 1 DFS:1 EFB; 1 DFS:1 CPH; and DFS:EFB:CPH in ratio of 1:1:1, 2:1:1, and 2:2:1, respectively. The mixtures were moistened with water to about 60% moisture content following the method described by Estevez-Schwarz et al. (2012) and laid in windrows. Each of the five treatment formulations was replicated three times in a completely randomized design. Average size of a pile was 2 m long, 1.5 m wide, and 0.6 m high. Piles were turned manually every 3 days for the first 2 weeks and then once a week for the rest of the 10 weeks. Throughout the experiment, the moisture contents of the piles were maintained as recommended by Day and Shaw (2000). Daily temperature readings were taken by inserting a thermometer probe into three perforated bamboo pipes of approximately equal sizes mounted in each pile. Samples of the co-composting material were taken every 2 weeks for physicochemical analyses (pH, EC, TOC, NH_4-N, NO_3-N, Total N, P, and K). *E. coli,* total coliform, and helminth egg populations particularly *Ascaris* eggs were determined in samples every 3 weeks.

Feedstock and co-compost analyses

The feedstock and co-compost samples were air dried and milled (<0.5 mm) to homogenize the samples. Total N was determined by the modified Kjeldahl method described by Black (1965). Ammonium (NH_4-N), Nitrate (NO_3-N), Total P, Avail-P, and K were determined by methods, as described in Okalebo et al. (2002). The pH and electrical conductivity (EC) were measured using methods described in USDA and USCC (2001). Total organic carbon (TOC) was determined by the methods described in Walkley and Black (1934). For all samples, total coliform and helminth egg counts were determined using the spread plate method (APHA-AWWA-WEF 2001) and the flotation and sedimentation method following a modified USEPA method (Schwartzbrod and Gaspard 1998), respectively. Compost samples were prepared for total coliform and *E. coli* population estimates as follows: each sample was thoroughly mixed and a subsample of 10 g (dry weight) was weighed into 90 ml of sterile phosphate-buffered saline and shaken vigorously for about 2 min. Further tenfold serial dilutions were made, and Chromocult coliform agar (MERCK1 KgaA 64271 Darmstadt, Germany) was inoculated from each dilution. The inoculated media were incubated at 35–37 °C for 24–48 h (APHA-AWWA-WEF 2001). All colonies with purple-to-blue–black colour were counted as *E. coli.* Temperature, pH, C:N ratio, and CO_2 evolution rates of co-compost were used to determine co-compost stability and maturity. Carbon dioxide (CO_2) evolution rates were determined by methods, as described in Okalebo et al. (2002). Criteria for selecting the best co-compost formulation were based on subjective performance star ratings in terms of pathogen reduction, total nutrient content, and amount of N-mineralised and N-loss. The ratings were as follows: ***** = Excellent, **** = Very Good, *** = Good, ** = Fair and * = Poor. The formulation with the highest number of stars was the best.

Greenhouse experiment

The best co-compost produced above was used as the soilless medium together with rice husk biochar to grow tomato transplants. The matured compost was first milled and sieved through a 2 mm sieve to obtain a uniform texture before mixing with the rice husk biochar to prepare three growing medium types: a 100% FS-based compost (FSC), a 100% rice husk biochar (RHB), and a 50% FS-based compost: 50% rice husks biochar mix (FSC-RHB). Tomato seeds (*Lycopersicon esculentum var. M2*) were sown in the three different growing medium at three seeds/plug and irrigated with tap water until the seeds germinated. Germinated seeds were counted starting from the fourth to tenth days after seeding (DAS). The percentage emergence and mean days to emergence (MDE) were calculated as described by Gerson and Honma (1978). The young transplants were continuously irrigated with tap water until most of them were showing second true leaves after which they were thinned to one plant per plug. Compost tea (CT) was brewed from the same co-compost for fertigation by steeping 2.5 kg of the compost in 15 L of distilled water (1:3 *v/v*) for 6 days following the bucket-fermentation method described by Diver (2001). For application, the CT stock solution was further diluted to 1:5 (CT-1) and 1:10 (CT-2) (stock solution:distilled water). The two compost tea dilutions were prepared in 20L batches and analysed for NPK content before applying to the transplants. To compare with the CT solutions, inorganic fertilizer solutions containing 400 mg N/L:176 mg P/L:332 mg K/L (In-fert-1) and 200 mg N/L:176 mg P/L:332 mg K/L (In-fert-2) were also prepared. After thinning the transplants, irrigation with tap water was terminated. The transplants were then sub-irrigated with the fertilizer solutions (CT-1, CT-2, In-fert-1, or In-fert-2) until the growth medium was saturated. Ordinary tap water served as control solution. Sub-irrigation with the fertilizer solutions was carried out for 21 days after which the experiment was terminated. The experiment was carried out in a complete randomized design with three replications.

Transplant sampling and analyses

Total N in the growth medium and compost tea solutions were determined by the semi–micro-Kjeldahl method (Black 1965). Total phosphorus (P) and potassium (K) were determined using the spectrophotometer (Philip PU8620 UV/VIS/NIR model) and Jenway flame photometer (PFP7), respectively. Plant available N (NH_4 and NO_3) were determined according methods described by Bremmer (1965). Electrical conductivity (EC) and pH of the medium and nutrient solutions were determined following standard procedures, as described in USDA and USCC (2001). Bulk density and water holding capacity of the medium were determined according to methods described in Okalebo et al. (2002) and Vengadaramana and Jashothan (2012), respectively. Transplants were sampled (five plants/treatment rep) every 7 days and the following growth parameters measured: leaf chlorophyll content, number of leaves, plant height, root length, stem diameter, shoot dry matter (DM), root DM, and shoot/root ratio. The leaf chlorophyll content was measured using Apogee chlorophyll content meter (Model CCM- 200), plant height and root length (in cm) were measured with a rule, and stem diameter was measured 1 cm above growth medium surface with a pair of Vernier callipers. The plant samples were then severed and separated into root and shoot parts. They were oven dried at 68 °C for 48 h to obtain dry weight/dry matter.

Statistical analysis

Analysis of variance (ANOVA) was performed on all data using Genstat 9th edition (Release 9.2) statistical package. Data for the microbial analysis were log transformed, while N-mineralised and N- and P-losses were transformed into percentages before ANOVA. Treatment means which were found to be significantly different from each other at ($p < 0.05$) were separated by the Least Significant Differences (LSD) tests. Two tailed correlation analysis between selected characteristics of the compost were determined using SPSS for windows software (version 16). Significant correlations at the 1% level ($p < 0.01$) have been considered.

Results

Co-composting DFS with EFB and CPH

General characterisation of the co-compost feedstock saw total N to be the highest (3.1%) in DFS followed by CPH and then EFB. Ammonium (NH_4-N) in the DFS was about fivefold higher than that in the EFB and CPH (Table 1). The C:N ratios were 43.5, 25.1, and 11.1 for EFB, CPH,

Table 1 Physico-chemical characteristics of feedstock used in co-composting ($x \pm$ SD) ($n = 3$)

Parameters	DFS	EFB	CPH
pH (1:5)	7.7 ± 0.0	8.9 ± 0.1	8.6 ± 0.1
TOC (%)	34.3 ± 2.0	64.6 ± 4.4	41.1 ± 0.8
C:N	11.1 ± 0.2	43. 5 ± 1.4	25.1 ± 1.0
Total N (%)	3.1 ± 0.2	1.5 ± 0.1	1.6 ± 0.0
Total P (%)	1.5 ± 0.5	0.1 ± 0.0	0.3 ± 0.0
Total K (%)	0.8 ± 0.0	1.3 ± 0.0	2.2 ± 0.1
NH_4-N (mg kg^{-1})	2062.8 ± 205.2	378.0 ± 5	399.5 ± 3.0
NO_3-N (mg kg^{-1})	662.4 ± 86.4	363.6 ± 126.0	320.4 ± 54.0
Avail-P (%)	NA	NA	NA

EFB empty fruit bunches, *CPH* cocoa pod husks, *DFS* dewatered faecal sludge, *NA* not available

and DFS, respectively. Potassium (K) level in CPH was about three and two times higher than that in the DFS and EFB, respectively.

Total coliform populations in the DFS ranged between 2.0×10^3 and 3.0×10^5 CFU g^{-1} with an average of $6.2 \times 10^5 \pm 7.1 \times 10^5$ CFU g^{-1}, while *E. coli* levels ranged between 0 and 6.0×10^3 CFU g^{-1}. The average helminth egg population was 1/10 g. The characteristics of the various formulations at the start of co-composting process are shown in Table 2. At the end of co-composting, total N was highest in piles 1 DFS:1 CPH and 2 DFS:2 EFB:1CPH followed by pile 2 DFS:1 EFB:1CPH (Table 2). The differences were, however, not significant. Pile treatment 1 DFS:1 EFB:1 CPH had the lowest total N, but it was not significantly different from pile 1 DFS:1 EFB. Total P did not vary significantly between pile treatments 1 DFS:1 CPH, 1 DFS:1 EFB:1 CPH, and 2 DFS:1 EFB:1 CPH. However, the highest concentration of P was measured in pile 2 DFS:2 EFB:1 CPH and the lowest in pile 1 DFS:1 EFB:1 CPH. Total K was the highest in pile 2 DFS:2 EFB:1CPH and the lowest in pile 1 DFS:1 EFB. Electrical conductivity (EC) was significantly higher in pile 1 DFS:1 CPH compared to the other treatment piles. There were, however, no significant differences between piles 1 DFS:1 EFB, 1 DFS:1 EFB:1 CPH, 2 DFS:1 EFB:1CPH, and 2 DFS:2 EFB:1CPH. In all cases, once the piles were formed, the temperature rose rapidly (Fig. 1). Temperature in all the piles peaked on Day 3, with 1 DFS:1 EFB; 2 DFS:1 EFB:1CPH, and 2 DFS:2 EFB:1 CPH reaching thermophilic conditions (Fig. 1). The maximum temperature attained by each pile ranged between 46.8 and 54.5 °C and the minimum temperature ranged between 31.7 and 32.9 °C.

In general, the CO_2 evolution rates indicating microbial respiration decreased with time over the 12-week period for all the piles (Fig. 2). Within the first week, the CO_2 evolution rates were the highest (8.4 mg g^{-1} compost per day) in pile 2 DFS:2 EFB:1CPH and the lowest (2.7 mg g^{-1} compost per day) in pile 1 DFS:1 CPH. By the fifth week, the evolution rates of pile 1 DFS:1 CPH became the highest (1.7 mg/g compost per day) amongst the treatments with pile 2 DFS:2 EFB:1CPH being the lowest (0.8 mg g^{-1} compost day^{-1}). At the end of the 12 week, all the piles recorded stable CO_2 evolutions which were less than 1 mg CO_2 g^{-1} compost day^{-1}.

The C:N ratios decreased drastically over the first 2–4 weeks of co-composting for all the piles, and remained low through the week 8 when concentrations began to increase (Fig. 2). Table 2 shows the levels of NH_4-N and NO_3-N and org-N after the 12-week composting period. The nitrogen was found predominantly in its organic form, with levels ranging between 89 and 95%, while the inorganic forms (NH_4-N and NO_3-N) ranged between 5 and 11%.

Table 2 Physico-chemical characteristics of co-compost piles at start and end of co-composting ($p < 0.05$) ($n = 3$)

Piles	EC (μS cm^{-1})	C/N ratio	Total N (%)	NH_4-N (mg kg^{-1})	NO_3-N (mg kg^{-1})	Total P (%)	Avail-P (%)	Total K (%)	*N*-mineralised			*N*- and *P-losses*	
									% Org-N	% NH_4-N	% NO_3-N	% N	% P
At start													
1DFS:1EFB	2150	38.2	2.2	1828.8	284.4	2.6	0.3	1.3	90.5	8.2	1.3	–	–
1DFS:1CPH	2875	32.2	3.1	2368.8	392.4	3.1	0.3	1.9	90.9	7.8	1.3	–	–
1DFS:1EFB:1CPH	2275	45.2	1.8	1569.6	241.2	2.4	0.3	1.6	89.8	8.9	1.4	–	–
2DFS:1EFB:1CPH	1750	61.0	1.5	954.0	252.0	2.4	0.1	1.5	92.1	6.2	1.7	–	–
2DFS:2 EFB:1CPH	2525	51.4	2.0	1785.6	190.8	2.4	0.2	1.5	90.2	8.8	0.9	–	–
LSD	551.4	10.37	0.38	604.2	66.79	0.54	0.05	0.21	1.46	1.57	0.35		
At end													
1DFS:1EFB	1975	20.3	1.2	367.2	849.6	0.8	0.2	0.7	89.4	3.2	7.4	73	84
1DFS:1CPH	2725	12.2	1.8	334.8	1000.8	0.7	0.3	0.9	92.4	1.9	5.4	63	86
1DFS:1EFB:1CPH	2175	19.7	1.1	381.6	561.6	0.6	0.1	0.9	91.0	3.6	5.4	71	88
2DFS:1EFB:1CPH	1925	17.9	1.4	334.8	594.0	0.7	0. 2	0.8	93.5	2.3	4.2	44	82
2DFS:2 EFB:1CPH	2025	19.7	1.8	327.6	568.8	0.9	0.2	1.0	94.9	1.9	3.2	56	82
LSD	417.3	4.20	0.17	30.09	185.36	0.12	0.04	0.14	1.82	0.40	1.55	9.2	3.0

EFB empty fruit bunches, *CPH* cocoa pod husks, *DFS* dewatered faecal sludge, and *LSD* least significant difference

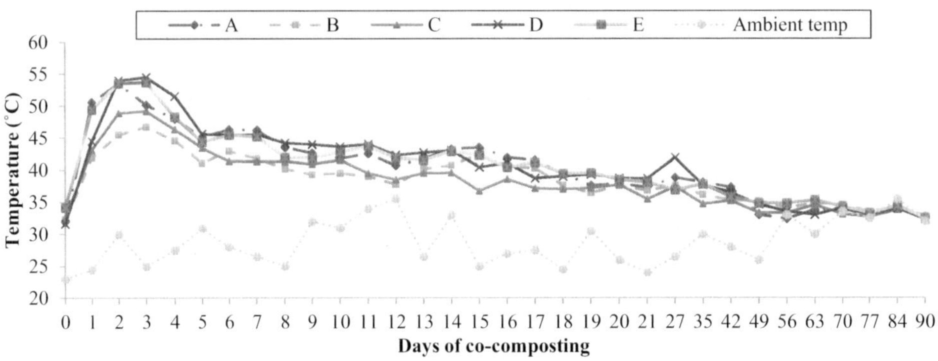

Fig. 1 Changes in temperature in the different piles during the co-composting process (*A* = 1 DFS:1 EFB, *B* = 1 DFS:1 CPH, *C* = 1 DFS:1 EFB:1 CPH, *D* = 2 DFS:1 EFB:1CPH and *E* = 2 DFS:2 EFB:1CPH)

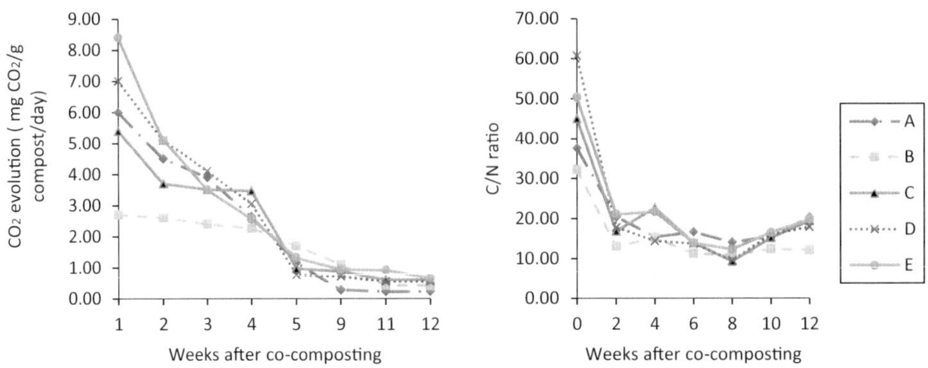

Fig. 2 Changes in CO_2 evolution and C/N in the different piles during the co-composting process (*A* = 1 DFS:1 EFB, *B* = 1 DFS:1 CPH, *C* = 1 DFS:1 EFB:1 CPH, *D* = 2 DFS:1 EFB:1CPH and *E* = 2 DFS:2 EFB:1CPH)

Significant ($p < 0.05$) losses of N and P were observed for all the treatment piles. The percentage nitrogen losses ranged from 44 to 73%, while phosphorus losses ranged from 82 to 88% (Table 2). There were, however, no significant differences in N-losses between piles 1 DFS:1 EFB and 1 DFS:1 EFB:1 CPH and then between piles 1 DFS:1

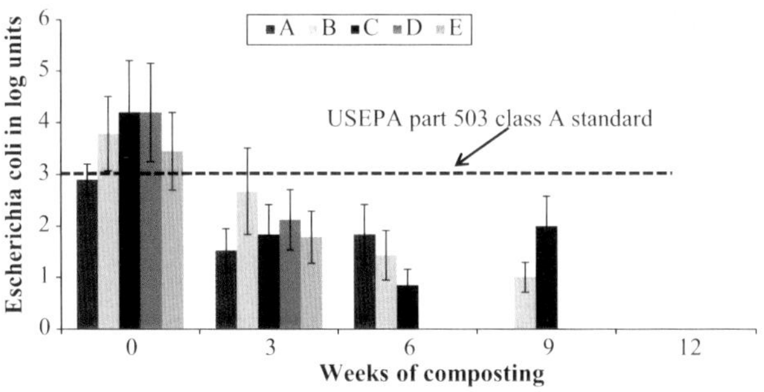

Fig. 3 Changes in *Escherichia coli* population in compost piles during composting process (*A* = 1 DFS:1 EFB, *B* = 1 DFS:1 CPH, *C* = 1 DFS:1 EFB:1 CPH, *D* = 2 DFS:1 EFB:1CPH and *E* = 2 DFS:2 EFB:1CPH)

CPH and 1 DFS:1 EFB:1 CPH. The change in levels of *E. coli* is represented in Fig. 3. *E. coli* levels were the highest in co-compost piles DFS:1 EFB:1 CPH and 2 DFS:1 EFB:1CPH (4.2 log units) and this was significantly ($p < 0.05$) higher than that of 1 DFS:1 EFB, 1 DFS:1 CPH, and 2 DFS:2 EFB:1 CPH. Apart from co-compost pile 1 DFS:1 EFB, all the other compost piles recorded *E. coli* concentrations above the USEPA part 503 class A exceptional quality (EG) compost standard (USEPA 1993) at the beginning of the co-composting process. For this standard to be achieved, *E. coli* and faecal coliform concentrations must be <1000 MPNg^{-1} or <3 log units CFUg^{-1}. This standard was adopted, because Ghana is yet to have a standard of its own and USA for many years has championed the concept of pathogen reduction as a number one compost quality standard (USEPA 1993). No *E. coli* was, however, recorded in the co-compost piles at the end of the composting period. The total coliform levels, ranging between 3.1×10^4 and 8.9×10^4 CFUg^{-1} at the beginning of the composting process, reduced to levels between 3.1×10^2 and 8.0×10^3 CFUg^{-1} after the period. Helminth eggs were not found in any of the co-compost piles at the end of the co-composting process.

The formulation 2 DFS:2 EFB:1 CPH was found to be the best mixing ratio for co-composting these three organic wastes. This formulation had the highest star ratings (15*), and it contained the highest nutrient (N, P, and K) content and least pathogen content among others, as shown in Table 3, and was used for the greenhouse experiment.

Greenhouse experiment

Electrical conductivity (EC) increased with increasing compost concentration in the growing medium (Table 4). The pH was significantly higher in FSC-RHB (7.8) than in FSC (7.5) and RHB (7.3). Total N, Total P, and Total K were higher in FSC than in FSC-RHB, though the differences were not significant at $p < 0.05$. Nitrate (NO_3) levels showed the following decreasing order: FSC > FSC-RHB > RHB. Available P was significantly higher in FSC-RHB than in the other growing media.

The dilution factors did not influence the pH of the compost tea solutions (CT-1 and CT-2) and the inorganic fertilizer solutions (In-fert-1 and In-fert-2) (Table 5). Higher concentrations of Total P and K were recorded in the compost tea solutions than in the inorganic fertilizer solutions.

The different growing media formulation affected tomato seed germination, emergence, and early vegetative growth. Germinated tomato seeds emerged in RHB 24 h earlier than in both FSC and FSC-RHB (Table 6). Percentage emergence was also higher in RHB medium than in the other two growing media.

Young transplants were taller in FSC than in RHB medium (Table 7); however, plant height was not significantly different between FSC and FSC-RHB media. Transplants in the RHB growth medium were significantly thinner than transplants growing in both FSC and FSC-RHB media. Transplants in the FSC medium recorded the highest shoot/root ratio, whereas those growing in the RHB medium recorded the least shoot/root ratio.

However, after 21 days of treatment, plant height, stem diameter, total (shoot and root) DM, leaf chlorophyll content, and number of leaves were significantly higher in both FSC and FSC-RHB media than in the RHB medium (Table 8). Shoot/root ratio was higher in FSC than in the FSC-RHB. Leaf chlorophyll content and root DM were significantly higher in FSC-RHB than in FSC.

The different nutrient solutions significantly affected growth of the transplants in the greenhouse. Growth response of transplants to the inorganic fertilizer solutions (In-fert-1 and In-fert-2) was significantly higher than that of the compost teas (CT-1 and CT-2) and the control (water) (Table 9). However, there were no significant differences in growth between In-fert-1 and In-fert-2 except in root DM, where it was significantly higher in In-fert-2 than in In-fert-1. Similarly, shoot DM, number of leaves, stem diameter, and plant height were higher in CT-1 than in CT-2. However, there were no significant differences in chlorophyll content, root DM, and shoot/-root ratio between the two compost tea solutions. Interestingly, the control solution recorded higher plant height, higher root DM, and higher root volume than in the CT-2 solution.

Increasing concentrations of FS-based compost in the growing medium (from 0 to 100%) and decreasing dilutions of the compost tea solutions (CT-2 to CT-1) resulted in significant increases in tomato shoot DM and shoot/root ratio (Table 10). Stem diameter, root length, leaf

Table 3 Selecting the best compost ratio based on selected parameters

Compost treatment	Faecal coliform reduction	N- and P-losses	N-mineralization	Total NPK	Total
1DFS:1EFB	*	*	*****	**	9 (*)
1DFS:1CPH	**	***	****	****	13 (*)
1DFS:1EFB:1CPH	***	**	***	*	9 (*)
2DFS:1EFB:1CPH	****	*****	**	***	14 (*)
2DFS:2EFB:1CPH	*****	****	*	*****	15 (*)

Table 4 Physico-chemical characteristics of growth medium types ($n = 3$ for each media)

Parameters	FSC	FSC-RHB	RHB	LSD ($p < 0.05$)
pH (1:5)	7.5	7.8	7.3	0.03
EC (1:10) (dS/cm)	2.1	1.6	0.0	0.08
Bulk density (g/cm^3)	0.7	0.6	0.3	0.02
Water holding capacity (per gram)	1.2	1.7	3.1	0.13
Total N (%)	1.8	1.7	1.1	0.28
NH_4 (mg/Kg)	205.0	223.0	158.0	96.10
NO_3 (mg/Kg)	1156.0	423.0	187.0	142.10
Total P (%)	0.5	0.4	0.1	0.16
Avail. P (%)	0.1	0.2	0.0	0.01
Total K (%)	1.1	0.9	0.5	0.05
C/N	19.8	13.3	11.0	3.82

FSC = 100% FS-based compost, FSC-RHB = 50% FS-based compost: 50% rice husk biochar mix and RHB = 100% rice husks biochar

Table 5 Physico-chemical characteristics of compost teas and inorganic fertilizer solutions ($n = 3$)

Parameters	CT-1	CT-2	In-fert-1	In-fert-2	LSD ($p < 0.05$)
pH	8.1	8.1	5.6	5.6	0.16
EC (dS/cm)	0.6	0.1	0.9	0.2	0.10
Total N (mg/L)	74.8	72.8	400.0	200.0	N/A
NH_4 (mg/L)	41.3	30.0	200.0	100.0	N/A
NO_3 (mg/L)	41.4	41.7	200.0	100.0	N/A
Total P (mg/L)	1085.0	560.0	176.0	176.0	N/A
Avail-P (mg/L)	51.5	51.4	176.0	176.0	N/A
Total K (mg/L)	7420.0	3340.0	332.0	332.0	N/A

CT-1 = 1:5 (compost tea stock:water), CT-2 = 1:10 (compost tea stock:water), In-fert-1 = 400 mg N/L:176 mg P/L:332 mg K/L, In-fert-2 = 200 mg N/L:176 mg P/L:332 mg K/L), and N/A = not available

Table 6 Effect of growing medium on tomato seed germination and emergence

Substrate type	Percentage emergence (%)	Mean days to emergence (MDE)
FSC	68.7	6.6
FSC-RHB	78.7	6.6
RHB	84.0	5.6
LSD ($p < 0.05$) ($n = 15$)	10.90	0.59

FSC = 100% FS-based compost, FSC-RHB = 50% FS-based compost: 50% rice husk biochar mix and RHB = 100% rice husks biochar

Table 7 Early vegetative growth of tomato transplant before treatment (sub-irrigation)

Substrate type	Plant height (cm)	Stem diameter (mm)	Root length (cm)	No. of true leaves	Shoot DM (mg)	Root DM (mg)	Shoot/root ratio
FSC	7.4	0.2	2.3	1.9	12.0	3.0	4.0
FSC-RHB	6.8	0.2	2.8	1.9	11.0	3.0	3.7
RHB	5.7	0.1	3.4	0.9	7.9	3.0	2.7
LSD ($p < 0.05$) ($n = 15$)	1.19	0.03	0.57	0.37	0.70	0.63	0.62

FSC = 100% FS-based compost, FSC-RHB = 50% FS-based compost: 50% rice husk biochar mix and RHB = 100% rice husks biochar

Table 8 Effect of substrate type on vegetative growth of tomato transplants after 21 days of treatment ($n = 3$)

Vegetative growth parameters	Growing medium			
	FSC	FSC-RHB	RHB	LSD ($p < 0.05$)
Plant height (cm)	27.1	26.7	14.2	0.98
Stem diameter (mm)	0.4	0.4	0.3	0.01
Shoot DM (mg/plant)	441.3	447	118.0	28.48
Root DM (mg/plant)	74.3	101.9	34.7	8.17
Chlorophyll content (CCI)	9.6	11.5	6.3	0.68
No. of leaves	6.5	6.3	4.5	0.23
Shoot/root	6.2	4.5	3.3	0.63

FSC = 100% FS-based compost, FSC-RHB = 50% FS-based compost: 50% rice husk biochar mix and RHB = 100% rice husks biochar

Table 9 Effect of compost teas and inorganic fertilizer solutions on vegetative growth of tomato transplant after 21 days of treatment (sub-irrigation) ($n = 3$)

Vegetative growth parameters	Nutrient solutions					
	CT-1	CT-2	In-fert-1	In-fert-2	*W*	LSD ($p < 0.05$)
Plant height (cm)	19.9	17.2	29.2	28.5	18.6	1.27
Stem diameter (mm)	0.3	0.3	0.4	0.4	0.3	0.02
Shoot DM (mg/plant)	279.4	213.3	490.0	506.1	221.1	36.77
Root DM (mg/plant)	51.7	44.4	86.1	102.8	66.4	10.55
Chlorophyll content (CCI)	6.3	6.0	14.2	13.2	6.1	0.87
No. of leaves	5.3	4.7	6.8	7.0	4.9	0.30
Shoot/root	4.8	4.2	5.7	5.2	3.3	0.80

CT-1 = 1:5 (compost tea stock:water), CT-2 = 1:10 (compost tea stock:water), In-fert-1 = 400 mg N/L:176 mg P/L:332 mg K/L, In-fert-2 = 200 mg N/L:176 mg P/L:332 mg K/L) and W = control

chlorophyll content, and root DM reduced with increasing compost concentration but were hardly significant. Similarly, there were no significant differences between FSC and FSC-RHB substrates in terms of root DM and plant height.

Interactions between medium type and inorganic fertilizer solutions show that total (shoot and root) DM were significantly reduced at the highest compost concentration (FSC) and the highest nitrogen (N) concentration in the inorganic fertilizer solutions (In-fert-1). Leaf chlorophyll content was significantly higher in FSC-RHB medium than in the others, but no significant difference was observed between In-fert-1 and In-fert-2.

Discussions

The temperature of the co-compost treatment piles was generally above 45 °C for the first few days but did not exceed 55 °C. According to Singh et al. (2011), the primary process criteria used for ensuring the microbiological safety of composts have been narrowly defined as time—temperature conditions. For complete sanitation of composts, the USEPA standard demands temperature of >55 °C for 2 weeks with five turns (USEPA 1993). Though the temperatures in this study did not meet any of the international requirements, they were higher than the maximum 45 °C temperature observed by Kala et al.

Table 10 Combined effect of growing medium and nutrient solutions on vegetative growth of tomato transplants after 21 days of treatment

	Nutrient solutions				
	CT-1	CT-2	In-fert-1	In-fert-2	*W*
Plant height (cm)					
FSC	25.6	22.4	31.6	30.7	25.5
FSC-RHB	24.7	20.4	32.9	33.3	22.2
RHB	9.4	8.9	23.2	21.4	8.0
LSD ($p < 0.05$) ($n = 3$)	2.20				
Stem diameter (mm)					
FSC	0.4	0.4	0.4	0.4	0.4
FSC-RHB	0.4	0.4	0.4	0.5	0.4
RHB	0.2	0.2	0.4	0.4	0.2
LSD ($p < 0.05$) ($n = 3$)	0.01				
Chlorophyll content (CCI)					
FSC	7.2	7.1	14.4	12.4	7.0
FSC-RHB	8.2	7.3	16.9	16.9	8.5
RHB	3.5	3.6	11.2	10.3	2.7
LSD ($p < 0.05$) ($n = 3$)	1.51				
Shoot dry matter (mg/plant)					
FSC	455.0	368.3	503.3	515.0	365.0
FSC-RHB	358.3	253.3	728.3	711.7	281.7
RHB	25.0	18.3	238.3	291.7	16.7
LSD ($p < 0.05$) ($n = 3$)	63.69				
Root dry matter (mg/plant)					
FSC	71.7	65.0	66.7	71.7	96.7
FSC-RHB	75.0	61.7	121.7	153.3	97.7
RHB	8.3	6.7	70.0	83.3	5.0
LSD ($p < 0.05$) ($n = 3$)	18.260				

FSC = 100% FS-based compost, FSC-RHB = 50% FS-based compost: 50% rice husk biochar mix, RHB = 100% rice husks biochar, CT-1 = 1:5 (compost tea stock:water), CT-2 = 1:10 (compost tea stock:water), In-fert-1 = 400 mg N/L:176 mg P/L:332 mg K/L, In-fert-2 = 200 mg N/L:176 mg P/L:332 mg K/L) and W = control (water)

(2012). The lower temperatures observed in this study could be due to the composting methods employed (open windrows) or to the dissipation of heat due to the small volume of the piles. This confirms earlier reports by Kala et al. (2012) that heap sizes affect the temperature generation during composting. The possible implication of the pile temperatures on pathogen reduction saw the level of *E. coli* in all piles declining below the USEPA class A compost standard by the third week of composting. The complete deactivation of *E. coli* in all piles at the end of composting (week 12) indicates that complete deactivation might have occurred during the cooler curing phases of composting and not during the active thermophilic phases. This confirms earlier reports by Droffner and Brinton (1995) that *Salmonel*la and *E. coli* were found to survive for 59 days at about 60 °C, although pathogens were destroyed during the cooler curing process in industrial compost. Results from this study suggest that high temperatures might not be the only factor responsible for *E. coli* deactivation. This phenomenon is in agreement with Dumontet et al. (1999) that other factors, such as pH, presence of metabolic, and antagonistic compounds produced by indigenous micro flora, accumulation of toxic NH_3, and microbiological competition for nutrients, are also involved in the deactivation. However, it is not clear from this study which of the measured factor or set of factors might have led to the inactivation of *E. coli* as none of the factors tested showed any significant correlation with the *E. coli* counts. Deactivation might, however, be due to competition among the microbes in the system. According to Kone et al. (2007), exposure to temperature over 45 °C for at least 5 days is known to inactivate *Ascaris* eggs which are very predominant in faecal sludge in developing countries. Similarly, the same reason could account for the complete elimination of Ascaris eggs from all compost piles at the end of the composting process. The decrease in carbon dioxide (CO_2) evolution rates over the 12 weeks period showed the microbial activities subsiding as the co-compost piles neared stability. This was as a result of a decrease in the amount of biodegradable substrate and N available (Hsu and Lo 1999). However, the differences in CO_2 production rates observed for the different piles show that the rates may have been affected by the differences in the easily biodegradable materials and the feedstock ratios. This confirms an earlier report by Ayuso et al. (1996) which stated that microbial activities to a large extent depend on the nature of the organic matter added to soil or the compost mixtures. A C:N ratio less than 20 could be considered as a satisfactory maturation level of compost (Heerden et al. 2002). This assertion was confirmed by this study, where the final C:N ratios of the piles ranged between 12 and 20. Ammonium (NH_4-N) concentration decreased as compost piles approached maturity. In all cases, the Ammonium (NH_4-N) levels showed negative correlations (correlation coefficients ranging between 55% and 71% (R^2) with time. This is in agreement with Laos et al. (2002) that NH_4-N concentration decreases towards stable values at the end of the thermophilic stage of composting. However, a reverse situation in the case of nitrate (NO_3- N) concentration with positive correlation coefficients ranging between 65 and 90% (R^2) was observed. These observations also agree with observations by other authors (Sun 2006; Huang et al. 2004; Banegas et al. 2007). The production of NO_3-N is very important, because it is the most available N for plant uptake. Hence, N-mineralization is a crucial process, because it converts organic N into ammonium (NH_4^+) and nitrates (NO_3^-). In this study, N-mineralisation was very low (<11%), making organic N the major form of nitrogen in the piles. This makes the compost a slow release nitrogen source (as also expected in other compost), contrary to the mineral fertilizer, where all the nitrogen is in the plant available form. However, the co-compost may have a longer residual effect on the soil as the microbes continue to breakdown organic N to the available form. In general, the piles with higher percentages of the DFS had significantly higher total N than in the other treatment piles at the end of co-composting. This indicates that the increase in DFS component during composting with EFB and CPH could result in higher N content of the final co-composts. This confirms earlier reports by Kala et al. (2009). In the present study, total N varying from 1.05 to 1.76% falls within an earlier report by Kulhman et al. (1989). Nitrogen losses during composting depend on the materials (feedstock) used and on the pH values of the mixtures (Sanchez-Monedero et al. 2001). It may also vary depending on several environmental factors, such as aeration, moisture content, and temperature (Bishop and Godfrey 1983), and also the frequency of turning. High total nitrogen losses during composting—as seen in this study—are a common trend, and such as the results of this research are largely attributable to decreases in NH_4-N (Rao Bhamidimarri and Pandey 1996). de Bertoldi et al. (1985) reported that the initial C/N ratio of the material also affects loss of N during composting. This could probably explain why pile 2DFS:1EFB:1CPH recorded the least losses in N at the end of composting. P-losses in this study were surprisingly higher (82–88%) than the N-losses. The feedstock ratio did not seem to have any effect on P loss. The losses were probably due to the higher solubility and subsequent leaching of phosphoric compounds in the compost materials as the process met with the minor raining season in the region. Thus, it may not be advisable to do co-composting of such or similar feedstock in the open without sheltering from rains as this might lead to a significant loss in nutrients. Total K on the other hand, which ranged between 0.7 and 1.0%, was higher than the recommended level (0.3%) suggested by Council of the European Communities (1986) in compost. The total K level recorded indicated that the major sources of K in the co-compost may have been CPH followed by EFB. 2 DFS:2 EFB:1CPH co-compost was selected as the best combination ratio, because it had relatively higher NPK content, had most pathogen reduction, though it had the least N-mineralised percentage. Even though this assertion may be subjective, the primary aim of composting DFS is to make it safe for handling and land application followed by the other agronomic benefits, such as organic matter and nutrient content.

Tomato seeds germinated and emerged faster in the RHB medium than in the FS-based compost amended media (FSC and FSC-RHB), indicating that the RHB medium may be preferred for seed germination. In addition, RHB had the lowest bulk density and the highest water holding capacity (WHC) amongst the three media types considered in this study. These may have provided the most favourable air and water infiltration for seed emergence. This agrees with an earlier report by Ball (1985) who reported that, a loose, porous medium is preferred for seed germination. Faecal sludge-based compost (FSC) recorded the lowest seed germination and emergence in this study and this may have been due to presence of certain inhibitory properties in the compost such as the high electrical conductivity (EC). This also confirms an earlier report by Wong and Chu (1985) that chemical constituents and properties of compost extracts, such NH_3, salts, ethylene oxide, heavy metals, and pH, affected seed germination and root elongation. Early vegetative growth of the tomato transplants after emergence was observed to be higher in the FS-based compost amended media (FSC and FSC-RHB) than in the RHB medium. This can be attributed to the relatively higher plant nutrients in the compost than the rice husks. According to Roe et al. (1997), plants that have adequate nutrition and water generally tend to have a higher shoot/root ratio than plants that are deficient in either. In this study, the lowest shoot/root ratio of transplants was recorded for RHB, while relatively higher ratios were recorded for the FS-based compost amended media, indicating that conditions in terms of water and nutrient availability for growth of transplants were more favourable in the compost amended media. However, between FSC and FSC-RHB, an interesting interaction may have occurred between the physical characteristics of the medium and its nutrient supplying capabilities. Transplants in the FSC were taller and slender compared to those in the FSC-RHB which had thicker stems. However, the other selected growth parameters measured were not significantly different. Stem diameter and root systems, according to Arenas (1999), are variables deemed most important by Florida Transplant Growers when it comes to transplant quality. After 21 days of treatment (sub-irrigation), transplants in the compost amended medium still showed superior growth qualities over transplants in RHB. This could be attributed to the relatively higher nutrient contents in the compost compared to RHB as discussed above. Similar observations were reported by Diaz-Perez et al. (2006). However, between transplants in FSC-RHB and FSC, there were no significant differences between the measured growth parameters except in the case of leaf chlorophyll content and root DM which were significantly ($p < 0.05$) higher in the FSC-RHB. This could mean that the relatively low bulk density of the FSC-RHB compared to the FSC medium may have provided lower resistance to the spread of the transplant root system to take up much water and nutrients. In addition, the FSC-RHB substrate may have provided better plant anchorage and nutrient uptake hence the more chlorophyll content found in the leaves. The direct cause of improved plant growth following compost extract (compost tea) application is often not clear (Shrestha et al. 2012). According to Merrill et al. (1997), the mineral nutrients extracted from compost can improve soil fertility directly. Other claims include a role for the extracted microbiota in improved mineralisation of soil organic matter and solubilisation of soil minerals, chelation of ions (Janzen et al. 1995), suppression/biocontrol of certain plant root and foliar diseases (Bernal-Vicente et al. 2008), and microbial production of plant growth promoting hormones, such as auxins (Garcia et al. 2002), or cytokinin-like substances (Arthur et al. 2001). In this study, transplants fertilized with CT-1 performed significantly ($p < 0.05$) better in terms of vegetative growth than CT-2. This can be due to higher nutrient content (nutritional effect) of CT-1. Similar results were also observed by Shrestha et al. (2012), where it was noted that growth benefit of compost tea was not directly biological in nature. They further observed that the high dose of compost extract may have had a positive impact on plant growth. Furthermore, responses to CT-2 were in most cases not significantly different ($p < 0.05$) from that of the control (water) indicating that CT-2 may have been too diluted resulting in lower nutrient content. Both inorganic fertilizer solutions (In-fert-1 and In-fert-2) had a higher positive effect on tomato transplant growth and leaf chlorophyll content than was observed in the compost tea treatments (CT-1 and CT-2) and control (water), indicating that a relatively higher nitrogen (N) concentrations may be preferred by the tomato transplants for growth and chlorophyll formation. However, there were no significant differences ($p < 0.05$) in transplants growth responses to In-fert-1 and In-fert-2 except for root DM, which was significantly higher in In-fert-2. This could probably mean that increasing the N concentration beyond 200 mg N/L may not necessarily result in any significant transplant growth and hence may not be cost effective. Similar results were also obtained by Kang and van Iersel (2004), where increasing concentration of N beyond 210 mg N/L did not further increase growth of *Salvia splendens*. In general, the root systems of the transplants in the compost tea treatments (CT-1 and CT-2) were poorly developed compared to transplants in the inorganic fertilizer solutions and control (water). This might be attributable to some sort of toxicity as a result of microbial activities around the root region. Comparing interactions between growing medium and inorganic fertilizer solution showed superior transplant qualities over that of the compost tea solutions and growth media. Interactions between compost tea solutions (CT-1 and CT-2) and growing medium showed that the highest total DM (shoot DM + root DM) was observed in the FSC medium fertilized with CT-1. This could be attributed to the higher nutrient contents in both FSC medium and CT-1 solution. A different observation was, however, made for the interactions between the inorganic fertilizer solutions and the medium types. Total DM was significantly ($p < 0.05$) reduced beyond FSC-RHB and In-fert-2 solution. This reduction in total DM could be as a result of the relatively higher EC levels arising from both the FSC medium and In-fert-2 solution, resulting in some form of toxicity or inhibition. Electrical conductivity (EC) levels approaching 3.0 dS/cm are considered higher than desirable (Davidson et al. 2000). This confirms earlier reports by Lumis et al. (2000) that reduced growth at the high fertilizer rate is the result of substrate EC and nitrogen levels being too high.

Conclusion

In conclusion, the study showed that co-composting DFS with EFB and CPH can be used as an alternative option for managing these wastes and to produce a suitable soil amendment. However, the quality of the compost could be further improved by reducing nutrient losses. Available options which were not tested in this study include carrying out the composting under sheds to protect the compost from rains and reduce leaching of nutrients, adding more bulking agents, such as biochar to trap nitrogen which could be lost through volatilization. Increasing the heap sizes could also increase the heap temperatures and further improve the sanitization of the compost. The fact that 2DFS:2EFB:1CPH was the best combination according to the criteria used indicates that this optimal ration also supports the idea of using this as a waste management solution as more human waste is used which subsequently can improve sanitation. Tomato seeds in this study germinated faster in RHB medium than in the other two FS-based compost amended growth medium, i.e., FSC and FSC-RHB. However, early vegetative growth of the transplants was higher in quality in the FSC and FSC-RHB growth media than in the RHB medium. FS-based compost is, therefore, a suitable growing medium for tomato. Though the compost tea solutions (CT-1 and CT-2) affected the growth of tomato transplants positively, its effects could not be compared with the inorganic fertilizer solutions. Indeed, transplants fertilized with In-fert-2 showed the highest growth response. Doubling the nitrogen concentration in the inorganic fertilizer did not result in any significant growth of tomato transplants. The compost tea solution CT-1 is, however, recommended for organic production of tomatoes. However, further studies into the optimal rate and frequency of application of compost teas on tomato are recommended.

Acknowledgements This study was made possible by financial support from African Water Facility (AWF)—the African Developments Bank (AfDB), the University of Berkeley, the International Water Management Institute (IWMI), and the CGIAR Research Program on Water, Land, and Ecosystems (WLE) through the project dubbed "Design for reuse—harvesting the value of effluents for sustaining the nutrients for sustaining the operation of sanitation facilities".

References

Abu Qdais H, Al-Widyan M (2016) Evaluating composting and co-composting kinetics of various agro-industrial wastes. Int J Recycle Org Waste Agricult 5:273–280. doi:10.1007/s40093-016-0137-3

Adamtey N (2005) Evaluation of agricultural and agro-industrial residues for composting for agricultural use in Ghana: a case study in the Kwaebibirim District. Dissertation, University of Ghana

APHA-AWWA-WEF (2001) Standard methods for the examination of water and wastewater, 22nd edn. American Public Health Association, Washington D.C., p 1360. ISBN: 978-087553-013-0

Zhang L, Sun XY, Tian Y, Gong XQ (2013) Effects of brown sugar and calcium superphosphate on the secondary fermentation of green waste. Bioresour Technol 131:68–75. doi:10.1016/j.biortech.2012.10.059

Arenas M (1999) Coconut (*Cocus nucifera*) pith or coir as an alternative media for tomato transplant production. Dissertation, University of Florida

Arthur GD, Jäger AK, Van Staden J (2001) The release of cytokinin-like compounds from *Gingko biloba* leaf material during composting. Environ Exp Bot 45:55–61

Atiyeh RM, Edwards CA, Subler S, Metzger JD (2001) Pig manure vermicompost as a component of a horticultural bedding plant medium: effects on physicochemical properties and plant growth. Bioresour Technol 78:11–20

Ayuso M, Pascual JA, García C, Hernandez T (1996) Evaluation of urban waste for agricultural use. Soil Sci Plant Nutr 42(1):105–111

Ball V (ed) (1985) Ball redbook: greenhouse growing, 14th edn. Reston Publishing Co., Reston

Banegas V, Moreno JL, Moreno JI, Garcı'a C, Leon G, Herna'ndez T (2007) Composting anaerobic and aerobic sewage sludge using two proportions of sawdust. Waste Manag 27:1317–1327. doi:10.1016/j.wasman.2006.09.008

Bernal-Vicente A, Ros M, Tittarelli F, Intrigliolo F, Pascual JA (2008) Citrus compost and its water extract for cultivation of melon plants in greenhouse nurseries. Evaluation of nutriactive and biocontrol effects. Bioresour Technol 99:8722–8728

Bishop PL, Godfrey C (1983) Nitrogen variations during sludge composting. BioCycle 24:34–39

Black CA (ed) (1965) Methods of soils analysis, Part 2 - chemical and microbiological properties, Agronomy monograph No.9. American society of Agronomy, Madison, Wisconsin USA

Boot NLD, Scott RE (2008) Faecal sludge management in Accra, Ghana; Strengthening links in the chain. In: 33rd WEDC International Conference, Accra, Ghana

Bremmer JM (1965) Inorganic forms of nitrogen. In: Black CAA (ed) Methods of soil analysis - chemical and microbiological properties, Part 2. American society of Agronomy, Madison, Winsconsin USA, pp 1179–1206

Bustamante MA, Paredes C, Moral R, Aguló E, Pérez-Murcia MD, Abad M (2008) Composts from distillery wastes as peat substitutes for transplant production. Resource Conserv Recycle 52:792–799

Chang NB, Davila E (2007) Minimax regret optimization analysis for a regional solid waste management system. Waste Manag 27(6):820–832

Cofie OO, Kranjac-Berisavljevic G, Drechsel P (2004) The use of human waste for peri -urban agriculture in Northern Ghana. Renew Agric Food Syst 20(2):73–80

Cofie O, Kone D, Rothenberger S, Moser S, Zubruegg C (2009) Co-composting of faecal sludge and organic solid waste for agriculture: process dynamics. Water Res (Oxf) 43(18):4665–4675

Council of the European Communities (CEC) (1986) Council directive on the protection of the environment, and in particular of the soil, when sewage sludge is used in agriculture (86/278/EEC). Official Journal of the European Communities No. L 181/16-27

Davidson H, Mecklenburg R, Peterson C (2000) Nursery Management: administration and culture. Prentice Hall, Upper Saddle River

Day M, Shaw K (2000) Biological, chemical and physical processes of composting. In: Stofella PJ, Kahn BA (eds) Compost utilization in horticultural cropping systems, Lewis Publishers, Boca Raton, USA, PP 17–50

de Bertoldi M, Vallini G, Pera A (1985) Technological aspects of composting including modeling and microbiology. In: Gasser

JKR (ed) Composting of agricultural and other wastes. Elsevier Applied Science Publishers, New York, pp 27–41

Diaz-Perez JC, Granberry DM, Germishuizen P (2006) Transplant growth and stand establishment of Bell pepper (*Capsicum annuum* L.) plants as affected by compost-amended substrate. International symposium on seed, transplants and stand establishment of horticultural crops; Translating seed and seedling physiology into technology. ISHS Acta Horticulturae 782: IV

Dick WA, McCoy EL (1993) Enhancing soil fertility by addition of compost. In: Hoitink HAJ, Keener HM (eds) Science and engineering of composting: design, environmental, microbial and utilization aspects. Wooster, Ohio, pp 622–633

Diver S (2001) Notes on compost teas: a supplement to the ATTRA publication: compost teas for plant disease control. Appropriate Technology Transfer for Rural Areas (ATTRA). Available: http://www.attra.ncat.org. Accessed 4 May 2013

Droffner ML, Brinton WF (1995) Survival of *E. coli* and Salmonella populations in aerobic thermophilic composts as measured with DNA gene probes. Int J Hygiene. Environ Med 197(5):387–397

Dumontet S, Dinel H, Baloda SB (1999) Pathogen reduction in sewage sludge by composting and other biological treatments: a review. Biol Agric Hortic 16:409–430

Eklind Y, Ramert B, Wivstad M (2001) Evaluation of growing media containing farmyard manure compost, household waste compost or chicken manure for the propagation of lettuce (*Lactuca sativa* L.) transplants. Biol Agric Hortic 19:157–181

Estevez-Schwarz I, Seoane-Labandeira S, Nunez-Delgado A, Lopez-Mosquera ME (2012) Production and characterization of compost made from garden and other waste. Pol J Environ Stud 21(4):855–864

Garcia MI, Cruz Sosa F, Saavedra AL, Hernandez MS (2002) Extraction of auxin-like substances from compost. Crop Res 24:323–327

Gerson R, Honma S (1978) Emergence response of pepper at low soil temperature. Euphytica 27:151–156

Grigatti M, Giorgioni ME, Ciavatta C (2007) Compost-based growing media: influence on growth and nutrient use of bedding plants. Bioresour Technol 98:3526–3534

Heerden IV, Cronje C, Swart SH, Kotze JM (2002) Microbial, chemical and physical aspects of citrus waste composting. Bioresour Technol 81:71–76

Hsu JH, Lo SL (1999) Chemical and Spectroscopic analysis of organic matter transformations during composting of pig manure. J Environ Pollut 104:189–196

Huang GF, Wong JWC, Wu QT, Nagar BB (2004) Effect of C/N on composting of pig manure with sawdust. Waste Manag 24:805–813. doi:10.1016/j.wasman.2004.03.011

Janzen RA, Cook FD, McGill WB (1995) Compost extract added to microcosms may simulate community-level controls on soil microorganisms involved in element cycling. Soil Biol Biochem 27:181–188

Jeong Y, Kim J (2001) A new method for conservation of nitrogen in aerobic composting process. Bioresour Technol 79:129–133

Kala DR, Rosenani AB, Fauziah CI, Thohirah LA (2009) Composting oil palm wastes and sewage sludge for use in potting media of ornamental plants. Malays J Soil Sci 13:77–91

Kala DR, Rosenani AB, Thohirah LA, Fauziah I, Ahmad SH (2012) Oil palm waste – sewage sludge compost as a peat substitute in a soilless potting medium for Chrysanthemum. Glob J Sci Front Res Agric Biol 12(2): Version 1

Kang J, van Iersel MW (2004) Nutrient solution concentration affects Shoot: root ratio, Leaf Area ratio, and growth of sub irrigated Salvia (*Salvia splendens*). HortScience 39(1):49–54

Kone D, Cofie O, Zurbrugg C, Gallizzi K, Moser D, Drescher S, Strauss M (2007) Helminth eggs inactivation efficiency by faecal sludge dewatering and co-composting in tropical climates. Water Res 41(19):4397–4402

Kulhman LR, Dale B, Groenhof AC, Young TFE (1989) Windrow composting of sewage sludge and other wastes. Presented at the international recycling congress, Geneva, Switzerland, 19–22 Jan 1993

Laos F, Mazzarino MJ, Walter I, Roselli L, Satti P, Moyano S (2002) Composting of fish offal and biosolids in northwestern Patagonia. Bioresour Technol 81:179–186. doi:10.1016/S0960-8524(01)00150-X

Lumis G, Purvis P, Taurins L (2000) Flood irrigation of container-grown Euonymus and Thuja as affected by fertilizer rate and substrate. J Environ Hort 18(1):13–17

Mann HT (1976) Sanitation without sewers-the aqua privy. Building research establishment, Overseas building notes No. 168, water research centre, Stevenage, UK, p 6

Merrill R, Hoberechi K, McKeon J (1997) Organic teas from composts and manures. In: Organic farming research foundation project report. Cabrillo Community College

Okalebo JR, Gathua KW, Woomer PJ (2002) Laboratory methods of soil and plant analysis: a working manual, 2nd edn. TSBF-CIAT and SACRED Africa, Nairobi

Ostos JC, Lopez-Garrido R, Murillo JM, Lopez R (2008) Substitution of peat for municipal solid waste and sewage sludge-based composts in nursery growing media: effects on growth and nutrition of the native shrub *Pistacia lentiscus* L. Bioresour Technol 99:1793–1800

Pant AP (2009) Vermicompost extracts influence growth, mineral nutrients, phytonutrients and antioxidant activity in pak choi (*Brassica rapa* cv. Bonsai, Chinensis group) grown under vermicompost and chemical fertilizer. J Sci Food Agric 89:2383–2392

Papafotiou M, Phsyhalou M, Kargas G, Chatzipavlidis I, Chronopoulos J (2004) Olive-mill wastes compost as growing medium component for the production of poinsettia. Sci Hortic 102:167–175

Perez-Murcia MD, Moral R, Moreno-Caselles J, Perez-Espinosa A (2006) Use of composted sewage sludge in growth media for broccoli. Bioresour Technol 97:123–130

Rao Bhamidimarri SM, Pandey SP (1996) Aerobic thermophilic composting of piggery solid wastes. Water Sci Technol 33(8):89–94. doi:10.1016/0273-1223(96)00265-X

Ribeiro HM, Romero AM, Pereira H, Borges P, Cabral F, Vasconcelos E (2007) Evaluation of a compost obtained from forestry wastes and solid phase of pig slurry as a substrate for seedlings production. Bioresour Technol 98:3294–3297. doi:10.1016/j.biortech.2006.07.002

Roe NE, Stoffella PJ, Graetz D (1997) Composts from various Municipal solid waste feedstock affect vegetable crops. I. Emerge and seedling growth. J Am Soc Hort Sci 122(3):427–432

Sanchez-Monedero MA, Roig A, Paredes C, Bernal MP (2001) Nitrogen transformation during organic waste composting by the Rutgers system and its effects on pH, EC and maturity of the composting mixtures. Bioresour Technol 78:301–308

Scheuerell JS (2004) Compost tea production practices, microbial properties, and plant disease suppression. I International conference on soil and compost eco-biology, Leon–Spain, 15–17 Sept 2004

Scheuerell S, Mahaffee W (2002) Compost tea: principles and prospects for plant disease control. Compost Sci Util 10:313–338

Schwartzbrod J, Gaspard P (1998) Quantification and viability determination for helminth eggs in sludge (Modified EPA France, 1993 Method). Faculty of Pharmacy, University "Henri Poincare" Department of Microbiology B.P.403 F-54001 Nancy Cedex/France, pp 3–6

Shrestha K, Walsh KB, Midmore DJ (2012) Microbially enhanced compost extract: does it increase solubilisation of minerals and mineralisation of organic matter and thus improve plant nutrition? J Bioremed Biodegrad 3:149. doi:10.4172/2155-6199.1000149

Singh R, Kim J, Sheperd MW Jr, Luo F, Jiang X (2011) Determining thermal inactivation of *Escherichia coli* 0157:h7 in fresh compost by simulating early phases of the composting process. Appl Environ Microbiol 77(12):4126–4135

Strauss M, Montangero A (2002) Faecal sludge treatment. A report of Swiss Federal institute for environmental science and technology. IHE Delft 2002:9–10

Sun X (2006) Nitrogen transformation in food-waste composting. Dissertation, University of Regina

Toledano J, Kumar S, Danielou M (2004) Government of Ghana-ministry of food and agriculture. The World Bank-Africa Region-Rural Development, Tree Crops Development Initiative, Reconnaissance Mission Report, Pretoria

USDA, USCC (2001) Test methods for the examination of composting and compost (TMECC). The United States Department of Agriculture and the United States Composting Council. Accessed 31 May 2013. http://tmecc.org/tmecc//

USEPA (1993) Standards for the use or disposal of sewage sludge 40 code of federal regulations part 503. US Environmental Protection Agency, Washington, DC

Vengadaramana A, Jashothan PTJ (2012) Effect of organic fertilizers on the water holding capacity of soil indifferent terrains of Jaffna peninsula in Sri Lanka. J Nat Prod Plant Resour 2(4):500–503

Walkley A, Black AI (1934) An examination of the degtjareff method for determining soil organic matter and proposed modification of the chromic acid titration method. Soil Sci 37:29–38

Wong MH, Chu LM (1985) The responses of edible crops treated with extracts of refuse compost of different ages. Agric Waste 14:63–74

18

Comparison of traditional and modern deficit irrigation techniques in corn cultivation using treated municipal wastewater

Sayyed-Hassan Tabatabaei[1] · Rohollah Fatahi Nafchi[1] · Payam Najafi[2] · Mohammad Mehdi Karizan[1] · Zohreh Nazem[1]

Abstract

Purpose Declining water resources, increasing drought period and increasing irrigated area cause a shift to deficit irrigated production which is not based on full water requirement. This study was conducted to investigate the impacts of two different deficit irrigation methods, including traditional deficit irrigation (TDI) and partial root-zone drying (PRD) on water use efficiency (WUE) in corn cultivations located at the lands close to the Shahrekord wastewater treatment plant.

Methods A factorial design was employed, consisted of fifteen treatments and three replications. The first factor was three water applications, including 60, 80 and 100 (control) percent of plant water requirement. The second factor was three water sources, including 100% fresh water (FW), 50% water and 50% wastewater (5050), 100% wastewater (WW). T-Tape irrigation type system was used for irrigation of corn (*Zea mays* L.). The third factor was water management in three levels: full irrigation (FI), PRD and TDI.

Results The result showed that dry and fresh weight, leaf area index (LAI), dry biomass percentage and WUE were affected by water requirement at 5% probability level expect of height plant that is affected at 1% probability level. Water quality was affected on all the study indicators of corn plant expect of WUE. WUE among PRD by deficit irrigation treatments were higher than TDI and it was the minimum in FI. The dry and wet weight and LAI were maximum at FI and then PDR_{80} than TDI_{80}. The height of corn plants were high at FI and TDI_{80} than PRD_{80}.

Conclusion Finally, wastewater application in treatment named 5050-PRD_{80} and 5050-TDI_{80} compensated water deficit in WUE, LAI and dry biomass percentage. PRD method recognized more suitable than TDI for corn plants.

Keywords Deficit irrigation · Partial root-zone drying · Full irrigation · Wastewater

✉ Sayyed-Hassan Tabatabaei
Tabatabaei@agr.sku.ac.ir

1 Department of Water Engineering, Faculty of Agriculture, Shahrekord University, Shahrekord, Iran

2 Department of Water Engineering, Faculty of Agriculture, Islamic Azad University, Isfahan Branch (Khorasgan), Isfahan, Iran

Introduction

Water shortage is assumed as major agricultural production constraint. Therefore, the relationship between crop yield and water use efficiency (WUE) has constantly been the focus of many agricultural researchers in the arid and semi-arid regions (Zhang and Oweis 1999).

In all agricultural systems, low WUE can occur when soil evaporation is high (considering crop evapotranspiration), early growth rate is slow, and water application does not correspond to crop demand and when shallow roots are unable to utilize deep water in the soil profile (Gallardo et al. 1996). Also, in cases of drought, nutrients are not easily available from inorganic sources of soil nutrients for plant (Chemura 2014). New innovations are particularly important in water-scarce regions with the aim of saving irrigation water, and thereby increasing crop WUE (Gencoglan et al. 2006). To do so, one way is to develop new irrigation scheduling techniques such as traditional deficit irrigation (TDI), which are not necessarily based on full water requirement.

Traditional deficit irrigation method is a practice of applying less water than crop evapotranspiration demand with the intent of imposing a managed level of water stress to the crop (Grant et al. 2008). The strategy is to maintain plants under certain water deficit for a prescribed duration of the growth season. The aim is to control reproductive and vegetative growth and to improve WUE and crop quality. In addition, correct application of TDI requires a thorough assessment of yield reduction's economic impact caused by drought stress (Geerts and Raes 2009). Maximum yield could be obtained with the fulfillment of complete crop water requirements. However, practicing TDI would increase in the irrigated area or frequency of cultivation. TDI provides a possible solution to the dilemma of sustained and/or increased productions (Geerts et al. 2008).

The effects of deficit and supplemental irrigation are diverse at different crop growth stages and in different soil moisture conditions. So, it is indispensable to supply enough water for crops at the stages; that they are most sensitive to water stress, and thereafter, irrigation water volume could be reduced in other stages. It is then important to monitor soil moisture and supply water based on the soil condition. Partial root-zone drying (PRD) is now considered as a general technique which is applied various methods at the field. The PRD is a more recent irrigation technique developed from TDI (Yazar et al. 2009). These include the controlled alternative drip irrigation of a part of root zone, applied in a vertical soil profile, controlled alternative border irrigation, controlled alternative furrow irrigation (FI), controlled alternative surface irrigation, and subsurface irrigation.

Zegbe-Domínguez et al. (2003) illustrated that leaf water potential, total plant fresh weight and total dry weight of tomato crop were lower in the TDI than the FI and PRD. In the PRD irrigation, WUE improved by 83%. In the TDI, dry mass that classified into stems and leaves was higher than dry mass of FI and PRD; nevertheless, dry mass that categorized into fruits was lower in PRD and TDI than the dry mass of FI. Various methods of water conservation have been investigated in agriculture, such as alternative PRD, TDI, and drip irrigation. In that respect, researchers have shown that WUE can be enhanced (Li et al. 2010). In general, these techniques are a trade-off: a lower yield for a higher WUE. This suggests that the PRD is a better water saving option than TDI. Saved water through PRD can be used for irrigation of additional cultivated areas (Patel and Rajput 2013). Optimum WUE in TDI is obtained by an economic analysis using yield and cost functions as described by English and James (1990).

Mohammad and Mazahreh (2003) studied the effect of wastewater irrigation on soil chemical properties and found that secondary treated wastewater could improve soil fertility parameters. However, more efficient treatment is recommended to reduce the salt content. In addition, proper irrigation management and periodic soil quality parameters' monitoring are required to minimize negative effects on the soil.

Sahin et al. (2014) showed root and sugar yields reduced as 22 and 24% in PRD treatment than FI, respectively. But irrigation water saving was about 42% by PRD treatment. They suggested that PRD technique could be useful strategy by drip-irrigated sugar beet in areas with water shortage regions.

Cantore et al. (2014) obtained enough yield and high quality tuber from 50% less water-applied treatment by TDI as like the FI treatment.

In Isfahan region, Iran, municipal and industrial treated wastewater used as a water resource or irrigation of different production (Tabatabaei et al. 2012). So, it is more necessary to select the best and new irrigation techniques with the management to improve efficiency of irrigation system (Najafi et al. 2016).

Therefore, the objective of this paper was to evaluate the type of irrigation system including PRD and to compare it with FI system as TDI which they both are water-saving irrigation practices. Furthermore, their effects on plant's indicators, such as dry and fresh weight, plant height, dry matter percentage and WUE were investigated.

Materials and methods

Site description

The Shahrekord wastewater treatment station is located 7 km south-west of the Shahrekord city at 32°20′N latitude and 50°51′E longitude. Shahrekord has a semi-arid climate using combination climatologic methods (Kham-chin Moghadam and Rezaei Pajand 2009). The average height of this region above the sea level and rainfall are about 2060 m and 321.5 mm, respectively (without any rainfall during experimental period). The average temperature of this city is 12.5 °C and the maximum and minimum relative humidity are 71 and 16%, respectively.

Soil and wastewater characteristic

Field experiments were carried out during the July–October 2010 (growth period). In general, the soil texture is clay in which its characteristics and also chemical analysis of treated wastewater by sequence batch reactor (SBR) method and water sources are presented in Tables 1 and 2 (Karizan 2011).

Experimental design

The field selected for the research was plowed and fertilized using macro- and micro-fertilizers in July 2010 based

Table 1 Chemical and physical properties of soil

Depth (cm)	EC (dS/m)	pH	Soil texture	ρ_b (g/cm^3)	FC (%)	PWP (%)
0–45	1.8	7.9	Clay	1.19	37	18

Table 2 Chemical analysis of pretreated municipal wastewater and water sources using as irrigation water

Sources	EC (dS/m)	pH	TDS (mg/l)	TSS (mg/l)	SAR (mol/l)$^{1/2}$	BOD (mg/l)	N–NO_3^- (mg/l)	K (mg/l)	PO_4^{-3} (mg/l)
Wastewater	0.75	7.8	450	30	2.2	18.75	16	14	30.5
Water	0.3	7.45	38	0	0.13	3.96	2.61	0.21	15

on the soil tests. Each treatment was designed in the format of furrow irrigation style. Experimental plot size was 3 × 4 m with 1 m distance between each plot. The distance between the planting rows was 75 cm and the distance between corn plants was 10 cm (Fig. 1). Research was implemented based on the factorial (split plot) design in three replicates and fifteen treatments. Quality and water management were the two factors, so that three water quality levels made up the main plots (Table 3). Water used for irrigation contained three quality levels: fresh water (FW), a mixture of 50% fresh water and 50% wastewater (5050), and finally 100% wastewater (WW).

Water management was the third factor determined in three levels: full irrigation (FI), partial root-zone drying (PRD) and the traditional deficit irrigation (TDI). Three levels of water application factor were implemented in 100, 80 and 60 percent of estimated crop water requirement (creating water stress of zero, 20 and 40 percentages, respectively). The reference evapotranspiration (ET_0) (mm/day) was estimated using the FAO Penman–Monteith equation and employing daily data from a nearby automatic weather station. Crop water requirements were calculated by multiplying the ET_0 values with the corn crop coefficients (K_c) given by Kheirabi et al. (2009) and E_{pan} as 5–8.5 mm/day. Concerning irrigation system, drip irrigation (T-tape) was used and was conducted 8 days after planting in late July. Corn crops were irrigated every three days. After the growth period, the plant physiological parameters such as shoot's fresh and dry weight and leaf area index

Table 3 Fifteen treatments applied in this study

Treatments		
WW-FI_{100}	5050-FI_{100}	FW-FI_{100}
WW-PRD_{80}	5050-PRD_{80}	FW-PRD_{80}
WW-TDI_{80}	5050-TDI_{80}	FW-TDI_{80}
WW-PRD_{60}	5050-PRD_{60}	FW-PRD_{60}
WW-TDI_{60}	5050-TDI_{60}	FW-TDI_{60}

FW 100% fresh water, *5050* 50% fresh water plus 50% wastewater, *WW* 100% wastewater, *FI* full irrigation, *PRD* partial root-zone drying, *TDI* traditional deficit irrigation, *60 AND 80* 80 and 60 percent of the estimated crop water requirement

(LAI) were measured. Plant height was measured in each treatment using a meter scale and the biomass (the above part of the ground). Thereafter, it was cut into small pieces and dried at 70 °C to reach a constant weight for the estimation of dry matter. Dry biomass percentage (DBP) was calculated using the following equation.

$$\text{DBP} = (\text{Sample total dry weight}/\text{Sample total fresh weight}) \times 100$$

Soil samples were collected from 0 to 45 cm soil depth. Air-dried soil samples were sieved through a 2-mm sieve to determinate physicochemical soil properties. Electrical conductivity (EC) and acidity of soil (pH), wastewater and water were identified by Jenway EC and pH meters (Page et al. 1991), respectively. Cation like as Ca and Mg was determined by titration method with 0.01 N EDTA (Klute 1986) and sodium (Na) and potassium (K) were measured

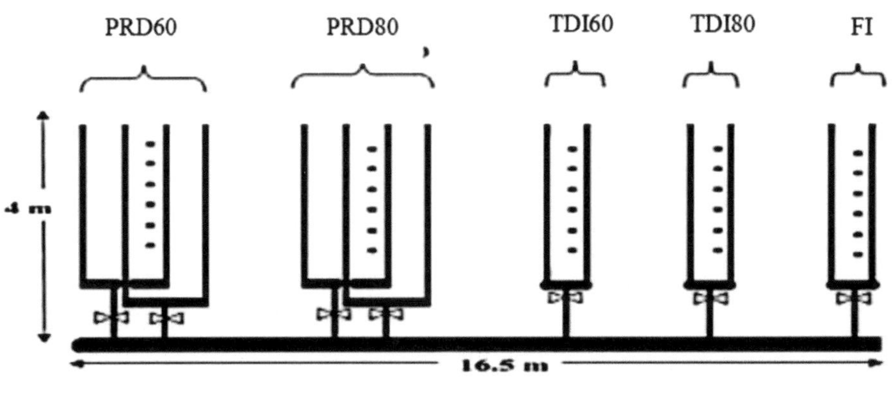

Fig. 1 Layout of the system. *PRD* partial root-zone drying, *TDI* traditional deficit irrigation, *FI* full irrigation. ⋈ Valve; *asterisks* corns; ▬ lateral

by flame photometry. Sodium adsorption ratio (SAR) was defined by the relation.

Soil's phosphor was measured using the method described by Olsen et al. (1954), soil texture was determined by hydrometer (Lee and Bauder 1986), total nitration by Kjeldahl digestion (Page et al. 1991), bulk density by clod and particle density by pycnometer (Klute 1986), total suspended solids (TSS) by gravimetric analysis (APHA 1975), total dissolved solids' (TDS) by Canadian Council of Ministers of the Environment (2011).

Afterwards, analysis of variance was performed and treatments' averages were compared with the least significant difference test (LSD) with 5% probability level. Statistical analysis was performed using "Statistical software" and also "Excel software" was applied for data management.

Results and discussions

Dry and fresh weight of corn shoots

The results showed that fresh weight of corn shoots in different treatments had significant difference at 5% level of probability. Fresh weight' mean of crop that irrigated by FI (5940 g/plant) was significantly higher than the fresh weight' mean of crop planted under TDI_{60} irrigation treatment (492.33 g) (Fig. 2). Water stress in TDI was the main reason to reduce shoot' fresh weight at 5% probability level. Although, wet and dry weight of corn plans of TDI_{60} plus PRD_{60} and TDI_{80} plus PRD_{80} together in each group were in the same probability level. Although, both PRD treatments (PRD_{60} and PRD_{80}) were higher than TDI treatments (TDI_{60} and TDI_{80}) (Fig. 2).

Nagaz et al. (2013) indicated that shoot lost its fresh weight rapidly in water stress situation. On the other hand, the corn plants irrigated with PRD had greater fresh weight in comparison with traditional irrigation system (Li et al. 2010); nevertheless, it was not significant at 5% probability level.

More dry weight of the corn shoots (119.7 g/plant) were observed in FI comparing TDI_{60} (91.44 g/plant). Also, dry weight of TDI_{80} with 108. 65 g/plant decreased to 91.44 (g/plant) in TDI_{60} (Fig. 2). Water deficiency primarily affects crops by reduction of dry matter accumulation (Karam et al. 2003) in which happens through TDI diminishing dry matter development (Lopez et al. 1996) because of water tension, stomata conductance and photosynthesis rate (Nazemosadat and Kazemini 2008). In this experiment, irrigation with 5050-treatment reached a peak of 556.3 g fresh weight/plant. It was then followed by WW-treatment used as irrigation water. FW irrigation treatment with 505.33 g fresh weight/plant was applied at the end of grouping (Fig. 3). The rate of fresh weight was raised from 4 to 10% using 5050-treatment as the first treatment and using WW-treatment as the second treatment, respectively ($p < 0.01$). Also, it was seen that 5050-treatment and WW-treatments resulted in 30.3 and 17.2% increase in dry weight ($p < 0.01$).

Also, Rousan et al. (2008) reported that dry weight was increased more under FI with fresh water and wastewater than deficit irrigation. This trend was perceived to be similar in the yield response of Dorjia et al. (2004). In another study implemented by Fulai et al. (2006), the results showed that deficit irrigation method reduced leaves' dry matter and in contrast increased roots' dry matter of potato' stems. Dorjia et al. (2004) in the course of their research on hot pepper find that total fruit's fresh mass per plant in PRD was 586.5 g higher than its amount in TDI; whereas it was 724.1 g lower than its amount in control irrigation treatment. However, there was no considerable difference in total fruit' dry mass among the treatments. These results were consistent with our findings as well. The ratio of $N{-}NO_3$ (mg/l) in wastewater was six times bigger than their volumes in FW-treatment; therefore, using WW-treatment as irrigation water increased more $N{-}NO_3$ concentration of corn' shoots. On the other

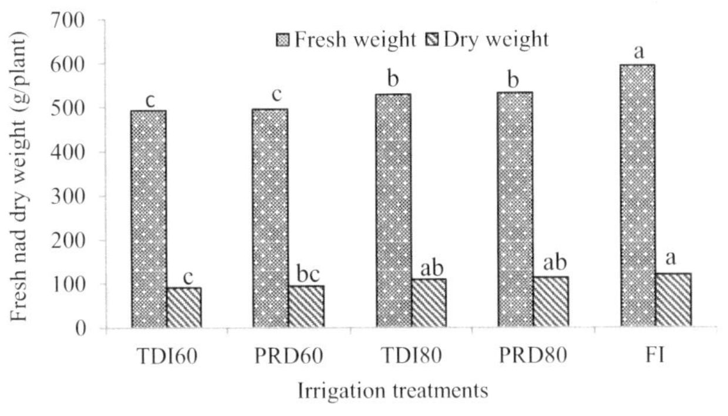

Fig. 2 Effect of irrigation treatments on dry and fresh weight of corn shoots ($p < 0.05$)

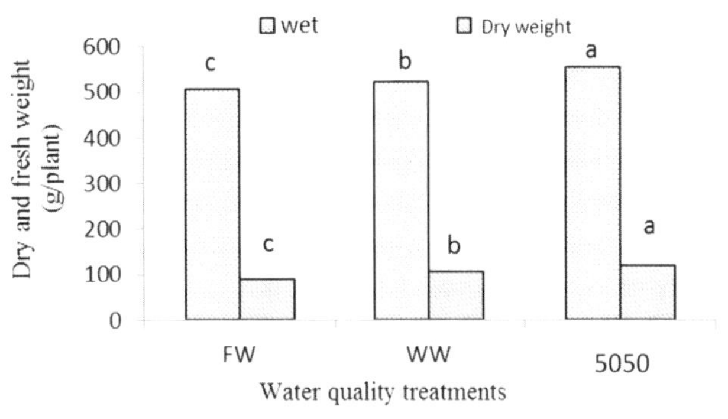

Fig. 3 Effect of water quality on dry and fresh weight of corn shots ($p < 0.01$)

hand, analysis of variance shoots' fresh weight demonstrated that interaction between water quality and TDI method was significant at 1% probability level. The results uncovered that the least and the most of yield were achieved using TDI_{60} (585.3 g) and 5050-FI (468.7 g) treatments, respectively. Also, WW-FI reduced two percent of yield in comparison with FW-FI due to nutrient toxicity. However, there was a slight rise in yield of WW-PRD_{80} and WW-PRD_{60} treatments.

Overall, PRD irrigation method using wastewater as irrigation water brought about greater yield than using FW-TDI method. Unlike fresh weight, irrigation with wastewater did not have any significant difference in dry weight instead; it indicated that there was only an increase in water percentage. Findings showed that the biggest dry weight was occurred in 5050-FI. FW-TDI_{60} was located at the end of grouping. Available nutrients of wastewater transmitted a signal from roots to shoots for strengthening of plant stomata (Li et al. 2010).

Dry biomass percentage (DBP)

In this study, the ratio of samples' total dry weight to samples' total fresh weight of samples is DBP. The amounts of DBP in all treatments were significantly different at 1% probability level. Concerning this parameter, PRD_{80} had the biggest DBP (21.29%) and TDI_{80} (with 18.51%) was placed at the end of this category (Fig. 4) because of its soil moisture stress and photosynthesis rate. Also, the DBP of TDI_{80} and FI treatments were 20.56 and 20.23 and there was no significant difference with the maximum DBP in PRD_{60} (Fig. 4). As it was already said, water deficiency primarily affects crops through prevention of dry matter accumulation (Karam et al. 2003) due to reduced dry matter development (Lopez et al. 1996). Dorjia et al. (2004) represented that the DBP to roots was equal among all different treatments; nevertheless, a higher allocation was revealed in stems and leaves in TDI and PRD plants comparing the control plants. In addition, wastewater irrigation might have an impact on the dissolution of the nutrient may be due to reduce the microbial activity in the soil and resulting in slower release of plant available nutrients (Shammi et al. 2016). In this research, applying lower irrigation water amount by TDI than FI techniques cause to increase more DBP.

In addition, analysis of variance showed that DBP had a significant difference at 5% probability level based on different irrigation water methods (Fig. 5).

The DBP means the ratio of total dry weight to total fresh weight of samples. The highest and lowest rates were reported applying 5050-WW with 21.47% and FW-treatment with 18.01%, respectively.

Rousan et al. (2008) illustrated that regardless of the water quality, TDI results in lower DBP. In their view, the reason was that amount of applied irrigation water was significantly less than the tomato crop's water requirement. Other researchers attributed improved plant growth with wastewater to improved soil condition, and in particular to enhanced soil organic matter content (Mohammad and Mazahreh 2003; Yonts 2010).

Leaf area index (LAI)

Leaf area is greatly changed with soil water availability. The effect of soil dryness on physiological indices was small over a certain range of soil water deficit started from 0 to approximately 40% (Patanè 2011). Reducing leaves' number and size are a result of cell behavior sensitive to water shortage (Sawia et al. 2002; Atti et al. 2004).

The findings indicated that water irrigation volume had a significant impact on LAI at 5% probability level (Fig. 6). According to the graph, a reduction of 20 and 40% in irrigation water happened as a result of TDI consequently diminished 9.3 and 25% of LAI. In addition, statistics demonstrate that PRD_{60} method decrease more effectively the LAI than the TDI_{60} method (Fig. 6). This could be due to transmitted signal from the roots to the shoots ordering them to decrease leaf growth and increase production of abscisic acid (ABA), and finally to close leaf stomata.

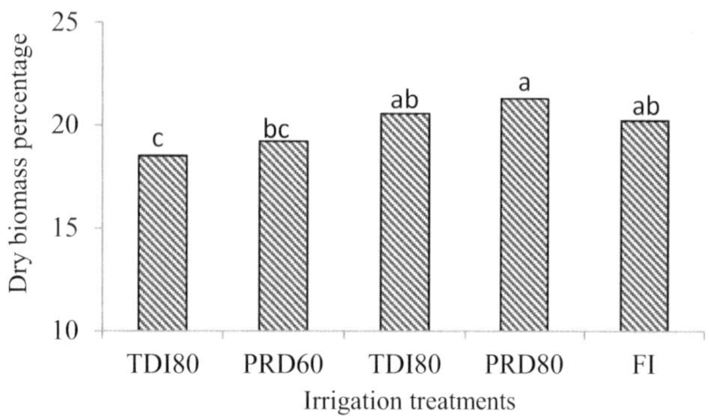

Fig. 4 Effect of irrigation treatments on dry biomass percentage ($p < 0.05$)

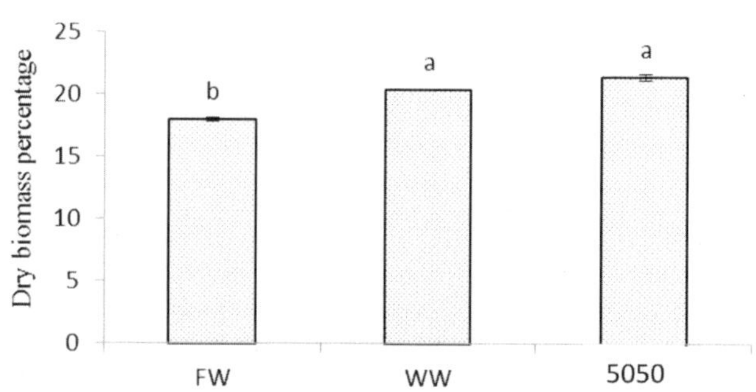

Fig. 5 Effect of water quality on dry biomass percentage ($p < 0.05$)

Similar results were reported by Liu et al. (2006) and Shahnazari et al. (2008). The findings also represented that in FI treatment the maximum and minimum LAI across the period were 7.17 and 5.2, respectively.

According to the research and based on the statistical comparison, water quality did not have any special effect on LAI. Nonetheless, the largest and least LAI belonged to 5050-treatment and FW-treatment with 6.24 and 6.13, respectively (Fig. 7).

Also, this finding is similar to the results of Hirich et al. (2011) in which a local variety of chickpea was planted under TDI (50% full irrigation) with treated domestic wastewater. The TDI applied for corn plants during vegetative growth lead to a greater proportion of its root systems developed deeper in the soil profile, which could lead to better use of stored soil water. Corn responded to drier soil conditions by increasing the proportion of roots deeper in the soil (Benjamin and Nielsen 2006). On the other hand, terminal water stress decreased the total plant dry matter including LAI and leaf weight (Sharma et al. 2007; Labidi et al. 2009; Behboudian et al. 2001).

Based on the results, a decrease in LAI was seen in all treatments expect FW-FI. This result similar to Yonts (2010) results that indicated the leaf area index of sugar beet planted under TDI was approximately 14% less than FI.

Also, the most and least LAI were distinguished in 5050-FI (7.5) and 5050-PRD_{60} treatments (4.8).

Plant height

One of the growth indexes that are affected strongly by water shortage is plant height (Atti et al. 2004; Sawia et al. 2002). In this research, plant height was measured three times (stages) but the results of two stages of height of corn plants were showed in Fig. 8.

Analysis of variance showed that height of corn plants had a significant difference at 1% probability level based on the irrigation methods. The TDI_{60} and PRD_{60} were at the same statistical level in the end of experiment.

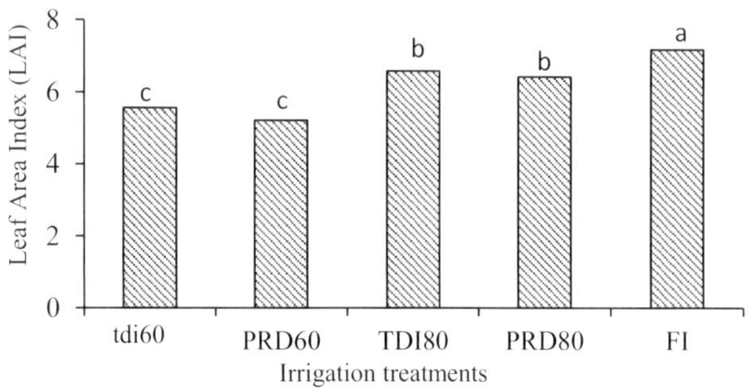

Fig. 6 Effect of irrigation treatments on leaf area index (LAI) ($p < 0.05$)

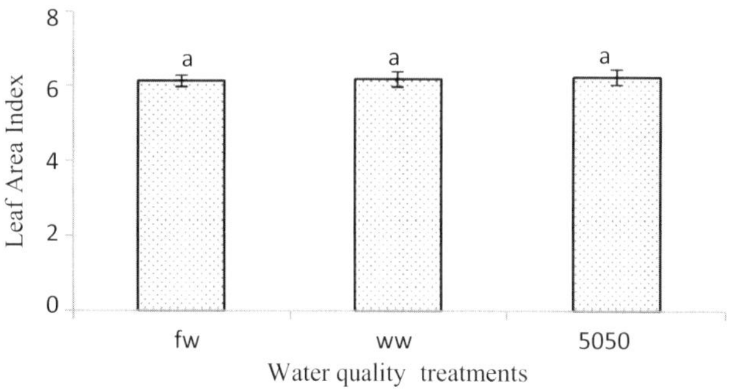

Fig. 7 Effect of water quality on leaf area index (LAI)

Although there is significant difference between these treatments on height of corn plants on 15 days before applied irrigation treatments (Fig. 8). On the other hand, three treatments included TDI_{80}, PRD_{80} and FI did not have significant difference in the first stage of measuring height of corn plants but using FI method caused to increase height of corn plants than TDI and PRD irrigation techniques.

Full irrigation had the maximum height of corn plants among irrigation treatments (Fig. 8). Also, PRD_{80}, TDI_{80}, PRD_{60} and TDI_{60} caused 7.1, 6.4, 12.8 and 12% of height reduction, respectively, in comparison with FI. The reduced plant height in all irrigation treatments was significant expect TDI_{80} at 1% probability level. However, there was not seen any significant difference between PRD and TDI irrigation methods. This outcome is similar to the results of Du et al. (2006) on cotton plants. It seems there is a genetic correlation between yield and plant height, cultivated plants with water stress resistance is able to maintain against reduced plant height.

Yonts (2010) reported that sugar beet plant height was 127–305 mm less in deficit irrigation method than full irrigation treatments. Also, variance analysis of interactive effects between water quality and water quantity of Yonts' study showed that there was no significant difference at 5% probability level.

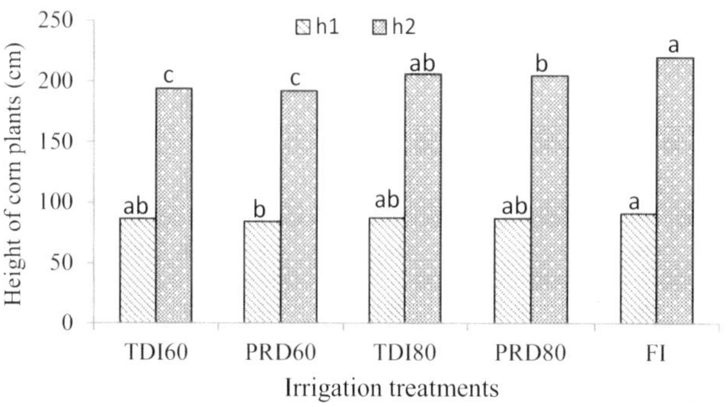

Fig. 8 Effect of irrigation treatments on height of corn plants ($p < 0.01$). h1: 15 days before applied irrigation treatments. h2: 50 days in the end of experiment

The analysis of variance showed that height of corn plants had a significant difference at the 5% probability level based on water quality. The results indicate that using 5050-FI was the best composition of irrigation water to increasing height of corn plant (Fig. 9).

Chemura (2014) showed in the research on coffee, irrigation water amounts in terms of height have significant effect ($p < 0.05$). Coffee plants were tallest and thickest plant stems when supplied the more irrigation water levels (1000 ml) and shortest and thinnest under the lowest irrigation amounts (500 ml) because of inorganic fertilizers are the most effective at high irrigation levels while organic manure performs better than inorganic fertilizers under low irrigation water amount.

Water use efficiency (WUE)

Water use efficiency, if defined as the biomass accumulation over water consumed may be fairly constant for a given species in given climate. WUE can be enhanced by less irrigation. However, such enhancement is largely a trade-off against lower biomass production (Zhang and Yang 2004). The maximum WUE was belonged to PRD_{60} with 3.09 (kg dry matter/m^3 water using) compared to the minimum WUE in FI treatment with 2.47 (kg dry matter/m^3 water using). Sahin et al. (2014) indicated that PRD technique increased by 34.9% WUE compared to FI of sugar beet (*Beta vulgaris* L.) because of increase root yield during 2-year study.

Overall, PRD had the WUE much more than TDI and it was significant at 5% probability level. It was similar to the results of Shahnazari et al. (2008) and Liu et al. (2006) on potato; Gencoglan et al. (2006) on green bean; De la Hera et al. (2006) on fruit quality in field-grown wine grapes. On the other hand, Hu et al. (2009) and Wang et al. (2008) reported that there was a significant rise in WUE for potato plant, when used the PRD to TDI method in pot.

Water use efficiency like other indicators dry and fresh weight shoots; plant height was being more affected by the amount of irrigation water than TDI methods. It is clear from the data that the amount of WUE in TDI_{60} to TDI_{80} and then TDI_{80} to FI were significant at 5% probability level (Table 4).

In other research, WUE was higher in TDI and PRD plants than in control plants (Zegbe-Domínguez et al. 2003). Acar et al. (2014) illustrated by use of DI by drip system average 25% volume of water saving can be made.

The analysis of variance showed that WUE had a significant difference at the 5% probability level based on irrigation water quality (Table 4). The 5050-treatment with 3.17 kg dry matter/m^3 using water was leader in this group. Irrigation with FW-treatment with 2.435 kg dry matter/m^3 water was in the end. The results indicated that irrigation with treatment included WW-treatment than 5050-treatment as another quality irrigation water has a significant decline on fresh water because of two main reasons: first of all, nutrient toxicity using WW-treatment as irrigation water and second, water stress by increasing osmotic potential due to high level of salts in this irrigation water source. The soil conditions such as arid climate and heavy soil texture caused to increase these effects if using wastewater as irrigation water (Table 1). On the other hand, wastewater has high amount of essential nutrients like N and P, while they are useful for plant growing.

It is clear that the minimum WUE belonged to use FW-FI and then irrigation with WW-FI. According to analysis of variance of interactive effects between water quality and water quantity, 5050-PRD_{80} and 5050-PRD_{60} with 3.36 and 3.3 kg/m^3 has the biggest WUE. Whereas FW-FI and FW-TDI_{80} with 2.31 and 2.26 kg/m^3 in contrast to 5050-PRD_{80} and 5050-PRD_{60} had the smaller WUE (Table 4). The results showed that using 5050-treatment caused to compensation of 40% water deficit for WUE because of higher concentration of $N{-}NO_3$ (16 mg/l) in wastewater in comparison with of $N{-}NO_3$ lower concentration (2.61 mg/l) in FW-treatment, increased dry matter development.

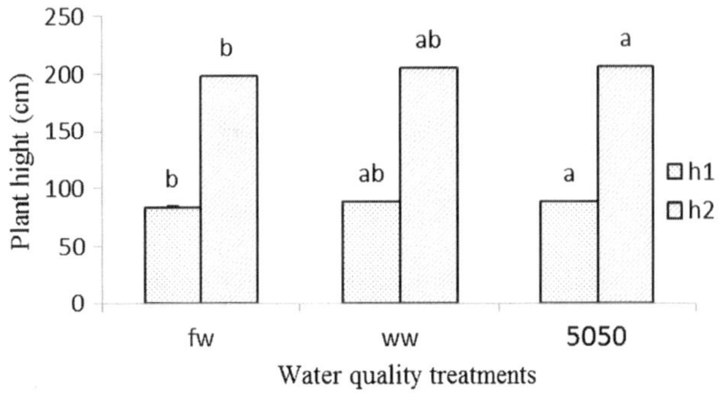

Fig. 9 Effect of water quality on height of corn plant ($p < 0.05$). h1: height of corn plants on 15 days before applied irrigation treatments. h2: height of corn plants on 50 days in the end of experiment

Table 4 Effects of irrigation treatments and water quality on water use efficiency (kg/m^3)

	TDI_{60}	PRD_{60}	TDI_{80}	PRD80	FI
Irrigation treatments	2.97ab	3.1a	2.74c	2.86bc	2.47d
	FW		WW		5050
Water quality	2.44c		2.88b		3.17a

FW 100% fresh water, *5050* 50% fresh water plus 50% wastewater, *WW* 100% wastewater, *FI* full irrigation, *PRD* partial root-zone drying, *TDI* deficit irrigation; 80 and 60 percent of the estimated crop water requirement

Conclusions

This experiment was done to assess TDI and PRD as water-saving irrigation techniques. In both TDI and PRD treatments than FI, dry and, fresh weight, height of plant and LAI of plants were reduced because of water stress but PRD and TDI method were caused to increase WUE and DBP. Using PRD_{80} method is more suitable than TDI method. After FI technique, corn plant that irrigated by PRD_{80} method has the bigger dry and fresh weight, DBP and WUE than TDI. So, it outcome in PRD_{80}, we can store irrigation water during study with less effects on corn plants yield because of better water distribution in hemispheric wetted volume than TDI as other water-saving irrigation technique. It seems transmitted signal from roots to shoots to decline the rate of leaf growth and increasing to produce ABA and to close stomata. The height of corn plants was more affected by declining the volume of irrigation water, therefore FI > PRD_{80} > TDI_{80} > -PRD_{60} > TDI_{60}. WUE as the main aim was the maximum by PRD_{60} to FI as the minimum WUE. Based on the results, it was parented that applying PRD method has higher performance than TDI. Also, in comparison of three water qualities, using 5050-treatment caused to compensation of water deficit for WUE, LAI, and yield indices but irrigation with WW-treatment alone had unsuitable effects because of nutrient toxicity and undesirable effects of increasing osmotic potential due to high level of salts. Changing of water quality has significant difference on dry and fresh weight, DBP, height of plant and WUE expect LAI. Although PRD and TDI methods were found to be feasible for corn cultivation but using PRD system was considered as a better strategy for water-saving than TDI.

Acknowledgements The authors would like to thank the faculty of Agriculture at Shahrekord University for its financial supports of this research project.

References

Acar B, Topak R, Yavu D, Kalender MA (2014) Is Drip irrigation technique sustainable solution in agriculture for semi-arid regions? A case study of Middle Anatolian region. Turk Int J Agric Econ Dev 2(2):1–8

APHA (1975) Method 208 D. Total non-filterable residue dried at 103–105 °C (total suspended matter) in standard methods for the examination of water and wastewater, 14th edn. American Public Health Association, Washington, DC, p 460

Zhang J, Yang J (2004) Crop yield and water use efficiency. In: Bacon MA (ed) Water use efficiency in plant biology. Blackwell Publishing, Oxford, pp 189–218

Atti S, Bonnell R, Smith DL, Prasher S (2004) Response of an indeterminate soybean to chronic water deficit during reproductive development under greenhouse conditions. Can Water Resour J 29(4):209–224

Behboudian H, Ma MQ, Palta CNT, Jairo A (2001) Reactions of chickpea to water stress: yield and seed composition. Sci Food Agric J 81(13):1288–1291

Benjamin JG, Nielsen DC (2006) Water deficit effects on root distribution of soybean, field pea and chickpea. Field Crops Res J 97:248–253

Canadian Council of Ministers of the Environment (2011) Protocol manual for water quality sampling in Canada. Canadian Council of Ministers of the Environment, P 180

Cantore V, Wassar F, Yamaç SS, Sellami MH, Albrizio R, Stellacci AM, Todorovic M (2014) Yield and water use efficiency of early potato grown under different irrigation regimes. Int J Plant Prod 8(3):409–428

Chemura A (2014) The growth response of coffee (*Coffea arabica* L.) plants to organic manure, inorganic fertilizers and integrated soil fertility management under different irrigation water supply levels. Int J Recycl Org Waste Agric 3:59

De la Hera ML, Romero P, Gomez-Plaza E, Martinez A (2006) Is partial root-zone drying an effective irrigation technique to improve water use efficiency and fruit quality in field-grown wine grapes under semi-arid conditions? Agric Water Manag J 87:261–274

Dorjia K, Behboudiana MH, Zegbe-Dominguez JA (2004) Water relations, growth, yield, and fruit quality of hot pepper under deficit irrigation and partial root-zone drying. Sci Hortic J 104:137–149

Du T, Kang SH, Zhang J, Li F, Hu X (2006) Yield and physiological responses of cotton to partial root-zone irrigation in the oasis field of northwest China. Agric Water Manag J 84:41–52

English M, James L (1990) Deficit irrigation. II: Observation in Colombia basin. J Irrig Drain Eng 116(3):413–426

Fulai L, Shahnazari A, Andersen MN, Jacobsen SE, Jensen CR (2006) Physiological responses of potato (*Solanum tuberosum* L.) to partial root-zone drying: ABA signaling leaf gas exchange and water use efficiency. Exp Bot J 57:3727–3735

Gallardo M, Jackson LE, Schulbach K, Snyder RL, Thompson RB, Wyland LJ (1996) Production and water use in lettuces under variable water supply. Irrig Sci J 16:125–137

Geerts S, Raes D (2009) Review: deficit irrigation as an on-farm strategy to maximize crop water productivity in dry areas. Agric Water Manag J 96:1275–1284

Geerts S, Raes D, Garcia M, Condori O, Mamani J, Miranda R, Cusicanqui J, Taboada C, Yucra E, Vacher J (2008) Could deficit irrigation be a sustainable practice for quinoa (*Chenopodium quinoa* Willd.) in the Southern Bolivian Altiplano? Agric Water Manag J 95:909–917

Gencoglan C, Altunbey H, Gencoglan S (2006) Response of green bean (*P. vulgaris* L.) to subsurface drip irrigation and partial root-zone-drying irrigation. Agric Water Manag J 84:274–280

Grant OM, Stoll M, Jones HG (2004) Partial rootzone drying does not affect fruit yield of raspberries. J Hortic Sci Biotechnol 79(1):125–130

Hirich A, Choukr-Allah R, Jacobsen Sven-Erik, Hamdy A, Youssfi L, El-Omari H (2011) Improving water productivity of chickpea by the use of deficit irrigation with treated domestic wastewater. World Acad Sci Eng Technol 59:1352–1357

Hu T, Kang S, Li F, Zhang F (2009) Effects of partial root-zone drying irrigation on the nitrogen absorption and utilization of maize. Agric Water Manag J 96:208–214

Karam F, Breidy J, Stephan C, Rouphael J (2003) Evapotranspiration, yield and water use efficiency of drip irrigated corn in the Bekaa Valley of Lebanon. Agric Water Manag J 63:125–137

Karizan MM (2011) Use of deficit irrigation techniques (DI and PRD) in irrigated with municipal wastewater branch under drip irrigation (T-tape) on corn silage. M.Sc. thesis, Department of Water Engineering, Faculty of Agriculture, Shahrekord University, Shahrekord, Iran

Kham-chin Moghadam F, Rezaei Pajand H (2009) Realize maximum daily rainfall by using De Martonne method and compare by linear combination method. Tech Eng J Islam Azad Univ Mashhad Branch 2:93–103

Kheirabi J, Darbandi S, Entesari M, Salamat A, Moghadesi F (2009) Evaluation and comparison Penman–Monteith equation with 24-FAO methods in Iran. Iranian National Committee on Irrigation and Drainage Publication. Tehran, Iran, p 208

Klute A (1986) Physical and mineralogical methods In: Methods of soil analysis. Part 1. American Society of Agronomy-Soil Science Society of America pubication, 2nd edn. p 1189

Labidi N, Henda M, Messedi D, Slama I, Chedly A (2009) Assessment of intervarietal differences in drought tolerance in chickpea using both nodule and plant traits as indicators. Plant Breed Crop Sci J 1:80–86

Lee GW, Bauder JW (1986) Particle size analysis. In: Klute A (ed) Methods of soil analysis, Part 1. American Society of Agronomy-Soil Science Society of America publication, pp 383–411

Li F, Yu J, Nong M, Kang S, Zhang J (2010) Partial root-zone irrigation enhanced soil enzyme activities and water use of maize under different ratios of inorganic to organic nitrogen fertilizers. Agric Water Manag J 97:231–239

Liu F, Shahnazari A, Andersen MN, Jacobsen SE, Jensen CR (2006) Effects of deficit irrigation (DI) and partial root drying (PRD) on gas exchange, biomass partitioning, and water use efficiency in potato. Sci Hortic J 109:113–117

Lopez FB, Johansen C, Chauhan YS (1996) Effects of timing of drought stress on phenology, yield and yield components of short-duration pigeonpea. Agron Crop Sci J 177:311–320

Mohammad MJ, Mazahreh N (2003) Changes in soil fertility parameters in response to irrigation of forage crops with secondary treated wastewater. Soil Sci Plant Anal J 34:1281–1294

Nagaz K, Mokh FE, Masmoudi MM, Mechlia NB (2013) Soil salinity, yield and water productivity of lettuce under irrigation regimes with saline water in arid conditions of Tunisia. Int J Agron Plant Prod 4:892–900

Najafi P, Asgari K, Samadi N (2016) Heavy metal elimination from industrial wastewater using natural substrate on pitcher irrigation. Int J Recycl Org Waste Agric 5:333–337

Nazemosadat SMJ, Kazemini SA (2008) The effects of water stress and photosynthesis active radiation on photosynthesis rate of wheat. Agric Sci Nat Resour J 14(6):34–44

Olsen SR, Cole CV, Watunable FS, Dean LA (1954) Estimation of available P in soil by extraction with Na-bicarbonate. Circular, No.939 USDA. US. Government, Washington DC

Page AL, Miller RH, Keeney DR (1991) Chemical and biological properties. In: Methods of soil analysis, part 2. American Society of Agronomy-Soil Science Society of America publication.

Patanè C (2011) Leaf area index, leaf transpiration and stomata conductance as affected by soil water deficit and VPD in processing tomato in semi-arid mediterranean. Clim Agron Crop Sci J 197:165–176

Patel N, Rajput TBS (2013) Effect of deficit irrigation on crop growth, yield and quality of onion in subsurface drip irrigation. Plant Prod J 7:417–436

Rousan LM, Rusan MJ, Amayreh J (2008) Irrigation with treated wastewater under full and deficit soil moisture conditions. In: Eds: Al-Mattareneh H, Sidek LM, Youssef MZ (eds) Proceeding of the international conference on construction and building technology (ICCBT): innovations in water resources and environmental engineering, Kuala Lumpour, Malaysia, vol 38, pp 405–416

Sahin U, Ors S, Kiziloglu FM, Kuslu Y (2014) Evaluation of water use and yield responses of drip-irrigated sugar beet with different irrigation techniques. Chil J Agric Res 74:302–310

Sawia M, Omaima U, EL-Beharya A, Abdolhamid AF (2002) Effect of irrigation level and N source on green bean productivity. Egypt Appl Sci J 47:326–347

Shahnazari A, Ahmadi SH, Lærke PE, Liu F, Plauborg F, Jacobsen SE, Jensen CR, Andersen MN (2008) Nitrogen dynamics in the soil-plant system under deficit and partial root-zone drying irrigation strategies in potatoes. Eur J Agric 28:65–73

Shammi M, Abul Kashem M, Mostafizu Rahman Md, Delwar Hossain Md, Rahman R, Khabir Uddin M (2016) Health risk assessment of textile effluent reuses as irrigation water in leafy vegetable Basella alba. Int J Recycl Org Waste Agric 5:113–123

Sharma KD, Pannu RK, Tyagi PK, Chaudhary BD, Singh DP (2007) Response of chickpea genotypes to plant water relations and yield under soil moisture stress. Agrometeo J 9:42–48

Tabatabaei SH, Najafi P, Mirzaei SMJ, Nazem Z, Heidarpour M, Hajrasoliha S, Afyuni M, Beigi Harchegani H, Landi E, Akasheh L, Zamanian M, Barani M, Amini H (2012) Compost' leachate recycling through land treatment and application of natural Zeolite. Int J Recycl Org Waste Agric 1:2. doi:10.1186/2251-7715-1-2. http://www.ijrowa.com/content/1/1/2

Wang J, Kang S, Li F, Zhang F, Li Z, Zhang J (2008) Effects of controlled alternate partial root-zone drying irrigation on soil microorganism and growth of maize. Plant Soil J 302:45–52

Yazar A, Gokgel F, Sezen MS (2009) Corn yield response to partial root zone drying and deficit irrigation strategies applied with drip system. Plant Soil Environ J 55:494–503

Yonts CD (2010) Development of season long deficit irrigation strategies for Suger-beets. The University of Nebraska Panhandle Research and Extension Center, Scottsbluff, p 69361

Zegbe-Domínguez J, Behboudian A, Lang MHA, Clothier BB (2003) Deficit irrigation and partial root zone drying maintain fruit dry mass and enhance fruit quality in 'Petopride' processing tomato (*Lycopersicon esculentum*, Mill.). Sci Hortic J 98:505–510

Zhang H, Oweis T (1999) Water-yield relations and optimal irrigation scheduling of wheat in the Mediterranean region. Agric Water Manag J 38:195–211

19

Dynamics of soil nitrogen availability indices in a sandy clay loam soil amended with animal manures

Habeebah Adewunmi Saka[1] · Jamiu Oladipupo Azeez[1] · Joy Nwakaego Odedina[1] · Shade John Akinsete[1]

Abstract

Purpose Transformation of nitrogen (N) from different manure sources to available forms to promote food security in context of rising climate change is vital. Effect of manure (types, nutrients, high rates and application time) on soil N transformation requires further investigation. This study evaluated effects of three animal manures on soil N mineralization of sandy clay loam.

Methods A 22-week field study in a Randomized Complete Block Design (three replicates) was conducted using dried Cattle, Goat and Poultry Manures (CGPM), applied at 5, 10, 20, 40, 60, 80, 120 and 150 t ha^{-1} once at onset of the study. Also, $N_{15}P_{15}K_{15}$ (0.4 t ha^{-1}) was incorporated in three splits of 2, 8 and 14 weeks after incorporation (WAI) of manures as reference. Soil NH_4^+–N and NO_3^-–N were determined bi-weekly.

Results The results showed N release peaked at 10 WAI with highest NH_4^+–N (830, 400, 253 mg kg^{-1}) and NO_3^-–N (316, 398, 250 mg kg^{-1}) at 150, 60 and 60 t ha^{-1} for CGPM, respectively. Initial rapid N release phase occurred at 0–4 WAI, NH_4^+–N and NO_3^-–N increased averagely by 182, 183, 139% and 131, 175, 144% for CGPM, respectively. Declines occurred at 8, 14 and 22 WAI but reduction observed at 22 WAI tripled 8 and 14 WAI.

Conclusions Application of NPK and all the manures provided highest soil NH_4^+–N and NO_3^-–N at 12 WAI. High rates of CGPM were not injurious to these soil properties; hence this has implication for N to maximize plants uptake as well as decrease losses to environment.

✉ Habeebah Adewunmi Saka
sakahabeebah.staad@gmail.com

[1] Federal University of Agriculture Abeokuta, Abeokuta, Nigeria

Keywords Animal manures · Rates · NPK fertilizer and nitrogen mineralization

Introduction

Agricultural growth as a result of population increase had contributed to intensification of land cultivation; this causes severe soil fertility depletion. However, losses of nutrients in the soil have been replenished through the use of animal manures (Ramsasa 2010). Hence, the potential of animal manure use in soil replenishment has received the global attention. Currently, to many soil fertility difficulties, animal manures offer an affordable and readily available solution (Spore 2006) and are important source of Ca, Mg, S and micronutrients (Gomez-Brandon et al. 2013); they contain low and highly variable amounts of N, P and K. Aside from being source of plant nutrients, the physical properties of soil are also improved (Akanni and Ojeniyi 2008). However, the use of animal manures can pose a major problem of excess nitrate in soil (Azeez and Van Averbeke 2010b; Navarro Pedreno et al. 1996). The difficulty in the prediction of manure nutrients to plants is due to turnover processes and losses in manures (Azeez and Van Averbeke 2010a; Sorenson 2001). Therefore, incorporation of animal manures to supply nitrogen in proper dose at the proper time without posing any injury to soil and plants is very important in soil replenishment to support crop productivity. Hence, in order to predict the net mineralization of nitrogen in animal manures, transformation of nitrogen from organic into inorganic form and immobilization processes need to be understood. These processes are mediated through microbial activities and naturally, they are biochemical (Vel and Swamam 2013; Bartholomow 1965). Large variability in manure qualities

(Chadwick et al. 2000) or types had posed difficulties for researchers in concluding on a specific rate of manure to support development of different crops. The consequence of these led to differences in the recommendations of 60 t ha^{-1} of cattle kraal manure for *Corchorus olitorius* L. (Masarirambi et al. 2012b), goat manure rate of 20 t ha^{-1} for sweet maize (Uwah and Eyo 2014); and poultry manure at the levels of 60 t ha^{-1} (Masarirambi et al. 2012a) and 30 t ha^{-1} (Enujeke et al. 2013) for lettuce and maize yield, respectively.

Animal manures at proper application time, adequate rates and proper nutrient contents when expanding the uses of organic matter in place of inorganic fertilizers for nitrogen mineralization to meet crop N requirement without adverse problem on soil, crop and the environment need to be investigated. This necessitated the study to (1) examine the release patterns of NH_4^+–N and NO_3^-–N from different rates of three animal manures and recommended inorganic fertilizer rate to soil, (2) evaluate the potential rates of mineralization of the three animal manures on soil and (3) determine the influence of high application rates of the three animal manures on soil.

Materials and methods

Location of the experiment

This research was conducted at the Teaching and Research Farm of the Federal University of Agriculture, Abeokuta, Ogun State, Nigeria, located between latitude 7°12′ N and longitude 3°20′ E (FUNAAB 2013). Irregularities of rainfall occurred during the period of the study (Table 1).

Experimental design

The experiment was a randomized complete block design (RCBD) with three replicates. Cattle, goat and poultry manures were applied at the rates of 5, 10, 20, 40, 60, 80, 120 and 150 t ha^{-1} separately; these were used to test the pattern of manure mineralization irrespective of the rates, to test the manure-induced soil factors that affect plant growth and also evaluate the residual effect of manure rates on soil properties. NPK 15–15–15 fertilizer was used as check for each manure treatment and applied at the rate of 400 kg ha^{-1} (Schippers 2000). Starting from 2 weeks after incorporation of manures but a day after application of NPK fertilizer, soil samples were collected at every 2-week intervals from the net portion of each bed at a depth of 0–15 cm with the exception of 6 and 16 WAI due to rainfall failure. The samples were air dried, prepared for laboratory analysis and analyzed for NH_4^+–N and NO_3^-–N.

Soil and manures analyses

Initial soil sample and those collected bi-weekly were analyzed using the following methods. Soil EC and pH were extracted with 1:2 soil:water ratio and measured potentiometrically using HANNA 215 electrical conductivity meter (Richards 1954) and glass pH meter electrodes (HANNA pHep), respectively, (Van Reeuwijk 1993). The NH_4^+–N and NO_3^-–N were extracted in 2 Molar K_2SO_4 and determined colorimetrically using the method of Okalebo et al. (1993). Available P was extracted by Bray P-1 extraction (Bray and Kurtz 1965), obtained colorimetrically (Murphy and Riley 1962). Exchangeable bases were extracted with 1 N Ammonium-acetate solution in 1:10 soil solution ratio, K^+ and Na^+ were analyzed with flame photometer, while Ca^{2+} and Mg^{2+} with Atomic Absorption Spectrophotometer (AAS) (Anderson and Ingram 1993). Organic carbon (OC) content was determined by Walkley–Black method (Nelson and Sommers 1990) and particle size by hydrometer method (Bouyoucos 1965).

The cattle, goat and poultry manures used were digested with nitric per chloric acid (2:1) (Watanabe et al. 2013; Silva and QueirÓz 2002). The digests were analyzed for

Table 1 Agrometeorological data for the period of the experiment

	2 WAI	4 WAI	6 WAI	8 WAI	10 WAI	12 WAI	14 WAI	16 WAI	18 WAI	20 WAI	22 WAI
Agrometeorological data for the period of the experiment											
Rainfall (mm)	119.4	33.3	8.8	15.1	62.5	147.8	3.5	NA	87.3	35.6	43.1
Mean Temp./day (°C)	27.9	28.1	26.9	27.5	47.3	54.6	39.1	24.9	26.1	26	26.9
Soil Temp./day (°C) (20 cm)	28.7	28.9	28.1	29.9	27.3	26.8	36.3	27.4	26.2	27.9	27.4
Average RH/day (%)	79.8	77.6	73.6	73.1	81.1	89.1	80.6	80.6	80.9	84.1	77.5
Mean sunshine (h/day)	5.5	6.7	5.2	5.4	3.7	2.7	2.9	3.4	3.1	4.1	5.2
Evaporation/day (mm)	3.1	4	3	2.9	1.5	NA	0.8	3.2	2.5	2.8	3.4

Source: Federal University of Agriculture, Abeokuta AGROMET STATION, May–October, 2013

WAI weeks after incorporation of manures, Soil temperature at the depth of 20 cm, *NA* not available, *Temp.* temperature, *mm* millimeter, *RH* relative humidity macronutrients and micronutrients using standard procedures (Kaira and Maynard 1991; Cater 1993). Distilled water was used for extraction to determine the electrical conductivity and pH of the manures.

Statistical analysis

Data collected were subjected to Analysis of Variance (ANOVA) using SAS (1999). Duncan Multiple Range Test at 5% level of probability was used to determine differences in the treatment rates means and also to show the significance effects on parameters measured in relation to the control.

Results

Agrometeorological data for the experimental period

Agrometeorological average data for rainfall, relative humidity, mean temperature, soil temperature, mean sunshine hour and evaporation at 2-week intervals during the experiment were presented in Table 1.

Soil characteristics

The result revealed that the soil used for the research had EC value of 0.69 dS m^{-1}, pH of 7.6 and total N value of 0.8 g kg^{-1}. The NH_4^+–N, NO_3^-–N, OC and available P values were 0.013, 0.014, 14.2 g kg^{-1} and 7.5 mg kg^{-1}, respectively. Exchangeable bases; K^+, Na^+, Ca^{2+} and Mg^{2+} had the values of (0.42, 0.82, 10.96 and 1.34) $cmol^+$ kg^{-1}, respectively. The carbon to nitrogen ratio (C: N) was 11.83. The micro-nutrients; Mn^{2+}, Fe^{2+}, Cu^{2+} and Zn^{2+} had the values of (55.1, 9.85, 1.2 and 5.1) mg kg^{-1}, respectively. The percentages of sand, silt and clay were 770, 68 and 162 g kg^{-1}, respectively.

Manures characteristics

The EC values of the cattle, goat and poultry manures were 12.4, 8.8 and 16.4 dS m^{-1}, respectively, while the pH values of the cattle, goat and poultry manures were 7.9, 7.7 and 7.8, respectively. The equivalent amount of total N, NH_4^+–N, NO_3^-–N, P, K, Na, Ca, Mg and OC in cattle manure added to the soil between the rates of 5–150 t ha^{-1} were 225.5–6765, 2.1–63, 2.2–66, 16–480, 37–1110, 46.5–1395, 148–4440, 47.5–1425 and 894–26,820 kg, respectively. However, addition of goat manure between the rates of 5 and 150 t ha^{-1} is equivalent to adding total N, NH_4^+–N, NO_3^-–N, P, K, Na, Ca, Mg and OC of 241–7230, 1.6–48, 1.8–54, 3.5–105, 18.5–555, 23–690, 90.5–2715, 50–1500 and 1115–33,450 kg, respectively, to the soil. Hence, the equivalent amount of total N, NH_4^+–N, NO_3^-–N, P, K, Na, Ca, Mg and OC in the poultry manure applied to the soil at the rates ranged from 5 to 150 t ha^{-1} estimated to the amount of 139–4170, 7.4–220.5, 7.5–223.5, 81–2430, 19.5–585, 24.5–735, 88.5–2655, 26.5–795 and 296.5–8895.0 kg, respectively (Tables 2, 3, 4).

Effect of cattle, goat and poultry manures on soil NH_4^+–N and NO_3^-–N (mg kg^{-1})

Effect of cattle manure on soil NH_4^+–N (mg kg^{-1})

Table 5 reveals that at 2 WAI, soil applied with cattle manure rate of 60 t ha^{-1} gave highest value of 62.06 mg kg^{-1} in NH_4^+–N and lowest of 27.69 mg kg^{-1} at 80 t ha^{-1} by 171 and 21%, respectively, above the control, whereas percentage decreases of 13, 9 and 0.04% were recorded in soil with cattle manure rates of 120, 150 t ha^{-1} and NPK, respectively. But there were no significant differences in control and the amendment rates at 4 WAI. While at 8 WAI, soil with cattle manure rate of 120 t ha^{-1} had NH_4^+–N of 51.66 mg kg^{-1} with the highest increase of 98%, soil applied with 20 t ha^{-1} of cattle manure had 36.50 mg kg^{-1} of NH_4^+–N with the lowest increase of 40% in relation to control. The soil with cattle manure rate of 150 t ha^{-1}gave highest NH_4^+–N value of 830.4 mg kg^{-1} by 636% but 5 t ha^{-1} resulted in the least value of 221.4 mg kg^{-1} by 96% relative to control at 10 WAI. However, the amendment rates were not significantly different from control soil at 12 WAI. At 14 WAI, highest increase of 61% and lowest increase of 14% above the control were obtained from soil applied with cattle manure rates of 150 and 40 t ha^{-1} with NH_4^+–N values of 29.32 and 20.88 mg kg^{-1}, respectively. While 5, 10, 20 t ha^{-1} and NPK decreased by 38, 29, 31 and 72%, respectively relative to control. At 18 WAI, soil with cattle manure rate of 150 t ha^{-1} with NH_4^+–N value of 86.55 mg kg^{-1} recorded the highest increase of 290%, but 10 t ha^{-1} had 23.33 mg kg^{-1} and recorded the least increase of 5% over the control, while 5 t ha^{-1} with 21.11 mg kg^{-1} of NH_4^+–N decreased by 5%. At 20 WAI, soil with cattle manure rate of 150 t ha^{-1} gave 49.53 mg kg^{-1} of NH_4^+–N and had the highest increase of 111% while the rate of 5 t ha^{-1} had 25.03 mg kg^{-1} of NH_4^+–N and recorded the lowest increase of 7% relative to control. But the rate of 60 t ha^{-1} with NH_4^+–N value of 5.26 mg kg^{-1} recorded the highest increase of 149%, while 20 t ha^{-1} and 120 t ha^{-1} with NH_4^+–N value of 2.99 mg kg^{-1} each had the least increase of 42% compared to control at 22 WAI. However, the level of soil NH_4^+–N decreased in relation to the corresponding rates at 20 WAI.

Table 2 Equivalent amount of nutrients present in cattle manure rates added to the soil

Rates (t ha^{-1})	Total N	NH_4^+–N	NO_3^-–N	P	K	Na	Ca	Mg	OC
Cattle manure nutrients (kg)									
5	225.5	2.1	2.2	16.0	37.0	46.5	148.0	47.5	894.0
10	451.0	4.2	4.4	32.0	74.0	93.0	296.0	95.0	1788.0
20	902.0	8.4	8.8	64.0	148.0	186.0	592.0	190.0	3576.0
40	1804.0	16.8	17.6	128.0	296.0	372.0	1184.0	380.0	7152.0
60	2706.0	25.2	26.4	192.0	444.0	558.0	1776.0	570.0	10,728.0
80	3608.0	33.6	35.2	256.0	592.0	744.0	2368.0	760.0	14,304.0
120	5412.0	50.4	52.8	384.0	888.0	1116.0	3552.0	1140.0	21,456.0
150	6765.0	63.0	66.0	480.0	1110.0	1395.0	4440.0	1425.0	26,820.0

Total N total nitrogen, *K* total potassium, NH_4^+–N ammonium–nitrogen, *Na* total sodium, NO_3^-N nitrate–nitrogen, *Ca* total calcium, *P* total phosphorus, *Mg* total magnesium, *OC* organic carbon

Table 3 Equivalent amount of nutrients present in goat manure rates added to the soil

Rates (t ha^{-1})	Total N	NH_4^+–N	NO_3^-–N	P	K	Na	Ca	Mg	OC
Goat manure nutrients (kg)									
5	241.0	1.6	1.8	3.5	18.5	23.0	90.5	50.0	1115.0
10	482.0	3.2	3.6	7.0	37.0	46.0	181.0	100.0	2230.0
20	964.0	6.4	7.2	14.0	74.0	92.0	362.0	200.0	4460.0
40	1928.0	12.6	14.4	28.0	148.0	184.0	724.0	400.0	8920.0
60	2892.0	19.2	21.6	42.0	222.0	276.0	1086.0	600.0	13,380.0
80	3856.0	25.1	28.8	56.0	296.0	368.0	1448.0	800.0	17,840.0
120	5784.0	38.4	43.2	84.0	444.0	552.0	2172.0	1200.0	26,760.0
150	7230.0	48.0	54.0	105.0	555.0	690.0	2715.0	1500.0	33,450.0

Total N total nitrogen, *K* total potassium, NH_4^+–N ammonium–nitrogen, *Na* total sodium, NO_3^-–*N* nitrate–nitrogen, *Ca* total calcium, *P* total phosphorus, *Mg* total magnesium, *OC* organic carbon

Table 4 Equivalent amount of nutrients present in poultry manure rates added to the soil

Rates (t ha^{-1})	Total N	NH_4^+–N	NO_3^-–N	P	K	Na	Ca	Mg	OC
Poultry manure nutrients (kg)									
5	139.0	7.4	7.5	81.0	19.5	24.5	88.5	26.5	296.5
10	278.0	14.7	14.9	162.0	39.0	49.0	177.0	53.0	593.0
20	556.0	29.4	29.8	324.0	78.0	98.0	354.0	106.0	1186.0
40	1112.0	58.9	59.6	648.0	156.0	196.0	708.0	212.0	2372.0
60	1668.0	88.2	89.4	972.0	234.0	294.0	1062.0	318.0	3558.0
80	2224.0	117.6	119.2	1296.0	312.0	392.0	1416.0	424.0	4744.0
120	3336.0	176.4	178.8	1944.0	468.0	588.0	2124.0	636.0	7116.0
150	4170.0	220.5	223.5	2430.0	585.0	735.0	2655.0	795.0	8895.0

Total N total nitrogen, *K* total potassium, NH_4^+–N ammonium–nitrogen, *Na* total sodium, NO_3^-–*N* nitrate–nitrogen, *Ca* total calcium, *P* total phosphorus, *Mg* total magnesium, *OC* organic carbon

Effect of goat manure on soil NH_4^+–N

Table 6 showed that aside the soil applied with goat manure rate of 150 t ha^{-1} that was significantly higher by 114% with NH_4^+–N of 48.27 mg kg^{-1} at 2 WAI, soil with other amendment rates were not statistically different from control. The soil with goat amendment rates followed similar trend as cattle treatment at 4 WAI. The soil NH_4^+–N increased in comparison with the corresponding rates at 2 WAI. At 8 WAI, soil with goat manure rate of 150 t ha^{-1} gave NH_4^+–N value of 44.17 mg kg^{-1} and recorded the highest increase of 53% while 20 t ha^{-1} had NH_4^+–N value of 34.60 mg kg^{-1} and with the least increase of 20% above the control. But reductions of 18 and 28% were obtained from 5 to 10 t ha^{-1}, respectively. The level of NH_4^+–N increased relative to corresponding rates at 4

Table 5 Influence of cattle manure rates and NPK fertilizer (t ha^{-1}) on mean soil NH_4^+–N (mg kg^{-1}) at 2 weeks intervals

Rates (t ha^{-1})	Means of bi-weekly analyses of soil NH_4^+–N (mg kg^{-1}) with applied cattle manure								
	2 WAI	4 WAI	8 WAI	10 WAI	12 WAI	14 WAI	18 WAI	20 WAI	22 WAI
0	22.88^{b}	68.11^{a}	26.10^{c}	112.90^{f}	79.90^{a}	18.245^{bc}	22.19^{g}	23.50^{f}	2.11^{e}
5	21.60^{b}	69.31^{a}	38.30^{b}	221.40^{d}	93.50^{a}	11.315^{de}	21.11^{g}	25.03^{e}	3.81^{c}
10	31.96^{b}	74.26^{a}	47.40^{bc}	270.30^{c}	118.80^{a}	14.43^{cd}	23.33^{f}	26.69^{d}	4.81^{b}
20	29.61^{b}	78.80^{a}	36.50^{bc}	246.80^{d}	133.60^{a}	12.63^{cde}	24.44^{f}	25.06^{d}	2.99^{d}
40	37.45^{ab}	75.52^{a}	38.29^{bc}	246.90^{d}	121.00^{a}	20.88^{b}	26.61^{f}	38.14^{c}	3.53^{c}
60	62.06^{a}	71.32^{a}	49.10^{bc}	219.20^{e}	107.90^{a}	23.94^{b}	36.66^{e}	39.53^{c}	5.26^{a}
80	27.69^{b}	73.13^{a}	49.10^{bc}	202.30^{d}	122.60^{a}	22.67^{b}	42.16^{d}	40.00^{c}	3.49^{c}
120	20.03^{b}	88.39^{a}	51.66^{a}	408.60^{b}	142.40^{a}	25.20^{ab}	66.58^{b}	40.59^{c}	2.99^{d}
150	20.79^{b}	75.35^{a}	40.50^{bc}	830.40^{a}	149.90^{a}	29.32^{a}	86.55^{a}	49.53^{a}	3.39^{c}
0.4	22.87^{b}	73.25^{a}	37.90^{bc}	270.40^{c}	149.40^{a}	5.2^{e}	50.30^{c}	46.52^{b}	3.59^{c}

Means followed by the same letter(s) within the same columns do not differ significantly at $P < 0.05$ according to Duncan multiple range test
WAI weeks after incorporation of manures

Table 6 Influence of goat manure rates and NPK fertilizer (t ha^{-1}) on mean soil NH_4^+–N (mg kg^{-1}) at 2 weeks intervals

Rates (t ha^{-1})	Means of bi-weekly analyses of soil NH_4^+–N (mg kg^{-1}) with applied goat manure								
	2 WAI	4 WAI	8 WAI	10 WAI	12 WAI	14 WAI	18 WAI	20 WAI	22 WAI
0	22.54^{b}	66.92^{a}	28.93^{d}	100.10^{i}	76.70^{a}	18.52^{b}	22.19^{j}	23.35^{f}	2.07^{e}
5	20.89^{b}	67.43^{a}	23.85^{d}	232.10^{g}	85.00^{a}	13.09^{cd}	26.63^{i}	24.45^{e}	3.04^{c}
10	21.50^{b}	75.99^{a}	20.75^{e}	225.70^{h}	93.50^{a}	16.12^{bc}	37.73^{h}	24.45^{e}	3.29^{c}
20	25.77^{b}	77.00^{a}	34.60^{c}	234.00^{f}	95.00^{a}	19.57^{ab}	47.71^{g}	29.79^{d}	3.95^{c}
40	27.57^{ab}	70.55^{a}	39.46^{b}	308.70^{c}	111.10^{a}	19.51^{ab}	49.93^{f}	30.89^{d}	3.28^{c}
60	27.75^{b}	77.68^{a}	37.87^{b}	400.30^{a}	112.20^{a}	18.67^{b}	68.80^{d}	30.89^{d}	4.12^{b}
80	27.98^{b}	80.27^{a}	37.58^{b}	298.00^{d}	134.60^{a}	19.42^{ab}	72.12^{c}	40.00^{c}	4.28^{b}
120	30.74^{b}	74.52^{a}	34.80^{c}	374.60^{b}	123.50^{a}	18.94^{b}	79.62^{b}	47.76^{a}	4.99^{a}
150	48.27^{a}	92.08^{a}	44.17^{a}	308.70^{c}	141.30^{a}	22.88^{a}	88.77^{a}	43.35^{c}	3.69^{c}
0.4	23.98^{b}	72.70^{a}	37.14^{b}	241.32^{e}	145.50^{a}	5.49^{e}	54.00^{e}	46.52^{b}	3.78^{c}

Means followed by the same letter(s) within the same columns do not differ significantly at $P < 0.05$ according to Duncan multiple range test
WAI weeks after incorporation of manures

WAI. The soil applied with goat manure rate of 60 t ha^{-1} resulted in NH_4^+–N value of 400.3 mg kg^{-1} with highest increase of 300% while 10 t ha^{-1} had 225 mg kg^{-1} and recorded the least increase of 126% relative to control at 10 WAI. But 12 WAI followed the same trend as soil applied with cattle amendment. Furthermore, soil NH_4^+–N decreased in comparison with the corresponding rates at 10 WAI. Highest increase of 24% was recorded from soil with goat manure rate of 150 t ha^{-1} which had NH_4^+–N value of 22.88 mg kg^{-1} at 14 WAI. The soil with rate of 60 t ha^{-1} resulted in 18.67 mg kg^{-1} and gave the lowest increase of 1% more than the control. However, 5, 10 t ha^{-1} and NPK reduced by 29, 13 and 70%, respectively. The level of soil NH_4^+–N decreased in relation to the corresponding rates at 12 WAI. However, soil applied with goat manure rate of 150 t ha^{-1} had highest NH_4^+–N value of 88.77 mg kg^{-1} by 300% while 5 t ha^{-1} resulted in 26.63 mg kg^{-1} and recorded the least increase of 20% at 18 WAI, compared with control. At 20 WAI, soil with goat manure rate of 120 t ha^{-1} recorded NH_4^+–N value of 47.76 mg kg^{-1} and had the highest increase of 105%. But the soil with goat manure rate of 5 and 10 t ha^{-1} with the NH_4^+–N value of 24.45 mg kg^{-1} each recorded the lowest increase of 5%. However, there was a decrease in soil NH_4^+–N level in relation to the corresponding rates at 18 WAI. Consequently, the soil with goat manure rate of 120 t ha^{-1} which gave NH_4^+–N value of 4.99 mg kg^{-1} recorded the highest increase of 141% at 22 WAI while 5 t ha^{-1} resulted in 3.04 mg kg^{-1} with the least increase of 47% relative to control. There were decreases in soil NH_4^+–N level compared with the corresponding rates at 20 WAI.

Effect of poultry manure on soil NH_4^+–N

At 2 WAI, soil applied with poultry manure rate of 60 t ha^{-1} gave NH_4^+–N value of 50.86 mg kg^{-1} and recorded the highest increase of 153%. The rate of 10 t ha^{-1} had soil NH_4^+–N value of 20.18 mg kg^{-1} and had the least increment of 0.5% more than the control. But the values at 4 WAI followed similar trend with soil on which cattle and goat manures were incorporated. The soil NH_4^+–N increased compared to the corresponding rates at 2 WAI. Meanwhile, soil applied with poultry manure rate of 150 t ha^{-1} recorded highest soil NH_4^+–N value of 47.85 mg kg^{-1} by 90% at 8 WAI. However, soil on which NPK was added with NH_4^+–N value of 37.21 mg kg^{-1} recorded the least increase of 47% over the control. The level of soil NH_4^+–N decreased in relation to the corresponding rates at 4 WAI. More so, soil with poultry manure rate of 60 t ha^{-1} resulted in 253.4 mg kg^{-1} of NH_4^+–N and had the highest increase of 138%, while the rate of 5 t ha^{-1} with soil NH_4^+–N value of 110.7 mg kg^{-1} recorded the least increase of 4% relative to control at 10 WAI. The soil applied with poultry amendment rates were not differerent significantly from control at 12 WAI but soil NH_4^+–N level decreased relative to corresponding rates at 10 WAI. At 14 WAI, soil with poultry manure rate of 150 t ha^{-1} had NH_4^+–N value of 28.06 mg kg^{-1} and recorded the highest increase of 49% but 40 t ha^{-1} resulted in 19.86 mg kg^{-1} of NH_4^+–N with lowest increase of 6% more than the control. However, at 14 WAI the level of soil NH_4^+–N decreased in relation to the corresponding rates at 12 WAI. At 18 WAI, soil with poultry manure rate of 150 t ha^{-1} gave highest NH_4^+–N value of 73.23 mg kg^{-1} by 230% increase while 5 t ha^{-1} had least NH_4^+–N value of 23.27 mg kg^{-1} by 5% increase compared with control, whereas, 80 t ha^{-1} resulted in soil NH_4^+–N value of 52.44 mg kg^{-1} and had the highest increase of 127% more than the control at 20 WAI while 20 t ha^{-1} with soil NH_4^+–N value of 27.41 mg kg^{-1} recorded the least increase of 19%. At 22 WAI, soil with poultry manure at the rate of 150 t ha^{-1} with NH_4^+–N value of 4.45 mg kg^{-1} recorded the highest increase of 119% while 5 t ha^{-1} gave soil NH_4^+–N value of 2.08 mg kg^{-1} and recorded the least increase of 3% above the control. In addition, soil NH_4^+–N values decreased in comparison with the corresponding rates at 20 WAI, Table 7.

Effect of cattle manure on soil NO_3^-–N

The soil applied with cattle manure rate of 60 t ha^{-1} resulted in NO_3^-–N value of 54.09 mg kg^{-1} at 2 WAI by 119% while the soil with NPK had the NO_3^-–N value of 25.98 mg kg^{-1} and recorded the least increase of 5% more than the control. The 4 WAI was in line with soil NH_4^+–N but the soil NO_3^-–N values increased compared with the corresponding rates at 2 WAI. However, at 8 WAI, soil with cattle manure at the rate of 120 t ha^{-1} gave NO_3^-–N value of 50.42 mg kg^{-1} with highest increase of 117% but 10 t ha^{-1} with soil NO_3^-–N value of 25.56 mg kg^{-1} recorded the least increase of 10% in relation to control. However, the soil NO_3^-–N values reduced in comparison with the corresponding rates at 4 WAI, Table 8. Moreover, soil with cattle manure rate of 150 t ha^{-1} with soil NO_3^-–N of 316.68 mg kg^{-1} resulted in highest increase of 201% at 10 WAI while 5 t ha^{-1} with 133.78 mg kg^{-1} had the least increase of 27% in relation to control. At 12 WAI, 80 t ha^{-1} with soil NO_3^-–N value of 139.60 mg kg^{-1}

Table 7 Influence of poultry manure rates and NPK fertilizer (t ha^{-1}) on mean soil NH_4^+–N (mg kg^{-1}) at 2 weeks intervals

Rates (t ha^{-1})	Means of bi-weekly analyses of soil NH_4^+–N (mg kg^{-1}) with applied poultry manure								
	2 WAI	4 WAI	8 WAI	10 WAI	12 WAI	14 WAI	18 WAI	20 WAI	22 WAI
0	20.08^{b}	67.22^{a}	25.25^{d}	106.50^{i}	71.50^{b}	18.79^{b}	22.19^{j}	23.13^{g}	2.03^{e}
5	19.94^{b}	79.79^{a}	23.50^{d}	110.72^{h}	99.17^{ab}	12.89^{cd}	23.27^{i}	28.60^{e}	2.08^{e}
10	20.18^{b}	77.08^{a}	24.72^{d}	244.86^{b}	120.20^{ab}	15.29^{bcd}	28.82^{h}	28.60^{e}	2.64^{d}
20	29.40^{ab}	76.61^{a}	23.57^{d}	219.16^{d}	127.43^{ab}	16.54^{bc}	32.18^{g}	27.41^{f}	2.89^{cd}
40	34.26^{ab}	74.74^{a}	35.25^{c}	217.03^{e}	121.00^{ab}	19.86^{b}	35.51^{f}	27.44^{f}	3.01^{c}
60	50.86^{a}	71.19^{a}	47.35^{a}	253.37^{a}	144.01^{a}	20.45^{b}	45.59^{d}	33.23^{d}	2.23^{de}
80	41.82^{ab}	80.70^{a}	43.37^{b}	227.75^{c}	156.33^{a}	21.95^{ab}	38.84^{e}	52.44^{a}	2.44^{d}
120	46.06^{a}	87.39^{a}	46.43^{a}	212.78^{f}	140.80^{a}	26.54^{a}	65.55^{b}	48.97^{b}	3.28^{b}
150	43.58^{ab}	75.44^{a}	47.85^{a}	200.07^{g}	153.40^{a}	28.11^{a}	73.23^{a}	47.44^{c}	4.45^{a}
0.4	21.75^{b}	72.98^{a}	37.21^{c}	212.24^{f}	144.50^{a}	5.88^{e}	57.70^{c}	46.52^{c}	3.02^{c}

Means followed by the same letter(s) within the same columns do not differ significantly at $P < 0.05$ according to Duncan multiple range test

WAI weeks after incorporation of manures

Table 8 Influence of cattle manure rates and NPK fertilizer (t ha^{-1}) on mean soil NO_3^-–N (mg kg^{-1}) at 2 weeks intervals

Rates (t ha^{-1})	Means of bi-weekly analyses of soil NO_3^-–N (mg kg^{-1}) with applied cattle manure								
	2 WAI	4 WAI	8 WAI	10 WAI	12 WAI	14 WAI	18 WAI	20 WAI	22 WAI
0	24.75^{b}	68.70^{a}	23.28^{d}	105.07^{f}	91.70^{b}	20.09^{bc}	28.25^{g}	20.00^{j}	3.18^{d}
5	21.81^{b}	75.61^{a}	29.17^{b}	133.78^{ef}	97.70^{b}	12.68^{d}	31.40^{fg}	29.24^{h}	4.50^{c}
10	35.95^{ab}	75.63^{a}	25.56^{cd}	179.56^{d}	99.28^{b}	13.89^{d}	38.51^{f}	28.74^{i}	6.58^{a}
20	31.29^{b}	71.55^{a}	26.00^{cd}	151.61^{e}	133.95^{a}	14.76^{d}	42.41^{e}	30.00^{g}	3.99^{c}
40	40.31^{ab}	73.86^{a}	27.24^{cd}	140.80^{e}	126.00^{a}	18.76^{cd}	45.31^{d}	37.66^{e}	4.64^{c}
60	54.09^{a}	69.75^{a}	48.69^{ab}	148.78^{e}	108.30^{ab}	24.78^{c}	48.39^{b}	38.95^{d}	6.58^{a}
80	30.96^{b}	76.65^{a}	47.09^{ab}	206.51^{c}	139.60^{a}	24.28^{c}	58.26^{a}	40.00^{b}	3.96^{c}
120	22.78^{b}	83.65^{a}	50.42^{a}	208.28^{c}	135.60^{a}	26.42^{a}	58.28^{a}	37.31^{f}	3.31^{d}
150	24.70^{b}	81.01^{a}	45.74^{ab}	316.68^{a}	123.40^{a}	25.74^{b}	54.19^{ab}	39.27^{c}	5.20^{b}
0.4	25.98^{b}	82.56^{a}	26.91^{cd}	267.77^{b}	145.50^{a}	10.76^{e}	46.82^{fg}	43.28^{a}	4.38^{c}

Means followed by the same letter(s) within the same columns do not differ significantly at $P < 0.05$ according to Duncan multiple range test
WAI weeks after incorporation of manures

gave highest increase of 52% but 5 t ha^{-1} had soil NO_3^-–N value of 97.70 mg kg^{-1} and had lowest increase of 7% relative to control. However, soil NO_3^-–N decreased relative to the corresponding rates at 10 WAI. The soil with cattle manure rate of 120 t ha^{-1} gave highest NO_3^-–N value of 26.42 mg kg^{-1} by 32% at 14 WAI while 80 t ha^{-1} with the value of 24.28 mg kg^{-1} resulted in least increase of 21% relative to control. But in comparison with the corresponding rates at 10 WAI, soil NO_3^-–N decreased, Table 8. At 18 WAI, soil applied with cattle manure at the rate of 120 t ha^{-1} gave NO_3^-–N value of 58.28 mg kg^{-1} by 106%, but 5 t ha^{-1} with NO_3^-–N value of 31.4 mg kg^{-1} recorded the least increase of 11%, relative to control. Consequently, at 20 WAI, soil applied with NPK had NO_3^-–N value of 43.28 mg kg^{-1} and recorded the greatest increment in soil NO_3^-–N by 116% followed by 80 t ha^{-1} with NO_3^-–N value of 40.00 mg kg^{-1} that had 100% increases, while 10 t ha^{-1} with soil NO_3^-–N value of 29.24 mg kg^{-1} had the lowest increase of 44% more than control. There was a decrease in soil NO_3^-–N for cattle treated soil at 20 WAI compared to the corresponding rates at 18 WAI. However, at 22 WAI, soil appiled with cattle manure rates of 10 and 60 t ha^{-1} with the same NO_3^-–N value of 6.58 mg kg^{-1} gave the highest increase of 107% while 150 t ha^{-1} with NO_3^-–N value of 3.31 mg kg^{-1} gave the least increase of 4% in comparison with control. The soil NO_3^-–N decreased in relation to the corresponding rates at 20 WAI.

Effect of goat manure on soil NO_3^-–N

At 2 WAI, soil applied with goat manure rate of 150 t ha^{-1} had NO_3^-–N value of 50.29 mg kg^{-1} which recorded the highest increase of 133% while 5 t ha^{-1} with soil NO_3^-–N value of 23.13 mg kg^{-1} recorded the least increase of 7% relative to control. At 4 WAI, the trend was in line with cattle manure treated soil as NH_4^+–N. However, there were increases of soil NO_3^-–N compared with the corresponding rates at 2 WAI. At 8 WAI, goat manure rate of 40 t ha^{-1} with NO_3^-–N value of 45.22 mg kg^{-1} recorded the highest increase of 102% but 10 t ha^{-1} with NO_3^-–N value of 24.95 mg kg^{-1} recorded the least increase of 11% above the control. In comparison of 8 WAI values with the corresponding rates at 4 WAI, soil NO_3^-–N decreased. The soil with goat manure rate of 60 t ha^{-1} which had NO_3^-–N value of 398.54 mg kg^{-1} recorded highest increase of 293%, 5 t ha^{-1} gave NO_3^-–N value of 227.02 mg kg^{-1} with the least increase of 124% in relation to control at 10 WAI. However, at 12 WAI, 80 t ha^{-1} resulted in soil NO_3^-–N value of 137.85 mg kg^{-1} and had highest increase of 51% but 5 t ha^{-1} with soil NO_3^-–N value of 100.70 mg kg^{-1} had lowest increase of 11% more than the control. The result revealed that soil NO_3^-–N decreased compared to the corresponding rates at 10 WAI. More so, at 14 WAI, 150 t ha^{-1} with soil NO_3^-–N value of 23.22 mg kg^{-1} was significantly higher while other goat manure treated soil rates decreased in soil NO_3^-–N compared with control. Consequently, the soil NO_3^-–N decreased relative to corresponding rates at 12 WAI. The soil applied with goat manure at the rate of 150 t ha^{-1} had NO_3^-–N value of 82 mg kg^{-1} and highest increase of 183%. The least increase of 4% was recorded from 5 t ha^{-1} with soil NO_3^-–N value of 30 mg kg^{-1} compared with control, at 18 WAI, while 120 t ha^{-1} with soil NO_3^-–N value of 44.69 mg kg^{-1} resulted in highest increment of 111% at 20 WAI. But 5 t ha^{-1} gave 28.75 mg kg^{-1} of NO_3^-–N and recorded lowest increase of 36% in relation to control. In comparison with the corresponding rates at 18 WAI, there were decreases in soil NO_3^-–N level. At 22 WAI, soil with goat manure rate of 120 t ha^{-1} had NO_3^-–

Table 9 Influence of goat manure rates and NPK fertilizer (t ha^{-1}) on mean soil NO_3^-–N (mg kg^{-1}) at 2 weeks intervals

Rates (t ha^{-1})	Means of bi-weekly analyses of soil NO_3^-–N (mg kg^{-1}) with applied goat manure								
	2 WAI	4 WAI	8 WAI	10 WAI	12 WAI	14 WAI	18 WAI	20 WAI	22 WAI
0	21.61^{b}	70.33^{a}	22.39^{d}	101.35^{f}	91.10^{d}	19.00^{ab}	29.00^{i}	21.22^{g}	3.26^{d}
5	23.13^{b}	78.52^{a}	25.49^{bcd}	227.02^{e}	100.70^{d}	14.68^{cde}	30.00^{h}	28.75^{f}	3.34^{d}
10	35.21^{ab}	70.40^{a}	24.95^{cd}	232.24^{e}	114.10^{b}	15.92^{cde}	31.00^{g}	30.00^{e}	4.27^{c}
20	30.42^{ab}	81.69^{a}	28.25^{c}	270.24^{d}	112.10^{bc}	18.14^{bcd}	49.00^{d}	30.08^{e}	5.34^{b}
40	27.64^{b}	74.28^{a}	45.22^{a}	309.51^{b}	113.50^{bc}	19.25^{ab}	36.00^{f}	33.55^{d}	5.55^{b}
60	27.11^{b}	83.51^{a}	33.84^{b}	398.54^{a}	115.80^{b}	18.95^{bc}	72.00^{b}	41.76^{c}	5.34^{b}
80	24.49^{b}	83.83^{a}	28.46^{c}	313.51^{c}	137.80^{a}	19.85^{ab}	70.00^{c}	40.00c	5.41^{b}
120	34.26^{ab}	80.55^{a}	31.81^{bc}	317.41^{c}	132.70^{b}	17.56^{bcd}	72.00^{b}	44.69^{a}	5.79^{a}
150	50.29^{a}	84.04^{a}	31.47^{bc}	374.18^{b}	117.80^{b}	23.22^{a}	82.00^{a}	40.77^{c}	5.51^{b}
0.4	26.32^{b}	78.15^{a}	27.59^{c}	261.51^{d}	141.05^{a}	10.82^{e}	46.68^{e}	42.22^{b}	4.53^{c}

Means followed by the same letter(s) within the same columns do not differ significantly at $P < 0.05$ according to Duncan multiple range test
WAI weeks after incorporation of manures

Table 10 Influence of poultry manure rates and NPK fertilizer (t ha^{-1}) on mean soil NO_3^-–N (mg kg^{-1}) at 2 weeks intervals

Rates (t ha^{-1})	Mean of bi-weekly analyses of soil NO_3^-–N (mg kg^{-1}) with applied poultry manure								
	2 WAI	4 WAI	8 WAI	10 WAI	12 WAI	14 WAI	18 WAI	20 WAI	22 WAI
0	22.59^{d}	68.51^{a}	22.46^{d}	104.19^{f}	91.40^{g}	19.11^{b}	28.18^{j}	20.28^{j}	3.19^{a}
5	22.38^{d}	72.95^{a}	24.23^{cd}	114.46^{f}	99.49^{f}	12.08^{e}	30.42^{i}	30.00^{i}	5.59^{a}
10	23.54^{d}	82.25^{a}	24.88^{cd}	239.48^{b}	145.95^{bc}	12.12^{d}	30.96^{h}	30.48^{h}	4.16^{a}
20	28.91^{bcd}	76.99^{a}	23.94^{c}	230.91^{b}	123.60^{de}	13.79^{d}	35.63^{g}	38.77^{f}	4.12^{a}
40	46.00^{abc}	68.69^{a}	25.55^{c}	240.08^{b}	131.15^{d}	17.87^{cd}	37.37^{f}	42.08^{e}	3.55^{a}
60	51.22^{a}	65.63^{a}	24.98^{cd}	250.14^{a}	144.35^{c}	16.50^{cd}	44.12^{d}	49.64^{c}	2.96^{a}
80	43.11^{abcd}	77.76^{a}	28.96^{b}	228.89^{c}	153.75^{a}	16.28^{cd}	40.22^{e}	56.38^{a}	3.18^{a}
120	47.37^{ab}	84.95^{a}	35.42^{a}	217.42^{d}	121.90^{e}	17.19^{cd}	46.55^{a}	54.19^{b}	4.73^{a}
150	37.65^{abcd}	75.35^{a}	28.05^{b}	211.43^{e}	145.95^{bc}	25.87^{a}	45.66^{b}	31.32^{g}	2.37^{a}
0.4	24.76^{cd}	84.32^{a}	28.96^{b}	261.76^{a}	145.40^{bc}	10.34^{f}	45.23^{c}	42.81^{d}	4.35^{a}

Means followed by the same letter(s) within the same columns do not differ significantly at $P < 0.05$ according to Duncan multiple range test
WAI weeks after incorporation of manures

N value of 5.79 mg kg^{-1} and gave highest increase of 78% but 5 t ha^{-1} with NO_3^-–N value of 3.34 mg kg^{-1} recorded least increase of 3% more than the control. The values also decreased compared to corresponding rates at 20 WAI, Table 9.

Effect of poultry manure on soil NO_3^-–N

Table 10 presented that at 2 WAI, soil applied with poultry manure rate of 60 t ha^{-1} had NO_3^-–N value of 51.22 mg kg^{-1} and recorded the highest increase of 127%. The rate of 10 t ha^{-1} gave NO_3^-–N value of 23.54 mg kg^{-1} and recorded the least increase of 7% more than the control. The 4 WAI was also similar to others above. The soil with poultry manure rate of 120 t ha^{-1} with the value of 35.42 mg kg^{-1} recorded the highest increase of 59% in soil NO_3^-–N while 20 t ha^{-1} with NO_3^-–N value of 23.74 mg kg^{-1} had the least increase of 7% relative to control at 8 WAI. Soil NO_3^-–N decreased in comparison with the corresponding rates at 4 WAI. Furthermore, soil applied with NPK gave highest NO_3^-–N value of 261.76 mg kg^{-1} with increment of 151% while 5 t ha^{-1} with NO_3^-–N value of 114.46 mg kg^{-1} presented least increase of 10% compared with control at 10 WAI. However, soil with poultry manure rate of 80 t ha^{-1} with 153.75 mg kg^{-1} of NO_3^-–N resulted to highest increment of 68% but 5 t ha^{-1} with NO_3^-–N value of 99.49 mg kg^{-1} had lowest increase of 9% relative to control at 12 WAI. There were decrements at 12 WAI in relation to the corresponding rates at 10 WAI. More so, soil with poultry manure rate of 150 t ha^{-1} recorded NO_3^-–N value of 25.87 mg kg^{-1} and gave the highest increase of 35% at 14

WAI. But soil NO_3^-–N decreased in relation to the corresponding rates at 12 WAI. The rate of 120 t ha^{-1} recorded highest NO_3^-–N value of 46.55 mg kg^{-1} by 62% and the least increase of 8% was recorded from 5 t ha^{-1} with NO_3^-–N value of 30.42 mg kg^{-1} relative to control, at 18 WAI. Consequently, at 20 WAI, 80 t ha^{-1} presented soil NO_3^-–N value of 56.38 mg kg^{-1} and this resulted in highest increment of 178% while 5 t ha^{-1} had NO_3^-–N value of 30 mg kg^{-1} by 88% over the control. Increments in relation to 18 WAI were erratic. At 22 WAI, the rates of soil amended with poultry manure were not significantly different from control but NO_3^-–N level reduced when compared to the corresponding rates at 20 WAI.

Discussion

Soil characteristics

The soil used for the research was sandy clay loam (USDA 2010), slightly saline (LAS 2014), and slightly alkaline (Pam and Brian 2007) and these could support plant performances. More so, the soil was deficient in total nitrogen (McBride 2015; USDA-SCS 1974) and had low available P content (ENDMEMO 2015; Mallarino 2000) which could retard crop development. It contained moderate OC (McBride 2015; USDA-SCS 1974), exchangeable K^+ and Mg^{2+} (Pam and Brian 2007); these were optimal for performance of the crop. The high exchangeable Na^+ and Ca^{2+} contents (Pam and Brian 2007) could be attributed to the decomposition of the organic matter content of the soil due to high temperature, evaporation rate, sunshine hours/day and low rainfall during the research. This confirmed the studies of some researchers (Davidson and Jannssens 2006; Friedlingstein et al. 2006). The NH_4^+–N and NO_3^-–N contents were low and C: N ratio was considered normal (Hill 2001) and could be as a result of low rainfall during this period. The copper, iron and zinc contents were very high (Enwezor et al. 1989).

Manure characteristics

The EC of the three manures applied were very strongly saline (LAS 2014). This could be as a result of high concentration of cationic salts in the manures. More so, the highest EC value of poultry manure compared with the cattle and goat manures support the findings of Azeez and Van Averbeke (2010a, b) and could be attributed to the highest NH_4^+–N, NO_3^-–N and total P contents in poultry manure compared to others. The pH of the pure goat and poultry manures was considered mildy alkaline, whereas that of cattle manure was moderately alkaline (Pam and Brian 2007). This would allow for nutrients availability in the soil after mineralization. It was observed from the result that cattle manure had the highest total K, Na and Ca over the goat and poultry manures. However, goat manure was higher in total N, Mg and OC in relative to cattle and poultry manures. Very low value of total P in goat manure could be a result of the nutrient concentrations of the feed given to the animals. This is because they were fed with elephant grasses and wheat offal under intensive system although the nutrient contents of the feed were not measured. Equivalent amount of total N contents present in goat manure applied was higher than cattle and poultry manures. High P content in poultry manure could also contribute to the yield quality of crops at lower rates than cattle and goat manures. Cattle manure recorded highest equivalent amount of pH, total K, Na and Ca than goat and poultry manures; this could be attributed to the higher EC value in cattle manure than goat manure and confirmed the report of Monica (2013). Highest contents of total N, Mg and OC in goat manure compared with cattle and poultry manures suggested the reason for lowest value of EC in goat manure applied during the studies. This is because Mg was the only salt forming cation that was highest in goat manure. Highest total N could be a result of N level in the feed given to the goat animals, although the nutrient contents of the feed were not determined.

Manure mineralization processes in the soil

Incorporation of cattle, goat and poultry manures increased the concentration of NH_4^+–N; this supports the findings of Sajal and Abul Kashem (2014) and NO_3^-–N depending on the rates (Eghball et al. 2002) compared with control. However, the low values of NH_4^+–N between 0 and 8 WAI of the manures could be attributed to conversion of NH_4^+–N to NO_3^-–N as supported in the study of (Hoskins 2015) that, between 2 and 4 weeks, NH_4^+–N converted relatively quickly to NO_3^-–N in a soil applied with broiler manure. However, mineralization of organic N into inorganic form of NH_4^+–N and NO_3^-–N occurred between 8 and 14 WAI of manures. Consequently, the low NH_4^+–N between 18 and 22 WAI could be attributed to ammonia volatilization (Meissinger and Jokela 2000) while low NO_3^-–N contents between 18 and 22 WAI could be attributed to leaching (Tom 2002; Marco et al. 2002) and plant root uptake. The low NO_3^-–N contents in soil applied with NPK fertilizer at 18 WAI compared with 8 and 14 WAI could be attributed to acidic properties of NPK fertilizer, since acidity reduces soil microbial activity and this would in turn reduce nutrient mineralization (Xu et al. 2002).

Mineralization of NH_4^+–N and NO_3^-–N contents occurred in the soil with the maximum values at 10 WAI for cattle, goat and poultry manures but was contrary to the results of Azeez and Van Averbeke (2010b) that sharp increase in N release was at 120 days; and (Ayeni 2011; Ayeni and Adeleye 2011), that release of poultry manure nutrients proportion is between 1 and 2 months of incubation. The variability of mineralization processes could be attributed to microbial activities as suggested by Vinten et al. (2002); and climatic variations. Hence, the turnover processes and losses have influence on availability of nutrients in manures; this is why the prediction of manure nutrients to plants is problematic (Azeez and Van Averbeke 2010b; Sorenson 2001).

Very low N mineralization at 14 WAI compared to 12 WAI could be attributed to very low rainfall during this period. This corroborated the report of Jonathan (2006) that in dry soils, N mineralization is low because soil micro-organism activity is limited by water availability. More so, low NH_4^+–N suggested to be as a result of volatilization of NH_3 as this is favored by warm temperature wet soils under drying conditions (Tom 2002; James 2001). While low NO_3^-–N could be as a result of leaching that might have occurred at 12 WAI due to very high rainfall at this period, this also corroborated the work of Tom (2002) that NO_3 in the soil can be lost through percolation of water below the active root zone.

Across weeks after incorporation of the three manures, application rates of 20 t ha^{-1} for cattle, 40 t ha^{-1} for goat and poultry manure at the rate of 10 t ha^{-1} were not widely significantly different when compared to higher rates. Therefore, addition of manures more than these rates to the soil could lead to wastage as these did not increase the amount of NH_4^+–N and NO_3^-–N in the soil sequentially according to this study. Although the high rates of cattle, goat and poultry manures were significantly higher relative to lower rates in this research, these suggested not to be encouraged as the increments were not regular across the weeks after manure application. The irregularity could be attributed to volatilization (Meissinger and Jokela 2000) and leaching (Tom 2002; Marco et al. 2002) of the NH_4^+–N and NO_3^-–N in the soil, respectively.

During this work, NPK fertilizer applied was able to supply equal amount of NH_4^+–N to the soil when compared with 10 t ha^{-1} of cattle manure treatment between 2 and 12 WAI. But goat manure treated soil at 60 t ha^{-1} supplied equal amount of NH_4^+–N compared to NPK fertilizer between 2 and 8 WAI, while 40 t ha^{-1} of poultry manure was able to supply equal amount of NH_4^+–N to the soil compared to NPK fertilizer at 0.4 t ha^{-1} between 2 and 8 WAI. This suggested that cattle, goat and poultry manures incorporated to the soil at 10, 60 and 40 t ha^{-1}, respectively, supplied equal amount of NH_4^+–N compared to NPK fertilizer at 0.4 t ha^{-1} within 6 weeks. The low NH_4^+–N and NO_3^-–N contents of the soils applied with NPK fertilizer at 14 WAI compared with cattle, goat and manure treated soils rates could be attributed to acidic properties of NPK fertilizer. This corroborated the study results of Xu et al. (2002) that acidity reduces soil microbial activity resulting in reduced nutrient mineralization. This could also be attributed to depletion of soil organic matter as a result of synthetic fertilization (Jonathan 2006).

Conclusion and recommendation

Generally, addition of NPK 15–15–15 fertilizer, cattle, goat and poultry manures improved the soil NH_4^+–N and NO_3^-–N compared with control. The NH_4^+–N and NO_3^-–N contents of the soil between 2 and 8 WAI increased with the application of cattle, goat and poultry manures. However, between 8 and 14 WAI mineralization of NH_4^+–N and NO_3^-–N occurred but peak values were obtained at 10 WAI irrespective of the manures types and rates. Finally, the effect of cattle, goat and poultry manures on NH_4^+–N and NO_3^-–N decreased between 18 and 22 WAI.

To increase soil NH_4^+–N and NO_3^-–N for optimum production of crops yield, application of cattle, goat and poultry manures is recommended. It is recommended that application rate of 20 t ha^{-1} for cattle, 40 t ha^{-1} for goat and poultry manure at the rate of 10 t ha^{-1} would provide optimum and improve the soil nutrients level which will be retained in the soil for longer period. To improve soil with low N contents, poultry manure is recommended. Meanwhile, addition of cattle, goat and poultry manures to supply NH_4^+–N and NO_3^-–N to the soil with regard to sowing dates is recommended to be 2 WAI as this would allow equilibration of the manures with the soil prior to planting, while incorporation of these three manures for crop needs is recommended to be 8 WAI.

Acknowledgements The assistance of people that contributed to the success of this research is gratefully appreciated while the expressions, views and conclusions attained are for the authors.

References

Akanni DI, Ojeniyi SO (2008) Residual effect of goat and poultry manures on soil properties nutrient content and yield of amaranthus in southwestern Nigeria. Res J Agric 2(2):44–47

Ayeni LS (2011) Integrated plant nutrition management: a panacea for sustainable crop production in Nigeria. Int J Soil Sci 6:19–24

Ayeni LS, Adeleye EO (2011) Soil nutrient status and nutrient interactions as influenced by agro wastes and mineral fertilizer in an incubation study in the South—west Nigeria. Int J Soil Sci 6:60–68. doi:10.3923/ijss.2011.60.68

Azeez JO, Van Averbeke W (2010a) Fate of manure phosphorus in a weathered sandy clay loam soil amended with three animal manures. Biores Technol 101:6584–6588. doi:10.1016/j.biortech.2010.03.073

Azeez JO, Van Averbeke W (2010b) Nitrogen mineralization potential of three animal manures applied on a sandy clay loam soil. Biores Technol 101:5645–5651. doi:10.1016/j.biortech.2010.01.119

Bartholomow WW (1965) Mineralization and immobilization of nitrogen in the decomposition of plant and animal residues. In: Bartholomow WW, Clark FE (eds) Soil nitrogen. American Society of Agronomy, Madison, pp 285–306

Bouyoucos GH (1965) A calibration of the hydrometer method for testing mechanical analysis of soils. Agric J 43:434–438

Bray RH, Kurtz LT (1965) Determination of total organic and available forms of phosphorus in soil. Soil Sci 59:225–624

Cater MR (1993) Soil sampling and methods of analysis. Canadian Society of Soil Science, Ontario

Chadwick DR, John F, Pain BF, Chambers BJ, Williams JC (2000) Plant uptake of nitrogen from the organic nitrogen fraction of animal manures: a laboratory experiment. J Agric Sci Camb 134:159–168

Eghball B, Wienhold BJ, Gilley JE, Eigenberg RA (2002) Mineralization of manure nutrients. J Soil Water Conserv 57(6):470–473

ENDMEMO (2015) Conversion of part per million to milligram per kilogram. Complete concentration percentage unit conversions. (Online). http://www.endmemo.com/sco

Enujeke EC, Ojeifo IM, Nnaji GU (2013) Residual effects of organic manure and inorganic fertilizer on maize grain weight and some soil properties in Asaba Area of Delta State. ISSN 2250–3579. Int J Adv Bio Res 3(3):433–442

Enwezor WO, Udo EJ, Usoroh NJ, Ayotade KA, Adepetu JA, Chude VO, Udegbe GJ (1989) Fertilizer use and management practices for crop in Nigeria (eds). Federal Ministry of Water Resources and Rural Development (FMAWRRF) Lagos, pp 52–56

FUNAAB (2013) Record of daily agrometeorological observation. Department of Water Resources Management and Agrometeorology. College of Environmental Management. Federal University of Agriculture, Abeokuta, pp 1–12

Gomez-Brandon M, Fernandez-Delgado MJ, Domiguez J, Insam H (2013) Animal manures: recycling and management technologies. INTECH Open Sci. doi:10.5772/53454

Hill laboratories (2001) Soil tests and interpretation. Technical Notes. http://www.hill-laboratories.com, vol 4, 3196, p 9

Hoskins B (2015) Nutrient content, availability, and release rates from natural fertilizers. University of Maine. http://www.newengladvfc.org/2015_conference/2, pp 1–3

James JC (2001) Nitrogen in soil and fertilizer. First Published in the: Science Turfgrass Foundation News 8(1):6–10

Jonathan D (2006) Nitrogen mineralization potential in important agricultural soils of Hawai, i. Cooperative extension service. Soil Crop Managt 15:1–5

Kaira YP, Maynard DC (1991) Method manual for forest soil and plant analysis. For, Can., Northwest Reg., North. For, Cent., Edmonton, Albert Inf. Rep. NOR-X-319

LAS (2014) soluble salts or electrical conductivity of soils and green house media. Litchfield Analytical Services, pp 1–2

Mallarino A (2000) Soil testing and available phosphorus. Integrated crop management. LOWA State University. University Extension, pp 164–166

Marco J, Vlado K, Hrvoje P, Jasna S, Mitra R (2002) Influences of Nitrogen fertilizer and irrigation on nitrogen leaching. In proceedings of the second international conference on sustainable agriculture for food, energy and industry, pp 1641–1647

Masarirambi MT, Dlamini P, Wahome PK, Oseni TO (2012a) Effects of chicken manure on growth, yield and quality of lettuce (*Lactuca sativa* L.) "Taina" under a lath house in a semi-arid sub-tropical environment. American- Eurasian. J Agric Environ Sci 12(3):399–406

Masarirambi MT, Sibandze N, Wahome PK, Oseni TO (2012b) Effects of kraal manureapplication rates on growth and yield of wild okra (*Corchorus olitorius* L.) in a sub-tropical environment. Accepted: November, 29, 2011. ISSN: 2041–3890. Asian J Agric Sci 4(1):89–95

McBride C (2015) How to calculate percentages in grams. Demand Media

Meissinger JJ, Jokela WE (2000) Ammonium volatilization from dairy and poultry manure. Natural Resource, Agriculture, and Engineering Service. http://www.nraes.org, pp 1–2

Monica ZB (2013) Water and soil characterization. pH and electrical conductivity. Microbial life. Educational Research. Montana State University, Bozeman, pp 1–5

Murphy J, Riley JP (1962) A modified single solution method for determination of phosphate in natural waters. Anal Chem 27:31–36

Navarro Pedreno J, Moral R, Gomez I, Mataix J (1996) Reducing nitrogen losses by decreasing mineral fertilization in horticultural crops in eastern Spain. Agric Ecosyst Environ 59:217–221

Okalebo JR, Gathua KW, Woomer PL (1993) Laboratory methods of soil and plant analysis. A working manual. Soil Science Society of East Africa (SSSEA). Technical Publication 1

Pam H and Brian M (2007) Interpreting soil test results. What Do all the numbers mean? Pam Hazelton and NSW Department of Natural Resources, pp 1–66

Ramsasa C (2010) How to replenish the nutrients in the soil of your garden. Ezine Article. http://EzineArticles.com

Sajal R, Abul Kashem M (2014) Effects of organic manures in changes of some soil properties at different incubation periods. Open J. Soil Sci. 4(3):81–86. doi:10.4236/ojss.2014.43011

Schippers RR (2000) African indigenous vegetables—an overview of the cultivated species, pp 103–118

Silva DJ, QueirÓz AC (2002) Analse de alimentos: metodos quimicos e biologicos, 3 (ed) Vicosa: Imprensa Universitaria/UFV

Sorenson P (2001) Short-term nitrogen transformations in soil amended with animal manure. Soil Biol Biochem 33:1211–1216

Spore (2006) Issue 123, June 2006, p 13

Tom D (2002) Nitrogen Sources. University of Nebraska Lincoln Extensoin. 402. Available through: 441–7180. http://Lancaster.uni.edu, p 288

USDA (2010) Soil texture calculator. United State Development of Agriculture Natural Resources Conservation Service Soils. http://www.nrcs.usda.gov/wps

USDA–SCS (1974) Total nitrogen and organic carbon ratings and interpretation. Interpretation Guides for Evaluating Analytical Data. USDA–SCS

Uwah DF, Eyo VE (2014) Effects of number and rate of goat manure application on soil properties, growth and yield of sweet maize (*Zea mays* L. saccharata strut). Sustainable Agriculture Research; ISSN 1927-050X. Accepted: September 2, 2014. doi:10.5539/sar.v3n4p75

Van Reeuwijk LP (1993) Procedure for soil analysis. International Soil Reference and Information Centre Technical paper, pp 9

Vel AM, Swamam TP (2013) Nitrogen release pattern from organic manures applied to an acid soil. J Agric Sci 5(6):174–184. ISSN 1916–9760. doi:10.5539/jas.v5n6p174

Vinten A, Whitmoore AP, Bloem J, Howard R, Wright F (2002) Factor affecting nitrogen immobilization or mineralization kinetics for cellulose-, glucose-, and straw amended sandy soils. Biol Fert Soils 36:190–199

Watanabe PH, Thomas MC, Pascoal LAF, Ruiz, US, Daniel E, Amorim AB (2013) Manure production and mineral excretion in faeces of gilts fed ractopamine. Acta Scientiarum. Animal

Sciences. Maringa. Received on September 19, 2012. Accepted on December 14, 2012, v35n3, 267–272. http://www.redalyc.org/articulo.oa?id=303128706007; http://www.ehow.com/how, pp 1–6

Xu RK, Coventry DR, Farhoodi A, Schultz JE (2002) Soil acidification as influenced by crop rotations, stubble management, and application of nitrogenous fertiliser, Tarlee, South Australia. Aust J Soil Res 40:483–496

Bioconversion of organic solid wastes into biofortified compost using a microbial consortium

Payel Sarkar[1] · Rounak Chourasia[1]

Abstract

Purpose Urban municipal solid waste in India are 75–85% organic. Uncontrolled dumping of this waste is a major health concern. Degradation of organic waste by use of a microbial consortium is safe, efficient and economic. Therefore, this study was taken up to recycle the organic solid waste into effective compost using a microbial consortium.

Methods Bacterial consortia were developed using antagonism assay. Concomitant enzyme production by the consortia was determined. The best consortium was further employed for degradation of 30 kg of organic solid waste. Compost analysis of 30 kg of wastes was done to determine the level of C, N, K, P and S.

Results In this study, of the four consortia proposed, consortia no. 2 had the highest degrading capability. It exhibited consistent degrading capabilities of 30 kg waste. The volume of the waste was reduced to 82%, with a reduction in mass and moisture content to 65 and 42%, respectively, after 30 days of degradation study. The compost produced after 30 days had a dark colour and grainy texture without any crustacean population and lacked foul smell. Compost analysis of 30kg wastes inoculated with consortium 2 showed C:N ratio of 22:1 compared to 32:1 in control, and increased percentage of K, P and S which are required for enhancement of soil fertility.

Conclusion Therefore, we can conclude that consortium 2 can serve as a biological tool for the removal of organic solid wastes from the environment, and the compost generated from the degradation can be applied to increase the fertility of the soil.

Keywords Municipal solid waste · Microbial consortium · Compost · Soil fertility

Introduction

Biodegradable waste refers to the amount of organic matter in waste. Biodegradable waste is commonly found in municipal solid waste (MSW) where the organic matter in waste ranges from 75 to 85% in urban MSWs of India. The varying content of organic matter in waste depends on the economic condition of the country from where it is generated (Bandara et al. 2007). Recent decades had seen rapid increase in human population, accelerating economy and urbanization, which have led to an exponential increase in waste generation every year (Gouveia and Prado 2009). According to Gupta and Arora (2016), it has been estimated that urban India generates approximately 1,88,500 tonnes of municipal solid waste per day (68.8 million tonnes per year). The most common and cheap waste disposal technique involves dumping of waste in a specified disposal yard. This results in the pollution of soil by toxic chemicals present in waste, leading to emergence of diseases in plants, human beings and cattle (Forastiere et al. 2011).

Thus, micro-organisms present in soils of dump sites are exposed to varied substrates and chemicals, which raises the possibility of screening effective bacterial strains with valuable applications. Rapid organic waste degradation

✉ Payel Sarkar
payal.jobs83@gmail.com

1 Department of Microbiology, University of North Bengal, Raja Rammohunpur, Siliguri, West Bengal 734013, India

requires the action of micro-organisms with enhanced enzyme activity to break down complex polymers into simpler degradable molecules. Therefore, inoculation of waste with microorganisms that produce extracellular enzymes such as cellulase, amylase, protease, pectinase and lipase at higher levels enhances waste degradation, thus helping in keeping up of the waste degradation rate to that of waste dumping (Saha and Santra 2014).

This requirement is fulfilled with the co-operation of many micro-organisms, each of which contributes in secreting different enzymes. The degradation of organic waste by use of a microbial consortium is considered highly efficient, and the capability of waste degradation by a consortium depends on its functional and structural stability (Mirdamadian et al. 2011). The stability of a potential waste-degrading consortium can be observed by testing antagonistic (competition for substrate among different microbes present in the consortium) and synergistic (co-operation between different microbes present in the consortium for better degradation of wastes) activities within the strains and testing the concomitant production of different enzymes by the bacterial strains within the consortium to study the consortium's functioning under natural conditions of substrate availability (Sarkar et al. 2011).

Therefore, the main objective of this present study was to develop a microbial consortium that can act as a biological tool for removal of organic solid wastes from the environment and application of the compost generated from the degradation to increase the fertility of the soil.

Materials and methods

Collection of samples

Soil samples were collected from the dumping land of Matigarahat (latitude—26°85′N, longitude—88°44′E, Siliguri, Darjeeling District, West Bengal, India) in sterile 'Whirl-Pak' plastic maintaining aseptic conditions and immediately transported to the laboratory for further tests.

Characterization of the soil sample

Determination of moisture content (%) in soil samples

The soil sample was taken in a pre-weighed container and the initial weight was recorded. The sample was dried in a hot air oven at a temperature of 110 °C while weighing the sample at regular intervals until a constant weight was achieved. The moisture content of the sample was calculated using the following Eq. (1) (AWPA 1986):

$$\mathrm{MC}\ (\%) = \frac{W - w}{w} \times 100, \tag{1}$$

where MC is the moisture content, W the original weight and w the constant weight after oven drying.

Chemical analysis of soil samples

The chemical characteristics of the soil sample were tested for the following parameters: pH, organic carbon (%), nitrogen (%), potassium (ppm), phosphorus (ppm), sulphur (ppm), organic matter (%) and electrical conductivity (EC). Organic carbon, nitrogen content, phosphorus, sulphur and potassium were determined using the rapid titration method (Walkey and Black 1934), Kjeldahl procedure (Bremner 1960), barium chloride colorimetric method (Jackson 1973) and flame photometric method after ammonium acetate extraction (Toth and Prince 1949). The pH was determined using a digital pH meter.

Isolation of bacteria from the soil sample

Isolation of bacteria from the soil sample was done from the waste dumping land of Matigarahat (latitude—26°85′N, longitude—88°44′E) Siliguri, West Bengal, India, by the serial dilution method. 1 g of the soil sample was added to a test tube containing sterile 10 ml deionized water. This tube was marked as stock. It was shaken vigorously to dissolve the soil and form a homogenized solution, followed by subsequent tenfold dilution by transferring 1 ml solution to sterile 9 ml deionized water-containing tubes until a dilution of 10^{-6} was obtained. 0.1 ml of the solution from last three dilution was transferred to sterile nutrient agar culture media plates (pH 7) via the spread plate method and incubated at 37 °C for 24 h. Bacterial colonies were selected on the basis of colony morphology and pigmentation. The isolated colonies were purified by streaking and preserved in slants at 4 °C.

Screening of bacteria for the production of extracellular enzymes and morphological characterization

The isolated strains were further characterized on the basis of their substrate specificity (starch, cellulose, casein, pectin, tributyrin) and gram character. The inoculated plates were incubated overnight at 37 °C and checked for a zone of clearing around each bacterial isolate. For starch agar, the zone of clearing was observed after flooding the plates with iodine. Cellulase-producing organisms were detected by flooding the plates with 0.1% Congo Red for 20 min, followed by rinsing with 1 M NaCl. Caseinase-positive bacterial isolates will show clear halo zone around the culture growth in skimmed milk agar. Pectinase-positive bacterial isolates will show clear zones around culture growth due to lack of formation of the pectin–iodine complex after flooding with iodine. Lipase-positive isolates show a clear zone around bacterial growth that was observed due to the hydrolysis of tributyrin.

The proposed combination of the consortium

Four different microbial consortia were prepared by permutation and combination, keeping in mind their concomitant enzyme production and gram character. The compatibility of the bacterial strains within the consortia was checked by antagonism assay.

Antagonism of bacterial strains

The cross streaking method was used to test the antagonism among the selected bacterial strains before consortium preparation. Each bacterial culture (24 h) was inoculated in sterile nutrient agar media plates as a 1.5 cm-wide streak (wider streak to obtain larger inhibitory zone in case of antagonism) diametrically across the plate and incubated at 37 °C for 24 h. After incubation, indicator organisms (to be present in the potential consortium) were streaked at right angle to the original inoculum. This process was followed to test each organism's antagonism against the other members of the consortium.

Concomitant enzyme production by the consortium

Each consortium was inoculated (1% inoculum load) in 50 ml synthetic media: modified Czapek mineral-salt broth [(g/l) sodium nitrate 2, potassium chloride 0.5, magnesium sulphate 0.5, di-potassium hydrogen phosphate 1, ferrous sulphate 0.01 supplemented with CMC, pectin, starch, skimmed milk and tributyrin (0.5% each)] and incubated at 37 °C at 150 rpm for 7 days (Sarkar et al. 2011). After every 24 h, 6 ml of the sample was withdrawn from the broth and centrifuged at 10,000 rpm. The supernatant was subjected to amylase, cellulase, protease, pectinase and lipase enzyme assays.

Amylase assay

Amylase activity was determined by the DNS method. 1% starch solution was prepared in 50 mM phosphate buffer (pH7) and used as substrate. The substrate, crude enzyme and buffer were added to tubes and incubated at 37 °C for 30 min. After incubation, the contents of the enzyme blank were added, 1 ml DNS reagent added to all test tubes and all the solutions in the tubes were boiled for 5 min. The tubes were allowed to cool to room temperature and then diluted with 5 ml distilled water. Absorbance was measured at 540 nm and the glucose standard curve was prepared using different glucose dilutions. One unit of enzyme activity is defined as the amount of enzyme that releases 1 μmol of reducing sugar (glucose) per minute under assay conditions (U/ml/min).

Cellulase activity

Cellulase activity was determined by the DNS method. 1% CMC solution was prepared in 50 mM phosphate buffer (pH7) and used as the substrate. The substrate crude enzyme and buffer were added to tubes similar to the amylase assay. The assay was conducted following a similar protocol to the amylase assay under the same incubation temperature and time. Absorbance was measured under 540 nm. Glucose standard curve was used as standard. One unit of enzyme activity is defined as the amount of enzyme that releases 1 μmol of reducing sugar per minute under assay conditions (U/ml/min).

Protease assay

Protease activity was determined spectrophotometrically following the protocol of Sigma's universal protease activity assay. Casein was used as a substrate and the ability of the enzyme to digest casein into tyrosine was tested. 1% casein was prepared in 50 mM potassium phosphate buffer (pH 7.5) and used as substrate. The substrate, crude enzyme and buffer were added to tubes and incubated at 37 °C for 10 min. After incubation, the contents of the enzyme blank were added and to all the tubes 0.5 ml stopping agent trichloroacetic acid (TCA) (20% w/v) was added. The tubes were incubated at 37 °C for 30 min. After the incubation period, the content of each tube was centrifuged at 10,000 rpm for 10 min to separate insoluble substances from the sample. 1 ml of the supernatant was taken in a fresh test tube to which 5 ml of 0.5 M sodium carbonate was added, followed by the quick addition of 0.5 ml Folin–Ciocalteu reagent. The tubes were kept in dark for 20 min. Absorbance was recorded at 660 nm. One unit of enzyme activity is defined as the amount of enzyme that releases 1 μmol of tyrosine per minute under assay conditions (U/ml/min).

Pectinase assay

Pectinase activity was determined titrimetrically following standard Sigma protocol using pectin as the substrate. 0.5% pectin solution was prepared in distilled water. This solution was used as the substrate. 4.90 ml of the substrate was taken in flaks marked as test and blank and kept at room temperature for 5 min. 0.1 ml of crude enzyme was added to test flasks, mixed thoroughly and the flasks along with the blank were incubated at room temperature for 5 min. After incubation, 5 ml 0.1 N iodine and 1 ml 1 M sodium carbonate were added to the solution, which was incubated at room temperature for 5 min. Then, 2 ml 2 N sulphuric acid was added, mixed well and the flasks were kept in dark for 20 min. Titration of the dark brown solution was done by 0.1 M sodium thiosulphate until the solution turned faint yellow in colour. Here, one drop of 1% starch solution was added to obtain a dark blue colour and then the solution was finally titrated until the solution turned colourless. One unit of enzyme activity is defined as the amount of enzyme that releases 1 μmol galacturonic acid per minute under assay conditions using Eq. (2):

$$\text{Units/ml enzyme} = \frac{(\text{ml of titrant for blank} - \text{ml of titrant for test}) \times 1 \times df \times 100}{0.100 \times 5 \times 2}, \tag{2}$$

where df = dilution factor, 1 = one micromole of galacturonic acid is oxidized by 1 micro-equivalent of I_2, 100 = micro-equivalents of S_2O_3 per millilitre of titrant, 0.100 = volume (in millilitre) of enzyme used in enzymatic reaction, 2 = micro-equivalents of S_2O_3 oxidized per micro-equivalent of I_2 reduced and 5.0 = time of incubation of assay in minutes per unit definition.

Lipase assay

Lipase activity was determined titrimetrically using tributyrin oil as the substrate. 1 ml of tributyrin was mixed with 3 ml 50 mM Tris–HCl buffer (pH8) and an emulsion was formed by swirling the contents. 1 ml of the crude enzyme was added to the emulsion (1 ml buffer in blank) and incubated at 37 °C for 30 min. After incubation of one drop of the indicator, 1% phenolphthalein solution was added to both test and blank and titrated with 50 mM sodium hydroxide until a pale pink colouration was developed. One unit of enzyme activity is defined as the amount of enzyme that releases 1 μmol of fatty acid per minute under assay conditions.

Determination of organic waste degradation (laboratory trials with 5 kg of wastes)

The ability of the consortia to degrade 5 kg of organic wastes was determined by individually inoculating 5% (v/v) consortium (24 h old cultures) in two sets of organic wastes collected from canteens around North Bengal University. The consortium was mixed evenly and kept in normal conditions for a month to observe the degradation. The heaps were mixed periodically to maintain aeration. A third set of uninoculated organic wastes was kept under similar experimental conditions that served as control. The sample was withdrawn periodically at a constant interval of 7 days till 28 days and waste degradation was measured on the basis of the following parameters.

Determination of waste volume

The buckets used for the trials have the shape of truncated cone and thus the volume of the bucket was calculated using Eq. (3):

$$\text{Volume} = \pi/3h(R^2 + \text{Rr} + r^2), \tag{3}$$

where h is the height of the bucket, R^2 the radius of the upper surface of the bucket, Rr the radius of the middle of the bucket and r the radius of the bottom of the bucket.

The volume of the waste was calculated accordingly at an interval of 7 days with respect to the bucket as the waste level decreased due to degradation.

Determination of waste density

Density of the waste was determined by filling a Petri dish full of the waste sample and weighing the content. The volume of the Petri plate was determined using the formula of the volume of the cylinder. Then the formula for density (ρ = mass/volume) was used.

Determination of waste mass

Mass of the degrading waste was determined after 7 days for a month with the data obtained from density and volume of the waste at the respective day, using the formula for density.

Change in temperature

Temperature changes were observed at three levels: top surface, centre and bottom. This was done by making holes in the wall of the bucket to observe temperature gradations of the three levels with respect to the degradation.

Determination of moisture content in waste

Waste degradation is associated with reduction in moisture content. Moisture content estimation was similar to the protocol followed during moisture content analysis of the soil.

Physical observation of waste

After 30 days of inoculation, wastes were observed for changes in colour and odour. After conducting the degradation study for 30 days with 5 kg organic wastes, the best consortium was selected to be employed in the upscaling of the organic waste degradation.

Scaling up of degradation study using the best consortium

On the basis of results obtained from laboratory trials, the best consortium was selected for further scaling up of the degradation study using 30 kg wastes. The degradation study was done using the composting process, because composting incurs lower cost and has lesser negative impact on the environment (Lim et al. 2016). In addition, composting of biodegradable solid wastes is deemed useful for its utilizable end product as nutrient-rich organic fertilizers or for land application (Wu et al. 2014). In this study, organic wastes, collected from different canteens of North Bengal University was inoculated individually with overnight grown the best consortium (5% v/v). The consortium was mixed evenly and kept under normal environmental conditions for a month to observe the degradation. The heaps were mixed periodically to maintain aeration. A second set of uninoculated organic wastes was kept under similar experimental conditions that served as the control. The sample was withdrawn periodically at a constant interval of 7 days till 28 days, and waste degradation was measured on the basis of the following parameters: volume reduction, density increase, mass reduction, temperature change and moisture content of the compost following the protocol explained in the previous section.

Chemical analysis of waste

The 30 kg organic waste inoculated with the best consortium was subjected to compost analysis after 30 days of degradation study. This included tests for the quantitative analysis of factors: pH, electrical conductivity (E.C.), organic matter, C:N ratio, phosphorus, potassium and sulphur. This test was performed to observe the ability of the degraded organic waste to be used as manure to increase the fertility of soil.

Results and discussion

Determination of moisture content of the soil sample

The soil sample collected from the waste dumping site of Matigarahat was dried at 110 °C and weighed until a constant weight was attained. The moisture content of the sample was calculated using the formula by AWPA and it was found to be 52%. High bacterial density is found at high moisture content and optimal activity of aerobic bacteria lies at the moisture content range of 50–75%. Aerobic soil commonly consists of bacteria *Pseudomonas* sp., *Bacillus* sp. and *Clostridium* sp., along with a higher population of Actinomycetes (Alexander 1977).

Chemical analysis of the soil sample

The chemical characteristics of the soil sample were determined with the following parameters: pH, organic carbon (%), nitrogen (%), potassium (ppm), phosphorus (ppm), sulphur (ppm), organic matter (%) and electrical conductivity (EC). The results are summarized in Table 1. Soil pH is hydrogen ion activity measurement and depends on relative amounts of absorbed metallic ions and hydrogen. The pH was determined with a soil to water ratio of 1:2.5. The pH of the soil sample was 5.49, which is slightly acidic. The reason for this can be attributed to the very high levels of available sulphur (30 ppm), resulting from organic matter degradation as the soil was collected from the organic waste-rich dumping site (Bezdicek et al. 2002).

Electrical conductivity is the measurement of the ability to transport electric charge. The EC of soil depends on salts and sodium content level. Uncontrolled discharge of wastes results in the significant increase of EC that can be toxic to plants, preventing their ability to obtain water from soil. The EC value of soil less than 0.5 m mho/cm is considered safe for plant growth. The result showed that the EC of the soil sample was 0.12 m mho/cm, which was considered safe for plant growth.

A proper C:N ratio must be present in soil for organisms to grow. Excess of carbon in soil means a surplus presence of energy, and microbial cells will draw larger amount of nitrogen to make use of available carbon. This is known as 'robbing' the soil of nitrogen and it delays the availability of nitrogen as fertilizer in soil (Meuser 2010). A C:N ratio of less than 20 is good for the growth of organisms the soil of nitrogen. The C:N ratio of the soil sample was 12.5, which is considered good for microbial growth. However, Rawat et al. (2008) found the C:N ratio of Indian dumpsites to vary between 11 and 30, which is in agreement with our result.

The level of nitrogen was 0.11%, which was fairly low. Meuser et al. (2011) found the nitrogen content of waste dumping site to range between 0.34 and 0.54%. However, higher levels of potassium and phosphorus were observed due to the high solubility of K and P in solid wastes that get readily leached to soil. The sources of these minerals are considered to be the organic materials in wastes.

The sulphur content of the soil sample was reported to be 30 ppm. Pillai et al. (2014) also reported the presence of 25 ppm sulphur content in the soil at the dumping site. This level is considered very high and is good for the growth of annual crops such as cabbage. High levels of sulphur

Table 1 Chemical analysis of soil

pH	Org carbon (%)	E.C (m mho/cm)	N (%)	K_2O (ppm)	P_2O_5 (ppm)	Sulphur (ppm)	Org. matter (%)
5.49	1.38	0.12	0.11	68.5	15	30	2.37

indicate the presence of decaying organic matter in the soil and results in low soil pH.

Isolation of bacteria from soil sample

19 different bacterial colonies were isolated from the soil sample by the serial dilution method after 24 h incubation at 37 °C. The bacterial isolates were screened according to colony morphology (shape, margin, elevation texture) and pigmentation. The results are summarized in Table 2. A similar method was used by for the successful isolation of organisms from municipal solid waste. The organisms were further utilized in consortia for fast and effective degradation of organic waste (Saha and Santra 2014).

Screening of bacteria for the production of extracellular enzymes

The isolated bacteria were screened for the production of five extracellular enzymes (amylase, cellulase, protease, lipase and pectinase) that are required for the effective degradation of organic solid wastes. The organisms were streaked on plates with the respective screening media and the zone of hydrolysis was observed after 24 h incubation. The gram characteristics of the isolates were also determined.

10 different bacterial cultures were screened out of the 19 bacterial isolates on the basis of extracellular enzyme production and gram characteristics (Table 3). Sarkar et al. (2011) reported 35 bacterial isolates out of 80 isolates that showed similar enzyme production. The soil of dumping sites is subjected to high concentration of organic material, which consists of various kinds of substrates for the indigenous organisms in the soil. Micro-organisms residing in soil with heightened supply of substrates develop mechanisms to secrete increased concentration of extracellular enzymes to utilize the available substrate (Raju and Divakar 2013).

Proposed composition of consortium

To prepare a successful microbial consortium, the bacterial cultures must be compatible with each other to concomitantly produce all these enzymes required for the degradation of organic wastes. Four different microbial consortia were prepared by permutation and combination (Table 4),

Table 2 Morphological characterization of the isolated strains from waste dumping site

Bacterial colony	Shape	Margin	Elevation	Pigmentation	Texture
W_1A	Filamentous	Filamentous	Flat	White	Dry
W_1B	Circular	Entire	Convex	Pale	Mucoid
W_1C	Circular	Entire	Flat	Pink	Dry
W_1D	Circular	Entire	Convex	White	Mucoid
W_1E	Irregular	Entire	Raised	Pale	Dry
W_1F	Circular	Entire	Raised	Orange	Glossy
W_1G	Circular	Entire	Raised	Yellow	Glossy
W_1H	Circular	Irregular	Convex	Pale	Glossy
W_1I	Rhizoid	Lobate	Raised	White	Mucoid
W_1J	Irregular	Rhizoid	Flat	White	Rough
W_1K	Irregular	Lobed	Flat	White	Rough
W_1P	Circular	Entire	Convex	Pale	Glossy
W_1Q	Punctiform	Entire	Convex	Yellow	Glossy
W_1R	Circular	Entire	Convex	Orange	Glossy
W_1S	Punctiform	Entire	Convex	White	Mucoid
W_1T	Circular	Entire	Convex	Oily yellow	Mucoid
W_1U	Circular	Entire	Convex	White	Glossy
W_1V	Irregular	Undulate	Raised	Opaque	Glossy
W_1W	Punctiform	Entire	Convex	Orange	Glossy

Table 3 Screening of the isolates on the basis of enzyme production and cell morphology

Bacteria	Amylase	CMCase	Protease	Lipase	Pectinase	Gram nature	Morphology
W_1A	+	–	–	++	+++	Positive	Rods
W_1B	–	–	++	+++	+++	Positive	Cocci
W_1C	+++	–	+	+++	+++	Positive	Rods
W_1D	–	–	–	+++	+++	Negative	Cocci
W_1E	++	–	++	+++	+++	Positive	Rods
W_1F	–	–	–	++	+++	Positive	Rods
W_1G	–	–	–	++	+++	Positive	Cocci
W_1H	–	–	–	++	+++	Positive	Cocci
W_1I	–	++	+++	+++	–	Negative	Rods
W_1J	–	–	–	+++	+++	Negative	Cocci
W_1K	–	+++	–	++	–	Positive	Rods
W_1P	–	–	–	++	–	Negative	Cocci
W_1Q	+	–	–	++	+++	Negative	Cocci
W_1R	–	–	++	++	+++	Positive	Cocci
W_1S	–	–	–	++	–	Positive	Rods
W_1T	+	–	++	+	+++	Negative	Cocci
W_1U	–	–	+	+++	–	Negative	Cocci
W_1V	+++	–	+	+++	+++	Negative	Rods
W_1W	+++	–	+	++	–	Negative	Cocci

– no enzyme production, + low, ++ moderate, +++ high enzyme production

Table 4 Proposed composition of the consortium

Combination	Composition
1	W_1C, W_1I, W_1E, W_1K
2	W_1R, W_1T, W_1V, W_1K.
3	W_1V, W_1K, W_1B, W_1W
4	W1B, W_1E, W_1K, W_1Q

keeping in mind their concomitant enzyme production and gram character. The compatibility of the bacterial strains within the consortia was checked by antagonism assay.

Antagonism assay

A cross streaking method was used to test antagonism among the bacterial strains within the proposed consortia. Plates were observed after incubation for zone of inhibition, which would indicate the presence of antagonism between organisms. Among the four combinations, two combinations of organisms (consortium 1 and 2) showed no antagonism among each other as shown in Fig. 1a, b.

Positive antagonism was observed in the case of combinations 3 and 4. The zone of inhibition of growth caused by antagonism is illustrated in Fig. 1c. Organisms W_1B inhibited the growth of W_1Q, W_1Q inhibited the growth of W_1E, and W_1K and W_1W inhibited the growth of W_1B. Therefore, they were not further considered for consortium preparation.

Preparation of a successful consortium requires the micro-organisms in the consortium to grow in the presence of each other without hampering the activities of other native micro-organisms. Antagonistic relations between micro-organisms of a consortium lead to instability of the consortium and the expected functioning of the consortium is not obtained. The result of the antagonism assay between organisms of individual combination is presented in Table 5.

Micro-organisms grow in various associations with other organisms, one of them being an antagonistic relation. Antagonist micro-organisms can employ various mechanisms to inhibit the growth of other microbes, including production of toxins, antibiotics and siderophores (Hibbing et al. 2010). Van Hamme et al. (2003) showed the presence of antagonistic relations between microbes in a population of mixed community.

Combinations 1 and 2 lacked antagonism and thus were selected for further consortium preparation. Consortia were prepared by inoculating 150 μl each of four overnight-grown cultures in 20 ml sterile nutrient broth for each consortium.

Concomitant enzyme production by the consortium

Czapek Dox mineral salt broth was selected for concomitant enzyme production, because it is a defined medium where all the constituents are known. This medium was

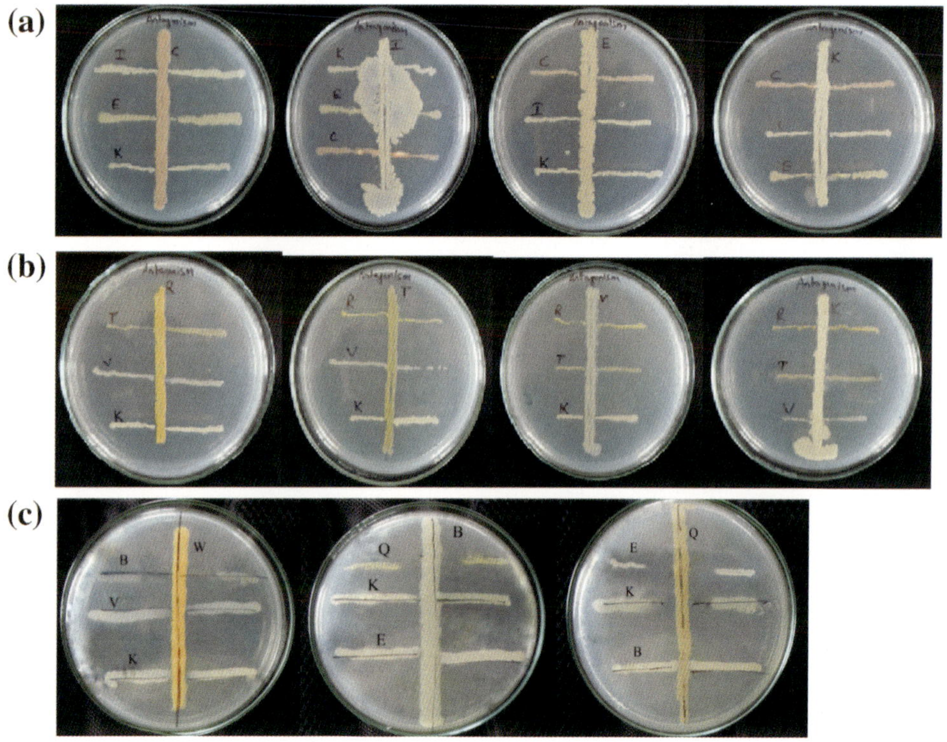

Fig. 1 Determination of antagonism among the bacterial strains within the proposed consortia. Antagonism assay for combination 1 (**a**), combination 2 (**b**), and combinations 3 and 4 (**c**)

supplemented with five substrates (CMC, starch, skimmed milk, pectin and tributyrin oil) and different extracellular enzyme productions were assayed after every 24 h interval till 7 days.

Consortium 2 showed the highest amylase activity on day 2 (Fig. 2a), followed by a gradual decline. A delay in cellulose activity was observed in both the consortia. The highest cellulase activity was noted on day 5 in consortium 1, whereas in consortium 2 it was highest on day 4. Consortium 1 had the highest cellulase activity (Fig. 2b). Consortium 1 showed greater protease activity than consortia 2, and the highest activity was observed on day 4 (Fig. 2c). Pectinase activity was significantly high for both the consortia from day 1 and kept increasing till day 4; consortia 2 showed slightly higher activity than consortium 1 (Fig. 2d).

The results of the lipase assay were different from those of the rest of the enzymes. No activity was observed for both the consortia on day 1. Enzyme activity was exhibited on day 2 only by consortium 2. Significant increase in lipase activity was observed in both the consortia after 72 h incubation with a constant increase till day 7. Consortium 2 showed higher activity than consortium 1 (Fig. 2e).

Among the two consortia, consortium 2 showed better concomitant enzyme production compared to consortium 1. Both the consortia were selected for laboratory trial with 5 kg organic waste. Similarly, Gopinath et al. (2014) constructed five different microbial consortia and the compatibility of the consortia to degrade organic waste was checked by enzyme assays of four substrates (starch, cellulose, casein and tributyrin). Sarkar et al. (2011) also reported three consortia out of 11 different microbial consortia that were concomitant producers of all three enzymes.

Determination of organic waste degradation (5 kg laboratory trials)

Consortia 1 and 2 were subjected to 5 kg of organic waste degradation for 28 days. The degradation capabilities were tested on the basis of various parameters at a constant interval of 7 days till 28 days.

Determination of physical changes in waste (volume, density and mass)

A rapid decrease in waste volume was observed in both the consortia as well as in the control after the first week. However, decrease in volume for consortium 2 was greater than that in consortium 1. After 28 days trial, 85% volume of the waste decreased in case of consortium 2 and 72% in consortium 1 (Fig. 3a), whereas only 55% volume reduction was observed in the control. Elango et al. (2009) observed degradation of MSW by thermophilic composting with 78% volume reduction of raw MSW in 40 days. The resulting compost provided good humus to build up soil characteristics and some basic plant nutrient.

Table 5 Antagonism assay for the proposed microbial combination

Combination 1				
Organisms	W_1C	W_1I	W_1E	W_1K
W_1C	X	–	–	–
W_1I	–	X	–	–
W_1E	–	–	X	–
W_1K	–	–	–	X
Combination 2				
Organisms	W_1R	W_1T	W_1V	W_1K
W_1R	X	–	–	–
W_1T	–	X	–	–
W_1V	–	–	X	–
W_1K	–	–	–	X
Combination 3				
Organisms	W_1V	W_1K	W_1B	W_1W
W_1V	X	–	–	–
W_1K	–	X	–	–
W_1B	–	–	X	–
W_1W	–	–	+	X
Combination 4				
Organisms	W_1B	W_1E	W_1K	W_1Q
W_1B	X	–	–	+
W_1E	–	X	–	–
W_1K	–	–	X	–
W_1Q	–	+	+	X

X identical organism, – no antagonism, + antagonism present

Density and volume are inversely related and a rapid decrease in volume indicates increasing density. As expected, the density of 5 kg waste increased in consortium 2, as it exhibited 85% decrease in volume of waste. Increase in density is important during composting, as it indicates particle size decrease due to waste degradation. Compost supplemented with consortium 2 showed significant density increase in comparison to consortium 1 and the control (Fig. 3b). Tom et al. (2016) reported a 50% density enhancement of bio-treated MSW resulting in compaction of the compost.

Mass reduction is a primary objective of waste treatment processes. Organic matter is utilized by degrading organisms and incorporated into cells and is also released as gas which causes a reduction in waste mass (Tom et al. 2016). Significant reduction in mass was observed for waste supplemented with consortium 2 in comparison to other composts, indicating better degradation amounting in reduction up to 70% of the initial mass (Fig. 3c). Higgins et al. (2006) also reported similar mass reduction of 70% during composting of municipal solid waste and emphasized the eligibility of the compost to be used as a fertilizer in soil for plant growth.

Determination of moisture content in waste

Water is a critical factor during composting. Compost moisture content above 80% leads to decrease of oxygen diffusion, resulting in the composting process to slow down and become anaerobic. A moisture content of 40–60% should be maintained during the composting process for proper microbial process. The moisture content reduction was uniform for the first 7 days for both test consortia and control composts, indicating moisture reduction due to drying and evaporation (Fig. 3d). 40% moisture content was found after 28 days of degradation study in consortium 2, whereas the moisture content was 45 and 65% in case of consortium 1 and control, respectively. Increase in moisture reduction in consortium 2 was more due to temperature rise, resulting from increased microbial degradation activities. Gautam et al. (2010) performed composting of MSW of Jabalpur City and observed a final moisture content of 36% in the final compost. This moisture content of compost is within the quality parameters set by international standards.

Change in temperature

Temperature changes of the three levels of the waste were observed after every 7 days interval. The temperature of the compost pile rose quickly and reached a peak at day 14 (Table 6), for all the test and control composts. Compost supplemented with consortium 2 attained the highest temperature among the test composts at all the three levels, and the highest temperature was observed in the centre layer (48 °C). Rise of temperature is associated with vigorous microbial activity and rapid organic matter degradation. Increase in temperature leads to degradation of waste by thermophilic organisms that supplement mesophilic degradation (Tom et al. 2016).

Temperature increase leads to reduction in pathogenic organism population, though turning of the compost pile at regular interval was employed to prevent increase of temperature to such extents that would kill the degrading organisms and stop the degradation process. This result supports the mass and volume reduction results, where consortium 2 demonstrated higher reduction in mass and volume than compost supplemented with consortium 1. Singh and Sharma (2003) observed the highest temperature (58 °C) at the bottom of the compost pile with maximum temperature increase from the 14th to the 21st day.

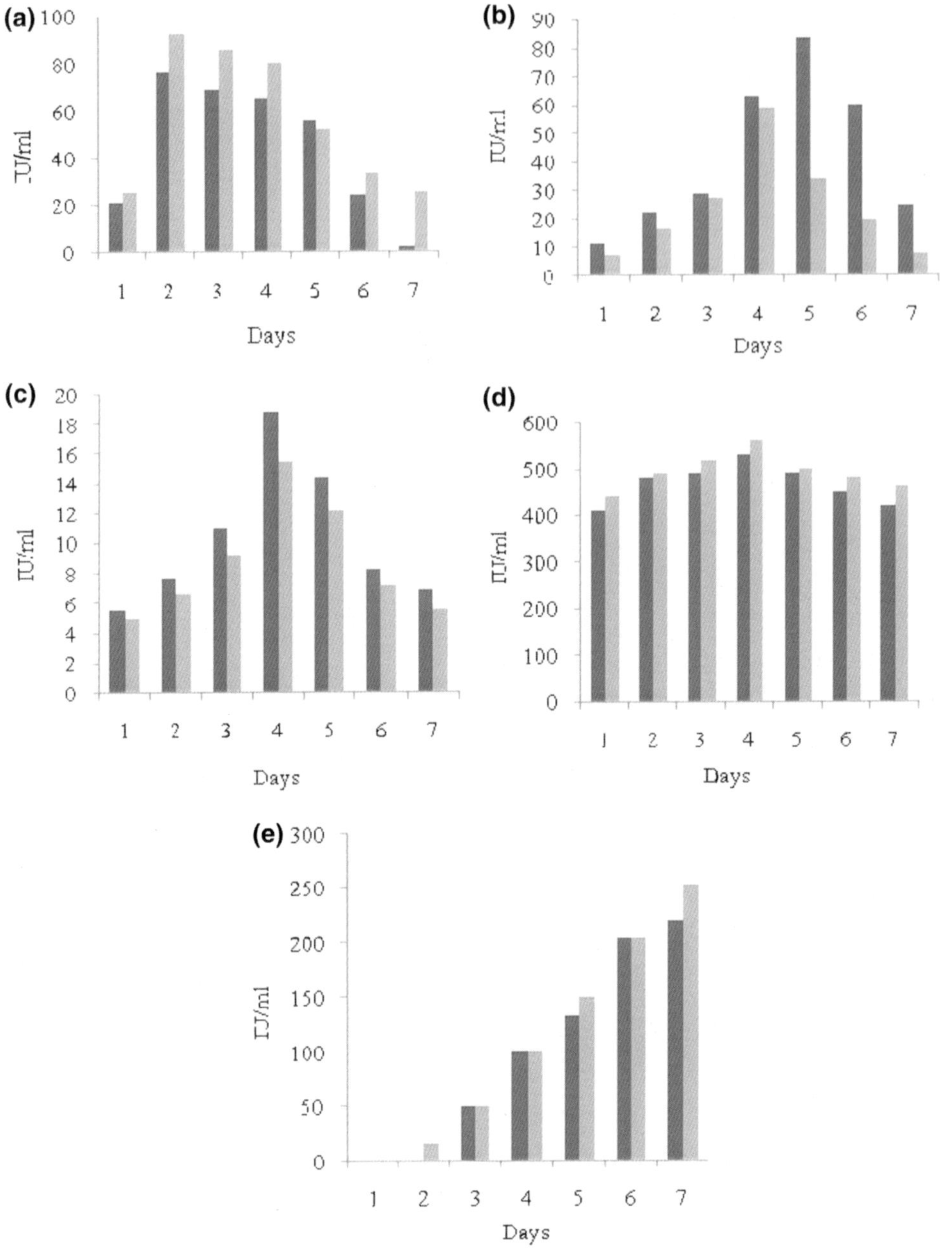

Fig. 2 Concomitant enzyme production by consortium 1 (black bar) and consortium 2 (grey bar) amylase activity (**a**), cellulase activity (**b**), protease activity (**c**), pectinase activity (**d**) and lipase activity (**e**)

Changes in colour and odour of the wastes

Wastes inoculated with consortia and control varied in colour, texture and other characteristics after 28 days. Waste inoculated with consortium 2 was dark (Fig. 4), had a grainy texture and had very low smell (Visilind and Spinosa 2001). It lacked the presence of crustaceans. Consortium 1-inoculated waste had a slight slimy texture and the presence of a small population of crustaceans, whereas the control had a sharp rotten stench. This might be a result of domination of anaerobic bacteria in the control waste which produce hydrogen sulphide and other sources of offensive odour such as cadaverine, in addition to organic acids and amines (Higgins et al. 2006).

In addition to the smell, the control consortium was teeming with small crustaceans including millipedes which are essentially harmless to the composting process; instead crustaceans break down the organic waste to help in further degradation (Shamaini and Fauzia 2014). But the presence of these organisms indicate that the temperature of the compost is not high enough as required for the degradation of the waste by thermophilic micro-organisms and thus the process of degradation of the control compost was slower than composts inoculated with the consortium.

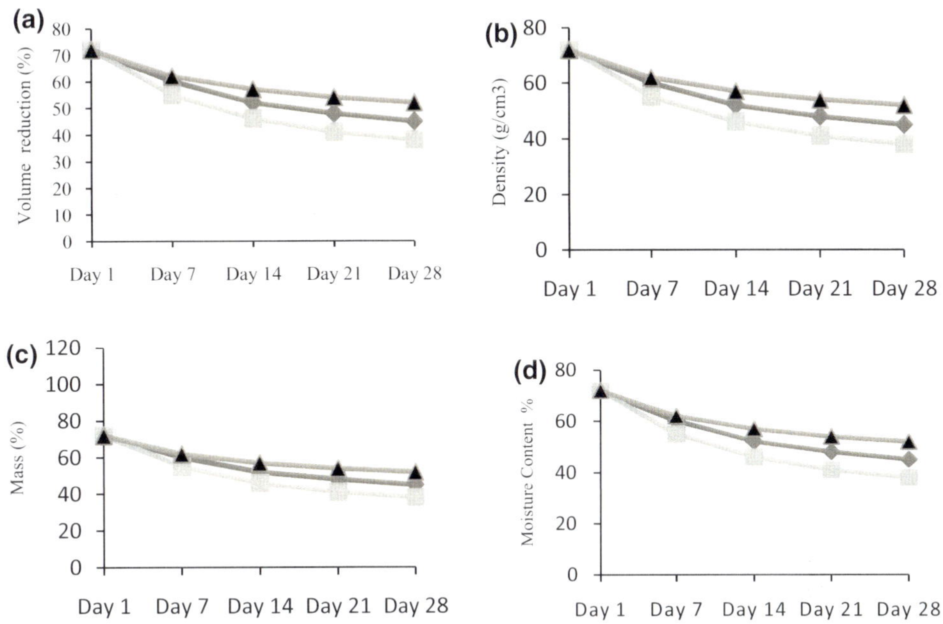

Fig. 3 Determination of various parameters of 5 kg organic waste degradation laboratory trials (consortium 1, consortium 2, control). **a** Volume reduction, **b** waste density, **c** mass reduction and **d** moisture content reduction

Table 6 Determination of temperature changes in the 5 kg compost

Day	Level	Temperature (°C)		
		Control	Con 1	Con 2
1	Surface	31	31	31
	Centre	37	39	37
	Bottom	36	36	35
7	Surface	34	37	35
	Centre	40	45	44
	Bottom	38	41	40
14	Surface	38	41	43
	Centre	40	46	48
	Bottom	39	41	43
21	Surface	37	39	42
	Centre	40	42	45
	Bottom	38	39	42
28	Surface	35	38	40
	Centre	39	40	43
	Bottom	37	38	42

Fig. 4 Determination of colour and texture changes in 5 kg waste after 28 days of degradation study

Scaling up of the degradation study using the best consortium

Among the two consortia used for laboratory trials with 5 kg waste, consortium 2 showed better degradation capabilities and thus was selected for further scaling up of the degradation study using 30 kg organic waste. Degradation and compost maturation with 30 kg waste were tested on the basis of the same parameters that were checked during 5 kg of laboratory trials for 30 days at an interval of 7 days. Reduction of volume in 30 kg waste inoculated with consortium 2 followed a similar pattern, as was observed during 5 kg trials. During the 30 days trial, the volume of the inoculated waste was decreased by 82% which was similar to 85% volume reduction of 5 kg waste (Fig. 5a).

A maximum density value of 0.85 g/cm^{3}was observed for the 30 kg waste, which was in accordance to the density value 0.90 g/cm^3 observed for 5 kg wastes inoculated with consortium 2 (Fig. 5b). The mass reduction observed for 30 kg waste was 65%, which correlated with 68.75% mass reduction observed for 5 kg waste. This difference is insignificant and indicative of the ability of consortium 2 to sustain its waste degradative capability of larger quantity of wastes (Fig. 5c). The final moisture content of 30 kg waste was 42% as compared to 40% observed during 5 kg trials. The maximum reduction in moisture content for 30 kg waste was observed during the 28th day as compared to

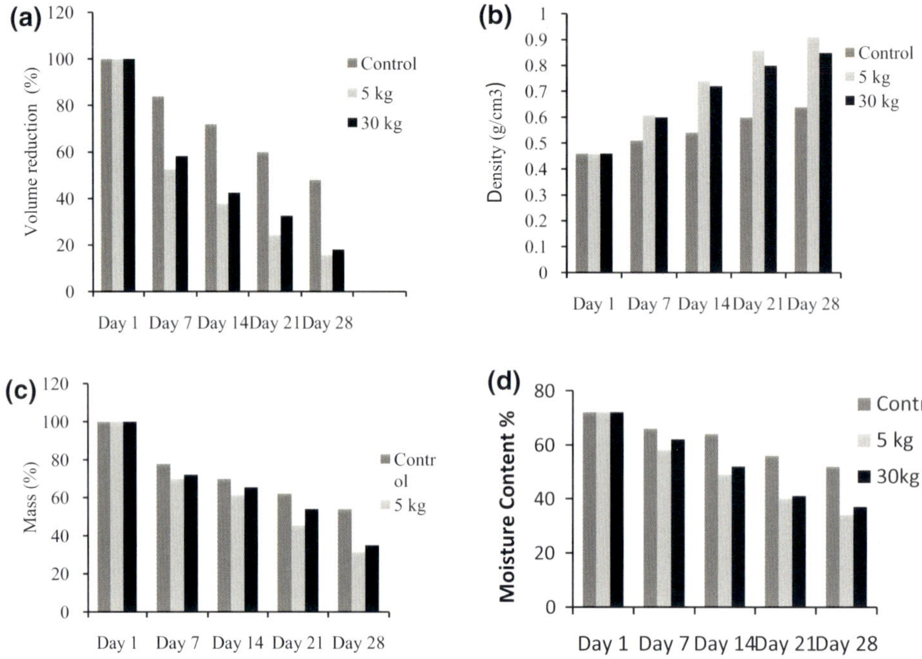

Fig. 5 Comparison of various parameters of 5 and 30 kg organic waste degradation by consortium 2. **a** Volume reduction, **b** waste density, **c** mass reduction and **d** moisture content reduction

Table 7 Comparison of temperature difference in the 5 kg and 30 kg waste

Day	Level	Temperature (°C)		
		Control	5 kg	30 kg
1	Surface	35	31	35
	Centre	38	37	39
	Bottom	36	35	35
7	Surface	40	35	42
	Centre	43	44	45
	Bottom	39	40	43
14	Surface	42	43	46
	Centre	44	48	49
	Bottom	33	43	47
21	Surface	44	42	51
	Centre	48	45	54
	Bottom	45	42	50
28	Surface	40	40	49
	Centre	44	43	56
	Bottom	43	42	50

maximum moisture reduction of 5 kg waste during the first 2 weeks (Fig. 5d).

Increase in temperature of the 30 kg waste was higher than the temperature rise for 5 kg wastes. A maximum value of 54 °C was observed in 30 kg waste, while a maximum temperature of 45 °C was observed in 5 kg waste at the 21st day (Table 7).

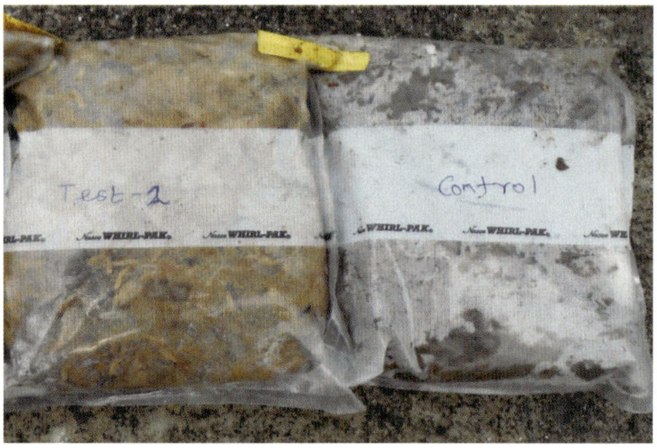

Fig. 6 Determination of colour and texture changes in 30 kg waste after 30 days of degradation study

The colour and odour of the 30 kg waste after 30 days were similar to the results obtained for the 5 kg waste. The waste had developed a dark colour, had a grainy texture and lacked crustacean population in comparison to the control waste; it also lacked the foul smell observed in control waste (Fig. 6).

Chemical analysis of the compost

The waste inoculated with consortium 2 was subjected to further compost analysis after the successful degradation of 30 kg wastes. This included tests for the quantitative analysis of the factors: pH, electrical conductivity (E.C), organic matter, C:N ratio, phosphorus, potassium, sulphur. The results of compost analysis are tabulated in Table 8.

Table 8 Chemical analysis of the final compost

Sample	pH	E.C (m mho/cm)	Org. matter (%)	C:N	K_2O (%)	P_2O_5 (%)	Sulphur (%)
Fresh waste	6.38	1.90	33.35	32:1	0.39	0.32	0.08
Control	5.91	1.96	21.17	29:1	0.48	0.41	0.08
Test compost	6.71	2.11	18.06	22:1	0.55	0.50	0.10

The pH of the control compost (5.91) was lower than that of the test compost (6.71), indicating the dominance of anaerobic organisms instead of thermophilic degraders in the control. This resulted in slow and less degradation of waste in the control compost compared to the test.

Reports also suggest that production of organic acids and other anaerobic degradation products lead to the generation of foul odour and a decrease in pH (Tejada and Gonzalez 2007). The pH of the test compost was 6.7, an ideal property of the compost to be used to increase soil fertility.

Compost EC typically ranges from values between 1 and 10 m mho/cm (Tambone et al. 2013). The EC of the test compost was 2.11 m mhos/cm. Reports suggested that the application of *E. eugeniae* increased the EC value and permitted a significant increase in salts such as Ca (42.3–91.6%), K (93.8–235%), Mg (25.4–84.6%) and P (37.1–129%) due to degradation of organic matter, which is important in delivering essential minerals to the plant (Lim et al. 2011). Janakrinam and Sridevi (2010) reported an increase of EC content from 7.39 to 10.77 m mho/cm during MSW composting.

The organic matter content of the test compost was 18.06% lesser than that of the control compost (21.17%) after 28 days. Organic matter content decreases with composting as the degrading organisms utilize the organic carbon for energy generation and biosynthesis. This reduction of organic matter in the test compost explains the reduction in volume and mass of the compost.

A C:N ratio of 30:1 before composting is considered desirable. Higher C:N ratio causes initial N immobilization and very low C:N ratio causes overabundance of N, resulting in accumulation of ammonia. The C:N ratio decreases during composting, as organic carbon is converted to CO_2 by degraders and lost in its gaseous phase, whereas the nitrogen is converted to other forms by organisms. The C:N ratio of the test compost was reported to be 22:1, suggesting it to be effective for the enhancement of soil fertility. Jeyapriya and Saseetharan (2007) reported a reduction in the C:N ratio of MSW from 30:1 to the final compost C:N ratio of 19:1. Jilani (2007) observed a huge reduction in the C:N ratio from 40 to 26 during MSW composting (64). A final C:N ratio of less than 26:1 is considered suitable for the compost for its applicability to increase the fertility of soil.

There was a slight increase in the percentage of ###phosphorus, potassium and sulphur in the test compost as compared to fresh waste. Phosphorus increased from 0.32 to 0.50%, potassium increased from 0.39 to 0.55% and sulphur increased from 0.02 to 0.10%. Similar results were obtained by Janakrinam and Sridevi (2010), where phosphorus increased from 0.29 to 0.40% and potassium increased from 1.61 to 1.78%.

Conclusion

We can conclude from the study that among the two consortia prepared for organic waste degradation, consortium 2 had the highest degrading capability and the compost produced after 30 days of degradation had a dark colour and lacked foul smell. Consortium 2 was consistent in degrading a greater quantity of waste (30 kg) and can be employed for composting a large quantity of wastes in future. Compost analysis of 30 kg wastes inoculated with consortium 2 showed C:N of 22:1 and increased percentage of K, P and S which are required for enhancement of soil fertility. Therefore, we can conclude from this study that consortium 2 can serve as a biological tool for removal of organic solid wastes from the environment and the compost generated from the degradation can be applied to increase the fertility of the soil.

References

Alexander M (1977) Introduction to soil microbiology, 2nd edn. Wiley, New York

AWPA (1986) American Wood Preservers Association Standards, Books of standards, Stevensville, Maryland, USA

Bandara NJ, Hettiaratchi JP, Wirasinghe SC, Pilapiiya S (2007) Relation of waste generation and composition to socio-economic factors: a case study. Environ Monit Assess 135(1):31–39. doi:10.1007/s10661-007-9705-3

Bezdicek D, Fauci M, Schwab G (2002) Impacts and management of soil acidity under direct seed system—effects of beneficial soil microbes and soil fauna. Direct seed conference. Department of crop and soil sciences, WSU, Pullman, WA

Meuser H, Grewal KS, Anlauf R, Malik RS, Narwal RK, Saini J (2011) Physical composition, nutrients and contaminants of typical waste dumping sites. Am J Environ Sci 7(1):26–34. doi:10.3844/ajessp.2011.26.34

Mirdamadian SH, Khayam-Nekoui SM, Ghanavati H (2011) Reduce

of fermentation time in composting process by using a special microbial consortium. World Acad Sci Eng Technol 52:475–479
Bremner JM (1960) Determination of nitrogen in soil by the Kjeldahl method. J Agric Sci 55(1):11–33. doi:10.1017/S0021859600021572
Elango D, Thinakaran N, Pannerselvam P, Sivanesan S (2009) Thermophilic composting of municipal solid waste. Appl Energy 86:663–668. doi:10.1016/j.apenergy.2008.06.009
Forastiere F, Badaloni C, Hoogh KD, Kraus MK, Martuzzi M, Mitis F (2011) Health impact assessment of waste management facilities in three European countries. Environ Health 10:53. doi:10.1186/1476-069X-10-53
Gautam SP, Bundela PS, Pandey AK, Jamaluddin AM (2010) Optimization of the medium for the production of cellulase by the *Trichoderma viride* using submerged fermentation. Int J Environ Sci 1(4):656–665
Gopinath LR, Christy PM, Mahesh K, Bhuvaneshwari R, Divya D (2014) Identification and evaluation of effective bacterial consortium for efficient biogas production. J Environ Sci Toxic Food Technol 8(3):80–86
Gouveia N, Prado RR (2009) Health risks in areas close to urban solid waste landfill sites. Rev de Saude Publica 44(5):1–8
Gupta B, Arora SK (2016) Municipal solid waste management in Delhi—the capital of India. Int J Innov Res Sci Eng Technol 5(4):5130–5138. doi:10.15680/IJIRSET.2016.0504106
Hibbing ME, Fuqua C, Parsek MR, Peterson SB (2010) Bacterial competition: surviving and thriving in the microbial jungle. Nat Rev Microbiol 8(1):15–25. doi:10.1038/nrmicro2259
Higgins Matthew J, Chen YS, Murthi DP, Sudhir N, Nick A, John T (2006) Cycling of volatile organic sulfur compounds in anaerobically digested biosolids and its implications for odours. Waste Environ Res 78(3):243–252
Jackson ML (1973) Soil Chemical Analysis, 1st edn. Prentice Hall of India Pvt. Ltd., New Delhi, India
Janakrinam T, Sridevi K (2010) Conversion of waste into wealth: a study in solid waste managent. Electron J Chem 7(4):1340–1345. doi:10.1155/2010/549185
Jeyapriya SP, Saseetharan MK (2007) Study on municipal solid waste reuse characteristics and leachate samples of Coimbatore city. J Nat Environ Pollut Technol 6(1):149–152
Jilani S (2007) Municipal solid waste composting and its assessment for reuse in plant production. Pak J Bot 39(1):271–277
Lim PN, Wu TY, Sim EYS, Lim SL (2011) The potential reuse of soybean husk as feed stock of *Eudrilus eugeniae* in vermicomposting. J Sci Food Agric 91(14):2637–2642. doi:10.1002/jsfa.4504
Lim SL, Lee LH, Wu TY (2016) Sustainability of using composting and vermicomposting technologies for organic solid waste biotransformation: recent overview, greenhouse gases emissions and economic analysis. J Clean Prod 111:262–278. doi:10.1016/j.jclepro.2015.08.083
Meuser H (2010) Contaminated urban soils, 1st edn. Springer, Verlag, UK, p 318
Pillai S, Peter AE, Sunil BM, Shrihari S (2014) Soil pollution near a municipal soil waste disposal site in India. International conference on biological, civil and environmental engineering (BCEE-2014), pp 148–152, Dubai (UAE), 17–18 Mar 2014. doi:10.15242/IICBE.C0314080
Raju EV, Divakar G (2013) Production of pectinase by using *Bacillus circulans* isolated from dump yards of vegetable wastes. Int J Pharm Sci Res 4(7):2615–2622. doi:10.13040/IJPSR.0975-8232.4(7).2615-22
Rawat M, Singh UK, Mishra AK, Subramaniam V (2008) Methane emission and heavy metals quantification from selected landfill areas in India. Environ Monit Assess 67(74):137. doi:10.1007/s10661-007-9729-8
Saha A, Santra SC (2014) Isolation and characterization of bacteria isolated from municipal solid waste for production of industrial enzymes and waste degradation. J Microbiol Exp 1(1):1–8. doi:10.15406/jmen.2014.01.00003
Sarkar P, Meghvanshi M, Singh R (2011) Microbial consortium: a new approach in effective degradation of organic kitchen wastes. Int J Environ Sci Dev 2(3):170–174
Shamaini K, Fauzia SH (2014) Enhanced vermicomposting for combination of organic waste through subsequent treatment with selected microorganisms. J Microbiol Res 4(2):54–67
Singh A, Sharma S (2003) Effect of microbial inoculation on mixed solid waste composting, vermicomposting and plant response. Comp Sci Util 11(3):190–199
Tambone F, Genevini P, Adani F (2013) The effects of short-term compost application on soil chemical properties and on nutritional status of maize plant. Compost Sci Util 15(3):176–183
Tejada M, Gonzalez JL (2007) Influence of organic amendments on soil structure and soil loss under stimulated rain. Soil Tillage Res 93(1):197–2305
Tom AP, Pawels R, Haridas A (2016) Biodrying process: a sustainable technology for treatment of municipal solid waste with high moisture content. Waste Manag 49:64–72. doi:10.1016/j.wasman.2016.01.004
Toth S, Prince AL (1949) Estimation of cation-exchange capacity and exchangeable Ca, K, and Na contents of soils by flame photometric techniques. Soil Sci 67(6):439–446
van Hamme JD, Singh A, Ward OP (2003) Recent advances in petroleum microbiology. Microbiol Mol Biol Rev 67(4):503–549. doi:10.1128/MMBR.67.4.503-549.2003
Visilind PA, Spinosa L (2001) Production and Regulations. In: Spinosa L, Visilind IA (eds) Sludge into biosolids: processing, disposal, utilization. IWA Publishing, London, pp 3–18
Walkey A, Black IA (1934) An examination of the Degtjareff method for determining soil organic matter and a proposed modification of the chromic acid titration method. Soil Sci 37(1):29–38
Wu TY, Lim SL, Lim PN, Shak KPY (2014) Biotransformation of biodegradable solid wastes into organic fertilizers using composting or/and vermicomposting. Chem Eng Trans 39:1579–1584. doi:10.3303/CET1439264

21

Rapid production of organic fertilizer from degradable waste by thermochemical processing

C. R. Sudharmaidevi[1] · K. C. M. Thampatti[1] · N. Saifudeen[1]

Abstract
Purpose Chemical decomposition was studied as a potential method for the rapid conversion of waste to organic fertilizer.
Methods Chemicals were screened, and process parameters were optimized. The physicochemical properties, phytotoxicity, and manurial efficiency of the product were assessed. A prototype machine was fabricated for the operation.
Results Chemical treatment of ground fresh waste with HCl (0.25 N, 50 ml kg^{-1}) for 30 min followed by KOH (0.5 N,100 ml kg^{-1}) for 30 min at 100 °C, and ambient pressure yielded a product that could be used in place of conventional organic manure. Only 8–14 h were required to complete the entire process. No by-product or leachate was produced. The quality of the product was comparable to that of conventional composts, except for the absence of microorganisms. The fortified organic fertilizer enhanced the yield of vegetables in pot trials. The process and the prototype machine were found beneficial by a public evaluation.
Conclusions The new thermochemical waste processing method provides a quick and sustainable solution for hygienic waste disposal and the production of organic fertilizer.

Keywords Quick recycling · Organic fertilizer · Pollution reduction · Thermochemical · Waste processing

✉ C. R. Sudharmaidevi
sudharmaidevi@gmail.com

1 Department of Soil Science and Agricultural Chemistry, College of Agriculture, Vellayani Kerala Agricultural University, Trivandrum, Kerala, India 695 522

Introduction

Scientific and hygienic waste disposal is a serious concern (Abu Qdais and Al-Widyan 2016) in developing countries, especially in urban areas, where the population density is high and the availability of land for waste processing and disposal is limited. The total municipal solid waste (MSW) generated in urban India is estimated to be 68.8 million tons per year or 188,500 tons per day (Annepu 2012). The current infrastructure available in the country is insufficient for the efficient handling and disposal of this huge quantity of waste on a daily basis (Baishya et al. 2016). Collected waste is often dumped in open land or used for landfilling. This leads to severe environmental hazards (Du et al. 2014). Due to the scarcity of land, existing dumping yards are sometimes overfilled. Although incineration is increasingly being used for waste disposal (Dube et al. 2014), it cannot be advocated widely due to the associated toxic gas emissions (Kim and Kim 2010). An analysis of solid waste in the major cities of India revealed that $\sim 51\%$ of it is composed of degradable organic material (Annepu 2012). Although this organic waste cannot be utilized as fuel because of its high moisture and low calorific value (Rana et al. 2013), the plant nutrient content makes it ideal for recycling as manure for crop production. The conversion of solid waste to organic fertilizer is a desirable option, in light of reports that severe depletion of soil organic matter is a major cause of declining crop productivity (Lal 2015). The use of organic fertilizers not only reduces the quantity of the organic fraction that ends up in landfills (Lim et al. 2016), but also reduces the use of inorganic fertilizers (Lim et al. 2015). Aerobic and anaerobic composting practices are popular in the country, but they are slow processes, limiting the turnover rate of waste recycling and disposal. Waste has to be collected from sources and transported to the composting yards, and large-scale dumping near the units for long periods leads to severe environmental pollution. In addition, improper separation of the inert materials, such as plastics and glass, may degrade the quality of the final compost for agricultural purposes (Wu et al. 2014). In this context, a quick method for the conversion of source-segregated waste to organic fertilizer will help to overcome these problems.

Hydrolytic degradation of biomass employing acids/alkalis is a popular practice to obtain fermentation feedstock for the production of ethanol and other chemicals (Sharma et al. 2013; Adela et al. 2014; Elemike et al. 2015; Zhu et al. 2015). Organic materials can be subjected to rapid hydrolytic degradation by treatment with acids and alkalis, and the reaction can be further accelerated by heat (Sharma et al. 2013; Zhu et al. 2015). However, the use of chemical digestion for the rapid conversion of waste to organic fertilizer has not been attempted. Therefore, a study was conducted at Kerala Agricultural University, India, from 2009 to 2014 to investigate the possibility of chemical decomposition as a quick method for the production of organic fertilizer from degradable waste.

Materials and methods

Waste: collection and characterization

Degradable waste, segregated at the source, was collected from the canteens, student hostels and agricultural farms of the College of Agriculture, Kerala Agricultural University, Vellayani, India. The composition of the waste was analyzed, and a representative 1 kg sample was taken from the collected waste, air-dried, oven-dried at 60 °C, and stored in an air-tight container for the determination of its physical and chemical properties.

Laboratory screening trials

Screening trials were conducted to select the most suitable chemical combination and optimum conditions for the degradation of organic waste. Two acids, i.e., HCl (Thys and Aires 2013; Elemike et al. 2015) and H_2SO_4 (Adela et al. 2014; Zhu et al. 2015), and an alkali, KOH (Bensah and Mensah 2013; Sharma et al. 2013), which are commonly used for hydrolytic degradation of biomass, were selected. Two concentrations (0.25 and 0.5 N) of the chemicals and two temperatures (100 and 150 °C) (Maeda et al. 2013; Elemike et al. 2015; Zhu et al. 2015) were tested. Reaction times of 30 min (15 min each for acid and alkali) and 60 min (30 min for the acid and alkali) were adopted (Maeda et al. 2013; Elemike et al. 2015). A total of 32 combinations were tested. The collected waste was homogenized, and 500 g of the fresh sample was transferred to a 2 L glass beaker and placed in a custom-made thermostatic digester with an automatic stirrer. Acid was added at a rate of 50 ml kg^{-1} waste and was allowed to react for the specified time. Then, the alkali was added and allowed to react for the remaining time. The quantity of alkali was adjusted (50 ml kg^{-1}, which was increased to 100 ml kg^{-1}) to obtain a neutral endproduct. The pH, color, electrical conductivity (EC), and total organic carbon (TOC) were recorded.

Phytotoxicity tests

The digest from the selected treatments was subjected to phytotoxicity bioassays in the laboratory using 1:10 (w/v) aqueous extract (Zucconi et al. 1985; Paradelo et al. 2010). Twenty seeds of *Amaranthus* sp. and ten seeds each of *Vigna unguiculata* (L.) Walp and *Abelmoscus esculentus* (L.) Moench were sown in Petri dishes lined with filter paper. The selection of the plant species was performed as described in OECD-OCDE (Organization for Economic Co operation and Development (2006). Distilled water was used as a control. Three replications of each species were kept in an incubation chamber at 28 °C for 5 days. The germination percentage and root length were recorded, and the germination index was calculated (Zucconi et al. 1985; Gariglio et al. 2002). Germination tests were also conducted using pro trays (OECD-OCDE (Organization for Economic Co operation and Development 2006). The potting mixture used to fill the trays was prepared by mixing sterilized coir pith and the digest in a 2:1 ratio. A growth medium of coir pith and vermicompost in a ratio of 1:1, which is the standard medium used on the Agricultural Farm of the college, was used as the control. The germination index was calculated according to Asgharipour and Sirousmehr (2012). The best treatment from these tests was selected for further evaluation.

Evaluation of the selected method with existing bioconversion methods

A comparative evaluation of the new process was performed with the existing composting methods to draw conclusions about the efficiency of the process and quality of the product. In this experiment, the selected thermochemical method, (1) mechanical disintegration and thermochemical decomposition at 100 °C for 60 min (MTC), was compared with the most commonly practiced methods in India, (2) mechanical disintegration followed by microbial composting (MMC), (3) mechanical disintegration followed by conventional aerobic composting (MCC), (4) vermicomposting (VC), and (5) conventional aerobic composting (CC). A 5 kg fresh waste sample was used in each treatment, and the experiment was performed in triplicate. The period required by each method for composting and the quality of the product obtained were studied. The maturity parameters prescribed for composts produced by bioconversion were taken as a reference in the absence of specifications for thermochemical products. Since the digest was free of microorganisms due to the intense physicochemical treatment, only the physical and chemical properties, such as color, odor, pH, EC, TOC, and C:N ratio, were assessed. After fortification with calculated doses of essential plant nutrients to make it balanced, the new product was denoted fortified organic fertilizer (FOF).

Evaluation of crop performance with FOF

The manurial efficiency of FOF was compared with that of commonly used manures, i.e., farm yard manure (FYM) and vermicompost (V), in pot culture experiments using vegetables (chili, tomato, and brinjal) as test crops. The experiment used a completely randomized design with four treatments and four replicates. Nitrogen, phosphorus, and potassium were applied to pots receiving FYM and V as per standard recommendations of Kerala Agricultural University. The growth and yield of the plants were recorded.

Physicochemical properties of the thermochemical digest with and without fortification

The moisture content of the samples was determined gravimetrically. The pH and EC was measured as described by Sasaki et al. (2003). pH was measured using a pH meter (Cyber Scan PC510, EuTech Instruments, Singapore), and EC was measured using an EC–TDS analyzer (CM183, Elico, India). TOC and N were determined using a CHNS analyzer (vario El cube, elementar, Germany). After nitric-perchloric (4:1) wet digestion (Abbruzzini et al. 2014) of the samples, total P was analyzed by the vanadomolybdic yellow color method (Greenberg et al. 1992). K, Ca, and Mg were analyzed by atomic absorption spectrophotometer (novAA 350 BU, AnalytkjenaAG, Germany), the heavy metals Pb, Cd, Ni, Cr, and As were analyzed by inductively coupled plasma optical emission spectrometer (ICP OES Optima 8000, Perkin Elmer Inc., USA), and B was analyzed by the Azomethine-H method (Roignavarro et al. 1996).

Fabrication of a prototype machine for the thermochemical processing of waste and the public evaluation trial

An automated machine with a 20 kg capacity was designed and fabricated to implement the new technology. A larger capacity machine (100 kg) was installed in June 2015, in a residential colony in Trivandrum, Kerala, India, for public evaluation.

Economic and statistical analyses

Since the unit installed in the residential colony represents the actual waste processing situation, cost analysis was performed based on the expenditure incurred and income generated by the unit. Statistical analyses of the data of the crop growth experiment were conducted using two-way analysis of variance (ANOVA). The F values for treatments were compared with the table values. If the values were significant, the CD values at the 5% significance level were calculated to compare the means and to interpret the results. For all other data, the means and standard deviations were calculated.

Results and discussion

The degradable waste collected from the campus had the average composition listed in Table 1. Food waste was the dominant fraction, contributing 78%. The raw waste was acidic (pH 4.2), because Indian cookery employs vegetables, spices, condiments, and other ingredients that are acidic in nature. Biowastes are universally reported to be acidic (Sundberg et al. 2013; Junoh et al. 2015). The moisture content was high (82.5%), and the color varied with the components. A foul smell developed within 24 h due to the decomposition of food particles and the release of volatile organic compounds and gases (Du et al. 2014). The mean organic carbon and nitrogen contents were 34.5 and 1.08%, resulting in a high C:N ratio of 31.94, which restricted its direct application to soil for crop production. The addition of organic materials with C:N ratios >30 affects plant growth due to immobilization of the available N pool by soil microbes, restricting N availability to plants (Ahmad et al. 2006).

Table 1 Composition of waste collected from the campus

Sl. no.	Components	Average %
1	Vegetable peelings	46.5
2	Food waste	18.0
3	Fish waste	5.5
4	Fruit waste	8.0
5	Garden litter	8.0
6	Agricultural wastes	12.0
7	Inert materials	2.0

Values are average of 12 lots

Selection of the process conditions

Laboratory screening aimed to develop a process for the fast and environmentally friendly conversion of stubborn and offensive organic waste materials into a product that can be conveniently used to promote the growth of crop plants. Since MSW is composed of plant and animal wastes, both acids and an alkali were screened. Although NaOH is (Maeda et al. 2013; Adela et al. 2014) an efficient hydrolytic agent for the degradation of biomass, it was not considered due to the undesirable effects it has on soil characteristics and plant growth (Suarez et al. 2006; Qadir et al. 2007). A reaction time of 30 min (15 + 15) was insufficient, since some of the original waste particles were identifiable after 30 min, indicating incomplete degradation. Therefore, 60 min (30 + 30) was selected. Alkali at a rate of 50 ml kg^{-1} was also not sufficient, since all the treatments except one had a pH less than 7 upon completion of the reaction time of 60 min, and the color ranged from greenish-brown to black. The physical characteristics, such as color, odor, and temperature, provided a general idea of the decomposition stage (Bernal et al. 2009). The acidic reaction might be due to the inherent acidic nature of the raw waste used for the study. Some components in the waste exhibited pH values <4, which required the addition of a higher quantity of alkali to attain a neutral pH at the end of the reaction. When the experiment was repeated with the addition of a higher quantity of alkali (100 ml kg^{-1}), the desired pH and color were recorded in several treatments. The EC and TOC were also recorded. Although all the treatments effectively reduced the organic carbon content to levels comparable to common organic manure, only those with a brownish-black color, viscous consistency, safe EC, permissible TOC, and near neutral pH at the end of the processing time were selected for further tests for suitability as organic fertilizers. Four treatments, i.e., HCl (0.25 N) + KOH (0.5 N) and H_2SO_4 (0.25 N) + KOH (0.5 N) at two temperatures (100 and 150 °C), were selected (Table 2). The pH of an organic fertilizer is crucial, because it alters the pH of the soil to which it is applied (Yao et al. 2007; Santillan et al. 2014). The soil root zone, especially the rhizosphere, is the active zone of microbial and root activity (Lagos et al. 2015). The pH of the rhizosphere zone is, therefore, critical in deciding microbial activity (Wang et al. 2016), which in turn regulates the innumerable chemical reactions involved in nutrient transformations and uptake by plant roots. The ideal pH range for active growth and development of plants and microbes is 6.5–7.5. The EC in these treatments was also safe for use as organic fertilizers. Values <3.5 dS m^{-1} are considered safe for soil application (Chen Yiqun et al. 2014). EC is a measure of soluble salts. A high salt content is not desirable, because very high concentrations of soluble salts damage plants through specific ion effects and plasmolysis (Turan et al. 2010; Chauhan et al. 2016). TOC is an important parameter with respect to the maturity of the compost/manure, because a decrease in TOC is desired to attain the optimum C:N ratio for soil application. The main aim of composting is to reduce the organic carbon content to achieve a C:N ratio of 10:1–25:1, which is ideal for beneficial microorganisms. Approximately 36–56% of the initial TOC is lost at the end of conventional composting (Nada 2015). Although these desired parameters were achieved by digestion at both temperatures, treatments at the lower temperature of 100 °C were selected due to the energy savings. Hence, the digest from two treatments, i.e., HCl (0.25 N) + KOH (0.5 N) and H_2SO_4 (0.25 N) + KOH (0.5 N) at 100 °C, were selected for the phytotoxicity tests.

Germination bioassay

The maturity and safety of conventional composts are generally evaluated by biological methods, such as germination tests (Paradelo et al. 2010) and plant growth analysis or a combination of the two (Kapanen and Itavaara 2001). These tests realistically and efficiently evaluate the compatibility of manure with plants, because they assess the combined effects of several phytotoxic factors present in the manure (Emino and Warman 2004). According to the UK Composting Association, mature compost is one that does not have a negative effect on seed germination or plant growth (Gilbert et al. 2001). Since in the present study, the digest was intended for use in place of compost

Table 2 Selected treatments from laboratory screening experiment

Sl. no.	Combination of chemicals	pH	EC (dS m^{-1})	TOC (%)
A	At 100 °C, 60 min			
1.	HCl (0.25 N) + KOH (0.5 N)	6.85 (±0.35)	1.91 (±0.13)	23.93 (±0.37)
2.	H_2SO_4 (0.25 N) + KOH (0.5N)	7.15 (±0.23)	2.10 (±0.06)	23.33 (±0.47)
B	At 150 °C, 60 min			
3.	HCl (0.25 N) + KOH (0.5 N)	6.85 (±0.23)	2.01 (±0.15)	23.10 (±0.05)
4.	H_2SO_4 (0.25 N) + KOH (0.5 N)	6.9 (±0.26)	2.03 (±0.04)	21.10 (±0.30)

(Mean ± SD, $n = 3$) (500 g ground waste, acid (50 ml kg^{-1}), and alkali (100 ml kg^{-1})

Table 3 Phytotoxicity tests with selected thermochemical treatments

Sl. no.		*Amaranthus* sp.	*Vigna unguiculata*	*Abelmoscus esculentus*	Recommended value
Germination index (%) in Petri dishes[a]					
1.	HCl (0.25 N) + KOH (0.5 N)	84.7 (±1.8)	89.0 (±2.0)	88.5 (±1.3)	>50 Emino and Warman (2004)
2.	H_2SO_4 (0.25 N) + KOH (0.5N)	60.4 (±2.5)	56.2 (±3.2)	62.6 (±1.6)	
Germination index (%) in pro trays[a]					
1.	HCl (0.25 N) + KOH (0.5N)	82.4 (±1.4)	84.2 (±3.3)	88.0 (±1.0)	Asgharipour and Sirousmehr (2012)
2.	H_2SO_4 (0.25 N) + KOH (0.5N	54.5 (±0.80)	63.3 (±2.7)	64.0 (±2.2)	

[a] Values are mean ± SD

Table 4 Important properties of composts produced by different methods

Compost	pH	EC (dS m^{-1})	TOC (%)	N (%)	C:N
MTC	6.8 (±0.35)	1.91 (±0.13)	23.93 (±0.37)	1.51 (±0.03)	15.79 (±0.48)
MMC	7.1 (±0.02)	1.77 (±0.09)	18.66 (±0.68)	1.36 (±0.01)	13.71 (±0.61)
MCC	7.7 (±0.16)	1.95 (±0.07)	21.83 (±0.25)	1.14 (±0.02)	19.16 (±0.32)
VC	8.4 (±0.11)	2.16 (±0.10)	16.8 (±0.35)	1.59 (±0.01)	10.58 (±0.15)
CC	7.6 (±0.32)	1.87 (±0.06)	21.16 (±0.25)	1.20 (±0.01)	17.54 (±0.26)

Values are mean ± SD

MTC mechanical disintegration followed by thermochemical decomposition, *MMC* mechanical disintegration followed by microbial composting, *MCC* mechanical disintegration followed by conventional aerobic composting, *VC* vermicomposting, *CC* conventional aerobic composting

for crop production, and phytotoxicity tests were essential. The results of the germination tests are presented in Table 3. Composts with a germination index (GI) ≥80% are considered free of phytotoxic compounds (Emino and Warman 2004). A GI <50% is considered immature and indicates high phytotoxicity (Zucconi et al. 1985; Emino and Warman 2004). The phytotoxicity of immature composts is attributed to the release of phenolic compounds (Madejon et al. 2001) and volatile organic compounds (Buss and Masek 2014) or the excess accumulation of salts (Salman et al. 2013). In the present experiment, both treatments had GI >50%, but the treatment of HCl (0.25 N) + KOH (0.5 N) at 100 °C and a 60 min reaction time had a GI value >80% in petri dishes and pro trays. Therefore, this treatment was selected as the best thermochemical method for processing the waste.

Waste processing efficiency of the selected method compared to existing methods

In the composting trial, the thermochemical treatment required only 8–14 h to produce a dried balanced organic fertilizer, including the time taken for sun-drying to attain moisture content suitable for packing. The MMC treatment took 19 days, and VC, MCC, and CC took 38, 54, and 90 days, respectively. In CC, the breakdown of complex and simple organic materials is accomplished through the activity of different types of aerobic microbes that become dominant during different stages of composting (Nada 2015); therefore, the method requires more time (Chan et al. 2016). In VC, the process is mediated by earthworms. They are voracious feeders and consume approximately their body weight per day and achieve higher mass reduction than conventional composting (Pattnaik and Reddy 2010). In MCC, the particle size of the waste material is reduced to provide a larger surface area for easy microbe access. Conventional composting can be made more efficient by size reduction (Mishra et al. 2003). In MMC, a consortium of microorganisms is inoculated to accelerate the composting process (Saravanan et al. 2013). Since all these processes are mediated through biological activity, they are slower than the thermochemical process. Chemical analysis of the composts (Table 4) showed that the quality of the organic fertilizer obtained using the new process was comparable to other composts. The pH of composts from CC, VC, and MCC was more alkaline than the thermochemical and microbial composts. In conventional composts and vermicomposts, the degradation process releases salts as a result of the mineralization of organic matter (Pattnaik and Reddy 2010). Compared to conventional composting, vermicomposting is more efficient for the reduction of the bulk dry matter of substrates; therefore, more soluble salts are released into the medium. Moreover, Ca salts, such as $CaCO_3$, are also synthesized and released into the medium by the activity of the calciferous glands in earthworms (Gago-Duport et al. 2008), making vermicomposts more alkaline than conventional composts. The release of exchangeable cations, such as Ca, Mg, and K, in vermicompost increases the EC of vermicompost (Tognetti et al. 2005). In contrast, in the thermochemical process, the stoichiometry of the acid–alkali reaction was adjusted to obtain a near neutral product that does not pose a threat to soil or plant growth. The TOC content of the thermochemical compost was higher than that of conventional composts. This is expected since in conventional composts, the loss of C as CO_2 occurs through microbial respiration during the degradative phases of composting (Nada 2015). However, this was not reflected in the C:N ratio, because the N content was also higher in this treatment. The longer duration of conventional composting may have resulted in the loss of some of the N already mineralized by leaching or volatilization (Chan et al. 2016), accounting for the lower values. The only limitation observed for the new method was the absence of microbes due to heating at 100°C. However, once the manure is applied to soil, it will stimulate microbial activity and improve soil structure, similar to any other organic matter (Zhao et al. 2016). Therefore, this could be treated as an advantage in terms of the absence of pathogenic organisms (FAI 2007).

Performance of vegetables with FOF

Although the germination tests showed that the FOF posed no threat to the germination of crop seeds, plant growth trials are essential to assess the efficacy and safety on a long-term basis. Germination tests provide an instant picture of phytotoxicity, and plant growth trials assess the prolonged effects of the compost (Zucconi and de Bertoldi 1987). In India, the most commonly used organic manure is FYM (Reddy et al. 2015). Since FOF is intended to be a substitute for FYM, its effect on plant growth is of the utmost importance. When test crops were raised in pots with FOF, the growth and yield attributes showed significant differences between treatments. An increase in yield was observed in the two treatments, where FOF was applied (Table 5). This may be due to the combined effect of the organic and mineral nutrients in FOF as a result of fortification. The mineral portion contributed an initial fast release of nutrients to the plants, followed by a slow release from the organic portion, ensuring nutrient availability for long periods. The beneficial effects of the combined application of organics and inorganics to enhance crop yield and soil health (Jing et al. 2016) have been reported. In the present experiment, the application of FOF at a rate of 5.54 t ha^{-1} produced the highest fruit yield, which was significantly superior to all other treatments. The yield

Table 5 Effect of FOF on growth and yield of vegetables

Treatments	Height (cm)	No. of branches	No. of fruits	Wt of fruit (g)	Yield	
					g $plant^{-1}$	t ha^{-1}
Chili						
$FYM_{20}+N_{75}P_{40}K_{25}$	40.1^{a}	18^{a}	50^{a}	7.9	395^{a}	10.95^{a}
$FOF_{2.77}$	46.6^{b}	25^{b}	66^{bd}	8.5	561^{b}	15.50^{b}
$FOF_{5.54}$	54.5^{c}	29^{c}	74^{c}	9.0	666^{c}	18.44^{c}
$V_{10}+N_{75}P_{40}K_{25}$	42.2^{d}	21^{d}	55^{d}	7.9	435^{d}	12.01^{a}
CD(0.05)	1.51	0.72	0.63	NS	16.21	1.08
Tomato						
$FYM_{20}+N_{75}P_{40}K_{25}$	58.2^{a}	3^{a}	45.1^{a}	20.2^{a}	911.0^{a}	25.23^{a}
$FOF_{2.77}$	68.2^{b}	8^{b}	46^{b}	25.6^{b}	1177.6^{b}	32.60^{a}
$FOF_{5.54}$	73.5^{c}	11^{c}	53^{c}	30.8^{c}	1632.4^{c}	45.20^{b}
$V_{10}+N_{75}P_{40}K_{25}$	55.8^{d}	4^{a}	45.7^{d}	21.5^{d}	982.5^{a}	27.20^{a}
CD(0.05)	1.06	1.27	0.21	0.13	96.82	8.73
Brinjal						
$FYM_{20}+N_{75}P_{40}K_{25}$	78.3^{a}	7^{a}	52^{a}	28^{a}	1449^{a}	40.16^{a}
$FOF_{2.77}$	102.2^{b}	12^{b}	59^{b}	38^{b}	2240^{b}	62.04^{ab}
$FOF_{5.54}$	106.4^{c}	15^{c}	67^{c}	42^{c}	2814^{c}	77.83^{b}
$V_{10}+N_{75}P_{40}K_{25}$	85.6^{d}	9^{a}	54^{d}	33^{d}	1781^{d}	49.30^{a}
CD(0.05)	3.21	2.13	1.42	2.36	239.14	23.63

FYM farm yard manure, *FOF* fortified organic fertilizer, *V* vermicompost in t ha^{-1}, *N, P and K* in kg ha^{-1}, *CD* critical difference, *NS* not significant

[a,b,c,d] Significant difference between data

increased by 68.6% compared to FYM and 58.5% compared to vermicompost. Tomato and brinjal also showed the same trend, with much greater yield increases (tomato: 79 and 71.3%; brinjal: 94.2 and 71.2%). The effects of FOF for increasing crop yield and improving the soil properties have been reported (Jayakrishna and Thampatti 2016; Leno et al. 2016). The yields of the vermicompost treatments, although lower than that of FOF, were higher than FYM. Vermicompost contains higher levels of nutrients than FYM (Venkatesha et al. 2015), in addition to secondary and micronutrients and growth stimulating hormones (Kundu et al. 2015), which contributed to enhanced growth and yield of plants. Thus, the results revealed that the FOF could act as a substitute for the combined use of organic manure and chemical fertilizer for crop production.

Details of the new process and characteristics of the product

In the new process, degradable waste segregated at the source was subjected to size reduction by wet grinding (Sun and Cheng 2002) to reduce the particle size to increase the specific surface area (Zhang et al. 2014). The waste was then allowed to react with HCl (0.25 N HCl, 50 ml kg^{-1} waste) followed by KOH (0.5 N, 100 ml kg^{-1}) for 30 min at 100 °C and ambient pressure. Thermal treatment helped the disintegration of the cell wall and membrane to solubilize the organic matter (Audrey et al. 2011). An initial drying step was performed by the addition of sterilized coir pith, wood charcoal or saw dust at the rate of 40 g kg^{-1} waste before completely drying the waste in the sun or in an electric drier. Since soils in general are depleted of minerals, fortification with calculated doses of essential plant nutrients was performed to create a balanced organic fertilizer. The sources of nutrients used for the fortification and their average nutrient contents are given in Table 6. A 1 kg sample of fresh waste would yield ~200–225 g of FOF. The product characteristics before and after fortification are presented in Table 7. The texture of the dried digest was lumpy, but the addition of coir pith made it free flowing. The reaction was neutral, EC within safe limit, and the C:N ratio was <20. A slight increase in TOC was observed after fortification, which might be due to the added coir pith. When applied to soil for crop production, organic fertilizers should improve the health of the soil and crop (Zhang et al. 2016). Hence, the manurial value of the added product is important. The nutrient content of the digest was comparable to those of ordinary composts in terms of primary major nutrients. The low contents of secondary and micronutrients were addressed by fortification with organic and inorganic sources (Table 6). The increase in the concentrations of S, Fe, Mn, and Cu in the FOF, although they were not added, may have come from the sources of the other nutrients. The contents of the heavy metals Pb, Cd, Ni, Cr, and As were far below the permissible limits (FAI 2007), indicating that the organic fertilizer was safe for use. Thus, the organic fertilizer produced by the new process (Fig. 1) can be widely used in agriculture to supply sufficient organic matter for the carbon-craving soil and balanced quantities of essential nutrients for plants. The fertilizer could also be used as a potting medium for container cultivation and as a growth medium for germinating seeds in pro trays (Jayakrishna and Thampatti 2016). There was no foul smell produced by the FOF, even when the raw waste was rotten and had a bad odor.

The process does not produce any type of effluent or pollutant, since the digest is utilized for the production of FOF. Therefore, the process does not require pollution control efforts. The new process provides a quick and sustainable solution for waste disposal and recycling of organic residue. These small-scale processing units prevent

Table 6 Sources used for fortification with their average nutrient contents

Properties	Groundnut cake	Rock phosphate	Wood ash	Zinc sulfate	Borax
N (%)	6.80	–	0.06	0.0005	–
P (%)	1.41	8.92	0.94	–	–
K (%)	1.14	–	12.2	–	–
Ca (%)	0.08	15.33	9.18	–	0.002
Mg (%)	0.34	0.21	1.42	0.11	–
S (%)	0.09	0.02	0.01	–	0.001
Fe (mg kg^{-1})	215.70	–	1.94	0.50	0.38
Mn (mg kg^{-1})	52.17	–	0.66	–	–
Cu (mg kg^{-1})	15.85	–	0.01	–	0.50
Zn	68.54 mg kg^{-1}	–	0.07 mg kg^{-1}	20.86%	–
B	17.50 mg kg^{-1}	–	–	–	10.5%
Mo (mg kg^{-1})	1.01	–	0.01	–	–

Table 7 Physical and chemical properties of thermochemical digest with and without fortification (mean ± SD)

Physical and chemical properties	Without fortification	With fortification	Recommended value (FAI 2007)
Color	Brownish black	Brownish black	Dark brown to black
Odor	Odorless	Odorless	Absence of foul odor
pH	Neutral	Neutral	6.5–7.5
EC (dS m^{-1})	1.91 (±0.13)	2.13 (±0.09)	<4
Texture	Lumpy	Free flowing	
Moisture (%)	18.5 ± 1.6	16 ± 1.1	15–25
TOC (%)	23.93 ± 0.37	24.42 ± 0.12	>16
C:N ratio	15.79 ± 0.48	12.16 ± 0.60	<20:1
Pathogens	Nil	Nil	Nil
Pb (mg kg^{-1})	1.38 ± 2.18	1.08 ± 0.76	<100
Cd (mg kg^{-1})	BDL	BDL	<5
Ni (mg kg^{-1})	3.6 ± 1.85	5.65 ± 2.59	<50
Cr (mg kg^{-1})	8.58 ± 2.80	11.68 ± 5.59	<50
As (mg kg^{-1})	BDL	BDL	<10
Manurial value			Rate and source of fortification
N (%)	1.51 ± 0.03	2.01 ± 0.11	0.5 (Ground nut cake)
P (%)	0.42 ± 0.22	0.95 ± 0.15	0.5 (Rock phosphate)
K (%)	0.66 ± 0.31	1.1 ± 0.20	0.5 (Wood ash)
Ca (mg kg^{-1})	255 ± 15	2700 ± 120	
Mg (mg kg^{-1})	300 ± 30	1200 ± 150	
S (mg kg^{-1})	299 ± 42	650 ± 122	–
Mn (mg kg^{-1})	10.6 ± 3.04	25.1 ± 3.2	–
Zn (mg kg^{-1})	8.25 ± 0.27	66.6 ± 4.5	50 (Zinc sulfate)
Cu (mg kg^{-1})	2.3 ± 0.20	14.5 ± 2.5	–
B (mg kg^{-1})	0.27 ± 0.08	0.55 ± 0.05	0.5 (Borax)

BDL below detectable level

the need for the collection and transportation of waste to centralized processing plants, resulting in huge transportation cost savings. The process could be scaled up and operated in factory mode in the case of huge demand.

The prototype machine

The prototype machine of 1.5 m^3 and a capacity of 20 kg (Fig. 2) occupy a floor space of only 2.5 m^2. It was specifically designed to meet the requirements of residential colonies/apartments in densely populated cities. The machine has two main parts: a grinding unit with a 2 HP motor and a reactor unit with temperature control and constant stirring facility. Segregated waste is dumped into the grinding unit. Uniformly homogenized mass is automatically transferred to the reactor unit and allowed to react with the chemicals at the specified temperature and time. The processed organic fertilizer can be collected by opening the valve in the reactor. The water requirement for the process is very low (only for grinding and cleaning), and the electricity consumption for the operation of the machine is only 3 kW h.

Fig 1 Organic fertilizer produced by the new process

Public evaluation of the new process

The unit is installed in a residential colony for public evaluation and is processing ~75–110 kg of waste daily, collected from 75 houses. Two women workers run the

Fig 2 Prototype waste processing machine (inside view)

unit. Degradable waste from each house is segregated and kept in separate buckets supplied to the residents for the purpose. Depending on the quantity of waste to be processed, one or two batches are operated per day. The unit works on all days except Sundays. The quantity of waste processed varies from 2000 to 2600 kg per month. The fortified manure produced by the unit is either sold as a value added, balanced organic fertilizer or as a potting medium for container cultivation at four times the price of ordinary organic manure.

Cost analysis

Since the method is new and remains to be popularized, it is essential to determine the economic feasibility for large-scale adoption. Hence, the cost analysis was performed on the basis of the actual operating costs and income generated during June 2015 to December 2015 by the waste processing unit installed in the residential colony in Trivandrum. The items of expenditure included collection charges, labor charges, and cost of materials needed for processing, including chemicals and other miscellaneous expenditures. The average total cost per month was US \$294 (US \$0.128 per kg fresh waste). The dried FOF was sold at a rate of US \$0.91 per kg. The average production per month was 575 kg FOF. Thus, the average total income generated per month was US \$523 (0.229 US \$ per kg fresh waste). Hence, a net profit of US \$229 per month (US \$0.101 per kg waste) is generated. This makes the new process even more attractive. Waste processing using this method is a solution for the safe disposal of waste and also for income generation for the unemployed.

Conclusions

Thermochemical processing of degradable waste with HCl (0.25 N, 50 ml kg^{-1}) for 30 min followed by KOH (0.5 N, 100 ml kg^{-1}) for 30 min at 100 °C and ambient pressure provides a quick and sustainable solution for hygienic waste disposal and the production of organic fertilizer. The method has the merits of processing waste on the day of generation at the source without the formation of by-products, thus avoiding environmental problems connected with open dumping and transportation to centralized processing plants. Large areas being used as landfill sites could be used for productive purposes. The process will lead to the large-scale generation of balanced organic fertilizer for soil application, which will enhance crop yields and improve soil health and environmental quality. This could be a replicable model of waste processing, especially for developing countries.

Acknowledgements The authors gratefully acknowledge the financial support provided for 2 years by the Department of Agricultural Research and Education, Indian Council of Agricultural Research, Ministry of Agriculture, Government of India for undertaking this project. Authors are thankful to Dr. Samuel Mathew, Professor and Head, Aromatic and Medicinal Plants Research Institute, Odakkali, Kerala and Dr. P. Sureshkumar, Professor and Head, Radiotracer Laboratory, College of Horticulture, Thrissur for their valuable suggestions in drafting the manuscript.

Conflict of interest The authors declare that they have no conflict of interest.

References

Abbruzzini TF, Silva CA, Andrade DA, Carneiro WJO (2014) Influence of digestion methods on the recovery of iron, zinc, nickel, chromium, cadmium and lead contents in 11 organic residues. Rev Bras Cienc Solo 38:166–176

Abu Qdais H, Al-Widyan M (2016) Evaluating composting and co-composting kinetics of various agro-industrial wastes. Int J Recycl Org Waste Agricult 5:273–280

Adela NB, Nasrin AB, Loh SK, Choo YM (2014) Bioethanol production by fermentation of oil palm empty fruit bunches pretreated with combined chemicals. J Appl Environ Biol Sci 4:234–242

Ahmad R, Khalid A, Arshad M, Zahir ZA, Naveed M (2006) Effect of raw (un-composted) and composted organic waste material on growth and yield of maize (*Zea mays* L.). Soil Environ 25(2):135–142

Annepu RK (2012) Sustainable solid waste management in India. M.S. Thesis, Columbia University

Asgharipour MR, Sirousmehr AR (2012) Comparison of three techniques for estimating phytotoxicity in municipal solid waste compost. Ann Biol Res 3(2):1094–1101

Audrey P, Julien L, Christophe D, Patrick L (2011) Sludge disintegration during heat treatment at low temperature: a better understanding of involved mechanisms with a multiparametric approach. Biochem Eng J 54(3):178–184

Baishya P, Mahanta DK, Singh SB (2016) Waste composting a sustainable practice. Int J Innov Res Sci Technol 2:537–543

Bensah EC, Mensah M (2013) Chemical pretreatment methods for the production of cellulosic ethanol: technologies and innovations. Int J Chem Eng 2013:21. doi:10.1155/2013/719607

Bernal MP, Alburquerque JA, Moral R (2009) Composting of animal manures and chemical criteria for compost maturity assessment. A review. Bioresour Technol 100(22):5444–5453

Buss W, Masek O (2014) Mobile organic compounds in biochar—a potential source of contamination—phytotoxic effects on cress seed (*Lepidium sativum*) germination. J Environ Manage 137:111–119

Chan MT, Selvam A, Wong JWC (2016) Reducing nitrogen loss and salinity during 'struvite' food waste composting by zeolite amendment. Bioresour Technol 200:838–844

Chauhan A, Rajput N, Kumar D, Kumar A, Chaudhry AK (2016) Effect of different salt concentration on seed germination and seedling growth of different varieties of oat (*Avena sativa* L.). Int J Inf Res Rev 3(7):2627–2632

Chen Yiqun Yu, Fang Liang Shengwen, Zongping Wang, Zizheng Liu, Ya Xiong (2014) Utilization of solar energy in sewage sludge composting: fertilizer effect and application. Waste Manage 34(11):2014–2021

Du Y, Feng H, Zhang K, Hu LF, Fang CR, Shen DS, Long YY (2014) Role of iron in H_2S emission behavior during the decomposition of biodegradable substrates in landfill. J Hazard Mater 272:36–41

Dube R, Nandan V, Dua S (2014) Waste incineration for urban India: valuable contribution to sustainable MSWM or inappropriate high-tech solution affecting livelihoods and public health? Int J Environ Technol Manag 17(2/3/4):199–214

Elemike EE, Oseghale OC, Okoye AC (2015) Utilization of cellulosic cassava waste for bio-ethanol production. J Environ Chem Eng 3:2797–2800

Emino ER, Warman PR (2004) Biological assay for compost quality. Compost Sci Util 12(4):342–348

FAI (2007) The Fertiliser (Control) Order 1985. The Fertiliser Association of India, New Delhi

Gago-Duport L, Briones MJI, Rodríguez JB, Covelo B (2008) Amorphous calcium carbonate biomineralization in the earthworm's calciferous gland: pathways to the formation of crystalline phases. J Struct Biol 162:422–435

Gariglio NF, Buyatti MA, Pilatti RA, Gonzalez Rossia DE, Acosta MR (2002) Use of a germination bioassay to test compost maturity of willow (*Salix* sp.) sawdust. NZ J Crop Hortic Sci 30:135–139

Gilbert EJ, Riggle DS, Holland FD (2001) Large scale composting: a practical manual for the UK. The Composting Association, Northamptonshire

Greenberg AE, Clesceri LS, Eaton AD (1992) Standard methods for the examination of water and wastewater, 18th edn. American Public Health Association, Washington

Jayakrishna, Thampatti KCM (2016) Standardisation of growth medium based on thermo chemical digest produced from degradable solid waste by rapid conversion technology. Int J Appl Pure Sci Agric 2(10):76–80

Jing D, Wang M, Zhang H, Li S (2016) Effects of chicken manure co-applied with nitrogen fertilizer on soil organic carbon oxidation stability and carbon pool management index in kidney bean/maize rotation soil. Trans Chin Soc Agric Mach 47(8):192–200

Junoh H, Palanisamy K, Yip CH, Pua FL (2015) Optimization of NaOH thermo-chemical pretreatment to enhance solubilisation of organicfood waste by response surface methodology. Intl J Chem Mol Nucl Mater Metall Eng 9(12):1282–1288

Kapanen A, Itavaara M (2001) Ecotoxicity tests for compost applications. Ecotoxicol Environ Saf 49:1–16

Kim MH, Kim JW (2010) Comparison through a LCA evaluation analysis of food waste disposal options from the perspective of global warming and resource recovery. Sci Total Environ 408(19):3998–4006

Kundu S, Chakraborty S, Roy D, Ghosh B, Dutta P (2015) Studies on organic nutrition in improving yield and quality of ber cv. BAU Kul-1. J Crop Weed 11(2):14–18

Lagos L, Maruyama F, Nannipieri P, Mora ML, Ogram A, Jorquera MA (2015) Current overview on the study of bacteria in the rhizosphere by modern molecular techniques: a mini—review. J Soil Sci Plant Nutr 15:504–523

Lal R (2015) Soil carbon sequestration in agro ecosystems of India. J Indian Soc Soil Sci 63(2):125–143

Leno N, Sudharmaidevi CR, Meera AV (2016) Fertility evaluation and manurial effect of organic manure produced from degradable solid waste by rapid conversion technology. Adv Life Sci 5(11):4433–4436

Lim SL, Wu TY, Lim PN, Shak KPY (2015) The use of vermicompost in organic farming: overview, effects on soil and economics. J Sci Food Agric 95(6):1143–1156

Lim SL, Lee LH, Wu TY (2016) Sustainability of using composting and vermicomposting technologies for organic solid waste biotransformation: recent overview, greenhouse gases emissions and economic analysis. J Clean Prod 111:262–278

Madejon E, Burgos P, Murillo JM, Cabrera F (2001) Phytotoxicity of organic amendments on activities of select soil enzymes. Commun Soil Sci Plant Anal 32:2227–2239

Maeda RN, Barcelos CA, Anna LMS, Pereira N Jr (2013) Cellulase production by *Penicillium funiculosum* and its application in the hydrolysis of sugarcane bagasse for second generation ethanol production by fed batch operation. J Biotechnol 163:38–44

Mishra RV, Roy RN, Hiraoka H (2003) On-farm composting methods. Food and Agricultural Organisation of the United Nation, Rome

Nada WM (2015) Stability and maturity of maize stalks compost as affected by aeration rate, C/N ratio and moisture content. J Soil Sci Plant Nutr 15(3):751–764

OECD-OCDE (Organization for Economic Co operation and Development (2006) OECD guidelines for the testing of chemicals test no. 208: terrestrial plant test: seedling emergence and seedling growth test. OECD Publishing, Paris

Paradelo R, Villada A, Gonzalez D, Barral MT (2010) Evaluation of the toxicity of heavy metals and organic compounds in compost by means of two germination–elongation tests. Fresenius Environ Bull 19:956–962

Pattnaik S, Reddy MV (2010) Nutrient status of vermicompost of urban green waste processed by three earthworm species—*Eisenia fetida*, *Eudrilus eugeniae*, and *Perionyx excavates*. Appl Environ Soil Sci 2010:13. doi:10.1155/2010/967526

Qadir M, Oster JD, Schubert S, Noble AD, Sahrawat KL (2007) Phytoremediation of sodic and saline-sodic soils. Adv Agron 96:197–247

Reddy KS, Mohanty M, Rao DLN, Singh M, Rao AS, Pandey M, Blamey FPC, Dalal RC, Dixit SK, Menzies NW (2015) Nutrient mass balances and leaching losses from a farmyard manure pit in Madhya Pradesh. J Indian Soc Soil Sci 63(1):64–68

Roignavarro AF, Lopez FJ, Hernandez F (1996) Application of the azomethine-H method to the determination of boron in work-place atmospheres from ceramic factories. Fres J Anal Chem 356:103–106

Salman G, Hameed A, Alatar A, Hegazy A, Khan MA (2013) Seed germination ecology of *Cyperus arenarius*—a sand binder from Karachi coast. Pak J Bot 45(2):493–496

Santillan MY, Moreno FP, Garcia FP, Sandooval AAO (2014) Effect of the application of manure of cattle on the properties chemistry of soil in Tizayuca, Hidalgo. M Int J Appl Sci Tech 4(3):67–72

Saravanan P, Kumar SS, Ajithan C (2013) Eco-friendly practice of utilization of food wastes. Int J Pharm Sci Innov 2013:14–17

Sasaki N, Suehara KI, Kohda J, Nakano Y, Yano T (2003) Effects of CN ratio and pH of raw materials on oil degradation efficiency in a compost fermentation process. J Biosci Bioeng 96:47–52

Sharma R, Palled V, Sharma-Shivappa RR (2013) Potential of potassium hydroxide pretreatment of switchgrass for fermentable sugar production. Osborne J Appl Biochem Biotechnol 169(3):761–772

Suarez DL, Wood JD, Lesch SM (2006) Effect of SAR on water infiltration under a sequential rain–irrigation management system. Agr Water Manage 86(1–2):150–164

Sun Y, Cheng J (2002) Hydrolysis of lignocellulosic materials for ethanol production: a review. Bioresour Technol 83(1):1–11

Sundberg C, Yu D, Franke-Whittle I, Kauppi S, Smars S, Insam H, Romantschuk M, Jonsson H (2013) Effects of pH and microbial composition on odour in food waste composting. Waste Manag 33(1):204–211

Thys RCS, Aires AG (2013) The effect of acid hydrolysis on the technological functional properties of pinhão (*Araucaria brasiliensis*) starch. Ciênc Tecnol Aliment Campinas 33:89–94

Tognetti C, Laos F, Mazzarino MJ, Hernandez MT (2005) Composting vs. vermicomposting: a comparison of end product quality. Compost Sci Util 13(1):6–13

Turan MA, Elkarim AHA, Taban A, Taban S (2010) Effect of salt stress on growth and ion distribution and accumulation in shoot and root of maize plant. Afr J Agric Res 5(7):584–588

Venkatesha MM, Krishnamurthy N, Tuppad GB, Venkatesh KT (2015) Yield, economics and nutrient uptake of aerobic rice cultivars as influenced by INM practices. Res Environ Life Sci 8(1):113–118

Wang J, Xue C, Song Y, Wang L, Huang Q, Shen Q (2016) Wheat and rice growth stages and fertilization regimes alter soil bacterial community structure, but not diversity. Front Microbiol 7:1207

Wu TY, Lim SL, Lim PN, Shak KPY (2014) Biotransformation of biodegradable solid wastes into organic fertilizers using composting or/and vermicomposting. Chem Eng Trans 39:1579–1584

Rana R, Ganguly R, Gupta AK (2013) An assessment of solid waste management system in Chandigarh city, India. Electron J Geotech Eng 20(Bundle 6):1547–1572

Yao LX, Li GL, Tu SH, Gavin S, He ZH (2007) Salinity of animal manure and potential risk of secondary soil salinization through successive manure application. Sci Total Environ 383:106–114

Zhang C, Su H, Baeyens J, Tan T (2014) Reviewing the anaerobic digestion of food waste for biogas production. Renew Sust Energ Rev 38(1215):383–392

Zhang P, Chen X, Wei T, Yang Z, JiaZ Yang B, Han Q, Ren X (2016) Effects of straw incorporation on the soil nutrient contents, enzyme activities, and crop yield in a semiarid region of China. Soil Tillage Res 160:65–72

Zhao B, Zhang J, Yu Y, Karlen DL, Hao X (2016) Crop residue management and fertilization effects on soil organic matter and associated biological properties. Environ Sci Pollut Res 23(17):17581–17591

Zhu S, Huang W, Huang W, Wang K, Chen Q, Wu Y (2015) Pretreatment of rice straw for ethanol production by a two-step process using dilute sulfuric acid and sulfomethylation reagent. Appl Energy 154:190–196

Zucconi F, de Bertoldi M (1987) Compost specifications for the production and characterization of compost from municipal solid waste. In: de Bertoldi M, Ferranti MP, L'Hermite P, Zucconi F (eds) Compost: production, quality and use. Elsevier, New York, pp 30–50

Zucconi F, Monaco A, Forte M (1985) Phytotoxins during the stabilization of organic matter. In: Gasser JKR (ed) Composting of agricultural and other wastes. Elsevier, New York, pp 73–86

Improving chemical and hydro-physical properties of semi-arid soils using different magnitudes of crumb rubber

Vahideh Sadeghizadeh[1] · Vahidreza Jalali[1]

Abstract

Purpose Iran is located on the desertification strip where it has faced water shortage and poor quality soils. So it is necessary to improve the soil structure and increase irrigation efficiency. In this context, the use of industrial wastes can increase soil's organic matter and improve its structure, also amend environmental problems related to disposal of industrial wastes. The objective of this study was to survey the effect of crumb rubber (CR) application as soil amendment on soil water retention curve indices, soil aggregate stability and soil micronutrients.

Methods In this study, different amounts of CR (0, 0.12, 0.24, 0.36 and 0.48 g) were added to every kg of the two different soil textures (loamy sand and silty clay). Pots were kept at FC moisture conditions for 6 months. At the end of incubation time, the different soil micronutrient availability, soil structural stability and soil hydraulic properties were measured.

Results The results showed that application of CR as soil amendment had significant effects ($p < 0.001$) on increasing soil organic matter content and soil micronutrient availability including Zn and Fe. Adding CR particles by reducing the aggregate fractal dimension, soil penetration resistance and bulk density, improves soil structure in both types of soil. The results also demonstrated that the use of CR particles can improve water holding capacity in sandy soil and increased plant available water content and air capacity in clay soil ($p < 0.001$).

Conclusions According to our results adding CR particles can be considered as a low-cost solution in order to improve the hydro-physical and chemical properties of soils in these areas.

Keywords Aggregate stability · Crumb rubber · Penetration resistance · Plant available water content · Soil water retention curve

Introduction

Soil properties are the crucial factors for crop production (Dotaniya et al. 2016). Soils in arid and semi-arid regions are often characterized by low clay and organic matter contents which results in low water holding capacity and inherently low fertility (Abdelfattah 2013). In order to improve soil structure and to increase under cultivation areas in arid and semi-arid regions, there are various management methods such as adding soil amendments, mulches and conservation tillage (Roper et al. 2013; Zhao et al. 2011; Bhatt et al. 2016).

Saline and sodic soils in arid and semi-arid areas cause unfavourable soil physical and chemical properties, which would impose restrictions on plant growth (Jalali et al. 2017). So the applications of appropriate soil amendments result in improving soil properties and reduced soil salinity and other stresses (Courtney and Harrington 2012; Bhattacharjya et al. 2016).

Different synthetic polymers are used for soil amendments in the agriculture which are safe and non-toxic and will eventually decompose to carbon dioxide, water, ammonia and potassium ions. Applying water-absorbing

✉ Vahidreza Jalali
V.Jalali@uk.ac.ir

Vahideh Sadeghizadeh
sadeghizadevahideh@yahoo.com

[1] Department of Soil Science, Faculty of Agriculture, Shahid Bahonar University of Kerman, Kerman, Iran

soil amendment may improve soil structure, arise water storage and decrease other water-related limitations (Huettermann et al. 2009).

Due to shortage of natural organic matters in arid and semi-arid regions, the use of organic polymers to improve soil structure was very important over the past years and these polymers have shown their ability as soil amendments. In fact, soil management practices encourage the addition of organic matter to the soil which has significant effect on soil physical and chemical properties (Weyers and Spokas 2014).

Previous studies demonstrated positive changes in several soil physical, chemical and biological properties after the addition of organic amendments to soil (Faust et al. 2015; Pandian et al. 2016; Adrover et al. 2017). The improvement of crop productivity has been attributed to the increase in the soil available nutrients (Lia et al. 2017; Chaudhuri et al. 2016; Rajaie and Tavakoly 2016) and enhanced soil physical properties such as reduction in soil bulk density and soil penetration resistance (Saglam and Dengiz 2017) and increase in water-holding capacity after applying soil amendments (Emami et al. 2014; Parihara et al. 2016).

Unfortunately, in some arid and semi-arid areas, natural organic matter is either unavailable or scarce; as an example more than 70 percent of Iran's agricultural soil contains less than 0.5 percent organic matters and ingredient (FAO 2012); thus, improvement of soil structure may be achieved using synthetic polymers (Masowa et al. 2016; Park et al. 2015).

Of the annual total global production of rubber material, which amounts to more than 10 million tons, approximately 60–70% is used in tyre. After they have been worn-out during their limited service life, millions of used tyres are discarded every year. As polymeric materials, rubbers do not decompose easily. Large amounts of waste rubber materials discarded pose two major problems: waste of valuable rubber and disposal of waste rubber leading to environmental pollution. Two major ways to solve the problem are recycling and reusing the waste rubber and reclaiming rubber's raw materials (Zhao et al. 2009).

Crumb rubber (CR) is a secondary resource recycled from scrap automobile tires. This amendment contains graded rubber granules and sometimes additional components, such as compost or mineral materials. The Application of waste polymer such as CR into soil behaves as reinforcement factor in improving soil function. Applying waste plastic for reinforcement of soil is an efficient and reliable technique for improving the strength and stability of soils (Sivakumar Babu et al. 2008). In this area, the amount and size of CR are the most effective factors which determine the soil structure stability from both agriculture and engineering point of view.

CR is not truly inorganic; it behaves much in the same way. Amendments with coarse grade CR can reduce compaction stress in soil, increase total porosity and enhance water movement. Finer particle sizes are being used for topdressing and can protect the crown tissue of plants from abrasion (Groenevelt and Grunthal, 1998). Vanini and Rogers (1995), used the rubber crumb as a soil amendment in different turfgrass situations.

Groenevelt and Grunthal (1998) pointed out that 10–20% of CR significantly reduced surface hardness of sports turf. Riggle (1994) also stated that the inclusion of rubber particles in soil could improve soil physical properties such as aeration and water infiltration rates. He concluded that the water use efficiency and fertilizers were improved by up to 30%. Intention to reuse the abandoned tires by grinding them into small particles (crumb rubber) and using them in sports field, pavement and construction materials may be an important approach to reduce waste rubbers in large quantities (Yilmaz and Degirmenci 2009).

Zhao et al. (2011) investigated the effect of CR waste as a soil conditioner on the nematode assemblage and some soil properties in a turfgrass soil. Their results demonstrated that CR incorporation significantly decreased plant parasite and omnivorous nematode populations, but increased the abundance of predatory nematodes, and also CR application decreased soil bulk density and pH value, while soil moisture content increased.

Taheri et al. (2011) reported that ground tyre rubber and rubber ash were very effective amendments to increase the DTPA-extractable Zn in a calcareous Zn-deficient soil. Khoshgoftarmanesh et al. (2012) investigated the effect of acid-washed shredded waste-tire rubber in hydroponic tomato production. They said shredded waste rubber might be used as a component of container media in production of hydroponically grown tomato; however, acid-washing of rubber is required to prevent potential Zn toxicity for plant.

Pérez et al. (2012) investigated the use of rubber crumbs as drainage layer in green roof establishment. The results of their research raised the possibility of using recycled rubber from tires as a drainage layer in green roofs, substituting the porous stone materials currently used.

Several studies have been carried out throughout the world to identify the effects of incorporating different soil amendment and organic matter into soil and survey their effect on soil's physical and chemical properties. However, various potential use of CR and its environmental impacts have been less investigated and there have been no published investigations about using CR effects on some soil's physical and chemical properties in arid and semiarid region soils. Therefore, the objective of this study was to assess the impacts of waste CR as soil amendment on the soil hydro-physical and chemical properties in arid and semiarid soils. The investigations may provide a basis for mitigating the worn out tires' pollution problems in the environment, while enhancing some soil's chemical and hydro-physical properties of semi-arid soils.

Materials and methods

The study site and soil properties

This study was carried out in the research greenhouse of the Shahid Bahonar University of Kerman, southeast of Iran (latitude of 30°14′N and longitude of 57°06′E). Kerman and the surrounding regions have a semi-moderate and dry climate with mean annual rainfall of 153 mm (Yazdanpanah et al. 2016). Two different soil textures [loamy sand (*S*) and silty clay (*C*)], were collected and air-dried and passed through a 2-mm sieve and stored at room temperature before use.

Physical and chemical properties of the soil samples were quantified in our laboratory. Soil particle size distribution was determined using the hydrometer method (Gee and Bauder 1986), and organic carbon (OC) was measured using the wet-digestion method (Walkley and Black 1934). Soil electrical conductivity (EC) and reaction (pH) were determined in a soil–water extract of 1:5 w:v.

EDTA-extractable elements (Zn, Fe and Cu) were determined by atomic absorption spectrometry (AAS) (Quevauviller et al. 1996) and bulk density was determined with core sample (Blake and Hartge 1986), respectively. The properties of CR particles (with a diameter of less than 2 mm) used in these experiments are presented in Table 1 (Asadollah zadeh 2013). The cation exchange capacity (CEC) of used CR particles was reported about 57 $cmol^{+}$ kg^{-} (Khoshgoftarmanesh et al. 2012).

Preparation of test pots

In this study, the pots with a capacity of 6 kg were used. The test substrates were prepared by adding 0, 0.12, 0.24, 0.36 and 0.48 g CR (with a diameter of less than 2 mm) to every kg of the two different soil textures. The soil and CR were manually blended and transferred to close-end pots without drainage. The pots were stored at greenhouse temperature conditions (23 ± 2 °C) and humidity of field capacity for 6 months before being used to prepare the test samples. Depending on the soil type, specific climate conditions and Intended land-use, different researchers have reported different incubation period from at least 3 months (Zhao et al. 2011), 8 months (Bidaki et al. 2012) and 11 months (Smolders and Degryse 2002), so in this study, the mean of the periods reported (6 months) were selected as the incubation period.

Measuring soil physical properties

Penetration resistance (PR)

Penetration resistance (PR) is a suitable indicator for identifying changes in soil resistance to penetration over a period and for assessing the effects of soil resistance to root development (O'Sullivan 1992). After incubation time, at the end of the test, the samples were irrigated to field capacity water content, and penetration resistance of the samples was determined by penetrometer device (33-T0165 model, CONTROLS Inc).

Depending on the type of soil texture, 0.2 square inch metal cap was used for loamy sand (*S*) and 0.1 square inch metal cap were used for silty clay (*C*) samples. Penetrometer probe was immersed to a certain extent (depth of 3 cm) at a nearly constant speed in soil samples. Soil penetration resistance was calculated using the function employed by Tormena et al. (1999) as:

$$\mathrm{PR} = \frac{0.5X}{Y} \times \alpha, \tag{1}$$

where PR is the soil penetration resistance (MPa), X is registered device number (mm), Y is cross-sectional area of used metal cap (square inch) and α is acceleration gravity (9.8 ms^{-2}).

Soil aggregate stability

Soil aggregate stability was determined by mean weight diameter index (MWD) and Fractal dimension by using Eqs. 2 and 3 (Calero et al. 2008).

$$\mathrm{MWD} = \sum_{i=1}^{n} x_i w_i, \tag{2}$$

where X_i is the mean diameter of any particular size range of aggregates separated by sieving (mm), and W_i is the weight of the aggregates in that size range as a fraction of the total dry weight of the sample. The summation accounts for all size ranges, including the group of aggregates smaller than the openings of the finest sieve.

Table 1 Selected physical and chemical properties of crumb rubber (CR) particles

EC (dS m^{-1})	pH (–)	Zn (mg kg^{-1})	Fe (mg kg^{-1})	Cu (mg kg^{-1})	Cd (mg kg^{-1})	ρ_b (g cm^{-3})
0.354	6.82	12,000	108	4.25	2.25	0.2

Fractal dimension index of soil aggregates was calculated based on the size-number equation of Turcotte (1989) as follows:

$$N_i = cX_i^{-D} \quad (3)$$

where X_i is the mean diameter of remaining aggregates on each sieve(mm) and c is constant coefficient. Fractal dimension (D) was obtained through regression analysis between $\log N_i$ and $\log X_i$. The amount of N_i is determined from stable aggregate distribution and it will be determined at the end of sieving as follows (Eq. 4 and 5):

$$N_i = \sum_{i=1}^{n} n_i, \quad (4)$$

$$n_i = w_i \rho^{-1} \times x_i \rho^{-3}, \quad (5)$$

where n_i, ρ and w_i are the number of aggregates, aggregate bulk density and the remaining aggregate mass on each sieves, respectively.

Soil hydraulic properties

Water retention curve is one of the main soil hydraulic properties. It can be considered as a soil's fingerprint, since the shape of the curve is related to various physical and chemical soil properties, which are unique for each soil (Cornelis et al. 2001). Therefore, any change in soil structure will appear in soil water retention curve. In this research, the widely used model of van Genuchten (1980) was fitted to the data obtained from pressure plate apparatus.

Different physical quality indicators derived from water retention curve have been proposed and validated by several authors. Reynolds et al. (2007) found that air capacity (AC) and relative water capacity (RWC) could be good indicators because they responded to different soil management practice, so in this research air capacity, relative water capacity and plant available water capacity (PAWC) were considered.

According to Reynolds et al. (2007), AC is the difference between saturation water content and water content at field capacity (FC). PAWC is the difference between FC and permanent wilting point (PWP). RWC is the ratio of water content at FC to water content at saturation.

Statistical analyses

Current research was carried out with five different CR application rates (0, 0.12, 0.24, 0.36 and 0.48 g kg^{-1}) for two texturally differentiated soil which were named C_0, $C_{0.12}$, $C_{0.24}$, $C_{0.36}$ and $C_{0.48}$ for silty clay soil and S_0, $S_{0.12}$, $S_{0.24}$, $S_{0.36}$ and $S_{0.48}$ for loamy sand (S). These pots were arranged in a randomized complete block design with three replicates.

Statistical assessments of possible difference in soil's physical and chemical properties due to the five rates of CR treatment were analysed with one way-ANOVA by using the SAS program (SAS Institute 1996). Duncan's multiple range test was used to compute LSD values for comparing differences between pairs of treatment means

Results and discussion

Amending the soil chemical properties with CR

Electrical conductivity (EC) and soil reaction (pH)

Table 2 shows physical and chemical properties of the studied soils. Statistical analyses of CR particles' addition on soil physical and chemical properties are presented in Table 3. Based on the data presented in this table, it becomes clear that increasing different amounts of CR has not significantly affected the EC and pH of both types of soil in this limited period of incubation.

Considering the table, it is uncovered that with expanding the utilization of CR, the electrical conductivity of loamy sand soil has slight expanding pattern while this pattern has been declining in silty clay soil. On the other hand, increasing the amount of CR, increased and decreased the salinity in the loamy sand and silty clay soils, respectively, but as mentioned this slight changes were not statistically significant.

pH changes were quite different in both types of soils so that, by increasing the amount of crumb rubber, loamy sand soil's pH value decreases, whereas silty clay soil's pH was increased.

As previously mentioned, the main feature of CR particles is their low biodegradability and long-term effectiveness in soil. So no statistically significant effect of these treatments on a limited time scale seems quite logical. Due to the abundance of lime in semi-arid soils, (such as Kerman-Iran), overcoming buffering capacity and changing chemical properties of these soils needs more incubation time.

Organic carbon content (OC)

CR is a complex vulcanized elastic which is essentially involved natural rubber (NR), like polyisoperene, and synthetic elastic (SR), like styrene-butadiene elastic as well as butadiene elastic, which are cross connected with sulphur and fortified with carbon dark (CB) (Ghavibazoo and Abdelrahman 2013). Najim and Hall (2012) reported that CR had at least 28 percent of carbon dark by weight. So as

Table 2 Selected physical and chemical properties of the studied soils

Texture	Sand(%)	Silt (%)	Clay (%)	O.C (%)	EC (dS m^{-1})	pH (−)	Zn (mg kg^{-1})	Fe (mg kg^{-1})	Cu (mg kg^{-1})	ρ_b (gr cm^{-3})
Loamy sand	85	4	11	0.0	1.37	7.5	1.33	3.22	0.32	1.64
Silty clay	3	42	55	0.01	1.16	7.9	0.92	4.83	0.84	1.45

expected, the addition of CR significantly increased the organic carbon content (OC) in both soils. Since the soil's organic carbon content is very small in the control sample of both soils, then increase in OC, was highly significant ($p < 0.001$) when the CR amount was increased in two soils (Table 3).

For loamy sand soil, organic carbon content from about zero percent (in control) was increased to 1.4% in the $S_{0.48}$ treatment. While in the silty clay soil due to weaker aeration, organic carbon content in highest CR application rate ($C_{0.48}$), reached to 2.1%.

Considering these results, it can be concluded that in areas where natural organic material is not available, addition of CR particles can be the convenient, inexpensive and a fast solution to improve soil organic carbon content.

Micronutrient availability

The investigation results indicated that using CR as soil modification had significant effect on the soil micronutrient such as Fe and Zn, but at different CR application rate, the copper content is not significantly changed (Table 3).

In the loamy sand substrate, the concentration of zinc increased significantly ($p < 0.001$), when the amount of CR was increased to 0.24 g kg^{-1} and in higher application values ($S_{0.36}$ and $S_{0.48}$), despite slight increases in zinc content, these changes were not statistically significant. Also in these substrates, the concentration of Fe increased ($p < 0.001$), when the added CR was 0.36 g kg^{-1} and between $S_{0.36}$ and $S_{0.48}$ treatments, significant changes were not seen (Table 3).

For silty clay soil, increasing the concentration of applied CR has a significant effect on all treatments, so that the Zn content in control treatment (0.91 mg kg^{-1}) has increased up to 17.97 mg kg^{-1} in $C_{0.48}$ treatment. In the case of Fe for silty clay soil, there was significant difference between the lowest and highest CR application value with control. Increasing Zn content by increasing the CR application rate can be attributed to pure Zn-Oxide which was used in tyre production process. This results demonstrated that these findings are in perfect agreement with Taheri et al. (2011) which said CR and rubber ash were very effective in improving DTPA-extractable Zn in a calcareous Zn-insufficient soil.

Unlike the iron and zinc elements, increasing the application of CR in different amounts had no significant effect on increasing copper content in all treatments. The main reason for this can be attributed that copper unlike iron and zinc, not involved directly in the tires production process.

However, in calcareous soils, such as our research soils, due to the stabilization of metal elements such as zinc in top soil, there are little concerns regarding contamination of these elements, but by reducing the soil's pH and in acidic soils, due to the abundance of zinc in CR, application of this modifier should be used with caution.

Effect of CR amendment on soil physical properties

Bulk density (BD)

The mean comparison results of different CR application rate indicated that the application of CR cause significant decrease of soil bulk density (BD) in both soils.

In loamy sand soil, soil bulk density was decreased significantly ($p < 0.001$) from 1.64 g cm^{-3} in control treatment to 1.40 g cm^{-3} by addition of 0.48 g kg^{-3} CR to the soil and also for silty clay soil, bulk density was reduced significantly ($p < 0.05$) from 1.44 g cm^{-3} in the control to 1.28 g cm^{-3} in the 0.48 g kg^{-3} CR treatment (Table 3). These results are consistent with the findings of Zhao et al. (2011) who said that CR application decreased soil bulk density in a sand-based medium.

Porous structure in tyres causes a significant reduction in both soil bulk densities. This property of CR has potential for usage in environments that have high traffic and have been subjected to density increases.

Mean weight diameter (MWD)

Based on the results presented in Table 3, the application of different amount of CR had significant effect on soil aggregate mean weight diameter (MWD). Duncan's multiple range test was employed for mean comparison of the different CR levels on aggregate MWD (Table 3). Considering mean comparison data in Table 3, the addition of CR significantly increased the MWD in both soils. In loamy sand soil, MWD increased significantly ($p < 0.001$)

Table 3 Substrate's Chemical and physical characteristics based on varying magnitudes of added crumb rubber

	EC (dS m^{-1})	pH	O.C %	Zn (mg Kg^{-1})	Fe (mg Kg^{-1})	Cu (mg Kg^{-1})	BD (g cm^{-3})	MWD (mm)	PR (MPa)	D (−)
S_0	1.37 ± 5.95^a	7.53 ± 5.7^a	0^e	1.40 ± 1.47^c	3.2 ± 0.7^c	0.32 ± 1.66^a	1.64 ± 0.47^a	0.11 ± 1.28^c	1.36 ± 6.19^a	4.7 ± 1.23^a
$S_{0.12}$	1.57 ± 0.2^a	7.52 ± 0.2^a	0.40^d	6.66 ± 0.06^b	4.5 ± 0.43^b	0.33^a	1.5 ± 0.08^{ab}	0.17 ± 0.02^b	1.21 ± 2.71^{ab}	4.9 ± 0.31^a
$S_{0.24}$	1.53 ± 0.21^a	7.41 ± 0.17^a	0.8 ± 0.22^c	16.21 ± 3.18^a	4.7 ± 0.73^b	0.31 ± 0.01^a	1.5 ± 0.07^b	0.17 ± 0.03^b	1.22 ± 0.096^{ab}	5.0 ± 0.25^a
$S_{0.36}$	1.57 ± 0.19^a	7.42 ± 0.18^a	1.1 ± 0.36^b	16.25 ± 6.58^a	5.8 ± 0.85^a	0.36 ± 0.02^a	1.50 ± 0.07^b	0.18 ± 0.03^b	1.15 ± 0.1^{ab}	4.2 ± 0.23^b
$S_{0.48}$	1.58 ± 0.19^a	7.39 ± 0.17^a	1.4 ± 0.45^a	17.02 ± 6.72^a	5.8 ± 1.05^a	0.30 ± 0.02^a	1.40 ± 0.07^c	0.23 ± 0.03^a	1.01 ± 0.12^b	4.2 ± 0.39^b
C_0	1.07 ± 1.57^a	7.9 ± 1.38^a	0.01^e	0.91^c	4.9 ± 0.08^c	0.53^a	1.44 ± 0.7^a	0.14 c	3.49 ± 0.73^a	4.34^a
$C_{0.12}$	1.01 ± 0.19^a	7.9 ± 0.2^a	0.58^d	16.72 ± 0.43^b	6.03 ± 0.46^{bc}	0.90 ± 0.09^a	1.39 ± 0.02^{ab}	0.16^c	2.96 ± 0.34^b	4.1 ± 0.03^a
$C_{0.24}$	1.01 ± 0.17^a	7.96 ± 0.17^a	1.2 ± 0.05^c	17.20 ± 1.16^{ab}	7.0 ± 2.36^{abc}	0.83 ± 0.2^a	1.3 ± 0.07^{abc}	0.22^b	2.36 ± 0.39^c	3.81 ± 0.1^b
$C_{0.36}$	0.99 ± 0.16^a	8.03 ± 0.17^a	1.8 ± 0.21^b	17.67 ± 0.04^{ab}	7.54 ± 1.28^{ab}	0.79 ± 0.1^a	1.32 ± 0.06^{bc}	0.26 ± 0.05^b	2.30 ± 0.53^{cd}	3.8 ± 0.18^b
$C_{0.48}$	0.88 ± 0.16^a	8.11 ± 0.17^a	2.1 ± 0.05^a	17.97 ± 0.04^a	8.74 ± 0.96^a	0.88 ± 0.13^a	1.28 ± 0.06^c	0.34 ± 0.03^a	1.94 ± 0.54^d	3.5 ± 0.06^c

S loamy sand, *C* silty clay, *BD* bulk density, *MWD* mean weight diameter, *PR* penetration resistance, *D* fractal dimension

The same letters in each group indicate no significant differences according to Duncan's Multiple Range Tests from 0.11 mm in control treatment to 0.23 mm at highest CR level ($S_{0.48}$). Also for silty clay soil, MWD was increased significantly ($p < 0.001$) from 0.14 mm in control treatment to 0.34 mm in the 0.48 g kg^{-1} CR treatment ($C_{0.48}$). These finding is strong evidence that the CR application in soil through its effect on SOC increased MWD significantly. The strong correlation between MWD and SOC were reported by different researchers around the world. Chaney and Shift (1984) revealed a direct relationship between aggregate stability and SOC content in 120 soils gathered from various soil arrangement in Scotland and England. Similarly, Nyamangara et al. (2001) detailed that a 10–38% addition in soil organic carbon with yearly utilization of cows compost at a rate of 12.5 Mg ha^{-1} for a long time brought about three to fourfold increment in soil MWD in a granitic sandy soil in Zimbabwe.

Fractal dimension (D)

There are many methods to describe the aggregate stability which the important ones include several sieving methods leading to mean weight diameter (MWD), geometric weight diameter (GMD) and in the last few decades fractal dimension index (D). These have been used to describe aggregate stability. In this research, the soil aggregate's fractal dimension was obtained from dry sieving data and using the formulas 3 and 4. The addition of CR significantly decreased ($p < 0.001$) the D in loamy sand soil when the amount of CR was increased up to 0.36 g kg^{-1}, whereas in silty clay substrate, the decrease in the D with increasing rate of amendment was significant ($p < 0.001$) when the CR amount reached to 0.48 g kg^{-1}. As it is clear from these findings, aggregate's fractal dimension (D) has inverse proportionality with mean weight diameter, significantly decreasing the fractal dimension of different treatments related to improving soil structure at various level of CR application.

Penetrometer resistance (PR)

One of the negative effects of the lack of organic matter in arid and semi-arid soils, like Iran, is increasing soil penetration resistance which is caused by weak soil structure.

High penetration resistance in these soils imposes restrictions during soil tillage and plowing, water infiltration into the soil, soil respiration and seed germination. This research revealed that the addition of CR significantly decreased soil penetration resistance in both types of soils.

In loamy sand soil, penetration resistance was decreased from 1.36 MPa in control to 1.01 MPa in $S_{0.48}$ treatment. While this decline rate was higher in silty clay soil, so that the control penetration resistance was 3.49 MPa ,it decreased significantly to 1.94 MPa in $C_{0.48}$ treatment. These results are consistent with the findings of Groenevelt and Grunthal (1998) who said that an increase of 10–20% of CR fundamentally decreases surface hardness of games turf.

Effect of CR amendment on soil hydraulic properties

One of the most important ways to deal with water shortage in arid and semi-arid regions is adding amendments to soil in order to increase water use efficiency and improve the physical properties of these soils. The non-linear relationship between soil water content (θ_V) and its matric potential (ψ_m) called soil water characteristic curve (SWCC) which expresses the effect of soil structure, total porosity and pore size distribution on soil water availability.

By studying the soil moisture characteristic curve at the difference treatment levels (0.12, 0.24, 0.36 and 0.48 g of crumb rubber/kg soil), it was revealed that by addition of CR application rate, water content increased at each suction point in both types of soils and maximum water content was gained at highest application rate of CR (0.48 g kg^{-1}) (Fig. 1).

Observing the overall considering of SWCC in low suction points (1, 100 and 330 cm) and high suction points (5000–15,000 cm) range, it can be concluded that significant amount of moisture absorbed by the CR particle is released in low suction points, so the release of moisture in low suction point is considered as an advantage for using CR, especially in moderate to heavy textural soils. Comparison of residual water content at high suction points (5000–15,000 cm) showed that the residual water content in the rubber-amended soil was also significant, especially at high levels of CR application. The reasons for the increase of moisture in the treatments containing CR can be attributed to increasing porosity, decreasing density and increasing surface area of rubber-amended soils. In this area, Riggle (1994) expressed that incorporating rubber particles in soil could enhance the water use efficiency by up to 30%. In general, polymers due to their hydrophilic properties are effective in improving soil properties especially in increasing soil moisture content.

Effect of CR amendment on soil water retention curve parameters

Plant available water content (PAWC)

Routinely, plant-available water content (PAWC) is obtained by subtracting moisture at field capacity and permanent wilting point. Based on the results as given in Table 4, CR particles significantly ($p < 0.001$) increased

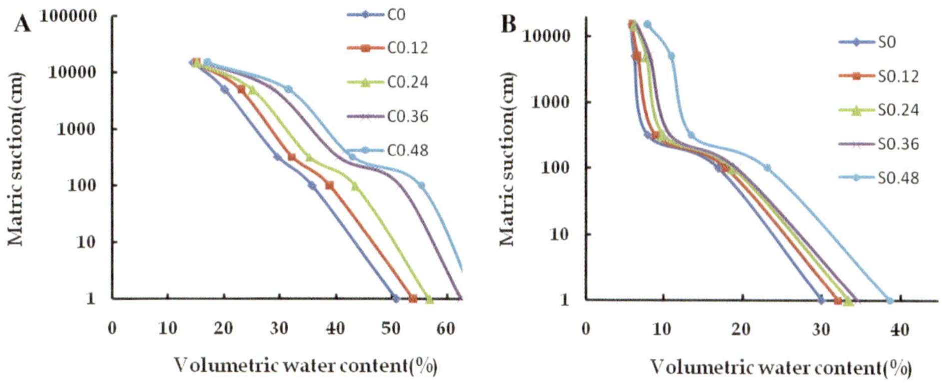

Fig. 1 Soil retention curves of silty clay (**a**) and loamy sand (**b**) soils as affected by different CR level

the PAWC in both types of soil, but the amount of PAWC increment in silty clay soil was greater than this increment in loamy sand soil, so that PAWC enhanced from 15.1% in C_0 (as control treatment of silty clay soil) to 25.7% in highest CR application rate ($C_{0.48}$), and this means an increase of more than 10 percent of PAWC compared to the control. In terms of availability of soil water for plants, adding CR particles in clay soils with very fine pores leads to the improvement of the availability of soil water for plants. For loamy sand soil, the maximum amount of PAWC is related to $S_{0.48}$ treatment and it was 5.45% and a difference with the control (S_0) is measured about 3%. These results confirmed that the CR addition in soil could improve pore size distribution and helped to increase the plant-available water content.

Relative water capacity (RWC)

Another indicator that shows the status of availability of soil water for plants is relative water capacity (RWC) index. Relative water capacity is obtained from the ratio of soil moisture at field capacity to soil saturation water content. These results indicated that CR affects the RWC in both soils significantly (Table 4).

Air capacity (AC)

The plant root needs both air and water simultaneously to survive. That portion of the soil porosity which is occupied by air is called the air capacity (AC). So soil air capacity index is one of the important indices for plant growth and its survival.

The volumetric difference between field capacity (θ_{fc}) and water saturated point (θ_s) considered as soil air capacity which both of these parameters (e.g. θ_{fc} and θ_s) are acquired from soil moisture characteristics curve. Based on statistical analysis, increasing the amount of CR in the loamy sand substrates significantly ($p < 0.05$) increased air capacity, while it did not have any effect on silty clay soil (Table 4).

Table 4 Different soil water retention curve indices based on varying magnitudes of added crumb rubber

Treatment	PAWC(%)	AC(%)	RWC (%)
S_0	2.08 ± 7.02^{d}	21.96 ± 5.65^{b}	0.26 ± 7.17^{c}
$S_{0.12}$	2.99 ± 0.5^{c}	23.02 ± 1^{b}	0.28 ± 0.02^{bc}
$S_{0.24}$	3.50 ± 0.67^{bc}	23.45 ± 1.06^{ab}	0.29 ± 0.01^{bc}
$S_{0.36}$	4.18 ± 0.75^{b}	23.71 ± 1.09^{ab}	0.31 ± 0.02^{b}
$S_{0.48}$	5.45 ± 0.9^{a}	25.07 ± 1.1^{a}	0.35 ± 0.02^{a}
C_0	15.15 ± 7.16^{e}	21.16 ± 2.93^{a}	0.58 ± 7.53^{c}
$C_{0.12}$	17.12 ± 1^{d}	21.83 ± 2^{a}	0.59 ± 0.02^{c}
$C_{0.24}$	20.13 ± 1.39^{c}	21.65 ± 1.82^{a}	0.61 ± 0.01^{bc}
$C_{0.36}$	23.62 ± 2.33^{b}	22.16 ± 1.75^{a}	0.64 ± 0.02^{ab}
$C_{0.48}$	25.74 ± 3.44^{a}	22.26 ± 1.74^{a}	0.65 ± 0.03^{a}

AWC available water content, *AC* air capacity, *RWC* relative water content

The same letters in each group indicate no significant differences according to Duncan's Multiple Range Tests

Conclusion

Soil organic matter is considered as the main factor of soil quality. Soil organic matter deficit is the most common problem in arid and semi-arid soils. In this study, amending two soils with distinct texture with different rates of crumb rubber modified some hydro-physical and chemical properties of soils.

Our results indicated that the application of CR has significant effect on decreasing the amount of soil compaction index (based on decreasing of bulk density and penetration resistance) and increasing volumetric water content at different suction points so that adding crumb rubber modified the water retention curve coefficients. Increasing plant-available water content (especially in silty clay soil) and soil relative water content are the examples of the positive impact of the crumb rubber application as a modifier.

According to different studies, the micronutrient deficits are very common in calcareous soils of arid and semi-arid regions like Iran. The results of this study showed that the addition of crumb rubber can be a good source of microelements, especially for zinc.

In general, the results of this research proved positive impact of crumb rubber particles on various soil properties. Since this research was carried out in controlled greenhouse conditions and for limited incubation time, it is recommended that the effect of adding crumb rubber particles be evaluated in field condition and for longer times.

References

Abdelfattah MA (2013) Pedogenesis, land management and soil classification in hyper-arid environments: results and implications from a case study in the United Arab Emirates. Soil Use Manag 29:279–294. doi:10.1111/sum.12031

Adrover M, Moyà G, Vadell J (2017) Seasonal and depth variation of soil chemical and biological properties in alfalfa crops irrigated with treated wastewater and saline groundwater. Geoderma 286:54–63. doi:10.1016/j.geoderma.2016.10.024

Asadollah zadeh MJ (2013) Effect of microbial inoculation and endophyte symbiosis on the Zinc (Zn) release from worn out crumb rubber and its uptake by plant. M.Sc. thesis Soil Science, Faculty of Agriculture, Isfahan University of Technology

Bhatt B, Chandra R, Ram S, Pareek N (2016) Long-term effects of fertilization and manuring on productivity and soil biological properties under rice (*Oryza sativa*)–wheat (*Triticum aestivum*) sequence in Mollisols. Arch Agron Soil Sci 62(8):1109–1122. doi:10.1080/03650340.2015.1125471

Bhattacharjya S, Chandra R, Pareek N, Raverkar KP (2016) Biochar and crop residue application to soil: effect on soil biochemical properties, nutrient availability and yield of rice (*Oryza sativa* L.) and wheat (*Triticum aestivum* L.). Arch Agron. Soil Sci 62(8):1095–1108. doi:10.1080/03650340.2015.1118760

Bidaki SMY, Hajabbasi MA, Khoshgoftarmanesh AH, Eshghizadeh HR (2012) Effect of waste tire rubber particles on some chemical properties of a calcareous soil. J Sci Technol Agric Nat Resour Water Soil Sci 16(59):101–115

Blake GR, Hartge KH (1986) Methods of Soil Analysis. Part 1-Physical and mineralogical methods, vol 9, 2nd edn. Agronomy Monograph. ASA and SSSA, Madison, pp 363–375

Calero N, Barron V, Torrent J (2008) Water dispersible clay in calcareous soils of southwestern Spain. Catena 74:22–30. doi:10.1016/j.catena.2007.12.007

Chaney K, Shift RS (1984) The influence of organic matter on aggregate stability in some British soils. Soil Sci 35:223–230. doi:10.1111/j.1365-2389.1984.tb00278.x

Chaudhuri PS, Paul TK, Dey A, Datta M, Dey SK (2016) Effects of rubber leaf litter vermicompost on earthworm population and yield of pineapple (*Ananas comosus*) in West Tripura, India. Int J Recycl Org Waste Agric 5:93–103. doi:10.1007/s40093-016-0120-z

Cornelis WM, Ronsyn J, Van Meirvenne M, Hartmann R (2001) Evaluation of pedotransfer functions for predicting the soil moisture retention curve. Soil Sci Soc Am J 65:638–648. doi:10.2136/sssaj2001.653638x

Courtney R, Harrington T (2012) Growth and nutrition of Holcus-lanatus in bauxite residue amended with combinations of spent mushroom compost and gypsum. Land Degrad Dev 23:144–149. doi:10.1002/ldr.1062

Dotaniya ML, Datta SC, Biswas DR, Dotaniya CK, Meena BL, Rajendiran S, Regar KL, Lata M (2016) Use of sugarcane industrial by-products for improving sugarcane productivity and soil health. Int J Recycl Org Waste Agric 5:185–194. doi:10.1007/s40093-016-0132-8

Emami H, Astaraei AR, Fotovat A, Khotabaei M (2014) Effect of soil conditioners on cation ratio of soil structural stability, structural stability indicators in a sodic soil, and on dry weight of Maize. Arid Land Res Manag 28(3):325–339. doi:10.1080/15324982.2013.856357

FAO (2012) Country report on history and status of soil survey in Iran. GSP regional workshop Jordan, Amman 1–5 April 2012. http://www.fao.org/fileadmin/user_upload/GSP/docs/Presentation_NEMA_Inception/The_Islamic_Republic_of_Iran_2.pdf

Faust S, Hanisch S, Buerkert A, Joergensen RG (2015) Soil properties under manured *Tamarindus indica* in the littoral plain of south-western Madagascar. Arid Land Res Manag 29(2):167–179. doi:10.1080/15324982.2014.944243

Gee GW, Bauder JW (1986) Particles size analysis. In: Kelute A (ed) Methods of soil analysis. Part 1. Physical and mineralogical methods. Soil Sci Soc Am J. Agronomy Monograph 9. 2nd edn, pp. 383–411

Ghavibazoo A, Abdelrahman M (2013) Composition analysis of crumb rubber during interaction with asphalt and effect on properties of binder. Int J Pavement Eng 14(5):517–530. doi:10.1080/10298436.2012.721548

Groenevelt PH, Grunthal PE (1998) Utilisation of crumb rubber as a soil amendment for sport turf. Soil Tillage Res 47:169–172. doi:10.1016/S0167-1987(98)00089-0

Huettermann A, Orikiriza LJB, Agaba H (2009) Application of superabsorbent polymers for improving the ecological chemistry of degraded or polluted lands. Clean Soil Air Water 37:517–526. doi:10.1002/clen.200900048

Institute SAS (1996) SAS user's guide: statistics. SAS Institute Inc., Cary

Jalali V, Asadi-Kapourchal S, Homaee M (2017) Evaluating performance of macroscopic water uptake models at productive growth stages of durum wheat under saline conditions. Agric Water Manag 180:13–21. doi:10.1016/j.agwat.2016.10.015

Khoshgoftarmanesh AH, Eshghizadeh HR, Chaney RL (2012) Using acid-washed shredded waste-tire rubber in soilless media for tomato production. J Plant Nutr 35(9):1341–1348. doi:10.1080/01904167.2012.684126

Lia R, Tao R, Ling N, Chu G (2017) Chemical, organic and bio-fertilizer management practices effect on soil physicochemical property and antagonistic bacteria abundance of a cotton field: implications for soil biological quality. Soil Tillage Res 167:30–38. doi:10.1016/j.still.2016.11.001

Masowa MM, Kutu FR, Shange PL, Mulidzi R, Vanassche FMG (2016) The effect of winery solid waste compost application on maize growth, biomass yield, and nutrient content under greenhouse conditions. Arch Agron Soil Sci 62(8):1082–1094. doi:10.1080/03650340.2015.1115018

Najim KB, Hall MR (2012) Mechanical and dynamic properties of self-compacting crumb rubber modified concrete. Constr Build Mater 27:521–530. doi:10.1016/j.conbuildmat.2011.07.013

Nyamangara J, Gotosa J, Mpofu SE (2001) Cattle manure effects on structural stability and water retention capacity of a granitic sandy soil in Zimbabwe. Soil Tillage Res 62:157–162. doi:10.

1016/S0167-1987(01)00215-X
O'Sullivan M (1992) Uniaxial compaction effects on soil physical properties in relation to soil type and cultivation. Soil Tillage Res 24:257–269. doi:10.1016/0167-1987(92)90091-O
Pandian K, Subramaniayan P, Gnasekaran P, Chitraputhirapillai S (2016) Effect of biochar amendment on soil physical, chemical and biological properties and groundnut yield in rainfed Alfisol of semi-arid tropics. Arch Agron Soil Sci 62(9):1293–1310. doi:10.1080/03650340.2016.1139086
Parihara CM, Yadavb MR, Jata SL, Singha AK, Kumara B, Pradhanb S, Chakrabortyb D, Jatc ML, Jatd RK, Saharawate YS, Yadava OP (2016) Long term effect of conservation agriculture in maize rotations on total organic carbon, physical and biological properties of a sandy loam soil in north-western Indo-Gangetic Plains. Soil Tillage Res 161:116–128. doi:10.1016/j.still.2016.04.001
Park CH, Li X, Jia RL, Hur JS (2015) Effects of superabsorbent polymer on cyanobacterial biological soil crust formation in laboratory. Arid Land Res Manag 29(1):55–71. doi:10.1080/15324982.2014.928835
Pérez G, Vila A, Rincón L, Solé C, Cabeza LF (2012) Use of rubber crumbs as drainage layer in green roofs as potential energy improvement material. Appl Energy 97:347–354. doi:10.1016/j.apenergy.2011.11.051
Quevauviller Ph, Lachica M, Barahona E, Rauret G, Ure A, Gomez A (1996) Interlaboratory comparison of EDTA and DTPA procedures prior to certification of extractable trace elements in calcareous soil. Sci Total Environ 178:127–132. doi:10.1016/0048-9697(95)04804-9
Rajaie M, Tavakoly AR (2016) Effects of municipal waste compost and nitrogen fertilizer on growth and mineral composition of tomato. Int J Recycl Org Waste Agric 5:339–347. doi:10.1007/s40093-016-0144-4
Reynolds WD, Drury CF, Yang XM, Fox CA, Tan CS, Zhang TQ (2007) Land management effects on the near—surface physical quality of a clay loam soil. Soil Tillage Res 96:316–330. doi:10.1016/j.still.2007.07.003
Riggle D (1994) Finding markets for scrap tires: utilization options. Biocycle 35:41–55
Roper MM, Ward PR, Keulen AF, Hill JR (2013) Under no-tillage and stubble retention, soil water content and crop growth are poorly related to soil water repellency. Soil Tillage Res 126:143–150. doi:10.1016/j.still.2012.09.006
Sağlam M, Dengiz O (2017) Spatial variability of soil penetration resistance in an alluvial delta plain under different land uses in middle Black Sea Region of Turkey. Arch Agron Soil Sci 63(1):60–73. doi:10.1080/03650340.2016.1178386
Sivakumar Babu GL, Vasudevan AK, Haldar S (2008) Numerical simulation of fiber-reinforced sand behavior. Geotext Geomembr 26(2):181–188. doi:10.1016/j.geotexmem.2007.06.004
Smolders E, Degryse F (2002) Fate and effect of zinc from tire debris in soil. Environ Sci Technol 36:3706–3710. doi:10.1021/es025567p
Taheri S, Khoshgoftarmanesh AH, Shariatmadari H, Chaney RL (2011) Kinetics of zinc release from ground tire rubber and rubber ash in a calcareous soil as alternatives to Zn fertilizers. Plant Soil 341(1–2):89–97. doi:10.1007/s11104-010-0624-7
Tormena CA, Da Silva AP, Libardi PL (1999) Soil physical quality of a Brazilian Oxisol under two tillage systems using the least limiting water range approach. Soil Tillage Res 52:223–232. doi:10.1016/S0167-1987(99)00086-0
Turcotte DL (1989) Fractals in geology and geophysics. Pure Appl Geophys 131(171–1):96
van Genuchten MTh (1980) A closed-form equation for predicting the hydraulic conductivity of unsaturated soils. Soil Sci Soc Am J 44:892–898
Vanini JT, Rogers JN (1995) Topdressing with crumb rubber from used tires in variation in a semi-arid catchment: implications for land and agricultural water management. Land Degrad Dev 25:163–172
Walkley A, Black IA (1934) An examination of digestion method for determining soil organic matter and a proposed modification of the chromic acid titration. Soil Sci 37:29–38
Weyers SL, Spokas KA (2014) Crop residue decomposition in Minnesota biochar-amended plots. Solid Earth 5:499–507. doi:10.5194/se-5-499-2014
Yazdanpanah N, Mahmoodabadi M, Cerdà A (2016) The impact of organic amendments on soil hydrology, structure and microbial respiration in semiarid lands. Geoderma 266:58–65. doi:10.1016/j.geoderma.2015.11.032
Yilmaz A, Degirmenci N (2009) Possibility of using waste tire rubber and fly ash with Portland cement as construction materials. Waste Manag 29:1541–1546. doi:10.1016/j.wasman.2008.11.002
Zhao S, Wang L, Duo L (2009) Effects of waste crumb rubber on medium characters and growth of *Lolium perenne* L. Pak J Bot 41(6):2893–2900
Zhao S, He T, Duo L (2011) Effects of crumb rubber waste as a soil conditioner on the nematode assemblage in a turfgrass soil. Appl Soil Ecol 49:94–98. doi:10.1016/j.apsoil.2011.06.011

23

Isolation and identification of bacteria from rotary drum compost of water hyacinth

Isha Vishan[1] · Senthilkumar Sivaprakasam[2] · Ajay Kalamdhad[3]

Abstract

Purpose The present work aims to isolate and identify bacterial community from the rotary drum compost of green waste such as the water hyacinth. Compost quality was also investigated with the physicochemical parameters and the heavy metal analysis.

Methods For rotary drum composting, a waste mixture of 150 kg with water hyacinth, cow dung and sawdust were prepared in the proportion of 6:3:1, respectively. The physicochemical parameters such as pH, temperature, volatile solids and the electrical conductivity were analyzed to detect the compost quality. Pb, Ni, Zn, and Cd were investigated for total heavy metals, toxicity characteristic leaching procedure (TCLP), diethylenetriaminepentaacetic acid (DTPA) and water solubility tests. Consistent and active bacterial community were isolated from the rotary drum compost of water hyacinth. Culture-dependent and culture-independent techniques were approached for the isolation process.

Results Twelve bacteria were isolated and identified by 16S rDNA sequencing and phylogenetic analysis; they majorly belonged to the *Bacillus* and *Enterobacter* family. The analysis of temperature, pH, EC, VS and heavy metals depicted the good quality of compost. Heavy metals concentration was in Pb > Ni > Zn > Cd; however, for Pb, Cd and Ni water solubility remained non-detectable, DTPA concentration for Pb and Cd was also found to be non-detectable.

Conclusions The microbes in water hyacinth compost are metabolically active degrading the organic matter, surviving in the heavy metal-loaded compost environment. Thus, green waste (water hyacinth) can be utilized to isolate bacterial species in anticipation of their application in heavy metal removal in liquid and solid waste sources for micro-bioremediation.

Keywords Rotary drum composting · *Bacillus* · *Enterobacter* · Heavy metals · PCR (polymerase chain reaction) · Micro-bioremediation

Introduction

Water hyacinth (*Eichhornia crassipes*) is an aquatic weed, originating in South America. It covers around 23.2% wetland area of the northeast part of India, it was introduced in West Bengal as an ornamental plant, but now it has become a nuisance throughout the country (Abbasi 1998; Husain 2003). Since then, water hyacinth growth has caused various problems in the whole area, e.g., reduction of fishes, blockage of shipping routes, and impeding other irrigational activities (Gunnarsson and Petersen 2007). There are many documentations in the literature regarding the utilization of water hyacinth as a substitute for animal

✉ Isha Vishan
reachisha01@gmail.com; isha@iitg.ernet.in

[1] Centre for the Environment, Indian Institute of Technology Guwahati (IIT G), Guwahati 781039, Assam, India

[2] Department of Biosciences and Bioengineering, Indian Institute of Technology Guwahati (IIT G), Guwahati 781039, Assam, India

[3] Department of Civil Engineering, Indian Institute of Technology Guwahati (IITG), Guwahati 781039, Assam, India

fodder (Tag El-Din 1992), fermentation feed and biogas production (Singhal and Rai 2003), pulp and paper industry and the vermicomposting and composting (Goswami and Saikia 1994; Gajalakshmi et al. 2001; Singh and Kalamdhad 2013a, b). Composting of water hyacinth followed by its use in the land application has come out to be the best method for treatment and utilization of this weed (Malik 2007). However, a huge amount of heavy metals has been reported in the compost of water hyacinth (Singh and Kalamdhad 2012). Additionally, a high count of actinomycetes, streptomycetes, bacteria and fungi was also observed throughout the whole period of compost formation of water hyacinth (Vishan et al. 2014). Contaminated regions such as industrial waste, wastewater, soil, municipal waste and sewage sludge have been used as source of the isolation of the most robust microbes (Affan et al. 2009; Bahig et al. 2008; Bautista-Hernández et al. 2012; Colak et al. 2011; Congeevaram et al. 2007; Jiang et al. 2008; Pal and Paul 2004).

Composting is a rigorous process of aerobic degradation of organic waste by the metabolic activities of the microorganisms, therefore it serves for the abundance of metabolically active microbes (Anastasi et al. 2005; Bhatia et al. 2013; Hassen et al. 2001; Pan and Sen 2013; Ryckeboer et al. 2003). The microbes are dependent on the surrounding conditions such as the temperature, availability of organic waste, pH, species diversity and availability of heavy metals. A large variety of mesophilic, thermotolerant microorganisms have been reported in composting of different types of waste materials, it includes bacteria, actinomycetes, streptomycetes and fungi (Hassen et al. 2001; Ryckeboer et al. 2003).

For the current study, the water hyacinth compost was utilized as a source of isolation of the most robust and persistent microbes. However, no work has been reported regarding the detection, isolation, and identification of bacteria present during the compost formation of water hyacinth using rotary drum composter. The hypothesis is that the bacterial community which is capable of sustaining the variable conditions of composting from the initial till the final day could be further utilized for the purpose of micro-bioremediation. Hence, there is an immense possibility in exploitation of microbes for bioadsorption, biotransformation, and/or bioaccumulation of the heavy metals in future (del Carmen Vargas-García et al. 2012; Anastasi et al. 2005). Accordingly, in the present study, basic physiochemical parameters and heavy metal content of compost were examined, along with this culture-dependent and culture-independent approaches were adopted to isolate and identify the bacterial strains.

Materials and methods

Rotary drum composting of water hyacinth

- Rotary drum composter

Figure 1 depicts the diagram of a rotary drum composter, with a capacity of 550 L; it works in batch mode operation. The pilot-scale rotary drum composter can hold around 150 kg of waste. The period of 20 days is sufficient for composting for proper degradation and stabilization of waste material as previously been studied during in-vessel composting (Kalamdhad and Kazmi 2009). The frequency of turning of the drum manually was per day (per 24 h) with a full rotation of the rotary drum ensuring that the waste moves from one portion to another, for higher temperature in order to degrade the waste properly (Kalamdhad and Kazmi 2009). To maintain the aerobic condition, the side doors at the ends of the rotary drum were left open.

- Feedstock materials

The waste mixture was prepared by mixing of water hyacinth, cow dung, and sawdust. Water hyacinth was obtained from the contaminated water in an industrial area called Amingaon near the campus of IIT Guwahati, Assam, India. Cow dung was brought from a dairy farm. Sawdust was purchased from a saw mill near to the campus. Fine powder of sawdust and cow dung was prepared by mixing them thoroughly. The water hyacinth was chopped down into small pieces of the order of 1 cm for maintaining proper aeration and moisture control. All the three components, i.e., 90 kg of water hyacinth, 45 kg of cow dung and 15 kg of sawdust were mixed thoroughly in the best proportion of 6:3:1 to the fine mixture as per the previous studies (Vishan et al. 2014).

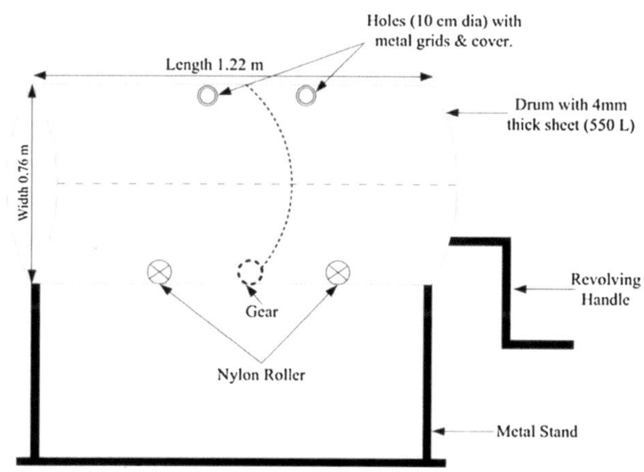

Fig. 1 Pilot-scale rotary drum composter schematic design

- Sampling and analysis of physicochemical parameters

Compost sample was grabbed from the end terminals and mid portion with the help of compost sampler in without disturbing the adjacent materials. On every 2nd-day, samples were collected in triplicates after manual turning of the rotary drum, air dried and separated through 0.2 mm of the sieve, followed by storing for physio-chemical analysis.

The observation of temperature was done thrice a day using digital thermometer throughout 20 days of the composting period. Thereafter, the samples were dried in an oven at 105 °C for 24 h to calculate the moisture content, other dried samples were used for analysis of pH (1:10 w/v, waste: water extract), volatile solids (loss on ignition at 550 °C for 2 h), total metal analysis was done by digesting 0.2 g of dried sample in 10 mL mixture of sulphuric acid perchloric acid in the ratio of 5:1 was fed in a block digester for 2 h at 300 °C (Pelican equipment Chennai-India). For water-soluble heavy metal analysis, 2.5 g of sample was extracted with distilled water of 50 mL volume, where sample and solution were in the ratio of 1:20 and kept in a shaker at 100 rpm for 2 h at room temperature. For DTPA (diethylenetriaminepentaacetic acid), 4 g of ground sample (sieved through 0.22 mm sieve) was mixed with 0.005 M DTPA (40 mL), 0.01 M $CaCl_2$ and 0.1 M (triethanolamine) (pH 7.3) and mechanically shaken at 100 rpm. The leachability of heavy metals was determined by TCLP (toxicity characteristic leaching procedure), following the EPA method 1311, 5 g of sample having size less than 9.5 mm was mixed with CH_3COOH (100 mL) at pH 4.93 ± 0.05. The sample and solution were taken in a ratio of 1:20 in 125 mL of reagent bottle and kept at 30 ± 2 rpm for 18 h at room temperature. The remaining suspensions were filtered by Whatman no. 42 filter paper after being centrifuged at 10,000 rpm for 5 min. The samples were stored in a specimen tube at 4 °C for heavy metal analysis. The heavy metal analysis was accomplished using atomic absorption spectrophotometer (AAS) (Brand Varian Spectra 55B). All the experiments were done in triplicates, and the result is depicted as the means of them.

For the calculation of the extraction efficiency of heavy metals extractable by DTPA method, the following equation was used (Singh and Kalamdhad 2013a):

$$\text{Extraction efficiency } (\%) = \frac{c_{\text{DTPA}}}{c_{\text{Total}}} \times 100,$$

where C_{DTPA} and C_{Total} denote the concentration of heavy metals extractable by DTPA, and the total heavy metal concentration, respectively.

Isolation of bacteria

- Sampling for microbial analysis

Samples were collected from mid-area, and end terminals on every fourth day such as 0, 4, 8, 12, 16, 20th day after manual turning of the rotary drum composter. The count of microbes was performed by compost addition (10 g) into a sterile distilled water (90 mL) having 0.85% (w/v) sterile NaCl solution into 250-mL Erlenmeyer flasks. Mechanical mixing of the solution was done at 150 rpm for 2 h and 25 °C. Finally, the serial dilution of waste suspensions was done for count of microbes on the different media.

- Media optimization

The first step towards the isolation of bacterial pure culture was by selecting appropriate growth media. The medium used for bacterial growth can have a major impact on the isolation efficiency. As the source of isolation was water hyacinth compost, the appropriate growth medium for bacteria was unknown. As mentioned in Table 1, four different types of media including two growth media and two minimal media were tested for a maximum number of bacterial species and consistency of performance in supporting the growth of targeted microbes. Cycloheximide (0.2 g/L) was added as antibiotic in all media to avoid contamination by fungi.

Table 1 Media compositions and their literature references

Media name	Ingredients	Reference
Nutrient agar	Peptone (5 g/L), meat extract (1g/L), yeast extract (2g/L), NaCl (5g/L), agar (20 g/L)	del Carmen Vargas-García et al. (2012)
Tryptone soy agar (HiMedia)	Casein (17 g/L), papaic digest of soya bean meal (3 g/L), NaCl (5 g/L), dextrose (2.5/L), K_2HPO_4 (2.5 g/L), agar (20 g/L)	Parungao et al. (2007)
Sucrose minimal low phosphate media	Sucrose (1%), $(NH4)_2SO_4$ (0.1%), K_2HPO_4 (0.05%), $MgSO_4$ (0.05%), NaCl (0.01%), yeast extract (0.05%), $CaCO_3$ (0.05%), agar (20 g/L)	Sheng et al. (2008)
Glucose minimal media	Glucose (1%), $(NH4)_2SO_4$ (0.1%), K_2HPO_4 (0.05%), $MgSO_4$ (0.05%), NaCl (0.01%), yeast extract (0.05%), $CaCO_3$ (0.05%), agar (20 g/L)	This media was investigated during the study by changing sugar of sucrose minimal low phosphate media to glucose

Different types of media are mentioned in the literature, on the basis of the goal for the isolation and its source. Growth can be affected by components of media, the difference can be observed in the same sample cultured in nutrient enriched and minimal media. Selection of suitable growth medium is a crucial step for the isolation of novel strains from a waste source, which is enriched with heterotrophic unidentified bacteria. However, a comparative study of bacterial growth and isolation in different nutrient media is missing in the literature.

- Culture-dependent bacterial identification

The sampling was done as described in the sampling procedure for microbial analysis. After the serial dilution of waste suspensions, diluted samples were used for microbial counts on all four media separately. Spread plate technique was performed to enumerate the bacterial colonies in the compost samples. The petri-plates were then incubated in an inverted position for 24–72 h at 25 °C. Plates with a lower number of distinguished bacterial colonies were picked up for purification and separation by quadrant streaking. Initially, streaking was done on every fourth day, and then the time interval was increased to 15 days. Plating was done till the final 20th day of composting with the observance of more number and sustainable bacterial colonies in different media. The serially diluted samples were plated on MacConkey agar for counting *Escherichia coli* number in plates kept for 24–48 h at 37 °C. Glucose minimal media depicted the maximum growth of bacterial colonies giving 12 distinguished robust bacterial isolates. Morphological testing of these 12 isolates was performed with HiMedia bacterial kit for the biochemical tests. The identifications were confirmed by using the Bergey's Manual of systematic bacteriology (Aneja 1999).

- Culture-independent bacterial identification
- Genomic DNA isolation from bacterial isolates

Genomic DNA was extracted from anexic culture pellets using the GeNei DNA Purification kit as per manufacturer's instruction. The purity of DNA was then checked by horizontal gel electrophoresis in 0.8% agarose gel.

- PCR amplification and 16S rDNA analysis

For PCR amplifications 8F (5′-AGAGTTTGATCCTG GCTCAG-3′) (forward) and 1492R (5′-CGGTTACCTTG TTACGACTT-3′) (reverse) primers were used. Reactions were performed in 50-μL volumes using the Ready mix Taq PCR reaction Mix (Sigma-Aldrich Ready Mix Taq PCR mixture) in PCR-peQLab thermocycler and following a protocol with a denaturation step at 94 °C for 5 min, followed by 35 amplification cycles (95 °C, 60 s; 52 °C, 60 s; 68 °C, 90 s) and a final extension step at 68 °C for 10 min. The amplification products were cleaned using the GenElute PCR Clean-Up kit (Sigma-Aldrich, St. Louis, MO, USA), sequenced and compared to the nucleotide database at GenBank by using BLAST (NCBI).

Results and discussion

Physiochemical analysis of compost

Microbial activities are inter-related with the variations of compost temperature; compost degradation takes place by the metabolic interventions of microbes with the organic wastes. The variability in temperature gives an index of the rate of decomposition by the biological process, and thus plays an important role in the succession of microbes throughout the composting period (Hassen et al. 2001). Figure 2 depicts the variability of the temperature throughout the 20 days of the composting period.

A maximum of 56.5 °C temperature was achieved on the fourth day. A proper combination of waste materials is important for the achievement of high temperature as it gives a proper combination of carbon and nitrogen to the microbial population for their growth and activities (Singh and Kalamdhad 2012). The temperature started rising within 24 h of feeding of the rotary drum with the feedstock and reached its maximum on the fourth day of composting, afterwards it became stabilized. As at the initial time of feeding, organic materials are readily available for microbes to degrade and decompose. In a composting process for proper degradation and maintenance of thermophilic activity, a temperature range of 52–60 °C is suitable (Ryckeboer et al. 2003).

Variability in pH causes changes in the availability of hydrogen ions which thereby affects the interaction and transport across the cell membrane of microbes, consequently affecting the composting process (Plette et al.

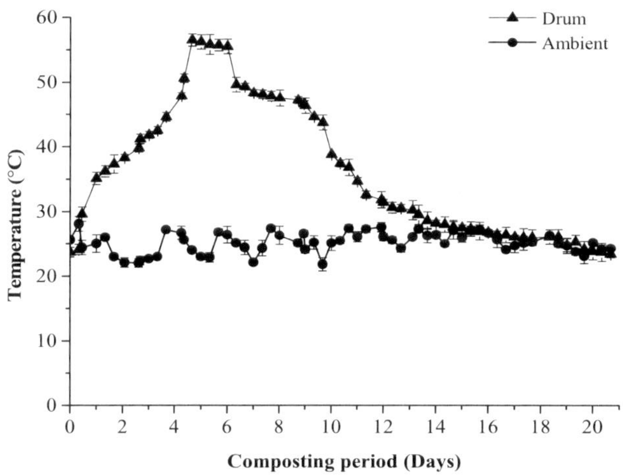

Fig. 2 Temperature variations during composting period

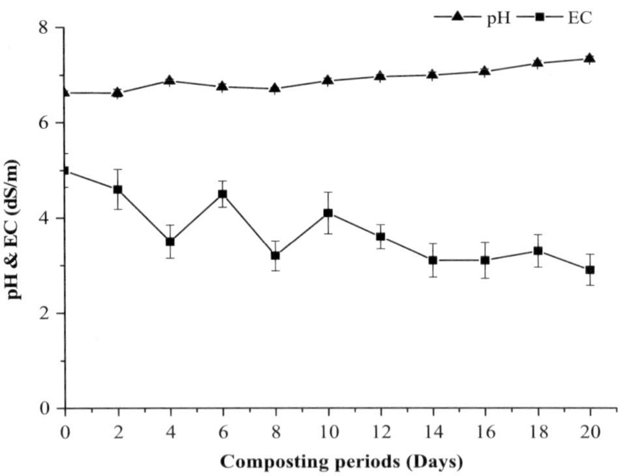

Fig. 3 pH and EC variations during the 20-day composting period

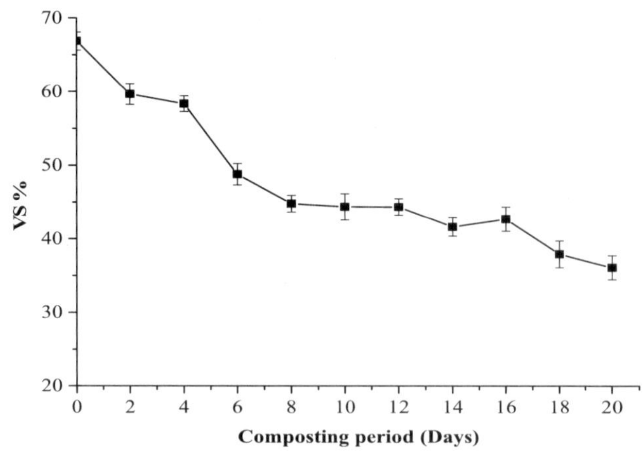

Fig. 4 VS% variations during the composting period

Table 2 Total metal concentrations during composting period (mg/kg dry matter)

Days	Pb	Cd	Zn	Ni
0	789 ± 6	42.6 ± 1	122.4 ± 2	251 ± 1.1
4	815.5 ± 4.5	46.3 ± 1.4	132.3 ± 1.4	243 ± 3.1
8	832.5 ± 2.5	48.7 ± 2.3	154.6 ± 2.5	250 ± 3.3
12	786 ± 4	49.2 ± 2.5	181.2 ± 2.1	231 ± 7.3
16	778 ± 7	49.7 ± 3.2	167.3 ± 2.3	242 ± 2.1
20	845.5 ± 4.5	53.2 ± 1.1	181.1 ± 3.1	254 ± 4.2

1995). Therefore, pH is also an important factor which influences the composting process. Bacteria can develop and survive in the optimal range of pH 6.0–7.0. A pH of 4.0 and 6.5 had been reported suitable for composting of food waste and garden waste, respectively. In composting of wheat straw, a pH range of 6.5–8.0 was reported to be favorable for microbial growth (Pan and Sen 2013). In the present study, initial pH characterization of water hyacinth, cattle (cow) manure, and sawdust was observed to be 6.5, 7.2 and 5.9, respectively. While the initial and final pH of the compost sample mixture was in the range of 5–6.6 and 7–7.6, respectively, as reported in Fig. 3. The pH value of the solution influences the precipitation, speciation, biosorption, hydrolysis and complexation process (Wang et al. 2009). These pH values favored the development of optimum conditions for bacterial growth and activities in the compost sample.

Electrical conductivity (EC) gives the chemical indication of soluble salts during composting process (Pan and Sen 2013). Transportation of nutrients and water into the plants can be reduced due to high EC, thereby affecting the growth of plants (Singh and Kalamdhad 2013a). As depicted in Fig. 3, the final EC value of compost was reduced down to 2.9 dS/m, similar values had been reported by Singh and Kalamdhad (2012). The decrease in EC at the later phase of composting can possibly be due to the precipitation of mineral salts and ammonia volatilization (Ganesh et al. 2012). Also, the EC reduction could be attributed to the release of humic substances which can reduce water solubility by interacting with the metal ions.

Volatile solids in the composting process account for the amount of organic matter oxidized due to the temperature rise. Microbes utilize the soluble organic matter as a primary source of nutrition for their growth and development. However, in the present study, the plant material, i.e., water hyacinth is rich in cellulose, hemicelluloses, reducing sugars, lignin, and lignocellulose. Where the lignocelluloses part is difficult to be solubilized (almost non-degraded) and forms humus-like substances, rest of the parts of the plant are easily degraded (Juan et al. 2013; Sarika et al. 2014). The volatile solids analyzed from 0th to 20th day reduced from 67 to 36%, i.e., an amount of 46% VS reduction was reported (Fig. 4). But this was an incomplete reduction which depicts that more time is needed for the degradation of the complex lignocellulosic material.

The total concentration of metals increases with the increase in organic matter degradation with the release of CO_2 and water (Singh and Kalamdhad 2013a). The presence of heavy metals in compost makes it too hazardous to be used in the field. Enhancement of heavy metals in soil laden with compost can increase the heavy metal concentration in plants. Table 2 demonstrates the variation in total concentration of metals (Pb, Zn, Ni, and Cd) during 20 days of the composting period. The order of heavy metals in the compost was Pb > Ni > Zn > Cd; whereas, Zn concentration was reported to be highest of about 220–370 mg/kg in the compost of sewage sludge (Chiang et al. 2007). However, all total metals concentration decreased for the compost of municipal solid waste (Castaldi et al. 2006). Total metal concentration gives information about the level of contamination, however, very less information is available on the bioavailability and leachability of heavy metals, whereby the present study focuses on the scenario of the toxicity of heavy metals in the compost and the microbes present in that contaminated environment.

Bioavailability of heavy metals during rotary drum composting process

- Water-soluble heavy metals

With the decomposition of organic matter, the fraction of heavy metals increases comparatively, but the water-soluble fraction of metals may decrease due to the changes of ionic and oxidizing conditions of surrounding compost environment. The extent of total metal content in compost is not sufficient to predict its potential toxicity, but the bioavailability and leachability of heavy metals are also important (Singh and Kalamdhad 2012). It is essential to determine the water-soluble fraction of metals as they can contaminate the food chain by being present in the compost (Singh and Kalamdhad 2013a). Table 3 shows the changes in water-soluble fractions of Pb, Zn, Ni and Cd during 20 days of the composting period. Water-soluble concentration of Zn was reported to decrease by 41% at the end of the composting period. Water-soluble fraction of Ni, Cd and Pb was not detected throughout the composting period. The increase in water-soluble concentration of Zn was reported by Singh and Kalamdhad (2012) in water hyacinth composting. A decrease of 95 and 60% water-soluble fraction of Pb, Cd and Zn, respectively, was reported in municipal solid waste composting by Castaldi et al. (2006).

Table 3 Water-soluble metal concentrations (mg/kg dry matter)

Days	Pb	Cd	Zn	Ni
0	ND	ND	2.9 ± 0.02	ND
4	ND	ND	3.1 ± 0.01	ND
8	ND	ND	4.0 ± 0.04	ND
12	ND	ND	2.4 ± 0.03	ND
16	ND	ND	2.0 ± 0.01	ND
20	ND	ND	1.7 ± 0.03	ND

ND not detected

- DTPA-extractable (plant available heavy metals)

DTPA-extractable heavy metal is the potentially available metal for the plant uptake (Chiang et al. 2007). It is a supplementary approach for finding out the potential ecotoxicity of heavy metals (Singh and Kalamdhad 2012). Table 4 shows the variations in DTPA-extractable fractions of Pb, Zn, Ni and Cd during 20 days of the composting period. Pb and Cd DTPA fractions were not detectable throughout the 20 days. The concentration of Ni was, however, reported to increase by 10% from initial till the final 20th day. Hydrolysis of metal ions due to alteration in pH contributes to the variability of metals. The concentration of Zn was observed to increase by 80% from its initial concentration. Similarly, not detectable Pb and Cd along with increased concentration of Ni and Zn were reported in water hyacinth composting (Singh and Kalamdhad 2013a).

- Leachable heavy metals during composting

TCLP depicts the mobility of the organic and inorganic components present in the solid and liquid waste materials (Singh and Kalamdhad 2013b). A heavy metal found in the leachable state can gradually percolate into the groundwater table causing contamination. For the compost, TCLP should be determined to detect its suitability of being non-hazardous for the land application. Table 5 denotes the leachability changes of Pb, Cd, Zn and Ni throughout the 20 days of the composting period. The concentration of Pb was reduced by 33% from the initial day of composting; however, an increase in Pb concentration was reported in water hyacinth composting by Singh and Kalamdhad (2012), whereas for Cd, the concentration was observed to be reduced by 57%. For Ni, 39% of reduction was reported, whereas 6% of the increase was observed for Zn from their initial till the final day concentration. The influence of pH on the solubility of metal ions is well known, either by solubility equilibria or by complexation pH increase affects

Table 4 DTPA-extractable metal concentrations (mg/kg dry matter)

Days	Pb	Cd	Zn	Ni
0	ND	ND	29.8 ± 0.4	2.0 ± 0.01
4	ND	ND	33.0 ± 0.1	1.6 ± 0.05
8	ND	ND	37.4 ± 2.4	1.4 ± 0.01
12	ND	ND	65.1 ± 0.2	1.6 ± 0.03
16	ND	ND	51.3 ± 0.2	1.7 ± 0.09
20	ND	ND	53.8 ± 0.2	2.2 ± 0.02

ND not detected

Table 5 TCLP-leachable metal concentrations (mg/kg dry matter)

Days	Pb	Cd	Zn	Ni
0	7.5 ± 0.1	2.1 ± 0.02	25.1 ± 0.5	7.8 ± 0.5
4	4.8 ± 0.2	2.3 ± 0.2	25.3 ± 0.4	5.3 ± 0.6
8	8.2 ± 0.6	2.0 ± 0.1	28.8 ± 1.7	4.6 ± 2.3
12	4.2 ± 0.4	1.8 ± 0.05	44.9 ± 0.6	5.5 ± 0.5
16	2.2 ± 0.2	1.6 ± 0.07	31.7 ± 1.1	6.6 ± 0.4
20	5 ± 0.4	0.9 ± 0.06	26.7 ± 0.2	4.7 ± 0.3

ND not detected

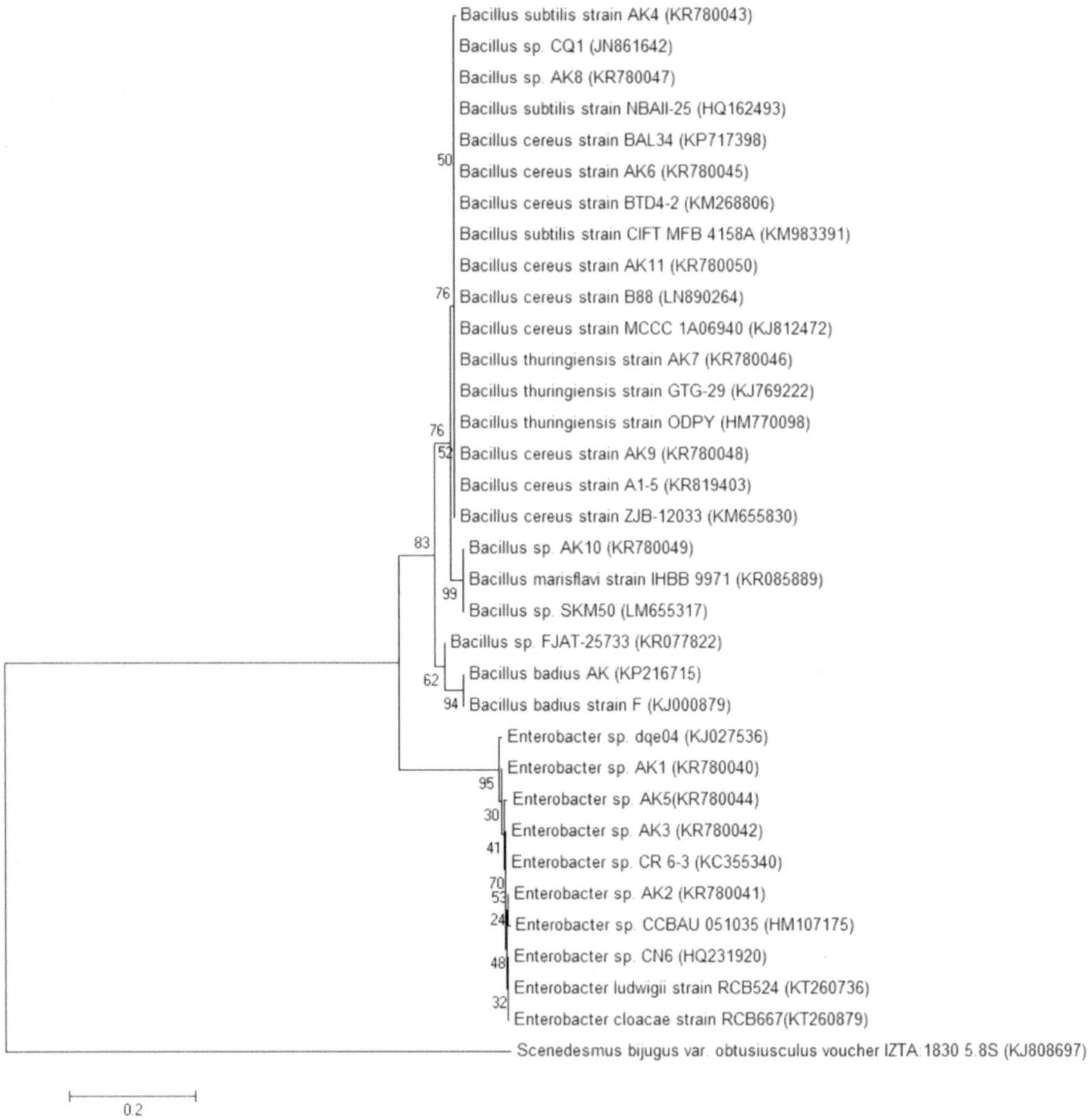

Fig. 5 Phylogenetic tree

the solubility of divalent trace metals (Cambier and Charlatchka 1999).

- Isolation and identification analysis

Media optimization was done on the basis of their maximum consistency and ability to support the growth of bacterial populations. For this, the different selective media were used for culturing bacterial populations during the whole composting period. From the previous study, it was found that about 10^4–10^7 dilutions consistently gave a good colony count in case of mesophilic bacteria. The initial count was as high as 50 for glucose minimal media which decreased to 18 gradually. This was synchronous with nutrient-rich media which yielded 41 colonies initially, reducing to 17 colonies in the end. The total number of colonies obtained from the compost was 656. The glucose minimal media gave the maximum 222 colonies followed by nutrient media with 214 colonies. Sucrose minimal media and tryptic soy agar media yielded a total of 102 and 118 colonies, respectively. The quadrant streaking was continued for 10 cycles which concluded with total 12 isolates from all four media. A total of 12 distinguishable isolates were obtained at the end. All the discrete isolates underwent morphological and biochemical tests as described in Table 5. Gram staining and phase contrast microscopy gave the shape identification of bacterial colonies, which were spotted to be in different forms such as, rod, circular and oval shapes (Bhatia et al. 2013). The colors of the strains were mostly, white, yellow, cream, orange and pink.

- Molecular identification

The sequence analysis of the 16S rRNA gene is an accurate and faster method to identify the phylogenetic position of bacteria. A full-length (about 1500 bp) 16S rDNA of all 12 bacterial isolates was sequenced and used for construction of a phylogenetic tree. The 16S rDNA gene sequence was used to carry out the BLAST alignment of NCBI GenBank database. On the basis of maximum identity score, the sequences were selected and aligned by using a multiple alignment software program such as Clustal W. Distance matrix was generated using RDP database and the phylogenetic tree was constructed using MEGA 5 (Tamura et al. 2013). Figure 5 shows that the evolutionary history was inferred using the neighbor-joining method (Saitou and Nei 1987). The bootstrap consensus tree inferred from 1000 replicates is taken to represent the evolutionary history of the taxa analyzed (Felsenstein 1985). Branches corresponding to partitions reproduced in less than 50% bootstrap replicates were collapsed. The percentage of replicate trees in which the associated taxa clustered together in the bootstrap test (1000 replicates) is shown next to the branches (Felsenstein 1985). The evolutionary distances were computed using the Kimura 2-parameter method (Kimura 1980) and are in the units of the number of base substitutions per site. The accession nos. received from NCBI were KR780040–KR780050 and KP216715. The bacterial groups were *Enterobacter, Bacillus subtilis, Bacillus cereus, Bacillus badius, Bacillus thuringiensis* (Fig. 5). Although composting is an aerobic process, members of facultative anaerobic family, i.e., *Bacillus* are much prevalent which signifies the versatility and survival capacity of these microbes.

Conclusion

Rotary drum composting of water hyacinth is enriched with numerous microorganisms. Bacteria in this environment are metabolically active which leads to degradation of the organic matter, also they are surviving in a heavy metal-loaded compost environment. 12 new robust bacterial strains were isolated as being the most consistent throughout the 20 days of the composting period. Huge degradation of organic matter, temperature, pH and heavy metals was observed. Both the culture-dependent and culture-independent approach yielded new bacterial strains. They survived not only the rigorous variations of composting process, but also the presence of toxic heavy metals in the compost. Reduction in water solubility, leachability and DTPA values of heavy metals indicated the efficient degradation of the organic matter by the microbial activities during the composting process. These bacterial species can be investigated further for their capability of heavy metal removal in liquid and solid waste sources. In this way, these bacteria can contribute to micro-bioremediation.

Acknowledgements The authors gratefully acknowledge the support by Central Instrument facility (CIF) IITG, Centre for the Environment Laboratory, IITG.

References

Abbasi SA (1998) Weeds of despair, and hope. Wetl India 3:12–21

Affan QUA, Shoeb E, Badar U, Akhtar J (2009) Isolation and characterization of bacterial isolates having heavy metal tolerance. J Basic Appl Sci 5:55–60

Anastasi A, Varese GC, Marchisio VF (2005) Isolation and identification of fungal communities in compost and vermicompost. Mycologia 97:33–44. doi:10.3852/mycologia.97.1.33

Aneja KR (1999) Experiments in microbiology plant pathology and biotechnology, 4th edn. New Age International Publishers, New Delhi. ISBN 81-224-1494-x

Bahig AE, Aly EA, Khaled AA, Amel KA (2008) Isolation, characterization and application of bacterial population from agricultural soil at Sohag Province, Egypt. Malays J Microbiol 4:42–50. doi:10.21161/mjm.11808

Bautista-Hernández DA, Ramírez-Burgos LI, Duran-Páramo E, Fernández-Linares L (2012) Zinc and lead biosorption by *Delftia tsuruhatensis*: bacterial strain resistant to metals isolated from mine tailings. J Water Resour Prot 4(4):207–216. doi:10.4236/jwarp.2012.44023

Bhatia A, Madan S, Sahoo J, Ali M, Pathania R, Kazmi AA (2013) Diversity of bacterial isolates during full scale rotary drum composting. Waste Manag 33(7):1595–1601. doi:10.1016/j.wasman.2013.03.019

Cambier P, Charlatchka R (1999) Influence of reducing conditions on the mobility of divalent trace metals in soils. In: Selim HM, Iskandar IK (eds) Fate and transport of heavy metals in the vadose zone. Lewis Publishers, Boca Raton/London/New York

Castaldi P, Santona L, Melis P (2006) Evolution of heavy metals mobility during municipal solid waste composting. Fresen Environ Bull 15(9):1133–1140

Chiang KY, Huang HJ, Chang CN (2007) Enhancement of heavy metal stabilization by different amendments during sewage sludge composting process. J Environ Eng Manag 17(4):249–256

Colak F, Atar N, Yazicioglu D, Olgun A (2011) Biosorption of lead from aqueous solutions by *Bacillus* strains possessing heavy-metal resistance. Chem Eng J 173:422–428. doi:10.1016/j.cej.2011.07.084

Congeevaram S, Dharani S, Park J, Dexillin M, Thamaraiselvi K (2007) Bio-absorption of chromium and nickel by heavy metal resistant fungal and bacterial isolates. J Hazard Mater 146(1–2):270–277. doi:10.1016/j.jhazmat.2006.12.017

del Carmen Vargas-García M, López MJ, Suárez-Estrella F, Moreno J (2012) Compost as a source of microbial isolates for the bioremediation of heavy metals: in vitro selection. Sci Total Environ 431:62–71. doi:10.1016/j.scitotenv.2012.05.026

El-Din Tag (1992) Utilization of water hyacinth hay in feeding of growing sheep. Indian J Anim Sci 62:89–992

Felsenstein J (1985) Confidence limits on phylogenies: an approach using the bootstrap. Evolution. doi:10.2307/2408678

Gajalakshmi S, Ramasamy EV, Abbasi SA (2001) Assessment of sustainable vermiconversion of water hyacinth at different reactor efficiencies employing *Eudrilus eugeniae* Kinberg. Bioresour Technol 80:131–135. doi:10.1016/S0960-8524(01)00077-3

Ganesh CD, Roshan SW, Khwairakpam M, Kalamdhad AS (2012) Composting of water hyacinth using sawdust/rice straw as a bulking agent. Int J Environ Sci 2(3):1223–1238

Goswami T, Saikia CN (1994) Water hyacinth—a potential source of raw material for greaseproof paper. Bioresour Technol 50:235–238. doi:10.1016/0960-8524(94)90095-7

Gunnarsson CC, Petersen CM (2007) Water hyacinth as a resource in agriculture and energy production: a literature review. Waste Manag 27:117–129. doi:10.1016/j.wasman.2005.12.011

Hassen A, Belguith K, Jedidi N, Cherif M, Boudabous A (2001) Microbial characterization during composting of municipal solid waste. Bioresour Technol 80:217–225. doi:10.1016/S0960-8524(01)00065-7

Husain Z (2003) Environmental issues of Northeast India. Regency Publication, New Delhi

Jiang CY, Sheng XF, Qian M, Wang QY (2008) Isolation and characterization of heavy-metal resistant bacteria *Burkholderia* sp. from heavy-metal contaminated paddy field soil and its potential in promoting plant growth and heavy metal accumulation in metal-polluted soil. Chemosphere 72(2):157–164. doi:10.1016/j.chemosphere.2008.02.006

Juan ALG, Lopez MJ, Vargas-Garcia MC, Suarez-Estrella F, Jurado M, Moreno J (2013) Tracking organic matter and microbiota dynamics during the stages of lignocellulosic waste composting. Bioresour Technol 146:574–584. doi:10.1016/j.biortech.2013.07.122

Kalamdhad AS, Kazmi AA (2009) Effects of turning frequency on compost stability and some chemical characteristics in a rotary drum composter. Chemosphere 74:1327–1334. doi:10.1016/j.chemosphere.2008.11.058

Kimura M (1980) A simple method for estimating evolutionary rates of base substitutions through comparative studies of nucleotide sequences. J Mol Evol 16(2):111–120. doi:10.1007/BF01731581

Malik A (2007) Environmental challenge vis a vis opportunity: the case of water hyacinth. Environ Int 33(1):122–138. doi:10.1016/j.envint.2006.08.004

Pal A, Paul AK (2004) Aerobic chromate reduction by chromium-resistant bacteria isolated from serpentine soil. Microbiol Res 159:347–354. doi:10.1016/j.micres.2004.08.001

Pan I, Sen SK (2013) Microbial and Physico-chemical analysis of composting process of wheat straw. Indian J. Biotechol 12:120–128

Parungao MM, Tacata PS, Tanayan CRG, Trinidad LC (2007) Biosorption of copper, cadmium and lead by copper-resistant bacteria isolated from Mogpog River, Marinduque. Philipp J Sci 136(2):155

Plette ACC, Van Riemsdijk WH, Benedetti MF, Van der Wal A (1995) pH dependent charging behavior of isolated cell walls of a gram-positive soil bacterium. J Colloid Int Sci 173:354–363

Ryckeboer J, Mergaert J, Vaes K, Klammer S, De Clercq D, Coosemans J, Insam H, Swings J (2003) A survey of bacteria and fungi occurring during composting and self-heating processes. Ann Microbiol 53:349–410

Saitou N, Nei M (1987) The neighbor-joining method: a new method for reconstructing phylogenetic trees. Mol Biol Evol 4(4):406–425

Sarika D, Singh J, Prasad R, Vishan I, Varma VS, Kalamdhad AS (2014) Study of physico-chemical and biochemical parameters during rotary drum composting of water hyacinth. Int J Recycl Org Waste Agric 3(3):1–10. doi:10.1007/s40093-014-0063-1

Sheng XF, Xia JJ, Jiang CY, He LY, Qian M (2008) Characterization of heavy metal-resistant endophytic bacteria from rape (*Brassica napus*) roots and their potential in promoting the growth and lead accumulation of rape. Environ Pollut 156(3):1164–1170. doi:10.1016/j.envpol.2008.04.007

Singh J, Kalamdhad AS (2012) Concentration and speciation of heavy metals during water hyacinth composting. Bioresour Technol 124:169–179. doi:10.1016/j.biortech.2012.08.043

Singh J, Kalamdhad AS (2013a) Assessment of bioavailability and leachability of heavy metals during rotary drum composting of green waste (water hyacinth). Ecol Eng 52:59–69. doi:10.1016/j.ecoleng.2012.12.090

Singh J, Kalamdhad AS (2013b) Reduction of bioavailability and leachability of heavy metals during vermicomposting of water hyacinth. Environ Sci Pollut Res 20:8974–8985. doi:10.1007/s11356-013-1848-x

Singhal V, Rai JPN (2003) Biogas production from water hyacinth and channel grass used for phytoremediation of industrial effluents. Bioresour Technol 86:221–225. doi:10.1016/S0960-8524(02)00178-5

Tamura K, Stecher G, Peterson D, Filipski A, Kumar S (2013) MEGA6: molecular evolutionary genetics analysis version 6.0. Mol Biol Evol 30(12):2725–2729. doi:10.1093/molbev/mst197

Vishan I, Kanekar H, Kalamdhad A (2014) Microbial population, stability and maturity analysis of rotary drum composting of water hyacinth. Biologia 69(10):1303–1313. doi:10.2478/s11756-014-0450-0

Wang SG, Sun XF, Gong WX, Ma Y (2009) Biosorption of metals on to granular sludge. Heavy metals in the environment. Taylor and Francis Group/CRC Press, Boca Raton, pp 201–223

24

Evolution of chemical and biological characterization during agitated pile composting of flower waste

Dayanand Sharma[1] · V. Sudharsan Varma[2] · Kunwar D. Yadav[1] · Ajay. S. Kalamdhad[2]

Abstract
Purpose Flower waste dumping in landfills and other disposal sites is causing major environmental issues due to its highly biodegradable nature. Instead, flower waste can be successfully composted for its higher nutrient value. Therefore, this study evaluated the pile composting of flower waste with different combinations of cow dung and saw dust to produce stabilized compost.
Methods Five agitated piles of flower waste mixture were formed in trapezoidal shape to perform the composting. Pile (P_0) was a control experiment, which contained 100 kg flower waste. The quantity of flower waste mixture with cow dung and saw dust in the other Piles was a total of 100 kg i.e. P_1 (50:40:10) kg, P_2 (60:30:10) kg, P_3 (70:20:10) kg and P_4 (80:10:10) kg, respectively.
Results Temperature above 50 °C was observed in all the piles except in P_0. Physico-chemical characteristics of compost showed a stabilized compost obtained by proportionate mix of flower waste, cow dung and saw dust. Pile P_2 was observed with pH 7.23 and electrical conductivity 3.98 mS cm^{-1}, total organic carbon 30.9%, total nitrogen 2.31% and C:N ratio 14, thus a good proportion for compost amongst all the piles.
Conclusions Appropriate proportion of waste mixture played an important role in providing favorable conditions for the microbial transformation of flower waste to stabilized compost. Temperature influenced the growth of microbes and maximum temperature was helpful in fast degradation of organic matter. Finally, flower waste with a combination of cow dung and saw dust was found to be successful during pile composting.

Keywords Agitated pile · C:N ratio · Co-composting · Flower waste · Nutrients

✉ Dayanand Sharma
dayanandsvnit@gmail.com

[1] Civil Engineering Department, Sardar Vallabhbhai National Institute of Technology, Surat, Gujarat 395007, India

[2] Civil Engineering Department, Indian Institute of Technology, Guwahati, Assam 781039, India

Introduction

In India, flowers are used for various purposes such as an offering to deities in religious places for worship, making garlands and bouquets, as decorative item for various religious ceremonies and social gatherings. As a result, about 300 MT of flower waste is generated per day and it has been found to be one of the major constituents of municipal solid waste (Singh et al. 2013). Most of the flower waste generated is disposed by directly dumping it in open land, water bodies. Alternatively, it is mixed with municipal solid waste for further treatment. Significant increase in leachate formation has been reported when flower waste/organic waste is mixed with municipal waste and is disposed off. Leachate production and percolation into ground water pollutes the ground of nearby landfill area (Varma and Kalamdhad 2014b). Flower waste disposed into rivers causes pollution of river water and is also harmful for both human and aquatic life. Thus, owing to the organic nature of flower waste and the significant presence of macro and micronutrients, composting offers a feasible solution to deal with flower waste.

Composting is defined as the natural biological decomposition of organic matter under self heating, aerobic and moist conditions to produce a stable nutrient enriched product which is used as organic manure (Bustamante et al.

2009). Organic wastes that have high moisture content generate the leachate during the initial stage of composting and are responsible for decrease in temperature and development of the anaerobic condition. Leachate problem can be solved by addition of bulking agent which also allows free passage of air during the process of composting. Bulking agents also act as buffer against the organic acids produced during the initial stage of decomposition. They also help to maintain the pH 6–8 (Sánchez-Monedero et al. 2001). Typically used bulking agents for composting process are saw dust and dry leaves (Varma and Kalamdhad 2014b), newspaper (Ball et al. 2000), Wheat bran (Wang et al. 2012), Pumice (Wang et al. 2014; Wu et al. 2015) and wood shaving (Adhikari et al. 2009).

Several studies have been reported for agitated pile composting of organic waste such as vegetable waste (Varma and Kalamdhad 2014a), water hyacinth (Singh and Kalamdhad 2015), municipal solid waste (Iyengar and Bhave 2006), chicken litter (Gao et al. 2010), sewage sludge and pig faeces (Huang et al. 2004), food waste and saw dust (Zhou et al. 2014). However, there is limited literature available on the composting of flower waste with cow dung and saw dust as bulking agent.

Therefore, this study was focused on physico-chemical and biological characterization during composting of flower waste with cow dung and saw dust in different proportions. The best proportion for flower waste composting along with the addition of cow dung and saw dust was evaluated based on maturity and final micro-macronutrient concentration during the different stages of the composting process.

Materials and methods

Feedstock

This study was conducted in Environmental Engineering laboratory, SVNIT, Surat, Gujarat (India). Surat is situated on the bank of the river Tapti. It is located at 21.17°N latitude and 72.83°E longitude. Flower waste was collected from temples in Surat area, India. The debris (plastic, threads, incense sticks, coconut, etc.) was removed from flower waste manually. Large quantities of marigold were found along with rose, lotus, and siroi lily. Fresh cow dung was collected from a dairy farm in a nearby village. Saw dust was collected from a saw mill in Bhatar market. Table 1 shows the initial characteristics of feedstock.

Agitated pile composting

Identical agitated piles of trapezoidal shaped were prepared. The bottom length was 2700 mm, width 450 mm, height 500 mm and top width was 100 mm and L/W ratio was kept as 6 with 100 kg of total weight. Cow dung and saw dust were mixed thoroughly followed by addition of the flower waste. The mixture was again made to blend properly. Studies were conducted in duplicate and average value was reported. Initially, all the piles were covered with plastic sheets for six days. The proportions of each pile are shown in Table 2.

All piles were manually turned upside down after three days. A sample of one kg was then collected from the top bottom and the centre at various locations of the pile and thoroughly mixed to form a homogenous sample. This process was repeated every three days for the first thirty days and then at an interval of seven days for the next thirty days. The collected samples were dried and crushed into fine powder. They were sieved through a 0.2 mm sieve and kept in desiccator for further analysis.

Analysis of physico-chemical parameters

The temperature was monitored throughout the composting period using digital thermometers. The temperature at different locations of each pile was measured and the average temperature was reported. Moisture content was monitored up to 30 days and after that moisture content was adjusted between 55 to 60%, if needed; water was added on dry weight basis during the turning of piles. Gravimetric method was used for determining the moisture content of the sample after pile turning (BIS, 1982). For determining the pH and conductivity, 10 g dried sieved sample was diluted by 100 mL distilled water (1:10 w/v) and kept through mixing in rotary shaker for 2 h. Then the sample was left to be settled for half an hour for settling and finally filtered through the whatman filter paper no 42 (Singh and Kalamdhad 2016).

Total nitrogen was determined by Kjeldahl method. Ammonical nitrogen (NH_4^+-N) was determined by KCl extraction method followed by phenate method (APHA 2005). Volatile solids were determined by keeping the oven dried sieved sample in a muffle furnace for 550 ± 5 °C for 2 h. The total organic carbon was calculated by dividing the volatile solids by 1.83 (Adhikari et al. 2009).

Fresh wet samples were used for the presence of CO_2 and determination of germination index. The CO_2 evolution rate was determined as described by Singh and Kalamdhad (2014). The Germination Index test was determined as described by Zhang and Sun (2015).

The macronutrients (P, K, Na, Ca and Mg) and the micronutrients (Fe, Mn, Zn and Cu) present in the compost were analyzed according to Singh and Kalamdhad (2012). The sample was digested before analysis by heating 0.2 g dry sample with a ratio of 5:1 of 10 mL H_2SO_4 and $HClO_4$ at 300 °C for 2 h. The digested sample was used for determining the total phosphorous present using stannous

Table 1 Initial characteristics of feedstock material

Parameters	Flowers waste	Cow dung	Saw dust
Moisture content (%)	82.01 ± 0.7	80.10 ± 2.1	16.71 ± 0.7
pH	5.26 ± 0.03	7.19 ± 0.04	5.85 ± 0.90
Electrical conductivity (mS cm^{-1})	4.51 ± 0.08	3.29 ± 0.08	0.98 ± 0.91
Total organic carbon (%)	44.70 ± 1.5	32.20 ± 1.8	54.20 ± 1.2
Total nitrogen (%)	2.13 ± 0.08	1.39 ± 0.20	0.68 ± 0.09
NH_4^+-N (%)	1.80 ± 0.08	0.37 ± 0.08	0.21 ± 0.07
C:N ratio	20.90 ± 0.3	23.10 ± 0.4	79.77 ± 2.1
Total phosphorous (g kg^{-1})	3.15 ± 0.08	2.52 ± 0.07	1.25 ± 0.09
Na (g kg^{-1})	0.90 ± 0.08	2.49 ± 0.05	0.65 ± 0.07
K (g kg^{-1})	13.75 ± 0.50	10.42 ± 0.31	1.31 ± 0.15
Ca (g kg^{-1})	6.31 ± 0.95	8.92 ± 0.63	2.94 ± 0.70
Cu (mg kg^{-1})	48.62 ± 0.3	36.24 ± 0.2	17.82 ± 0.2
Zn (mg kg^{-1})	129.53 ± 1.0	178.22 ± 1.3	6.21 ± 1.3
Mn (mg kg^{-1})	98 ± 0.8	258 ± 1.1	56 ± 1.0
Fe (g kg^{-1})	1.86 ± 0.57	1.88 ± 0.57	0.47 ± 0.51
Mg (g kg^{-1})	2.55 ± 0.19	5.83 ± 0.15	1.23 ± 0.12

NH_4^+-N ammoniacal nitrogen, *C:N ratio* carbon to nitrogen ratio, *Na* sodium, *K* potassium, *Ca* calcium, *Cu* copper, *Zn* zinc, *Mn* manganese, *Fe* iron, *Mg* magnesium

Table 2 Combination of flower waste, cow dung and saw dust

Name of pile	Flower waste (kg)	Cow dung (kg)	Saw dust (kg)
P_0	100	–	–
P_1	50	40	10
P_2	60	30	10
P_3	70	20	10
P_4	80	10	10

chloride method (APHA 2005; Singh and Kalamdhad 2014). The concentrations of Na and K were determined using flame photometer (Systronics 128μ) and the concentrations of calcium (Ca), copper (Cu), zinc (Zn), manganese (Mn), iron (Fe), chromium (Cr) and magnesium (Mg) were determined by Inductively Coupled Plasma Atomic Emission Spectroscopy method.

Statistical analysis

The value of measured Physico-chemical parameters reported for each trial is the mean of duplicate. A one way repeated measure ANOVA was used to test for differences in each of the measured Physico-chemical parameters during composting. SPSS 13.0 was used to perform the analysis of variance (ANOVA $P < 0.05$) for each parameter among five piles. The objective of statistical analysis was to determine the significant differences among the parameters analyzed for different combinations. Table 3 shows the Anova of physico-chemical parameters.

Results and discussion

Initial characteristic of waste material

The initial characteristic of the combination of different waste mixture content piles is shown in Table 1. Flower waste, cow dung and saw dust contained 82.01, 80.10 and 16.71% of moisture content, respectively. Singh and Kalamdhad (2014) reported that bulking agent regulates moisture content and provides free air passage. In pile P_0 (100 kg), flower waste lump formation was observed on the third day and it was possibly due to excess moisture content and lack of air passage between the particles. Leachate production was also observed in pile P_0. The covering of each pile with plastic sheet helped to increase the temperature inside the pile and it reached the thermophilic phase within one day. Due to the presence of moisture and the action of microbes there was an increase in humidity inside the pile. As a result, vapours were seen on the inner surface of the plastic sheets.

The bacterial activity increased due to the temperature inside the pile and presence of moisture (water) converted this heat into steam. Varma and Kalamdhad (2014b) reported that vapour provided was responsible for the breakdown of tough organic material (lignin or tannin). Turning or proper mixing of piles in every three days provided adequate oxygen to the piles, which reduced the stucking and enhanced the temperature.

Table 3 Anova of physicochemical parameters

Parameters	Anova	Sum of squares	Degree of freedom (*df*)	Mean square	*F* value	*P* value
Electrical conductivity	Between groups	75.780	4	18.945	209.660	0.0001
	Within groups	6.777	75	0.090		
	Total	82.557	79			
Total organic carbon	Between groups	279.835	4	69.959	5.004	0.001
	Within groups	1048.604	75	13.981		
	Total	1328.439	79			
C:N ratio	Between groups	269.943	4	67.486	4.677	0.002
	Within groups	1082.122	75	14.428		
	Total	1352.066	79			
Ash content	Between groups	937.140	4	234.285	5.004	0.001
	Within groups	3511.670	75	46.822		
	Total	4448.810	79			
Total nitrogen	Between groups	0.785	4	0.196	5.115	0.001
	Within groups	2.878	75	0.038		
	Total	3.663	79			
Sodium	Between groups	15.420	4	3.855	21.046	0.0001
	Within groups	13.737	75	0.183		
	Total	29.157	79			
Potassium	Between groups	205.705	4	51.426	54.613	0.0001
	Within groups	70.623	75	0.942		
	Total	276.329	79			

Temperature

Temperature is one of the key indicators for active microbial biomass. Figure 1 shows variation of temperature in each pile during the composting process. Each pile attained the mesosphilic phase (Temp <40 °C), thermophilic phase (>45 °C), cooling and maturation phase. However, pile P_0 could not attain proper thermophilic phase (>45 °C) because of the presence of excess moisture

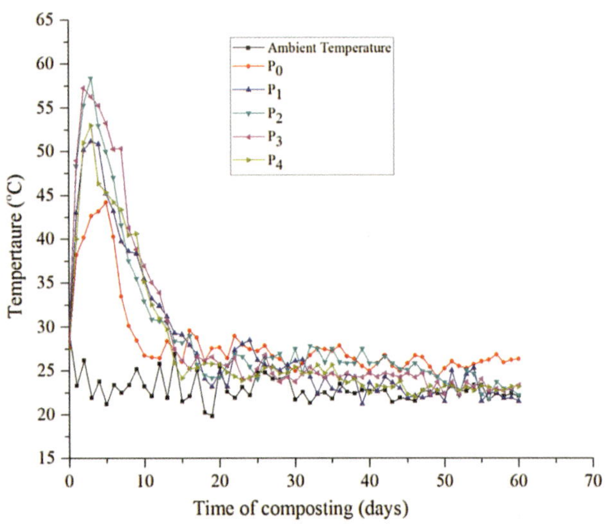

Fig. 1 Variation of temperature during the composting period

which was unfavourable for the growth of microbes. Piles P_1 and P_4 remained at temperature >50 °C for three and two days, respectively. Piles P_2 and P_3 remained at >50 °C for 4 and 6 days, respectively. Maximum thermophilic phase was observed in piles P_2 and P_3 due to the proper combination of flower waste and cow dung. The presence of higher and lower proportion of waste mixture was responsible for the duration of thermophilic phase and for achieving maximum temperature. The maximum temperature recorded in piles P_2 and P_3 was 58.40 and 57.32 °C, respectively. Temperature between 50 and 60 °C for two consecutive days kills the pathogens and this sanitizes the compost (Awasthi et al. 2015). For each pile, the observation was that within 15 days the temperature became near to the ambient temperature. Similar variations of temperature were reported by Abdullah et al. (2013), Singh and Kalamdhad (2014), Varma and Kalamdhad (2014b) for the composting of different organic waste.

Moisture content

Figure 2 illustrates moisture content variations in each pile. Initial moisture content of piles P_0, P_1, P_2, P_3 and P_4 were 82.0 ± 2.68%, 72.5 ± 3.01%, 73.7 ± 2.92%, 76.7 ± 3.21%, 81.2 ± 2.87%, respectively. Moisture content decreases linearly in P_1, P_2, P_3 and P_4. Whereas increase in moisture content and formation of leachate was observed in

P_0, decrease in moisture content was primarily because of the use of saw dust, which absorbed excess moisture, and thus maintained moisture content, whereas in absence of bulking agent, moisture content increased. Simultaneously, with the increase of cow dung, decrease in temperature was observed. Cow dung provides microbial inoculums which enhances microbial activity resulting in increasing the rate of degradation of organic matter. P_1 and P_4 took more time because of the presence of ratio of cow dung and flower waste i.e. 4:5 and 1:8, respectively. This indicates that the ratio of cow dung and flower waste significantly affects the composting and the adequate ratio must be maintained.

pH and Electrical conductivity (EC)

Figure 3 shows the variation in pH and electrical conductivity during composting. The increase in pH value was observed for the first 15 days and later a slight variation in pH value was observed for the next 15 days (15–30 days). After 30 days, pH of compost was stabilized at 7.5 ± 1.2 in each pile. The increase in pH was mainly due to the regular turning process which assists in maintaining adequate oxygen to keep aerobic condition (Awasthi et al. 2015).

The neutral pH value was observed after the 12th day as flower waste contained high water soluble K concentration which reacts with bi-carbonic acid (HCO_3^-) during the degradation of organic matter to form strong base KOH. The buffering capacity of humus and volatilization of ammonia result in a slight decrease in pH value. Similar results were reported for organic waste composting (Kalamdhad and Kazmi 2009; Mena et al. 2003). The final pH value between 6 and 8 shows the maturity of the compost (Varma and Kalamdhad 2014a).

Electrical conductivity represents the salinity of compost. The range of EC into compost is 0.14–12 mS cm^{-1} which is

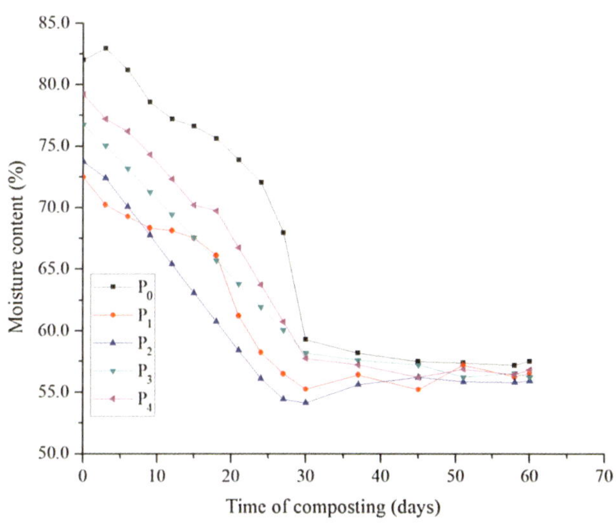

Fig. 2 Variation of moisture content during the composting period

suitable for the growth of plants (Huang et al. 2004). Initial low value in electrical conductivity in waste composition shows the suitability of flower waste for composting. It was observed that the increased electrical conductivity in the beginning of thermophilic phase became stable in cooling phase. However, this behaviour was not observed in pile P_0 where electrical conductivity increased. The degradation of organic matter took place in the beginning of the thermophilic phase which released mineral salts such as ammonium and phosphate and that increased the electrical conductivity. Stabilized electrical conductivity value attributes to the volatilization of mineral salts (Yadav et al. 2013). Inspite of electrical conductivity value in between 0.12 and12.2 mS cm^{-1} in P_0, it did not indicate quality compost due to the high moisture content and the low temperature which influence the degradation of the organic matter.

Total organic carbon (TOC), ash content and ammoniacal nitrogen (NH_4^+-N)

Changes in carbon concentration indicate compost maturity. Higher rate of carbon content degradation indicates stabilized compost at the end of the composting. Variation of total organic carbon is shown in Fig. 4a. Reduction of total organic carbon at the end of composting in piles P_0, P_1, P_2, P_3 and P_4 were 13.5, 24.5, 29.6, 30.9 and 22.4%, respectively. The low reduction of total organic carbon in P_0 was due to unavailability of inoculum such as cow dung and the absence of temperature >50 °C. Pile P_1 had easily digestible organic substrate for microbes. This reduction in total organic carbon was higher than that of P_4. The percentage reduction of total organic carbon in P_2 and P_3 was higher due to the availability of thermophilic bacteria which enhances the degradation rate of complex cellulose and protein. A similar result in the reduction of total organic carbon in organic waste was reported in the literature (Awasthi et al. 2015; Zhou et al. 2014). The reduction of the total organic carbon content indicates the mineralization of the organic matter and the maturity of the compost.

Figure 4b shows the variation of ash contents during the composting period. Initially the presence of ash content in pile P_0, P_1, P_2, P_3 and P_4 were 18.65 ± 2.63, 19.44 ± 3.12, 19.51 ± 2.75, 17.26 ± 1.95 and 17.39 ± 2.75%, respectively, which was increased, respectively, to 29.70 ± 3.11, 39.22 ± 2.42, 43.37 ± 1.58, 42.90 ± 1.75 and 35.95 ± 2.35% at the end of composting period. Higher rate of increase in ash contents was observed in pile P_2 (6:3:1) ratio of waste mixture due to the higher rate of degradation of organic matter at the time of thermophilic phase. In pile P_0, the increase in percentage of ash content was less as compared to the pile containing cow dung and saw dust due to lower temperature and presence of higher moisture content.

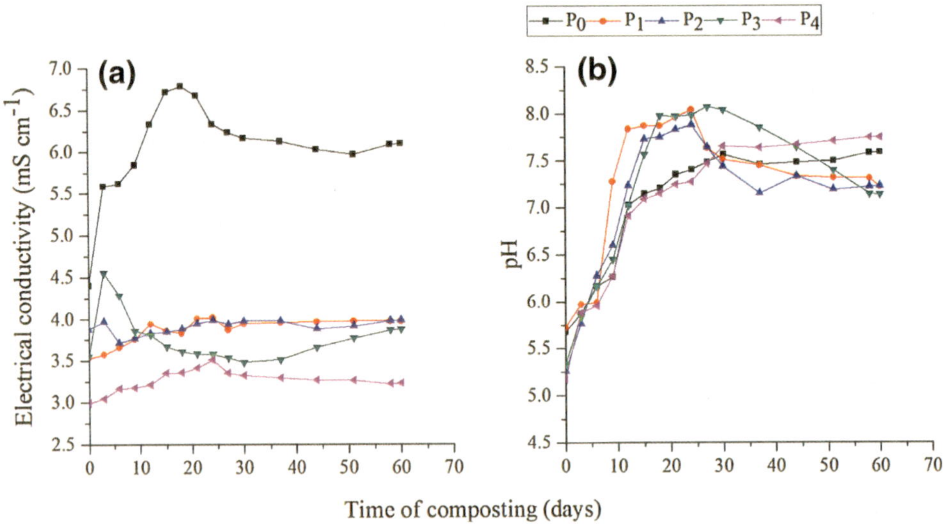

Fig. 3 Variation of pH and electrical conductivity during composting period

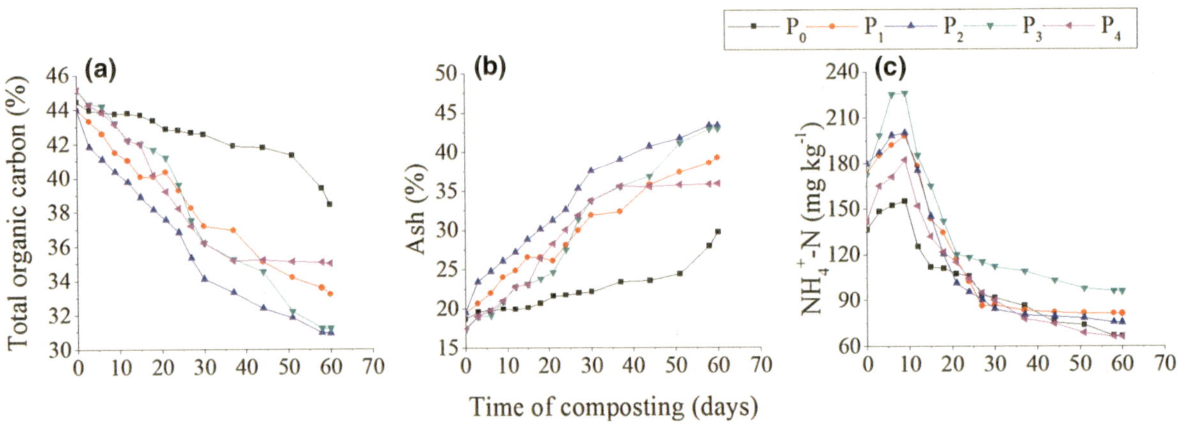

Fig. 4 Variation of total organic carbon ash content and ammoniacal nitrogen during composting period

Figure 4c shows the variation of ammoniacal nitrogen during the composting period. It was observed that in P_0 the ammoniacal nitrogen reduction was high when compared to other piles. This may be due to the low temperature in thermophilic phase and low degradation rate of the organic matter. In piles P_1, P_2, P_3 and P_4 the initial ammoniacal nitrogen were 73.0 ± 21.3, 179.8 ± 19.5, 172.3 ± 21.5 and 142.2 ± 16.5 mg kg^{-1}, respectively, which finally decreased to 81.8 ± 23.2, 75.3 ± 24.5, 95.8 ± 19.5 and 66.4 ± 18.9 mg kg^{-1}, respectively, during composting. The ammoniacal nitrogen concentration decreased due to the volatilization of NH_4^+-N by the microbes and the nitrification during the composting process.

Macro nutrients (N, P, K, Na, Ca and Mg) and micro nutrients (Fe, Mn, Zn and Cu)

The Macro nutrients in the organic matter are used by the microbes for body metabolism and also for the growth of plants. Figure 5a shows the variation of total nitrogen content. The initial concentration of total nitrogen in piles P_0–P_4 was between at 1.6 and 1.7% and increased to concentrations of 2.2–2.3% at the end of the composting process. The increasing trends of total nitrogen in this study was found to be similar to results by Jolanun and Tow-prayoon (2010). Total nitrogen increased due to the degradation of organic matter and loss of carbon in the form of CO_2 and the contribution of nitrogen fixing bacteria which were the responsible for the increase of total nitrogen content (Garcia et al. 1992).

Figure 5b shows the increase in concentration of sodium from the initial to the final day of composting. Initial concentrations of sodium in piles P_0–P_4 were 2.5–3.2 g kg^{-1} which increased to 5.3–4.9 g kg^{-1} within the 60 days composting period. The concentration of macronutrients P, K, Na, Ca and Mg increased gradually which indicates the net loss in dry mass due to the degradation of organic matter and the release of CO_2, NH_3 and H_2S during the composting process (Bustamante et al. 2009).

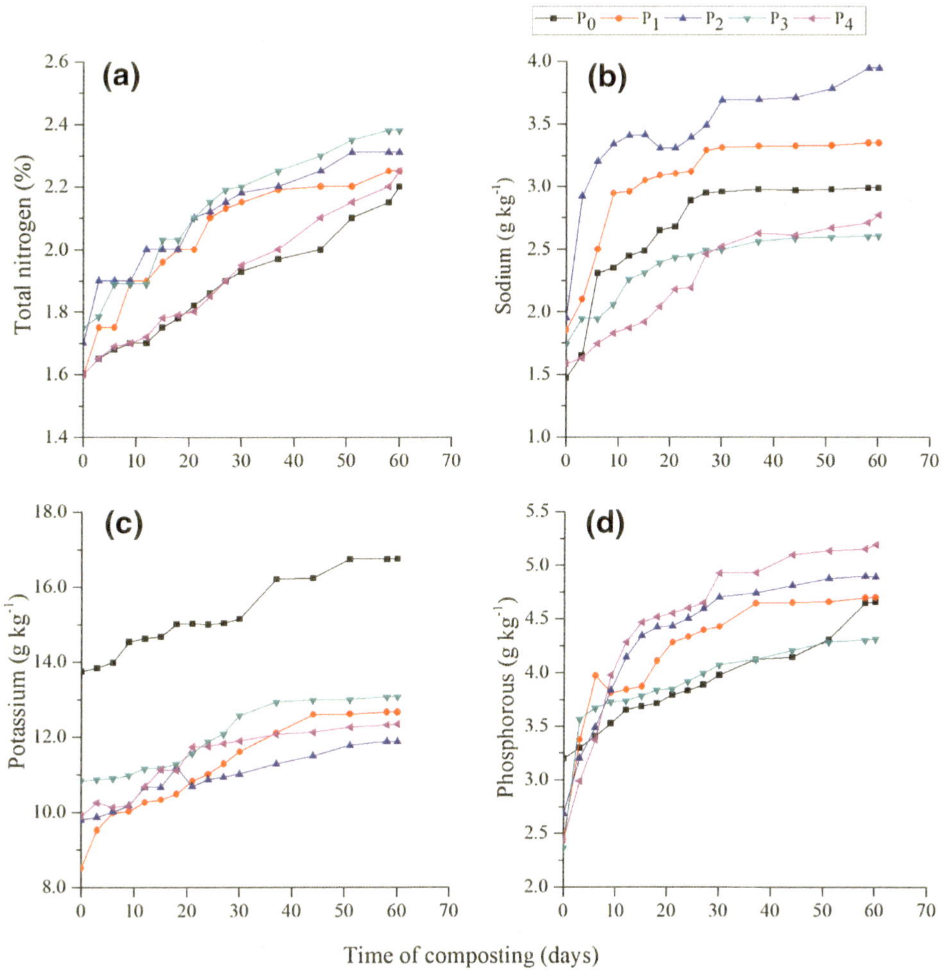

Fig. 5 Variation of total nitrogen, sodium, potassium, phosphorous during composting period

Initial total phosphorous content in the pile P_0 was 3.20 ± 0.09 g kg^{-1}, which finally increased to 4.65 ± 1.23 g kg^{-1}. In P_1, P_2, P_3 and P_4 the initial content of phosphorous were 2.44 ± 0.32, 2.69 ± 0.12, 2.37 ± 0.89 and 2.45 ± 0.85 g kg^{-1}, which increased, respectively, to 4.70 ± 0.12, 4.89 ± 0.52, 4.31 ± 0.74 and 5.19 ± 0.95 g kg^{-1} at the end of the composting process. The order of increase of the presence of phosphorous content in the piles was $P_4 > P_2 > P_1 > P_0 > P_3$. The increase in phosphorus concentration during composting can be attributed to the loss of organic matter. Similar increase in phosphorous content was observed by Singh and Kalamdhad (2014), Varma and Kalamdhad (2014b) during the composting of water hyacinth and vegetable waste.

Initial concentration of potassium in pile P_0 was 13.75 ± 0.50 g kg^{-1} and it increased to 16.75 ± 0.82 g kg^{-1} in 60 days. The concentration of potassium was rich in pile P_0. It shows the high inherent content in flower waste, suggesting that the compost is rich in potassium. The Initial potassium concentrations in piles P_1, P_2, P_3 and P_4 were 13.7 ± 0.54, 8.5 ± 1.2, 9.7 ± 0.94, 10.8 ± 0.63 and 9.9 ± 0.54 g kg^{-1}, and, respectively, increased to 16.7 ± 0.85, 12.6 ± 1.25, 11.8 ± 1.34, 13.0 ± 1.54 and 12.3 ± 1.24 g kg^{-1} at the end of the composting period. The increase in potassium concentration in all the piles is partly due to potassium assimilation and immobilisation by microbes (Singh and Kalamdhad 2012).

Micronutrients present in small amounts in the compost help in the growth of the plant and in plant reproduction (Wang et al. 2004). Manganese helps in nitrogen assimilation and photosynthesis in plant. Zinc helps in the secretion of enzymes and protein synthesis which regulate the growth of plants. Copper and iron help in the metabolism process of the plants. The initial and final concentration of micronutrient Fe, Mn, Zn and Cu present in the compost is shown in Table 4. In all the piles, an increase in micronutrients contents was observed. The increment order of micronutrient present on the final day of compost was Fe > Mn > Zn > Cu.

The concentration of iron in P_0 8.6 ± 0.49 g kg^{-1} was the highest due to the high iron content in flower waste. Initially the concentration of manganese in pile P_1 was more compared to the remaining piles. The concentration of manganese in cow dung and flower waste was 98 ± 0.8

Table 4 Macro and microelement contents in the initial and final day of the composting period

Elements	Day	Pile 0	Pile 1	Pile 2	Pile 3	Pile 4
Ca (g kg^{-1})	0	6.43 ± 0.77	10.66 ± 0.51	9.12 ± 1.37	7.83 ± 0.46	8.11 ± 0.88
	60	16.49 ± 0.76	29.67 ± 1.04	25.98 ± 0.50	18.09 ± 1.85	15.92 ± 1.03
Mg (g kg^{-1})	0	2.27 ± 0.02	3.55 ± 0.04	3.2 ± 0.01	2.83 ± 0.03	2.47 ± 0.01
	60	6.86 ± 0.02	6.53 ± 0.01	5.22 ± 0.02	6.05 ± 0.03	5.87 ± 0.02
Fe (g kg^{-1})	0	0.92 ± 0.46	1.64 ± 0.46	1.58 ± 0.52	1.57 ± 0.43	1.03 ± 0.20
	60	8.69 ± 0.49	6.8 ± 0.48	4.69 ± 0.33	6.77 ± 0.25	7.53 ± 1.01
Mn (mg kg^{-1})	0	99.63 ± 1.63	208.36 ± 0.50	143.17 ± 1.1	126.43 ± 0.70	110.24 ± 1.0
	60	126.36 ± 2.0	244.42 ± 0.50	171.32 ± 1.0	170.1 ± 1.12	152.71 ± 0.5
Zn (mg kg^{-1})	0	127.29 ± 1.0	130.76 ± 0.50	177.11 ± 1.1	152.99 ± 0.50	141.78 ± 0.5
	60	163.77 ± 0.7	177.42 ± 0.80	204.88 ± 0.3	192.06 ± 0.50	176.13 ± 0.7
Cu (mg kg^{-1})	0	36.42 ± 0.08	39.76 ± 0.11	38.18 ± 0.18	37.12 ± 0.12	35.11 ± 0.11
	60	48.14 ± 0.14	40.6 ± 0.25	42.73 ± 0.15	43.21 ± 0.03	45.26 ± 0.03

Ca calcium, *Mg* magnesium, *Fe* iron, *Mn* manganese, *Zn* zinc, *Cu* copper

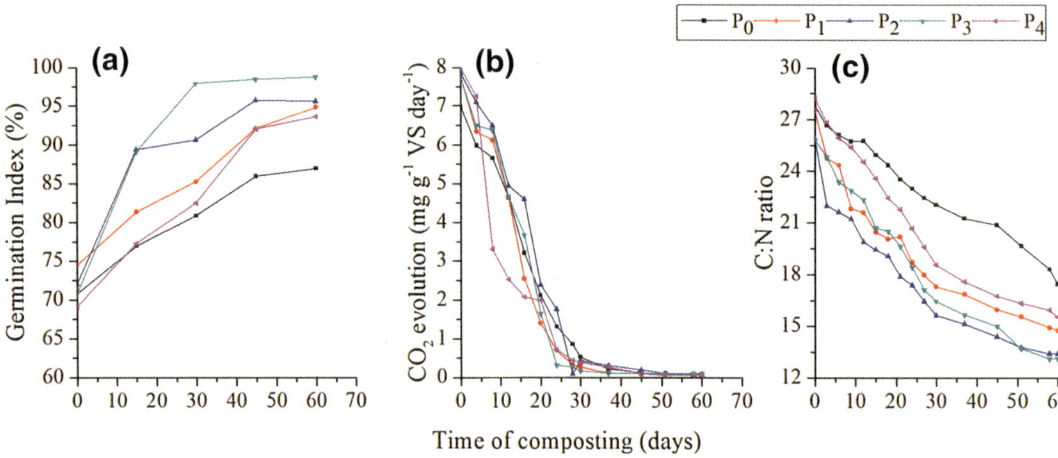

Fig. 6 Variation of germination index, CO_2 evolution rate, C:N ratio

and 258 ± 1.1 mg kg^{-1}, respectively. The increase in magnesium content in pile P_1 was due to increased quantity of cow dung. The concentration of zinc in pile P_2 was comparatively more than the remaining piles. This was due to the availability of zinc content in the flower waste 129 ± 1 mg kg^{-1} and cow dung 178 ± 1.3 mg kg^{-1}. The concentration of copper was more in the flower waste 48.62 ± 0.3 mg kg^{-1} whereas in cow dung it was 36.24 ± 0.2 mg kg^{-1}. The concentration of copper in pile P_0 was observed as 48.14 ± 0.14 mg kg^{-1} on the final day of composting. This study showed that the flower waste was rich in macro and micronutrients which are essential for the plants. The addition of cow dung helps in fast microbial activities. In each pile, it was observed that at the end of composting the concentration of micronutrients had increased due to the degradation of organic matter and the release of ammonia.

Germination index (GI), C: N ratio and carbon dioxide (CO_2) evolution

Germination index is necessary for the determination of maturity and phytotoxicity effects of compost (Gao et al. 2010; Huang et al. 2004). Germination index is widely used to check the suitability of compost for agricultural purposes. Figure 6a shows the germination index during the composting process of flower waste. The germination index which was initially 70, 74, 72, 70 and 69% in piles P_0–P_4 were increased to 86, 94, 95, 98 and 93%, respectively, at the end of the composting period. The germination index was more than 50% which shows that flower waste compost is a good source for the growth of plants. Zhang and Sun (2015) concluded that the optimum percentage of germination index suitable for mature compost is more than 50%. Similar trends of germination index were observed by Gao et al. (2010) and Singh et al. (2013) during composting of different proportions of organic waste.

CO_2 evolution is one of the best methods to determine the stability of the compost because it measures the carbon derived directly from decomposition or degradation of the organic matter. CO_2 evolution directly synthesizes aerobic respiration, which shows the direct measure of respiration and aerobic microbial activity (Kalamdhad and Kazmi 2009). CO_2 production decreased with time. Figure 6b depicts that the CO_2 production in piles P_1–P_4 was 6.9, 7.7, 7.6, 7.4 and 7.9 mg g^{-1} VS day^{-1} which decreased to 0.5, 0.2, 0.4, 0.1 and 0.3 mg g^{-1} VS day^{-1}, respectively, during the composting process. Similar decreasing trends of CO_2 production rate was observed by Varma and Kalamdhad (2014b) during the rotary drum composting of vegetable waste.

Figure 6c shows the variation of C:N ratio during the composting process. C:N ratio is a widely used parameter to indicate the maturity of the compost (Huang et al. 2004; Jiang et al. 2011). The initial C:N ratio of pile P_0 was 28 which reduced to 18 at the end of the composting period. The reduction in C:N ratio was due to the reduction of total organic carbon and increase of total nitrogen during the composting period. The initial C:N ratio of pile P_1–P_4 was 26–28 and it decreased to 14–16 at the end of composting period. The reduction of C:N ratio was due to the usage of carbon as a source of energy by the microorganisms while nitrogen was used for building the cell structure by the different group of microbes (Adhikari et al. 2009; Yadav et al. 2011; Yadav and Garg 2009). In P_2, due to fast degradation of cow dung, a 24.5% reduction of the C:N ratio was observed without reducing the carbon content, which indicates the quality of the matured compost. In pile P_0, due to rich content of flower waste and the absence of saw dust and cow dung, less reduction in total organic carbon content and an increase in the total nitrogen content was observed, all of which resulted in an increased C:N ratio. This research shows that the addition of cow dung plays an important role in the reduction of the total organic carbon and the duration of composting process.

Conclusions

A conclusion that the study derived is that the agitated pile composting of flower waste with cow dung and saw dust as bulking agent resulted in completely stabilized compost at the end of sixty days. The study also reported the leachate production during the process due to the absence of saw dust. The maximum degradation of organic matter was observed during the first 30 days of the composting process. The addition of the bulking agent helped in maintaining the aerobic condition, which reduced the excessive formation of leachate and maintained the thermophilic stage for a longer period. Macro and micronutrients contents after composting were observed in suitable ranges for agricultural purposes. The physicochemical analysis showed that the pile containing 60 kg of flower waste, 30 kg of cow dung, and 10 kg of saw dust would be the most appropriate combination for the composting of flower waste.

References

Abdullah N, Chin NL, Mokhtar MN, Taip FS (2013) Effects of bulking agents, load size or starter cultures in kitchen-waste composting. Int J Recycl Org Waste Agric 2:1–10. doi:10.1186/2251-7715-2-3

Adhikari BK, Barrington S, Martinez J, King S (2009) Effectiveness of three bulking agents for food waste composting. Waste Manag 29:197–203. doi:10.1016/j.wasman.2008.04.001

APHA (2005) Standard methods for the examination of water and wastewater, 20th edn. American Public Health Association Washington. https://www.standardmethods.org/

Awasthi MK, Pandey AK, Bundela PS, Khan J (2015) Co-composting of organic fraction of municipal solid waste mixed with different bulking waste: characterization of physicochemical parameters and microbial enzymatic dynamic. Bioresour Technol 182:200–207. doi:10.1016/j.biortech.2015.01.104

Ball AS, Shah D, Wheatley CF (2000) Assessment of the potential of a novel newspaper/horse manure-based compost. Bioresour Technol 73:163–167. doi:10.1016/S0960-8524(99)00169-8

BIS (1982) Methods for analysis of solid wastes (excluding industrial solid wastes). Indian Standard Institution, New Delhi. https://www.bis.org.in/

Bustamante M, Paredes C, Morales J, Mayoral A, Moral R (2009) Study of the composting process of winery and distillery wastes using multivariate techniques. Bioresour Technol 100:4766–4772. doi:10.1016/j.biortech.2009.04.033

Gao M, Li B, Yu A, Liang F, Yang L, Sun Y (2010) The effect of aeration rate on forced-aeration composting of chicken manure and sawdust. Bioresour Technol 101:1899–1903. doi:10.1016/j.biortech.2009.10.027

Garcia C, Hernandez T, Costa F, Ayuso M (1992) Evaluation of the maturity of municipal waste compost using simple chemical parameters. Commun Soil Sci Plant Anal 23:1501–1512. doi:10.1080/00103629209368683

Huang G, Wong J, Wu Q, Nagar B (2004) Effect of C/N on composting of pig manure with sawdust. Waste Manag 24:805–813. doi:10.1016/j.wasman.2004.03.011

Iyengar SR, Bhave PP (2006) In-vessel composting of household wastes. Waste Manag 26:1070–1080. doi:10.1016/j.wasman.2005.06.011

Jiang T, Schuchardt F, Li G, Guo R, Zhao Y (2011) Effect of C/N ratio, aeration rate and moisture content on ammonia and greenhouse gas emission during the composting. J Environ Sci 23:1754–1760. doi:10.1016/S1001-0742(10)60591-8

Jolanun B, Towprayoon S (2010) Novel bulking agent from clay residue for food waste composting. Bioresour Technol 101:4484–4490. doi:10.1016/j.biortech.2010.01.116

Kalamdhad AS, Kazmi A (2009) Effects of turning frequency on compost stability and some chemical characteristics in a rotary drum composter. Chemosphere 74:1327–1334. doi:10.1016/j.chemosphere.2008.11.058

Mena E, Garrido A, Hernandez T, Garcia C (2003) Bioremediation of sewage sludge by composting. Commun Soil Sci Plant Anal 34:957–971. doi:10.1081/CSS-120019102

Sánchez-Monedero M, Roig A, Paredes C, Bernal M (2001) Nitrogen transformation during organic waste composting by the Rutgers system and its effects on pH, EC and maturity of the composting mixtures. Bioresour Technol 78:301–308. doi:10.1016/S0960-8524(01)00031-1

Singh J, Kalamdhad AS (2012) Concentration and speciation of heavy metals during water hyacinth composting. Bioresour Technol 24:169–179. doi:10.1016/j.biortech.2012.08.043

Singh WR, Kalamdhad AS (2014) Potential for composting of green phumdi biomass of Loktak lake. Ecol Eng 67:119–126. doi:10.1016/j.ecoleng.2014.03.086

Singh J, Kalamdhad AS (2015) Assessment of compost quality in agitated pile composting of water hyacinth collected from different sources. Int J Recycl Org Waste Agric 4:175–183. doi:10.1007/s40093-015-0097-z

Singh WR, Kalamdhad AS (2016) Transformation of nutrients and heavy metals during vermicomposting of the invasive green weed Salvinia natans using *Eisenia fetida*. Int J Recycl Org Waste Agric, pp 1–16 doi:10.1007/s40093-016-0129-3

Singh A, Jain A, Sarma BK, Abhilash P, Singh HB (2013) Solid waste management of temple floral offerings by vermicomposting using *Eisenia fetida*. Waste Manag 33:1113–1118. doi:10.1016/j.wasman.2013.01.022

Varma VS, Kalamdhad A (2014a) Evolution of chemical and biological characterization during thermophilic composting of vegetable waste using rotary drum composter. Int J Environ Sci Technol, pp 1–10. doi:10.1007/s13762-014-0582-3

Varma VS, Kalamdhad AS (2014b) Stability and microbial community analysis during rotary drum composting of vegetable waste. Int J Recycl Org Waste Agric 3:1–9. doi:10.1007/s40093-014-0052-4

Wang P, Changa C, Watson M, Dick W, Chen Y, Hoitink H (2004) Maturity indices for composted dairy and pig manures. Soil Biol Biochem 36:767–776. doi:10.1016/j.soilbio.2003.12.012

Wang L-P, Shen Q-R, Yu G-H, Ran W, Xu Y-C (2012) Fate of biopolymers during rapeseed meal and wheat bran composting as studied by two-dimensional correlation spectroscopy in combination with multiple fluorescence labeling techniques. Bioresour Technol 105:88–94. doi:10.1016/j.biortech.2011.11.064

Wang K, Li X, He C, Chen C-L, Bai J, Ren N, Wang J-Y (2014) Transformation of dissolved organic matters in swine, cow and chicken manures during composting. Bioresour Technol 168:222–228. doi:10.1016/j.biortech.2014.03.129

Wu C, Li W, Wang K, Li Y (2015) Usage of pumice as bulking agent in sewage sludge composting. Bioresour Technol 190:516–521. doi:10.1016/j.biortech.2015.03.104

Yadav A, Garg VK (2009) Feasibility of nutrient recovery from industrial sludge by vermicomposting technology. J Hazard Mater 168:262–268. doi:10.1016/j.jhazmat.2009.02.035

Yadav KD, Tare V, Ahammed MM (2011) Vermicomposting of source-separated human faeces by *Eisenia fetida*: effect of stocking density on feed consumption rate, growth characteristics and vermicompost production. Waste Manag 31:1162–1168. doi:10.1016/j.wasman.2011.02.008

Yadav A, Gupta R, Garg VK (2013) Organic manure production from cow dung and biogas plant slurry by vermicomposting under field conditions. Int J Recycl Org Waste Agric 2:21. doi:10.1186/2251-7715-2-21

Zhang L, Sun X (2015) Influence of bulking agents on physical, chemical, and microbiological properties during the two-stage composting of green waste. Waste Manag 48:115–126. doi:10.1016/j.wasman.2015.11.032

Zhou Y, Selvam A, Wong JW (2014) Evaluation of humic substances during co-composting of food waste, sawdust and Chinese medicinal herbal residues. Bioresour Technol 168:229–234. doi:10.1016/j.biortech.2014.05.070

Permissions

All chapters in this book were first published in IJROWA, by Springer; hereby published with permission under the Creative Commons Attribution License or equivalent. Every chapter published in this book has been scrutinized by our experts. Their significance has been extensively debated. The topics covered herein carry significant findings which will fuel the growth of the discipline. They may even be implemented as practical applications or may be referred to as a beginning point for another development.

The contributors of this book come from diverse backgrounds, making this book a truly international effort. This book will bring forth new frontiers with its revolutionizing research information and detailed analysis of the nascent developments around the world.

We would like to thank all the contributing authors for lending their expertise to make the book truly unique. They have played a crucial role in the development of this book. Without their invaluable contributions this book wouldn't have been possible. They have made vital efforts to compile up to date information on the varied aspects of this subject to make this book a valuable addition to the collection of many professionals and students.

This book was conceptualized with the vision of imparting up-to-date information and advanced data in this field. To ensure the same, a matchless editorial board was set up. Every individual on the board went through rigorous rounds of assessment to prove their worth. After which they invested a large part of their time researching and compiling the most relevant data for our readers.

The editorial board has been involved in producing this book since its inception. They have spent rigorous hours researching and exploring the diverse topics which have resulted in the successful publishing of this book. They have passed on their knowledge of decades through this book. To expedite this challenging task, the publisher supported the team at every step. A small team of assistant editors was also appointed to further simplify the editing procedure and attain best results for the readers.

Apart from the editorial board, the designing team has also invested a significant amount of their time in understanding the subject and creating the most relevant covers. They scrutinized every image to scout for the most suitable representation of the subject and create an appropriate cover for the book.

The publishing team has been an ardent support to the editorial, designing and production team. Their endless efforts to recruit the best for this project, has resulted in the accomplishment of this book. They are a veteran in the field of academics and their pool of knowledge is as vast as their experience in printing. Their expertise and guidance has proved useful at every step. Their uncompromising quality standards have made this book an exceptional effort. Their encouragement from time to time has been an inspiration for everyone.

The publisher and the editorial board hope that this book will prove to be a valuable piece of knowledge for researchers, students, practitioners and scholars across the globe.

List of Contributors

Richard Adande, Hervé Kouessivi Janvier Bokossa, Mouhamadou Nourou Dine Liady and Emile Didier Fiogbé
Laboratory of Research in the Wetlands (LRZH), Faculty of Science and Techniques (FAST), University of Abomey- Calavi (UAC), 01 BP 526, Cotonou, Benin

Tarun Kumar Kumawat
Department of Biotechnology, JECRC University, Jaipur, Rajasthan, India

Anima Sharma
ICAR-Central Sheep and Wool Research Institute, Arid- Region Campus, Bikaner, Rajasthan, India

Seema Bhadauria
Department of Microbiology, JECRC University, Jaipur, Rajasthan, India

K. Azim, S. Komenane and B. Soudi
Research Unit of Integrated Crop Production, Regional Center of Agronomy Research, INRA-CRRA, B.P.124, CP: 80350 Inzegane, Agadir, Morocco
Department of Environment and Natural Resources, IAV Hassan II-Rabat, B.P. 6202, Madinat Al Irfane, Morocco

A. Embrandiri, P. F. Rupani, M. H. Ibrahim and M. O. Abd. Kadir
Environment Division, School of Industrial Technology, Universiti Sains Malaysia, 11800 Pulau Pinang, Malaysia

M. Shahadat
Department of Biochemical Engineering and Biotechnology, Indian Institute of Technology, IIT Delhi, New Delhi 110016, India

R. P. Singh
Institute of Environment and Sustainable Development, Banaras Hindu University, Varanasi 221005, India

S. A. Ismail
Eco-Science Research Foundation, Chennai 600041, India

Naseer Hussain, Tasneem Abbasi and S. A. Abbasi
Centre for Pollution Control and Environmental Engineering, Pondicherry University, Pondicherry 605014, India

A. I. Afe and F. Oluleye
Department of Crop Production, Kwara State University, Malete, Nigeria

Cinny Makkar and Chander Parkash
Department of Applied Sciences, I.K.G Punjab Technical University, Kapurthala, Punjab, India

Jaswinder Singh
Department of Zoology, Khalsa College, Amritsar, Punjab, India

Chris Fidelis and B. K. Rajashekhar Rao
Department of Agriculture, PNG University of Technology, Private Mail Bag, Lae, Morobe Province 411, Papua New Guinea

Amirhossein Malakahmad, Natasha Binti Idrus, Motasem S. Abualqumboz, Sara Yavari and Shamsul Rahman M. Kutty
Department of Civil and Environmental Engineering, Universiti Teknologi Petronas, 32610 Seri Iskandar, Perak Darul Ridzuan, Malaysia

Siamak Shirani Bidabadi
Department of Horticulture, College of Agriculture, Isfahan University of Technology, IUT, Isfahan 84156-83111, Iran

Safoora Dehghanipoodeh
Department of Horticultural Science, College of Agriculture and Natural Resources, University of Tehran, Karaj, Iran

Glenn C. Wright
The School of Plant Sciences, College of Agriculture and Life Sciences, University of Arizona, Tucson, AZ 85364, USA

A. G. Adebayo, H. A. Akintoye, A. O. Shokalu and M. T. Olatunji
Department of Vegetable and Floriculture, National Horticultural Research Institute, Ibadan, Nigeria

L. L. Mugivhisa and J. O. Olowoyo
Department of Biology, Sefako Makgatho Health Sciences University, Medunsa Campus, P.O Box 139, Pretoria 0204, South Africa

Oyeyemi A. Dada and Francis Imade
Department of Botany, Faculty of Life Sciences, Ambrose Alli University, Ekpoma, Edo State, Nigeria

E. M. Anifowose
Department of Botany, University of Ibadan, Ibadan, Oyo State, Nigeria

Zobia Anwar, Muhammad Irshad, Qaisar Mahmood, Farhan Hafeez and Muhammad Bilal
Department of Environmental Sciences, COMSATS Institute of Information Technology, Abbottabad, Pakistan

Hossein Karimi, Mehdi Mokhtari, Faezeh Salehi and Aliasghar Ebrahimi
Environmental Science and Technology Research Center, Department of Environmental Health Engineering, Shahid Sadoughi University of Medical Sciences, Yazd, Iran

Sajed Sojoudi
Department of Occupational Health Engineering, Shahid Sadoughi University of Medical Sciences, Yazd, Iran

Anteneh Argaw
School of Natural Resources Management and Environmental Sciences, College of Agriculture and Environmental Sciences, Haramaya University, Haramaya, Ethiopia

Eric Gbenatey Nartey, Philip Amoah and Surendra kumar Pradhan
International Water Management Institute (IWMI), PMB CT, 112, Cantonments, Accra, Ghana

Godfred Kwabena Ofosu-Budu
College of Agricultural and Consumer Sciences, University of Ghana Forest and Horticultural Crops Research Centre- Kade, Legon, Ghana

Ashley Muspratt
Pivot Ltd, Mombasa, Kenya

Sayyed-Hassan Tabatabaei, Rohollah Fatahi Nafchi, Mohammad Mehdi Karizan and Zohreh Nazem
Department of Water Engineering, Faculty of Agriculture, Shahrekord University, Shahrekord, Iran

Payam Najafi
Department of Water Engineering, Faculty of Agriculture, Islamic Azad University, Isfahan Branch (Khorasgan), Isfahan, Iran

Habeebah Adewunmi Saka, Jamiu Oladipupo Azeez, Joy Nwakaego Odedina and Shade John Akinsete
Federal University of Agriculture Abeokuta, Abeokuta, Nigeria

Payel Sarkar and Rounak Chourasia
Department of Microbiology, University of North Bengal, Raja Rammohunpur, Siliguri, West Bengal 734013, India

C. R. Sudharmaidevi, K. C. M. Thampatti and N. Saifudeen
Department of Soil Science and Agricultural Chemistry, College of Agriculture, Vellayani Kerala Agricultural University, Trivandrum, Kerala, India 695 522

Vahideh Sadeghizadeh and Vahidreza Jalali
Department of Soil Science, Faculty of Agriculture, Shahid Bahonar University of Kerman, Kerman, Iran

Isha Vishan
Centre for the Environment, Indian Institute of Technology Guwahati (IIT G), Guwahati 781039, Assam, India

Senthilkumar Sivaprakasam
Department of Biosciences and Bioengineering, Indian Institute of Technology Guwahati (IIT G), Guwahati 781039, Assam, India

Ajay Kalamdhad
Department of Civil Engineering, Indian Institute of Technology Guwahati (IITG), Guwahati 781039, Assam, India

Dayanand Sharma and Kunwar D. Yadav
Civil Engineering Department, Sardar Vallabhbhai National Institute of Technology, Surat, Gujarat 395007, India

V. Sudharsan Varma and Ajay. S. Kalamdhad
Civil Engineering Department, Indian Institute of Technology, Guwahati, Assam 781039, India

Index